Autodesk Fusion 360 Black Book (V 2.0.10027) - Colored

By
Gaurav Verma

Matt Weber

(CADCAMCAE Works)

D1091762

ISBN # 978-1-77459-022-5

NOTICE TO THE READER

DEDICATION

To teachers, who make it possible to disseminate knowledge
to enlighten the young and curious minds
of our future generations

To students, who are the future of the world

THANKS

To my friends and colleagues

To my family for their love and support

Table of Contents

Chapter 2 : Sketching

Chapter 3 : 3D Sketch and Solid Modeling

Chapter 4 : Advanced 3D Modeling

Chapter 5 : Practical and Practice

Chapter 12 : Sculpting

Chapter 13 : Sculpting-2

Chapter 14 : Mesh Design

Chapter 15 : Manufacturing

Chapter 16 : Generating Milling Toolpaths - 1

Chapter 17 : Generating Milling Toolpaths - 2

Chapter 18 : Generating Turning and Cutting Toolpaths

Chapter 20 : Introduction to Simulation in Fusion 360

Chapter 21 : Simulation Studies in Fusion 360

Chapter 22 : Sheetmetal Design

Chapter 23 : Generative Design New

Preface

Autodesk Fusion 360 is a product of Autodesk Inc. Fusion 360 is the first of its kind software which combine 3D CAD, CAM, and CAE tool in single package. It connects your entire product development process in a single cloud-based platform that works on both Mac and PC. In CAD environment, you can create the model with parametric designing and dimensioning. The CAD environment is equally applicable for assembly design. The CAE environment facilitates to analysis the model under real-world load conditions. Once the model is as per your requirement then generate the NC program using the CAM environment.

The **Autodesk Fusion 360 Black Book** (V 2.0.10027) is 4th edition of our series on Autodesk Fusion 360. The book is updated on Autodesk Fusion 360 Ultimate, Student V 2.0.10027. With lots of features and thorough review, we present a book to help professionals as well as beginners in creating some of the most complex solid models. The book follows a step by step methodology. In this book, we have tried to give real-world examples with real challenges in designing. We have tried to reduce the gap between educational use of Autodesk Fusion 360 and industrial use of Autodesk Fusion 360. This edition of book, includes latest topics on Sketching, 3D Part Designing, Assembly Design, Sculpting, Mesh Design, CAM, Simulation, Sheetmetal, 3D printing, Manufacturing, and many other topics. A new chapter of Generative Design has been added in this edition. The book covers almost all the information required by a learner to master the Autodesk Fusion 360. The book starts with sketching and ends at advanced topics like Manufacturing, Simulation, and Generative Design. Some of the salient features of this book are :

In-Depth explanation of concepts

Every new topic of this book starts with the explanation of the basic concepts. In this way, the user becomes capable of relating the things with real world.

Topics Covered

Every chapter starts with a list of topics being covered in that chapter. In this way, the user can easy find the topic of his/her interest easily.

Instruction through illustration

The instructions to perform any action are provided by maximum number of illustrations so that the user can perform the actions discussed in the book easily and effectively. There are about **2200** small and large illustrations that make the learning process effective.

Tutorial point of view

At the end of concept's explanation, the tutorial make the understanding of users firm and long lasting. Almost each chapter of the book has tutorials that are real world projects. Moreover most of the tools in this book are discussed in the form of tutorials.

Project

Free projects and exercises are provided to students for practicing.

For Faculty

If you are a faculty member, then you can ask for video tutorials on any of the topic, exercise, tutorial, or concept.

New

If anything is added or enhanced in this edition which is not available in the previous editions, then it is displayed with symbol **New** in table of content.

Formatting Conventions Used in the Text

All the key terms like name of button, tool, drop-down etc. are kept bold.

Free Resources

Link to the resources used in this book are provided to the users via email. To get the resources, mail us at ***cadcamcaeworks@gmail.com*** with your contact information. With your contact record with us, you will be provided latest updates and informations regarding various technologies. The format to write us mail for resources is as follows:

Subject of E-mail as ***Application for resources of _____book***.
Also, given your information like
Name:
Course pursuing/Profession:
E-mail ID:

Note: We respect your privacy and value it. If you do not want to give your personal informations then you can ask for resources without giving your information.

About Authors

The author of this book, Gaurav Verma, has authored and assisted in more than 16 titles in CAD/CAM/CAE which are already available in market. He has authored **AutoCAD Electrical Black Books** which are available in both **English** and **Russian** language. He has also authored books on various modules of Creo Parametric and SolidWorks. He has provided consultant services to many industries in US, Greece, Canada, and UK. He has assisted in preparing many Government aided skill development programs. He has been speaker for Autodesk University, Russia 2014. He has assisted in preparing AutoCAD Electrical course for Autodesk Design Academy. He has worked on Sheetmetal, Forging, Machining, and Casting designs in Design and Development departments of various manufacturing firms.

If you have any query/doubt in any CAD/CAM/CAE package, then you can contact the authors by writing at cadcamcaeworks@gmail.com

For Any query or suggestion

If you have any query or suggestion, please let us know by mailing us on *cadcamcaeworks@gmail.com*. Your valuable constructive suggestions will be incorporated in our books.

Page left blank intentionally

Chapter 1

Starting with Autodesk Fusion 360

Topics Covered

The major topics covered in this chapter are:

- **Overview of Autodesk Fusion 360**
- **Installing Autodesk Fusion 360 (Educational)**
- **Starting Autodesk Fusion 360**
- **Starting a New Document**
- **File Menu**
- **Preform**
- **Undo and Redo button**
- **User Account drop-down**
- **Help drop-down**
- **Data Panel**
- **Simulation Browser**
- **Navigation Bar**
- **Display Bar**
- **Customizing Toolbar and Marking Menu**

OVERVIEW OF AUTODESK FUSION 360

Autodesk Fusion 360 is an Autodesk product designed to be a powerful 3D Modeling software package with an integrated, parametric, feature based CAM module built into the software. Autodesk Fusion 360 is the first of its kind 3D CAD, CAM, and CAE tool. It connects your entire product development process in a single cloud-based platform that works on both Mac and Microsoft Windows; refer to Figure-1.

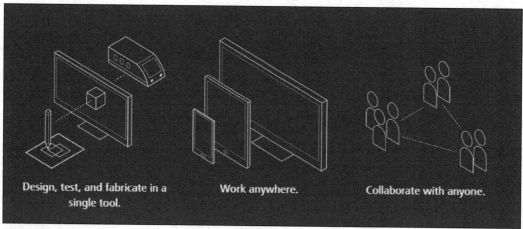

Figure-1. Overview

It combines mechanical design, collaboration, simulation, and machining in a single software. The tools in Fusion 360 enable rapid and easy exploration of design ideas with an integrated concept to production toolset. This software needs a good network connection to work in collaboration with other team members.

This software is much affordable than any other software offered by Autodesk. To use this software, one need to pay the monthly subscription of Fusion 360. You can work offline in this software and later save the file on Autodesk Server. User can access this software from anywhere with an internet connection. The user is able to open the saved file and also able to share files with anyone from anywhere as long as he/she has the software and good internet connection. Also, the pricing of this software is cost effective so anyone can use it for manufacturing of tools and parts. Autodesk Fusion 360 is build to work in multi body manner: both parts and assemblies build in a single file. The procedure to install the software is given next.

INSTALLING AUTODESK FUSION 360 (STUDENT)

- Connect your PC with the internet connection and then log on to **http://www.autodesk.com/products/fusion-360/students-teachers-educators** as shown in Figure-2.

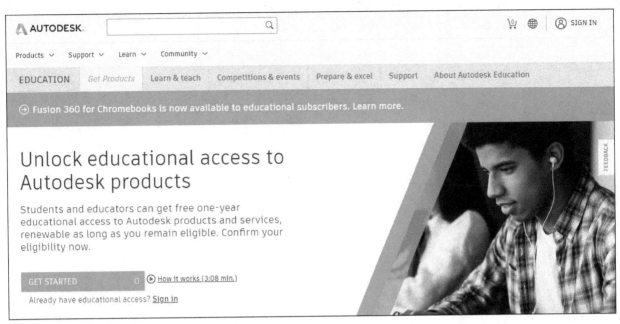

Figure-2. Autodesk Website

- Click on the **GET STARTED** button and on next page, select your Country, Education role, and Institution Type. In **Education role** drop-down, you need to select **Student**. In Institution Type drop-down, you need to select from either of the given two options as desired. (There is free subscription for students with license term of 3 years) as shown in Figure-3.

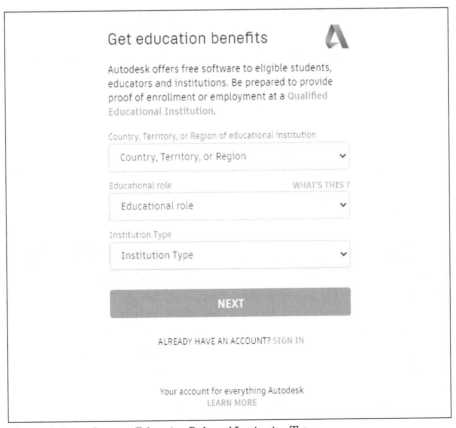

Figure-3. Select Country Education Role and Institution Type

- After filling the details, click on **NEXT** button. The next page of this website opens. In this page, you need to fill your personal details and then click on **Create Account** as shown in Figure-4.

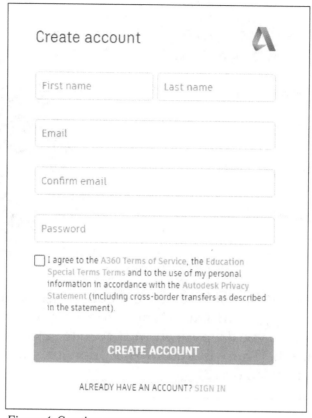

Figure-4. Creating account

- After creating account, you need to verify your E-mail address by login into your E-mail account. After verifying E-mail address, you need to give your **Education details**.
- After completion of these processes, you need to sign-in your Autodesk Account and search the **Autodesk Fusion 360 software**.
- Download the **Autodesk Fusion 360 software**.
- Open the download setup file and follow the instructions as per the setup instruction.
- The software will be installed in a couple of minutes.

STARTING AUTODESK FUSION 360

- To start **Autodesk Fusion 360** from **Start** menu, click on the **Start** button in the **Taskbar** at the bottom left corner and then select **Autodesk Fusion 360** option from the **Autodesk** folder. Select the **Autodesk Fusion 360** icon; refer to Figure-5.
- While installing the software, if you have selected the check box to create a desktop icon then you can double-click on that icon to run the software.
- If you have not selected the check box to create the desktop icon and want to create the icon on desktop now in Windows 8 or later, then drag the **Autodesk Fusion 360** icon from **Start** menu to Desktop.

Figure-5. Start menu

After clicking on the icon, the Autodesk Fusion 360 software window will be displayed; refer to Figure-6.

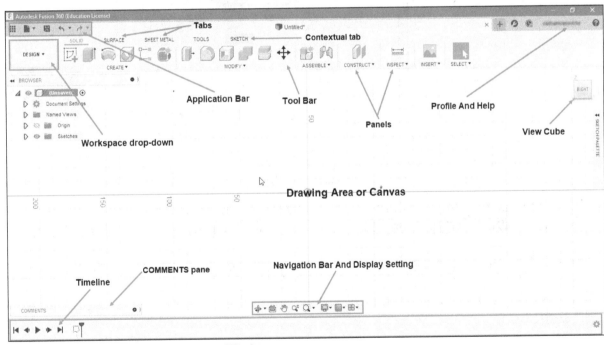

Figure-6. Autodesk Fusion 360 Application Window

STARTING A NEW DOCUMENT

• Click on the **File** drop-down and select the **New Design** tool as shown in Figure-7. A new document will open.

Figure-7. File menu

- Select desired workspace from the **Workspace** drop-down; refer to Figure-8.

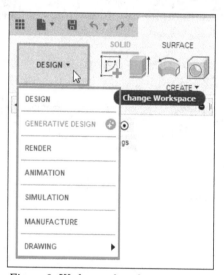

Figure-8. Workspace drop down

- There are six buttons available in **Workspace** drop-down in Educational version; **DESIGN, RENDER, ANIMATION, SIMULATION, MANUFACTURE,** and **DRAWING**.

- The **DESIGN** button is used to create solid, surface, and sheet metal designs.

- The **RENDER** button is used to activate workspace for rendering realistic model for presentation.

- The **ANIMATION** button is used to activate workspace for creating automatic or manual exploded views as well as direct control over unique animation of parts and assemblies.

- The **SIMULATION** button is used to perform Engineering Analyses.

- The **MANUFACTURE** button is used to generate G-codes for manufacturing processes like turning, milling, drilling, cutting, probing, and so on.

- The **DRAWING** button is used for generating drawings from model and animation.

If you are working on Professional version of software then another workspace **GENERATIVE DESIGN** is also available using which you can generate different iterations of designs based on specified limits and goals.

You will learn more about these work-spaces later in this book.

FILE MENU

The options in the **File** menu are used to manage files and related parameters. Various tools of **File** menu are discussed next.

Creating New Drawing

The tools in **New Drawing** cascading menu are used to initialize a new drawing from animation or design; refer to Figure-9. The methods to use these tools will be discussed later in the book.

Figure-9. New Drawing cascading menu

Opening File

The **Open** tool is used to open files earlier saved in cloud or local drive. You can also use this tool to import supported files of other software. The procedure to use this tool is given next.

- Click on the **Open** tool from the **File** menu or press **CTRL+O** from keyboard. The **Open** dialog box will be displayed; refer to Figure-10. Note that we are working in online mode of Autodesk Fusion.

Figure-10. Open dialog box

- By default, files saved on cloud are displayed in the dialog box. Select desired file and click on the **Open** button. The file will open in Autodesk Fusion.
- If you want to open a file stored in local drive then click on the **Open from my computer** button. The **Open** dialog box will be displayed; refer to Figure-11.

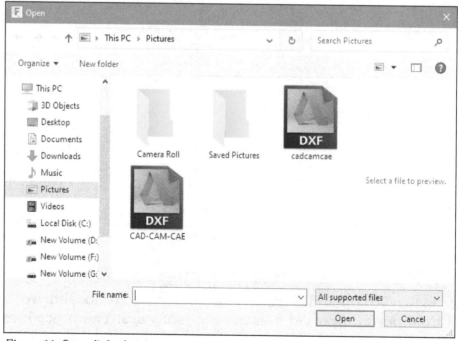

Figure-11. Open dialog box

- Select desired file and click on the **Open** button. The file will be displayed in the application window. If you have selected the file of different format from native format then the **Job Status** dialog box will be displayed notifying you the status of file import; refer to Figure-12.

Job Status			✕
Data	Generative Designs	Simulations	

Name	Status	Action
C02_Practical_1.par	Complete	Open

Close

Figure-12. Job Status dialog box

- Once the status of import is complete then click on the **Open** button from the **Action** column in the dialog box. The model will be displayed; refer to Figure-13 (Creo Parametric file imported in Fusion 360).

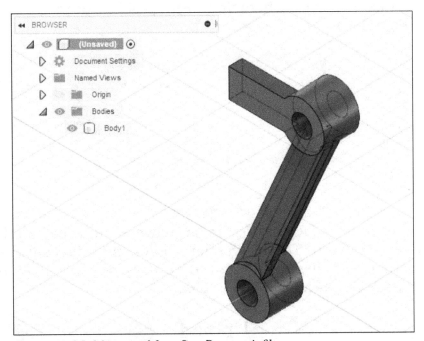

Figure-13. Model imported from Creo Parametric file

Recovering Documents

The **Recover Documents** tool is used to recover unsaved file versions which are created when software closes unexpectedly; refer to Figure-14. Note that there is an auto save feature in Autodesk Fusion which saves versions of file at a specific time intervals. If software has created the auto save version of your file then only you will be able to recover the file. By default, this time interval is 5 minutes and it can be changed in **Preferences** dialog box which will be discussed later. The procedure to use this tool is given next.

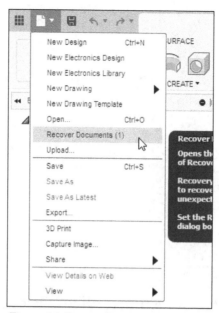

Figure-14. Recover Documents tool

- Click on the **Recover Documents** tool from the **File** menu. The **File Recovery** dialog box will be displayed where you need to select auto save version of your file; refer to Figure-15.

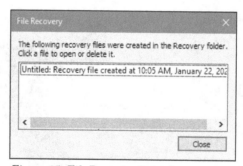

Figure-15. File Recovery dialog box

- Click on the file version that you want to be recovered and click on the **Open** button from the displayed menu. The file version will open in **Autodesk Fusion 360**. If the opened version is as desired then save it otherwise close it and open the other version from the **File Recovery** dialog box.

Upload

The **Upload** tool is used to upload files on cloud. The procedure to use this tool is discussed next.

- Click on **Upload** button from the **File** menu. The **Upload** dialog box will be displayed; refer to Figure-16. Click on the **Select Files** button or Drag & Drop the files to be uploaded in the **Drag and Drop Here** area of the dialog box.

Figure-16. Upload dialog box

- If you want to change the location where files will be uploaded then click on the **Change Location** button and select desired location from the **Change Location** dialog box; refer to Figure-17.

Figure-17. Change Location dialog box

- After selecting desired location, click on the **Select** button. You will return to **Upload** dialog box with updated location. Note that you can also create new folders and new locations by using the options in **Change Location** dialog box which will be discussed later in this book.
- Click on the **Select Files** button to upload files and select multiple files while holding the **CTRL** key from the **Open** dialog box displayed; refer to Figure-18.

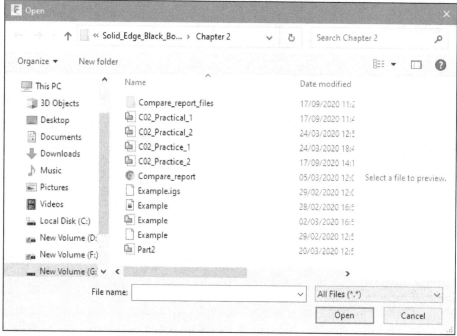

Figure-18. Open dialog box for file upload

- Click on the **Open** button from the dialog box to upload files. Or, drag the files in the dialog box and drop them. The options in the **Upload** dialog box will be updated; refer to Figure-19.

Figure-19. Updated Upload dialog box

- If you have selected a .PRT file of NX or Creo then select the **.PRT files are not assembly files** check box to explicitly tell software that these PRT files are not connected to any assembly.
- After specifying desired parameters, click on **Upload** button from **Upload** dialog box. Files will be uploaded and status will be displayed in **Job Status** dialog box.

Save

The **Save** tool is used to save the current file on cloud. You can press **CTRL+S** key from keyboard to save file. The procedure to save file is given next.

- Click on the **Save** tool from **File** menu or select the **Save** button from **Application bar** as shown in Figure-20. The **Save** dialog box will be displayed asking you to specify the **Name** and **Location** of your file; refer to Figure-21.

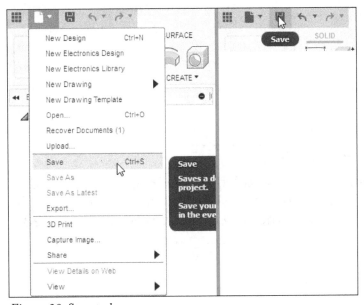

Figure-20. Save tool

Figure-21. Save dialog box

- Specify desired name in the **Name** edit box.
- Click on the down arrow next to **Location** edit box in the dialog box if you want to select a different location on cloud. A list of locations created for/by you in Autodesk Cloud will be displayed; refer to Figure-22.

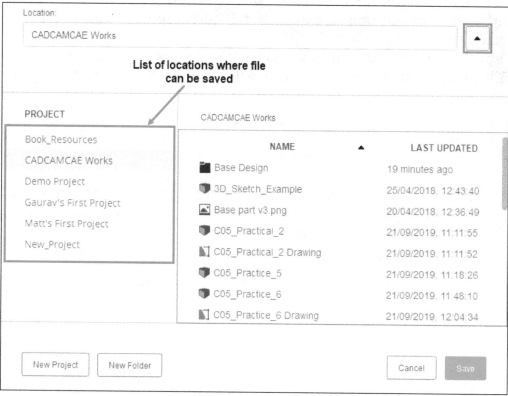

Figure-22. Locations to save files

- By default, Demo Project is added in the Project list. To add a new project, click on the **New Project** button at the bottom left corner of the dialog box. You will be asked to specify name of the new project; refer to Figure-23.

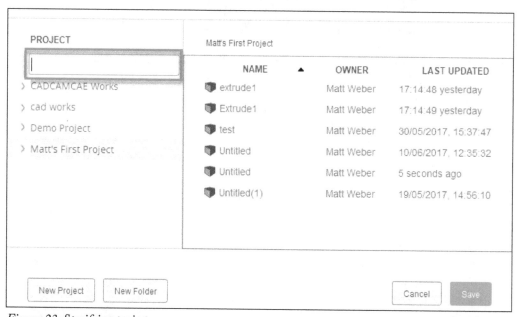

Figure-23. Specifying project name

- Specify desired name for the project in displayed edit box and click anywhere in blank area of the dialog box. The new location will be created. Select this new location to save the file; refer to Figure-24.

Figure-24. Location selected for saving file

- If you want to create folder in the project with desired name then click on the **New Folder** button from the dialog box. You will be asked to specify the name of the folder. Specify the name of folder and click in blank area of the dialog box.
- To save your file in the folder, double-click on folder name to enter the folder and then click on the **Save** button. The file will be saved at specified location.

Note that when next time you will use **Save** tool after making changes then the **Add Version Description** dialog box will be displayed asking you to specify description for this new version of same file; refer to Figure-25. Type any short description about what has changed in this version and click on the **OK** button. You will notice that file name has changed from xxxxx v0 to xxxxx v1 which describes that this is the version 1 of same file. These versions can be accessed by **Data Panel** which will be discussed later.

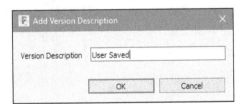

Figure-25. Add Version Description dialog box

Save As

Using the **Save As** tool, you can save the file with different name at different location. The procedure to use this tool is discussed next.

- Click on **Save As** tool from **File** menu; refer to Figure-26. The **Save As** dialog box will be displayed as shown in Figure-27.

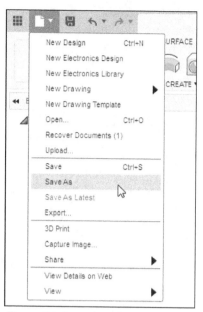

Figure-26. Save As tool

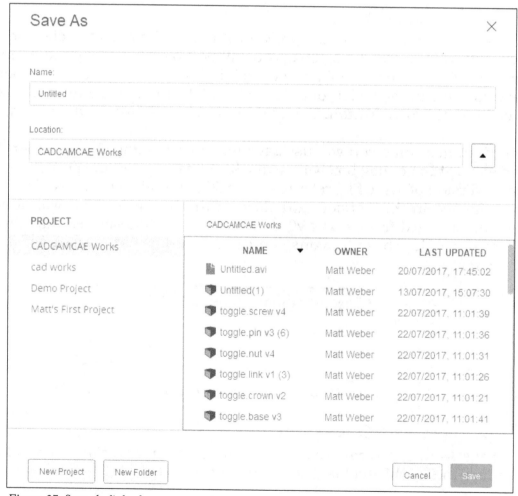

Figure-27. Save As dialog box

- Specify desired name and location for the file, and click on the **Save** button. The options in this dialog box are same as discussed in **Save** dialog box.

Save As Latest

When you are collaborating on a model with your team and after creating many versions of the model, you find that one of the previous versions is to be considered as final model for use. In such case, open the previous version file Autodesk Fusion and the **Save As Latest** tool will be active in **File** menu to save this version as latest; refer to Figure-28. On clicking the **Save As Latest** tool, an information box will be displayed showing how versions will be renumbered by making current version latest. Click on the **Continue** button from the information box. The **Save As Latest** dialog box will be displayed; refer to Figure-29. Specify desired description text about this version in **Version Description** edit box. Select the **Milestone** check box if you want to mark it as achievement of one of your design goals and click on the **OK** button from dialog box. The file will be saved on cloud.

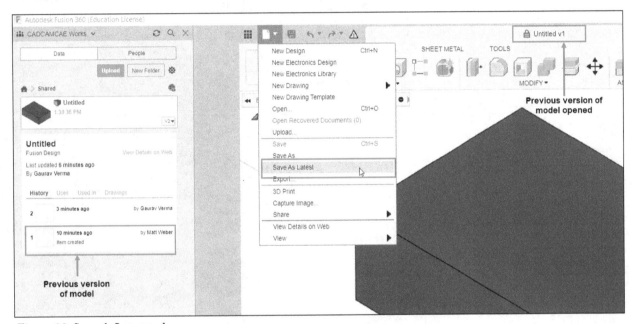

Figure-28. Save As Latest tool

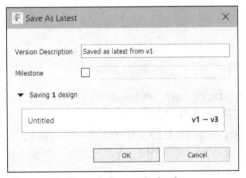

Figure-29. Save As Latest dialog box

Export

The **Export** tool is used to export the file in different formats. The procedure to use this tool is given next.

- Click on the **Export** tool from the **File** menu; refer to Figure-30. The **Export** dialog box will be displayed; refer to Figure-31.

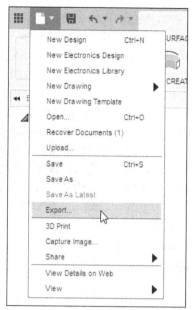

Figure-30. Export tool

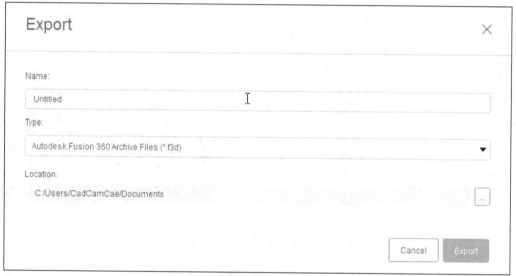

Figure-31. Export dialog box

- In the **Type** drop-down, there are various options to specify export format; **F3D**, **IGES**, **SAT**, **SMT**, **STEP**, **IPT**, **DWG**, **DXF**, and so on. Select desired format from the drop-down. Note that for some file formats, a cloud conversion will be required and take some time. You will be able to see progress of conversion in **Job Status** dialog box.

- Specify desired name and location for the file. Note that you can save the files only in the local drive by using this dialog box.
- Click on the **Export** button from the dialog box to save the file.

3D Print

The **3D Print** tool is used to prepare and send current model for 3D printing. This tool convert the selected body to mesh body and sends the output to 3D-Print Utility or converts it to **STL** format. The procedure to use this tool is given next.

- Click on the **3D Print** tool from the **File** menu or click on the **3D Print** tool from the **Make** panel in the **Tools** tab of the **Toolbar**; refer to Figure-32. The **3D PRINT** dialog box will be displayed; refer to Figure-33.

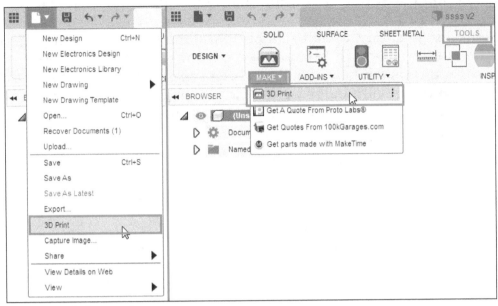

Figure-32. 3D Print tool

Figure-33. 3D Print dialog box

- By default, the **Selection** button is active in this dialog box and you are prompted to select model to be 3D printed. Click on the model to be 3D printed.
- Now, select the **Preview Mesh** check box to see the number of triangles forming in the selected body when it converts to mesh.

If you don't know why it is previewed in mesh then you need to know full name of standard file format used for 3D printing. STL format is used as standard for 3D printing which abbreviates **Standard Triangle Language**. So, your model will be broken into very small triangles and then these triangles will be arranged in the shape of your model in layers.

• Select desired level of resolution for printing in **Refinement** drop-down menu. There are four resolution in this drop-down. If you select the **High, Low,** and **Medium** command then the **Surface deviation, Normal deviation**, **Maximum Edge Length,** and **Aspect Ratio** parameters in the **Refinement Options** section will be adjusted automatically. (Click on the arrow before **Refinement Options** node to expand the section.) If you want to adjust these parameters manually then you need to select the **Custom** option in **Refinement** drop-down; refer to Figure-34. Note that your 3D printing machine should be able to print the model as set resolution. A higher resolution will take more time and power in 3D Printer. Unnecessarily high refinement should be avoided in 3D printing.

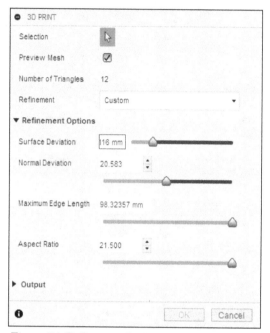

Figure-34. Refinement Options

Now, for the technical stuff :

The **Surface Deviation** is difference between real surface of model and triangulated tessellation of model. In simple language, it drives the number of triangles to be generated along the surfaces/faces in mesh; refer to Figure-35. The value specified in **Surface Deviation** edit box will be taken as maximum deviation allowed, the software will process the deviation lower than specified value but it will not go higher.

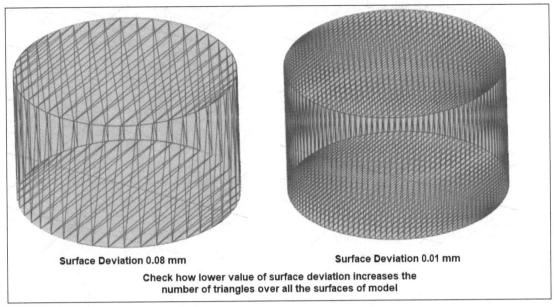

Figure-35. Surface Deviation in 3D Printing

The **Normal Deviation** is gap between two normals of consecutive triangles in tessellation. In simple words, a lower value of normal deviation will increase the number of triangles perpendicular to the edges of model in a given region; refer to Figure-36.

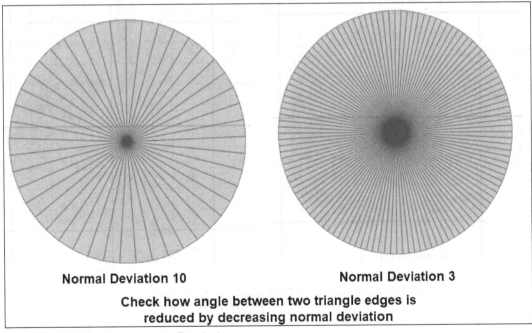

Figure-36. Normal Deviation in 3D printing

The **Maximum Edge Length** is used to define the maximum length of triangle edges. If you decrease the maximum edge length then more triangles of smaller edge length will be created in best possible orientations; refer to Figure-37.

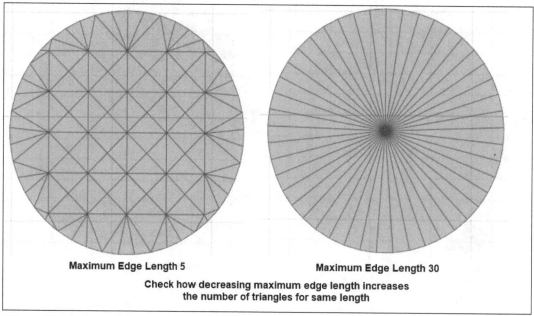

Figure-37. Maximum Edge Length in 3D Printing

The **Aspect Ratio** defines the ratio of number of triangles along width and height of the model; refer to Figure-38. (If you have irregular shape model then don't bother finding aspect ratio, your computer know better!!)

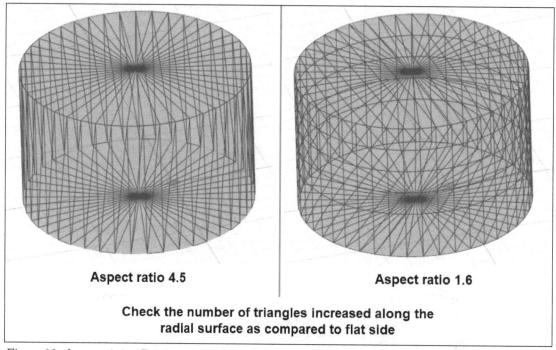

Figure-38. Aspect ratio in 3D printing

- Select the **Send to 3D Print Utility** check box in **Output** section to further edit the model in 3D printing utility. Select desired 3D Printing utility from **Print Utility** drop-down; refer to Figure-39. Note that 3D printing utilities are not installed automatically during Fusion 360 installation. On selecting the utility from **Print Utility** drop-down, an option to download and install the utility will be displayed in the dialog box. Download and install desired utility.

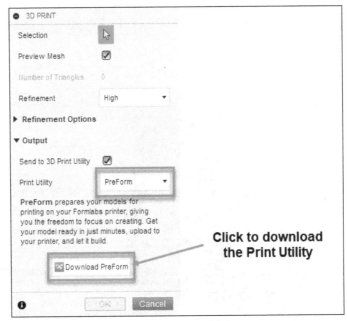

Figure-39. Option To Download Print Utility

- Clear the **Send To 3D Print Utility** check box if you want to save the project file in **STL** format so that you can open this file later in any 3D Printing software.
- In our case, we have selected **PreForm** option from the **Print Utility** drop-down (Needless to say, we have installed it already). Click on the **OK** button from the dialog box. The **PreForm** application window will be displayed; refer to Figure-40.

Figure-40. PreForm window

The tools of this application are discussed next.

PREFORM

Click on the **Apply** button from the **JOB SETUP** dialog box to enter the print utility, we will come back to these options later. On clicking **Apply** button, the model will be displayed in PreForm application window; refer to Figure-41.

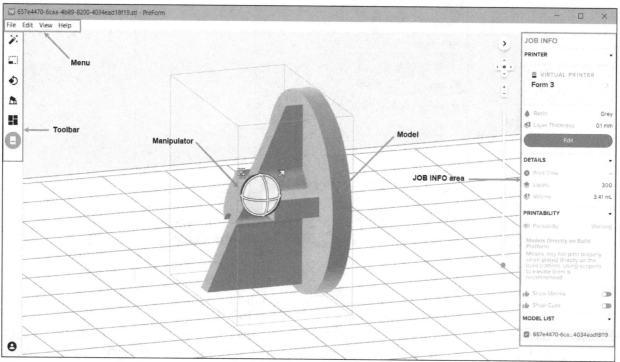

Figure-41. Interface of PreForm

Orienting the Model

There are two ways to orient a model in Preform, using **Orientation** tool from **Toolbar** or by using **Manipulator** displayed on the model. We will discuss both the ways here.

Orienting Model Using Manipulator

Click on the **Circle** handle in **Manipulators** for direction in which you want to rotate the model for orientation and drag the handle; refer to Figure-42. Note that software will automatically place the flat face of model on the bed of virtual machine.

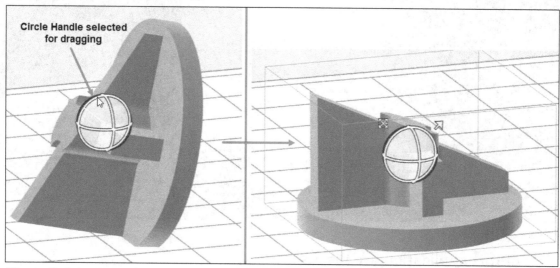

Figure-42. Orienting model using Manipulator

If you want to move the model at some other location then click on the **Move** handle in **Manipulators** and drag it to desired location; refer to Figure-43.

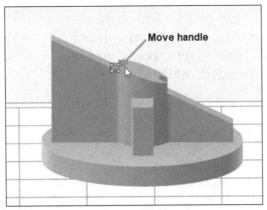

Figure-43. Move handle in Manipulator

Note that the **Manipulators** are displayed on model only when the model is selected. If the **Manipulators** are not displayed by default in your case or it is hidden then select the **Show Manipulators** option from the **View** menu; refer to Figure-44.

Figure-44. Show Manipulators option

Orienting Using Orientation Tool

• Click on the **Orientation** tool 🔄 from the **Toolbar** at the left in application window. The **ORIENTATION** toolbox will be displayed; refer to Figure-45.

Figure-45. ORIENTATION toolbox

• Click on the **Auto-Orient Selected** button to automatically orient the selected model on bed. (Note: this option was clumsy while writing this book and could not generate a proper orientation so we do not recommend this tool.)

- Click on the **Select Base** button to define the face which will be placed flat on the bed. After selecting this button, click on the face to be used as base of model; refer to Figure-46. The model will be automatically oriented to bed. Click on the **Done** button from the information box displayed in top-left corner. (This is the most used and recommended method to orient the part.)

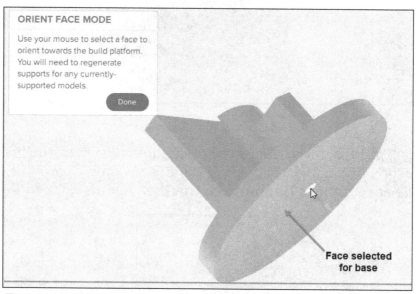

Figure-46. Selecting face for base

- You can use the spinners in **ORIENT AXES** area of the dialog box to define orientation angles for model. (These options are generally used to tweak the orientation; refer to Figure-47. You can also use handles in Manipulators to tweak the orientation.)

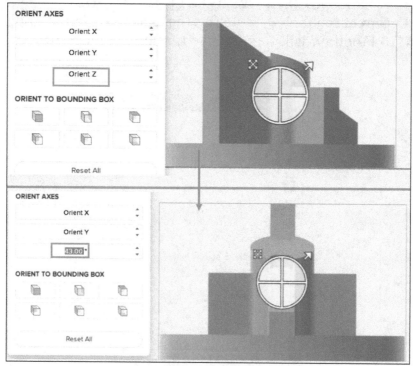

Figure-47. Tweaking the orientation of model

- The buttons in **ORIENT TO BOUNDING BOX** area of toolbox are used to orient the model along respective faces of the bounding box. (Bounding box is an imaginary cuboid which defines maximum length, width, and height occupied by the model.)

• At any time, you find that you have messed up the orientation then you can move back to original orientation by clicking on the **Reset All** button at the bottom in the toolbox.

Note that there are two sides of bed for 3D printing machine called FRONT and MIXER SIDE; refer to Figure-48. The front is the side from where you will be extracting the 3D printing model so this is the side of machine where you will be standing while controlling the machine. The Mixer side is the one where material cartridge of 3D printer will be located.

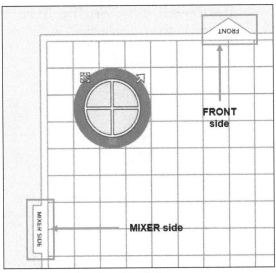

Figure-48. Sides of machine bed

Changing View

You can use the **Up**, **Down**, **Left**, **Right**, and **Home** buttons at the top-right corner of application window to switch between different views; refer to Figure-49. To dynamically change the view, right-click in modeling area, hold the button, and drag the cursor.

To zoom in and zoom out, click on the **+** and **-** buttons respectively which are available in the **View Switch** buttons or you can scroll up and scroll down with mouse.

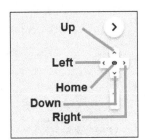

Figure-49. View Switch buttons

The shortcut keys for changing views are given next.

Up View	CTRL + Up key
Down View	CTRL + Down key
Left View	CTRL + Left key

Right View	CTRL + Right key
Home View	F
Zoom in	+ or Scroll up
Zoom out	- or Scroll down
Pan	SHIFT + Right-click drag
Dynamic View change	Right-click drag

Scaling Model

To scale up or scale down the model, click on the **Size** handle in **Manipulators**, hold the button and then drag in desired direction to scale up or scale down; refer to Figure-50.

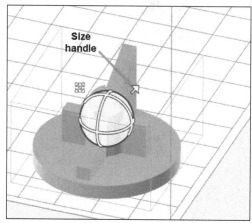

Figure-50. Size handle

OR

- Click on the **Size** button from the **Toolbar** at the left in application window. The **SIZE** toolbox will be displayed; refer to Figure-51.

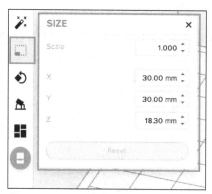

Figure-51. SIZE toolbox

- Specify desired scale value in the **Scale** edit box of toolbox or you can use the spinner to change its value.
- You can also use **X**, **Y**, or **Z** spinners to change the scale of model. Note that size of model will increase or decrease uniformly in all direction so it does not matter whether you use **Scale**, **X**, **Y**, or **Z** spinner.
- Click on the **Reset** button if you want to revert back to original scale. Close the toolbox after making desired changes.

JOB INFO Settings

The options in the **JOB INFO** area are used to set parameters related to printer. You can also check the warnings and details related to 3D printing process. Various tools and options are discussed next.

Selecting and Adding Printers

• Click on the **Edit** button from the **JOB INFO** area at the right in the dialog box. The **JOB SETUP** dialog box will be displayed; refer to Figure-52.

Figure-52. JOB SETUP dialog box

• Click in the **Select Printer** box of **Printer** area in the dialog box to select desired printer. The **PRINTER LIST** dialog box will be displayed; refer to Figure-53.

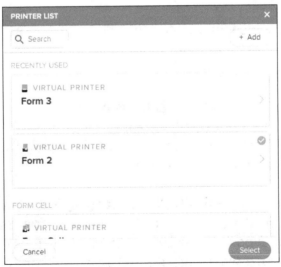

Figure-53. PRINTER LIST dialog box

• Select desired printer from the list and click on the **Select** button from the **PRINTER DETAILS** dialog box displayed.
• If you want to add a new printer in the list of **PRINTER LIST** dialog box, click on the **+Add** button from the dialog box. The **ADD PRINTER** dialog box will be displayed; refer to Figure-54. Specify the IP address of network printer in the **IP address** input boxes and click on the **Connect** button. The printer will be added in the list.

Figure-54. ADD PRINTER dialog box

- If you have selected a 3D printer which has options to select different types of materials then options to select material will be activated in the **Material** area of the dialog box; refer to Figure-55. Select desired resin and version of resins from the respective drop-downs in the dialog box.

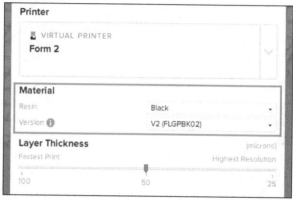

Figure-55. Material options

- Move the layer thickness slider to desired size for increasing print speed or resolution.
- Click on the **Apply** button to set the parameters.

Job Details

Various details of the printing job will be displayed in the **DETAILS** rollout of the **JOB INFO** area like estimated printing time, number of layers in 3D printed model, and total volume of material used in 3D printing.

PRINTABILITY Options

There are three options in the **PRINTABILITY** rollout of **JOB INFO** area; **Printability**, **Show Minima**, and **Show Cups**. The **Printability** icon defines whether the model is good for printing, there are warnings for 3D printing, or there are errors stopping from 3D printing like in Figure-56, a warning is displayed telling you that you need to elevate the model and add some supports for proper extraction.

Figure-56. Warning for 3D printing

- Select the **Show Minima** radio button to display the unsupported areas of model if the model is not oriented correctly; refer to Figure-57.

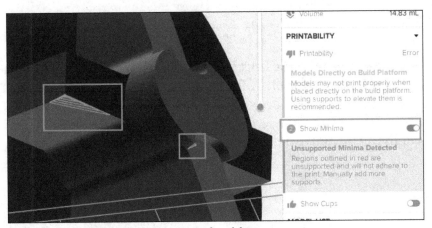

Figure-57. Unsupported minima areas of model

- Select the **Show Cups** radio button from the **PRINTABILITY** rollout to display areas on model that can trap air while 3D printing. Most of the time cups are generated when there are holes/cuts perpendicular to base plane.

Adding Supports

The **Supports** tool in the **Toolbar** is used to add supports to the 3d printing object. Supports are used when the model is unstable or extraction of 3d printed model from base plate can cause damage to the model. The procedure to add supports is given next.

- Click on the **Supports** tool from the **Toolbar** or press **C** from keyboard. The **SUPPORTS** toolbox will be displayed; refer to Figure-58.

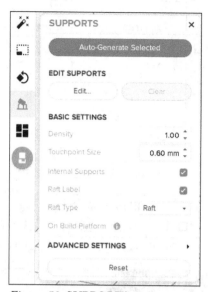

Figure-58. SUPPORTS toolbox

- Click on the **Auto-Generate Selected** button to automatically generate support features for the selected model; refer to Figure-59. Note that using this button, only positions of support elements are decided. It does not automatically decides the shape and size of support elements. So, you should click on this button after you have specified the parameters for supports in the toolbox.

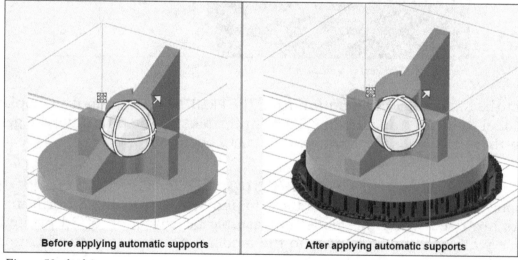

Figure-59. Applying supports

- Set desired density and touch point size in respective edit boxes in the toolbox. Density is the thickness of support base and touch point size is the thickness of upper section of support where it touches the model.
- Select the **Internal Supports** check box if you want to create internal supports whose start point and end point are on the model; refer to Figure-60.

Figure-60. Internal supports created

- Select the **Raft Label** check box if you want to 3D printing the name of file on the support raft.
- Select the **Mini-Rafts** option from the **Raft Type** drop-down if you want to mini-rafts for supports; refer to Figure-61.

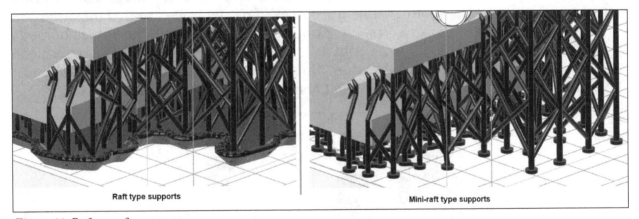

Figure-61. Raft types for supports

- Select the **On Build Platform** check box if you want to place the base of model on bed of machine directly; refer to Figure-62.

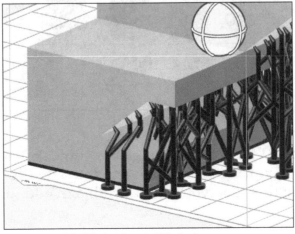

Figure-62. Model placed on build platform

- Specify desired parameters in the **ADVANCED SETTINGS** rollout of the toolbox if needed.

Creating Layout for Multiple Part Printing

- Click on the **Layout** tool from the **Toolbar**. The **LAYOUT** toolbox will be displayed; refer to Figure-63.

Figure-63. LAYOUT toolbox

- Set desired parameters like space between two instances of the model, raft overlap, rotation lock, and number of duplicate copies of model to be created on same bed.
- After setting desired parameters, click on the **Create** button. The copies of model will be created on the bed. Note that every time you click on the **Create** button, new copies of the model are created. So, it is better to set the duplicate count value carefully and then click on the **Create** button; refer to Figure-64.

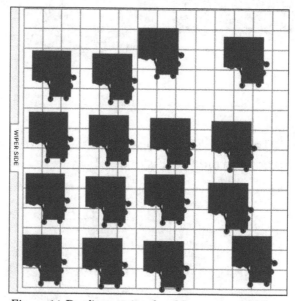

Figure-64. Duplicate copies of model

- Click on the **Layout All** button to arrange the copies of model in such a way that maximum number of copies can be accommodated on the bed; refer to Figure-65. Note that after using this tool, you might be able to add more copies of the model on bed.

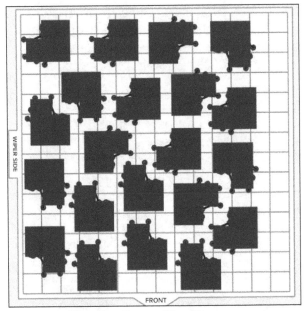

Figure-65. After using Layout All tool

- Click on the **X** button at top-right corner to exit the toolbox.

Printing

Now, we are ready to start 3D printing of model layout. If you have printer connected with your system then click on the **Start a print** button from the **Toolbar**. The **PRINT** dialog box will be displayed; refer to Figure-66. Click in the **Job Name** edit box and specify desired name for print job. If you have 3D Printer connected then it will be displayed in the Printer list. You need to select the printer from the list. If your printer is connected to a network on FORMLABS then you need to specify account information or log in to your FORMLABS account. After setting desired parameters, click on the **Upload Job** button. The file will be uploaded on FORMLABS account for printing.

Figure-66. PRINT dialog box

After performing desired operations, click on the **Quit PreForm** tool from the **File** menu in the application window to exit the PreForm application.

Capture Image

The **Capture Image** tool is used to capture the image of model in current state. The procedure is discussed next.

- Click on the **Capture Image** tool from **File** menu; refer to Figure-67. The **Image Options** dialog box will be displayed; refer to Figure-68.

Figure-67. Capture Image tool

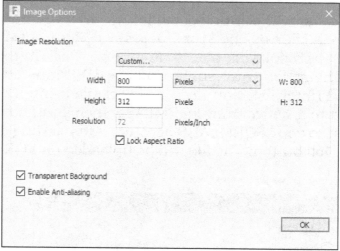

Figure-68. Image Options dialog box

- Select desired option from the drop-down in **Image Resolution** area to define size of image. All the other parameters of **Image Resolution** area will be defined automatically.
- If you want to set image resolution other than the standard one then select the **Custom** option from the drop-down and define the parameters. By default, you are asked to specify size in pixels but you can also define the image size in inches by selecting the **Inches** option from the drop-down next to **Width** edit box. Now, you will be able to define pixel density in the **Resolution** edit box manually. Clear the **Lock Aspect Ratio** check box to specify values of height and width individually.
- Select **Transparent Background** check box for setting the background of the image to be transparent.
- Select the **Enable Anti-aliasing** check box to smooth jagged lines or textures by blending the color of an edge with the color of pixels around it.
- After setting desired parameters, click on the **OK** button from **Image Options** dialog box. The **Save As** dialog box will be displayed; refer to Figure-69.

- Select the desired format from the **Type** drop-down for image. There are three formats available: PNG, JPG, and TIFF.
- Set the other parameters as discussed earlier and click on the **Save** button to save the image.

Figure-69. Save As dialog box for image capturing

Share

The **Share** tools are used to share the file or project on various portals. Click on the **Share** tool from **File** menu. Four options will be displayed in **Share** cascading menu; refer to Figure-70.

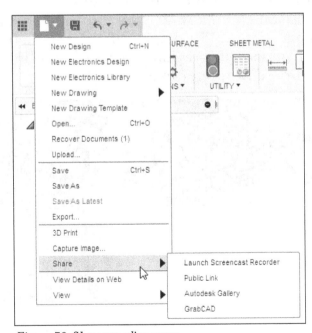

Figure-70. Share cascading menu

Launch Screencast Recorder

The **Launch Screencast Recorder** tool is used to record Autodesk Fusion 360 commands and entire product development process into the shareable videos. The procedure to use this tool is discussed next.

- Click on **Launch Screencast Recorder** tool from **Share** cascading menu.
- On selecting this tool, you will redirected to the **Autodesk** Website through which you can download **Autodesk Screencast Recorder** for Windows or Mac OS; refer to Figure-71.

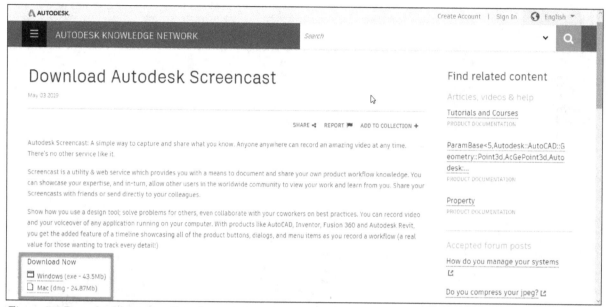

Figure-71. Download Screencast Recorder

- While installing this application, select the **Create a desktop icon** check box to create the icon of this application on the Desktop.
- After finishing the installation, double-click on the **Screencast Recorder** icon. The **Screencast** application will be displayed; refer to Figure-72.
- Click on **More** button and select a region on which you want to record a video from **Select a program** drop-down; refer to Figure-73.

Figure-72. Autodesk Screencast Recorder

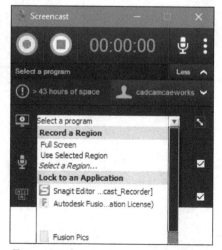

Figure-73. Select a program drop-down

- Select check boxes for **Microphone** and **Capture Keyboard Events** from the expanded dialog box to record voice and keyboard strokes while recording screencast.
- Note that if dimensions for recording are different from application window size then cross button will be displayed in the expanded dialog box; refer to Figure-74.

Figure-74. Mismatch in recording dimension and application size

- Click on the cross button to automatically resize the application according to recording dimensions.
- Switch on the **Record** button and start recording. Once you have completed the recording, click on the **Stop** button from **Screencast** application window. The **Screencast Recording Preview** dialog box will be displayed; refer to Figure-75.

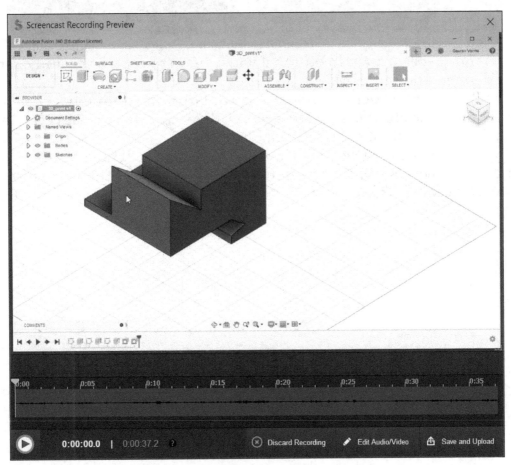

Figure-75. Screencast Recording Preview dialog box

- If you want to edit the audio and video of recording then click on the **Edit Audio/Video** button from the bottom of preview window. The time bar for recording will be displayed as shown in Figure-76.
- Move the time seeker at desired location and click on the **Set green marker on playhead** button to set start point for editing video/audio. Now, move the time seeker to position upto which you want to edit the video/audio and click on the **Set red marker on playhead** button; refer to Figure-77.

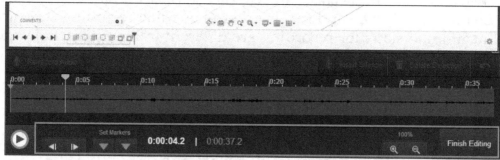

Figure-76. Updated options in timebar

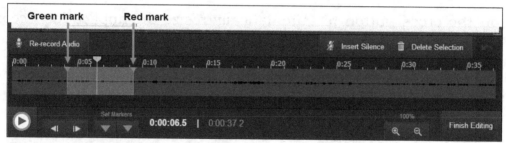

Figure-77. Marks for editing

- Click on the **Re-record Audio** button if you want to re-record the audio for section between green and red markers.
- Click on the **Insert Silence** button if you want to mute the audio for given section.
- If you want to delete the selected section then click on the **Delete Selection** button from the bottom of application window.
- After performing desired editing, click on the **Finish Editing** button.
- Click on the **Save and Upload** button to save and upload the video on Autodesk Knowledge Network. The **Screencast** dialog box will be displayed; refer to Figure-78.

Figure-78. Screencast dialog box

- Set desired title, description, and other parameters in the dialog box and then click on the **Upload** button to upload video on AKN.

Share Public Link

The **Public Link** tool is used to share the project link with anyone. The procedure to use this tool is discussed next.

- Click on **Public Link** tool of **Share** cascading menu from **File** Menu.
- On selecting this tool, the **Share Public Link** dialog box will be displayed, refer to Figure-79.

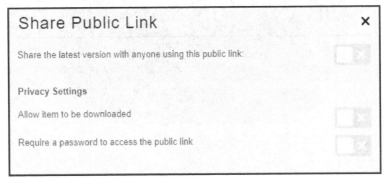

Share Public Link   ✕

Share the latest version with anyone using this public link:

Privacy Settings

Allow item to be downloaded

Require a password to access the public link

Figure-79. Share Public Link dialog box

- Click on the toggle buttons to activate or de-activate respective functions.
- After activating the sharing function, copy the link from **Copy** box and share the link via E-mail or other method with anyone. Set the **Allow item to be downloaded** toggle button to allow downloading of file.
- Activate the **Require a password to access the public link** option and specify desired password in the edit box displayed below it to secure your link with a password.

Share To Autodesk Gallery

The **Autodesk Gallery** tool is used to share the file online to the gallery of Autodesk where the file is accessible to everyone registered with the gallery. The procedure to use this tool is given next.

- Click on **Autodesk Gallery** tool from **Share** cascading menu. The **SHARE TO AUTODESK GALLERY** dialog box will be displayed; refer to Figure-80.

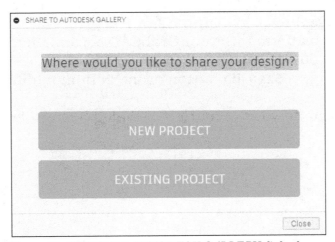

SHARE TO AUTODESK GALLERY

Where would you like to share your design?

NEW PROJECT

EXISTING PROJECT

Close

Figure-80. SHARE TO AUTODESK GALLERY dialog box

- Click on the **NEW PROJECT** button from the dialog box. The options will be displayed as shown in Figure-81.

Figure-81. Screen Capture Of 3D Model

- Scroll down in the dialog box and specify the information as required.
- Once all desired information is specified, click on the **Publish** button. **Your project has been successfully shared** message will be displayed.
- Click on the **Go and take a look!** button to check your model in gallery.

Publish to GrabCad

Autodesk Fusion 360 allows you to publish your model directly to **GrabCad** portal. The procedure to use this tool is given next.

- Click on **GrabCad** tool from **Share** cascading menu. The **PUBLISH TO GRABCAD** dialog box will be displayed; refer to Figure-82.
- Login to GrabCAD platform with your GrabCAD account details to upload or share the project. If you do not have an account on GrabCAD then you can create the account by clicking on **Sign Up** button at the bottom in dialog box and following the instructions.
- Once you have login to your account, the dialog box will be displayed as shown in Figure-83.

Figure-82. PUBLISH TO GRABCAD dialog box

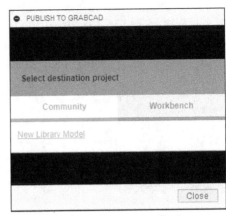

Figure-83. Select Destination Project

- Click on the **New Library Model** link button if you want to save model in community. If you want to share model in a workbench project then click on the **Workbench** tab in the dialog box and then click on the **New Workbench Project** link button. Based on your selection, the respective information will be displayed in the dialog box; refer to Figure-84.

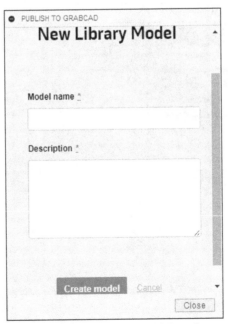

Figure-84. New Library Model Details

- Set desired parameters and click on the **Create model** button. In the next page, click on the **Publish Here** button. Your model will be uploaded to GrabCAD.

View details On Web

The **View Details On Web** tool is used to display the details of current file on Autodesk cloud in web browser; refer to Figure-85. You can use this option only after you have saved your file on Autodesk cloud.

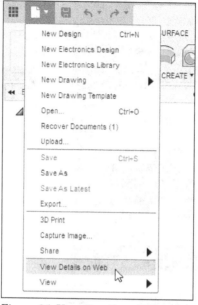

Figure-85. View Details on Web tool

View

The **View** tool is used to show and hide various elements of Fusion 360 application window.

* Click on the **View** tool from **File** menu. The **View** cascading menu will be displayed; refer to Figure-86.

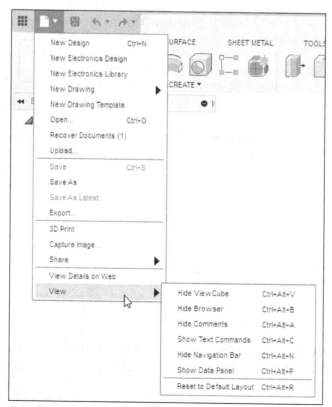

Figure-86. View cascading menu

Various tools in this cascading menu are given next.

Hide ViewCube

The **Hide ViewCube** tool is used to hide the **ViewCube.** If the **ViewCube** is already hidden then **Show ViewCube** tool will be displayed in place of it. On again selecting this tool, **View Cube** will be displayed. You can also show/hide **ViewCube** by pressing the **CTRL+ALT+V** keys.

Hide Browser

The **Hide Browser** tool is used to hide the **Browser** from the current screen. If the **Browser** is already hidden then **Show Browser** tool will be displayed in place of it. This tool can also be activated by pressing **CTRL+ALT+B** together.

Hide Comments

The **Hide Comments** tool is used to show and hide the **Comments** from Fusion 360 software. If the **Comment** is already hidden then **Show Comment** tool will be displayed in place of it. This tool can also be used by pressing **CTRL+ALT+A** together.

Show Text Commands

The **Show Text Commands** tool is used to show **Text Commands Bar**. If the **Text Commands Bar** is already displayed then **Hide Text Commands** tool will be displayed in place of it. This tool can also be activated by pressing **CTRL+ALT+C** together.

Hide Navigation Bar

The **Hide Navigation Bar** tool is used to hide the **Navigation bar**. If the **Navigation Bar** is already hidden then **Show Navigation Bar** tool will be displayed in place of it. This tool can also be activated by pressing **CTRL+ALT+N** together.

Show Data Panel

The **Show Data Panel** tool is used to show the **Data Panel**. If the **Data Panel** is already showing then **Hide Data Panel** tool will be displayed in place of it. This tool can also be activated by pressing **CTRL+ALT+P** together.

Reset To Default Layout

The **Reset To Default Layout** tool is used to reset user interface to default settings. This tool can also be used by pressing **CTRL+ALT+R** together.

Scripts and Add-Ins

The **Scripts and Add-Ins** tool is used to run and manage **Scripts** and **Add-Ins** in Autodesk Fusion 360. The procedure to use this tool is discussed next.

- Click on **Scripts and Add-Ins** tool from **ADD-INS** drop-down in **TOOLS** tab. The **Scripts and Add-Ins** dialog box will displayed; refer to Figure-87.

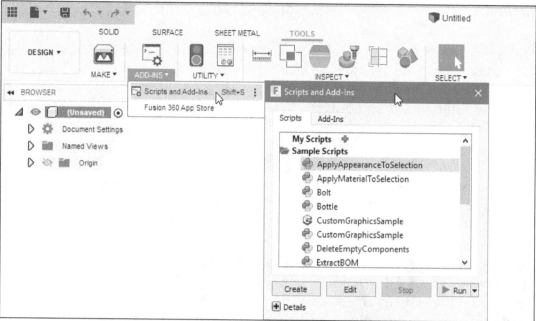

Figure-87. Scripts And Add-Ins

There are two tabs in this dialog box; **Scripts** and **Add-ins**. Select the **Script** tab to run Python scripts in Autodesk Fusion. Select the **Add-Ins** tab to run the third party applications inside Autodesk Fusion 360.

Creating Script and Add-Ins

- To create a script or Add-Ins, click on the **Create** button from **Scripts and Add-Ins** dialog box. The **Create New Script or Add-In** dialog box will be displayed; refer to Figure-88.

Figure-88. Create New Script or Add In dialog box

- Select **Script** radio button to create a script or select the **Add-In** radio button to create an Add-In.
- Select desired radio button from the **Programming Language** area to define the language in which you will be writing script/Add-in codes.

- Set the other parameters as required and click on the **Create** button. A new script/ Add-in will be added in the dialog box.
- Select the newly created script/Add-in and click on the **Edit** button. If the script/ Add-in is coded in **Python** then an information box will be displayed asking you to download and install VS Code. Download & install the application and once installation is complete, codes will be displayed in application when you edit; refer to Figure-89.

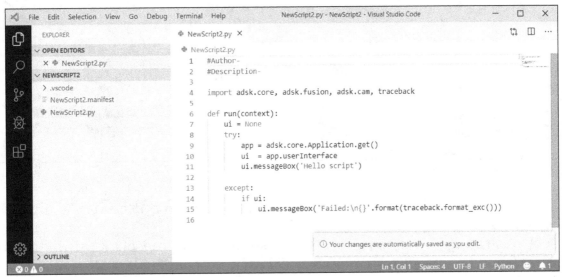

Figure-89. Visual Studio Code application window

- If you have opted for any other language then system will ask you for respective Programming software to display codes. If you like coding then you can check **Learning Python, 5th Edition by Mark Lutz** as reference book for Python programming.
- Click on the **Run** button from the **Scripts and Add-Ins** dialog box to run selected script.
- If you want to debug a script/Add-in then select it from the **Scripts and Add-Ins** dialog box and click on the **Debug** button in **Run** drop-down; refer to Figure-89. Programming interface will be displayed based on language of selected script/ Add-In.

Fusion 360 App Store

The **Fusion 360 App Store** tool in **File** menu is used to download and install third party Apps in Autodesk Fusion. On clicking this tool, a webpage is displayed in your default web browser; refer to Figure-90.

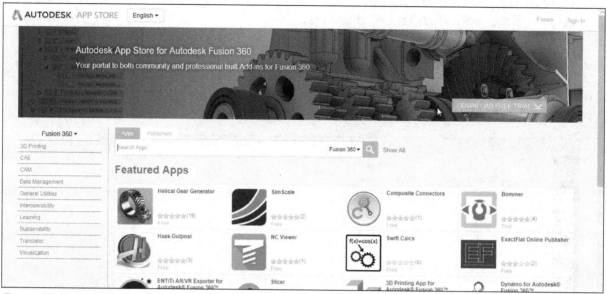

Figure-90. Fusion 360 Store Webpage

- Login to your Autodesk account after clicking **Sign In** button at the top right in the web page.
- Select desired app, set the operating system option and click on the **Download** button from web page; refer to Figure-91. The software will start downloading.

Figure-91. Downloading App

- Once downloaded, double-click on the setup file to install it and follow the instructions as displayed during installation.
- Once the app is installed, you can access it from **Add-Ins** tab of **Scripts and Add-Ins** dialog box; refer to Figure-92.

Figure-92. Installed App

UNDO AND REDO BUTTON

The **Undo** tool is used to revert back to condition before performing most recent action. The procedure to use this tool is discussed next.

* Click on **Undo** tool in **Application bar** or press **CTRL+Z** keys; refer to Figure-93.
* You can't undo some actions, like clicking commands on the **File** tab or saving a file.

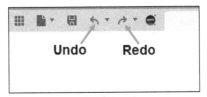

Figure-93. Undo And Redo Button

* The **Redo** tool is used to redo an action reverted by **Undo** tool. To use this tool, select the **Redo** button in **Application bar** or press **CTRL+Y** keys.
* The **Redo** button activates only after you have undone an action.

USER ACCOUNT DROP-DOWN

The tool in **User Account** drop-down are used to manage user account details and preferences for Autodesk Fusion 360; refer to Figure-94. The tools in this drop-down are discussed next.

Figure-94. User Account drop down

Preferences

The **Preferences** tool is uused to set the preferences for various functions of the software like you can set units, material libraries, display options, and so on. The procedure to use this tool is given next.

- Click on **Preferences** tool from **User Account** drop-down; refer to Figure-94. The **Preferences** dialog box will be displayed where you can specify various parameters for application; refer to Figure-95.

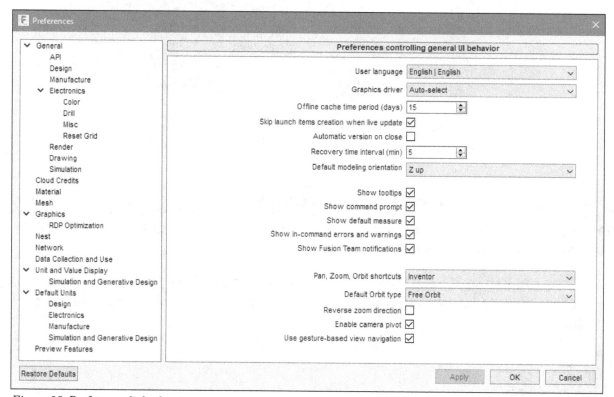

Figure-95. Preferences dialog box

General

In **General** node, you can specify the preferences for language, graphics, mouse functioning, default workspace, and so on. Click on the **General** node at the left in the dialog box to display general options. Various important options are discussed next.

- Click on the **User language** drop-down and select desired language for Autodesk Fusion interface.
- Click on the **Graphics driver** drop-down to select the graphics driver to be used by Autodesk Fusion.
- **Offline cache time period (days)** edit box/spinner is used to specify the number of days up to which your documents can be in the cache memory of Autodesk Fusion before you go back to online mode of Autodesk Fusion. If you specify high value then more local memory will be used by Autodesk Fusion to save temporary copy of your documents. If you specify a low value here then system will prompt you soon to go back online. The maximum number of days specified here can be 360 and minimum number of days can be 7.
- Whenever Autodesk Fusion is updated, launch icons are created automatically in various locations. Select the **Skip launch items creation when live update** check box to skip creating new launch icons.
- Select the **Automatic version on close** check box to automatically save the newer version of file when you close Autodesk Fusion.
- Specify desired time in minute after which a recovery copy of your model will be created automatically in the **Recovery time interval (min)** edit box.

- Click on the **Default modeling orientation** drop-down and select desired direction option to define default orientation of model. In most of the CAD software, Z axis upward is the default orientation of model. To set the common orientation, select the **Z up** option from the drop-down.
- Select the **Show tooltips**, **Show command prompt**, **Show default measure**, **Show in-command errors and warnings**, and **Show Fusion Team notifications** check boxes to display respective interface elements.
- Select desired software style from the **Pan, Zoom, Orbit shortcuts** drop-down to define which software style for shortcuts of pan, zoom, and orbit should be used. If you are switching from Alias, Inventor, Solidworks, or Tinkercad to Autodesk Fusion then you can select respective option from the drop-down to use familiar shortcuts.
- Select desired option from the **Default Orbit type** drop-down to define the default orbit type for rotating model view. There are two options available for orbit; **Constrained Orbit** and **Free Orbit**.
- Select the **Reverse zoom direction** check box to reverse the zoom direction of mouse scroll and other zoom methods.
- Select the **Enable camera pivot** check box to display camera pivot point used as center for orbiting.

Note that by default the pivot point is placed on center of mass point of model but if you want to change the position of pivot point then right-click in blank area of canvas and select the **Set Orbit Center** option. The green point will get attached to cursor and you will be asked to define the location of orbit point. Click at desired location to place the orbit center.

API Options

- Click on the **API** option under **General** node in the left of the dialog box and set the desired parameters to define default location and language of scripts/Add-Ins.

Design Options

- Click on the **Design** option under **General** node in the left of the dialog box to set the parameters related to model creation. The options in the dialog box will be displayed as shown in Figure-96.

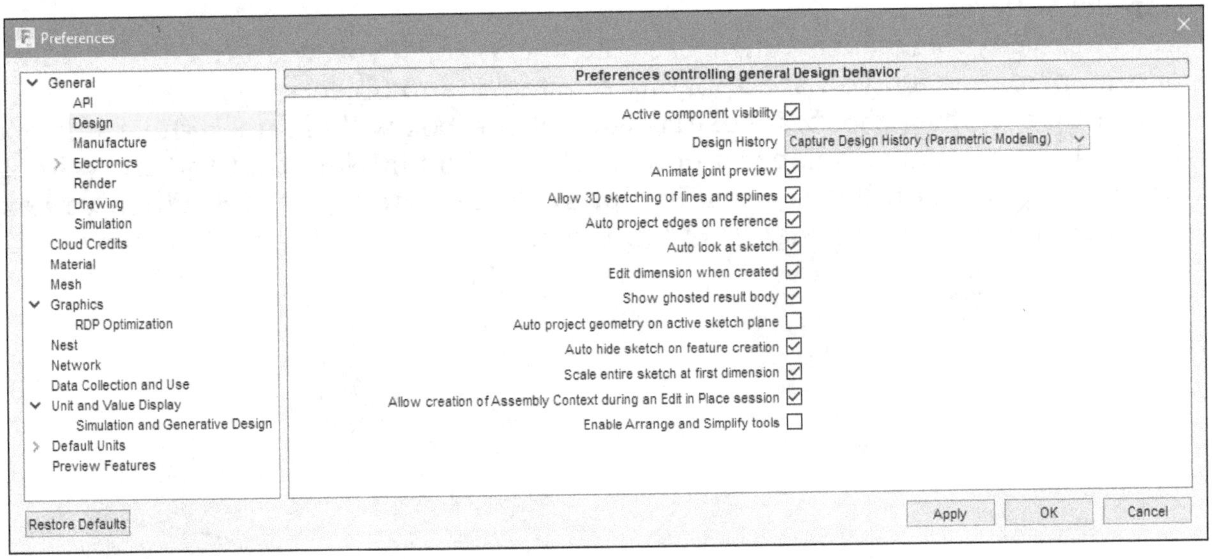

Figure-96. Design Page

- Select the **Active Component Visibility** check box to display only currently active component and hide rest of the components.
- Select the **Capture Design History** option from **Design History** drop-down if you want to keep detail of every operation you perform on the model in model history. If you want to save space or want to hide operation details while sharing file then create the model after selecting **Do not capture Design History** option from the **Design History** drop-down. Note that changes made in most of the options will be applied only after you have restarted Autodesk Fusion.
- Select the **Animate Joint Preview** check box if you want to check small animation of how joint will work. The animation can give you clues about degrees of freedom which are constrained by selected joint.
- Select the **Allow 3D sketching on lines and splines** check box if you want to create lines and splines which are not confined to single plane.
- Select the **Auto project edges on reference** check box if you want to automatically project edges of model when they are used as reference for dimensioning or constraining of other sketch elements. Refer to Figure-97.

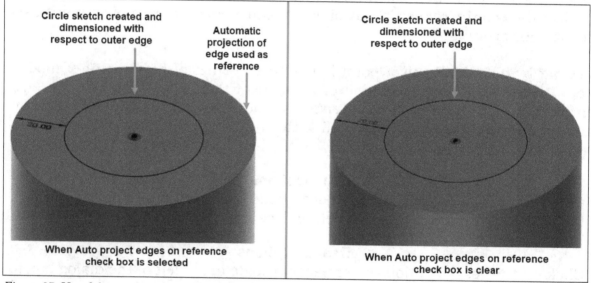

Figure-97. Use of Auto project edges on reference check box

- Select the **Auto look at sketch** check box to automatically make sketching plane parallel to screen.
- Select the **Edit dimension when created** check box if you want to edit the values of dimensions when you are applying dimensions to objects.
- Selecting the **Show ghosted result body** check box will create a ghost image of model before a sculpting operation is performed on it. Note that the ghost image will be displayed until you are in sculpt mode; refer to Figure-98. When you exit the model after making changes then the ghost image will not be displayed.

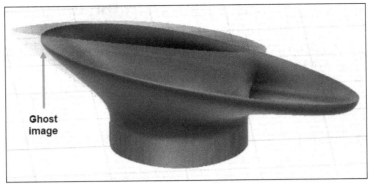

Figure-98. On selecting Show ghosted result body check box

- Select the **Auto project geometry on active sketch plane** check box if you want to project geometry on the sketching plane by default.
- Select the **Auto hide sketch on feature creation** check box to automatically hide the sketch after it has been used to create a feature.
- Select the **Scale entire sketch at first dimension** check box to automatically fit the sketch objects in view area with reduced dimensions after you specify first dimension of the sketch; refer to Figure-99. Note that all the other objects reduce in same ratio as the first dimension reduces its respective entity.

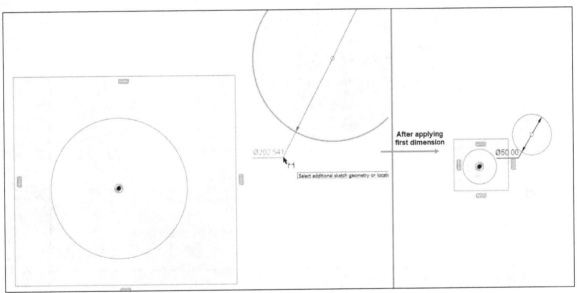

Figure-99. When scale entire sketch at first dimension check box is selected

- Similarly, set the other options as required on different nodes of the dialog box. We will work with options in this dialog box during rest of the book whenever a change in application settings is required.
- After setting the parameters, click on the **OK** button from the dialog box and restart Autodesk Fusion if needed.

Autodesk Account

The **Autodesk Account** tool in **User Account** drop-down is used to manage your Autodesk account profile in default web browser.

My Profile

The **My Profile** tool in **User Account** drop-down is used to manage your project data stored on cloud via web browser.

Work Offline/Online

The **Work Offline/Online** toggle button is displayed on clicking **Job Status** button at the left of **User Account** drop-down; refer to Figure-100. Click on this button to toggle between offline and online mode of Autodesk Fusion.

Figure-100. Work Offline Online

Extensions

Click on the **Extensions** button to display the list of current available extensions in the **Extension Manager** dialog box for adding more functionality to Autodesk Fusion; refer to Figure-101. If you want to use these extensions then select the extension to be used and click on **Purchase** button or activate button if available.

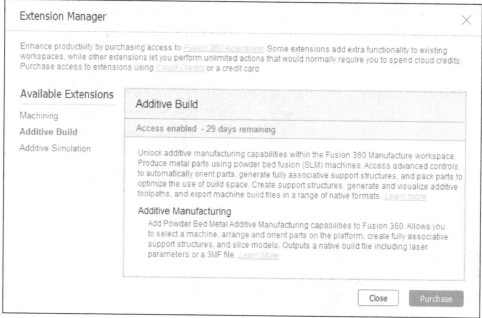

Figure-101. Extension Manager dialog box

HELP DROP-DOWN

When you face any problem working with Autodesk Fusion 360, the tools in **Help** drop-down can be useful; refer to Figure-102. Tools in the drop-down are given next.

Figure-102. Help drop down

Search Help Box

When you are facing any problem regarding tools and terms in the software, then you need to type keywords for your problem in **Search Box** and press **ENTER** to find a suitable solution to your problems; refer to Figure-103.

Figure-103. Search Box

Learning And Documentation

The tools in the **Learning and Documentation** cascading menu are used to display video tutorials, help files, and fusion 360 api related content in web browser; refer to Figure-104. On clicking the **Self-Paced Learning** button in the **Learning and Documentation** cascading menu of **Help** menu, the website of Autodesk Fusion 360 will be displayed; refer to Figure-105.

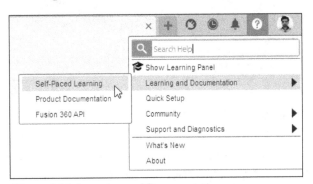

Figure-104. Learning and Documentation

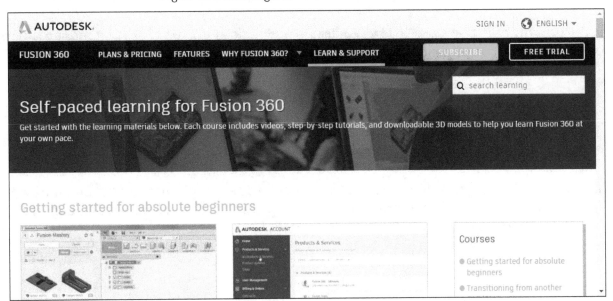

Figure-105. Fusion 360 Learning Website

Similarly, you can use **Product Documentation** and **Fusion 360 API** tools to learn about Fusion 360 interface, design methodology, and APIs.

Quick Setup

When you are new to Autodesk Fusion 360 then this tool will help you to understand this software terminology and basic settings.

- Click on **Quick Setup** tool from **Help** menu. The **QUICK SETUP** dialog box will be displayed; refer to Figure-106.

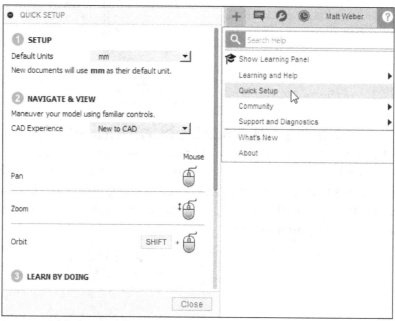

Figure-106. Quick Setup

- Set desired unit in the **Default Units** drop-down.
- The navigation functions of mouse can be set as per Autodesk Inventor, SolidWorks or Alias software by selecting respective option from the **CAD Experience** drop-down.
- If you are new to CAD then you can select the **New to CAD** option in **Cad Experience** drop-down. The mouse will function as per Autodesk Fusion 360.

Similarly, you can use **Community Forum, Feedback Hub, Gallery, Roadmap, Blog** tool to get help from Autodesk Fusion 360 users or provide feedback.

Support and Diagnostics

When you are having a problem in Fusion 360 and want **Technical support** team to look at software problem data, then you can create the log files and send it to the support team of Autodesk.

- To create log files, select **Diagnostic Log Files** tool under the **Support and Diagnostics** cascading menu in **Help** drop-down. The **Diagnostic Log Files** dialog box will be displayed; refer to Figure-107.

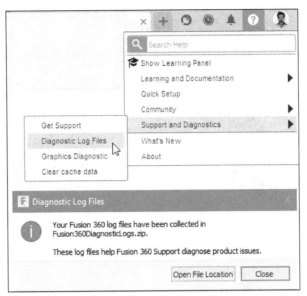

Figure-107. Diagnostic Log Files dialog box

- Click on **Open File Location** button and select the log file which you want to send to technical support team for the solution of problem.

Similarly, you can use **Graphics Diagnostic**, **Clear user cache data**, and other tools from **Help** drop-down in Autodesk Fusion 360.

DATA PANEL

The **Data** panel is used to manage the Autodesk Fusion 360 projects. The projects you save in cloud are shown in the **Data** panel. You can easily access the saved project files from anywhere and anytime with the help of internet access.

To show or hide the **Data** panel, click on the **Show Data Panel** or **Hide Data Panel** button ▦ at upper left corner of the Fusion 360 window; refer to Figure-108. A box will be displayed with files of current project. In this box, there are two sections; **Data** and **People**.

In **Data** section, the files which you have saved earlier in current project will be displayed. You can open any file by double-clicking on it.

In **People** section, the information of users will be displayed who have access the files of current project; refer to Figure-109.

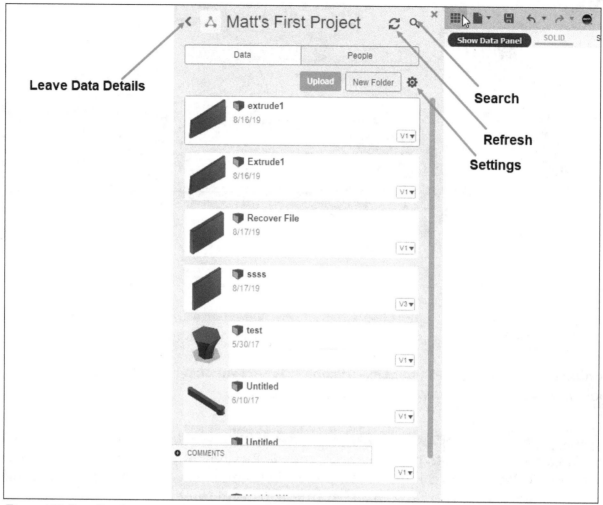

Figure-108. Data Panel

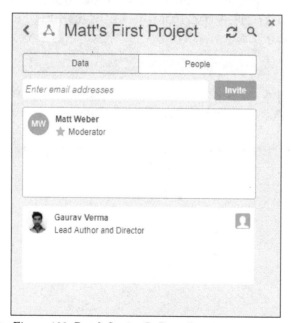

Figure-109. People Section In Data Panel

Working in a Team

Using the tools in **User** drop-down of **Data Panel**, you can create a new team or work in other's team to create designs in collaboration with others; refer to Figure-110. Various tools of this drop-down are discussed next.

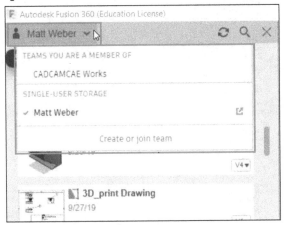

Figure-110. User drop-down

Creating a Team

- To create your own team for a given project, click on the **Create or join team** tool from the drop-down. An information box will be displayed about Fusion Team; refer to Figure-111. Note that this information box will be displayed for only one time when you are creating team using educational version of software as Educational version can create only 1 team for students & educators.

Figure-111. Create or join team information box

- Click on the **Next** button from the information box. The options to create or join teams will be displayed in a dialog box; refer to Figure-112.

Figure-112. Create a team or join existing team options

- Click on the **Create a Team** option from the dialog box displayed. You will be asked to specify name the team in next page of dialog box; refer to .

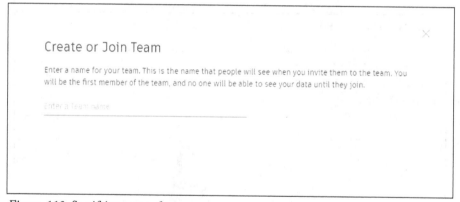

Figure-113. Specifying name of team

- Specify desired name for the team to be created and click on the **Next** button. The next page in dialog box will ask you whether to make the team discoverable on cloud server or make it private.
- Select desired radio button from the dialog box and click on the **Create** button. A message box will be displayed tell you that a team with your specified name is created. Click on the **Go to team** button from the information box. Name of the team will be added in the Data Panel; refer to Figure-114.

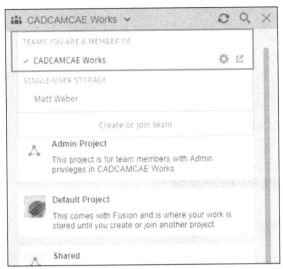

Figure-114. Team added in Data Panel

- Click on the **Open Administrator Console on the Web** button (similar to common settings button icon) displayed next to name of the team. The administration page for team will be displayed in the default web browser; refer to Figure-115.

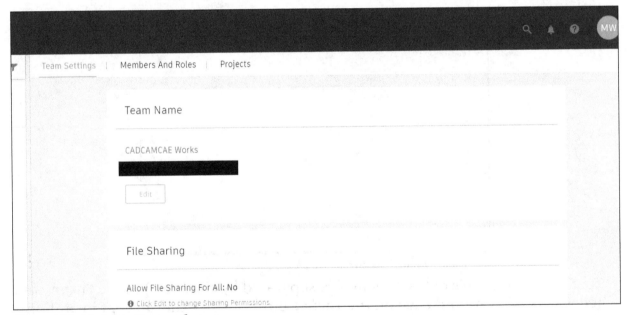

Figure-115. Administration page for team

- Specify desired parameters in this page to define settings related to file sharing, access, and collaboration. Note that you can set auto-approval settings for specified domains by using the **Add domain** button on this page. You will be asked to specify the text in email ID displayed after @ like for matt@cadcamcaework.com the @ cadcamcaework.com is domain.
- To add members in your team, click on the **Members and Roles** tab in the web page. The options will be displayed as shown in Figure-116.

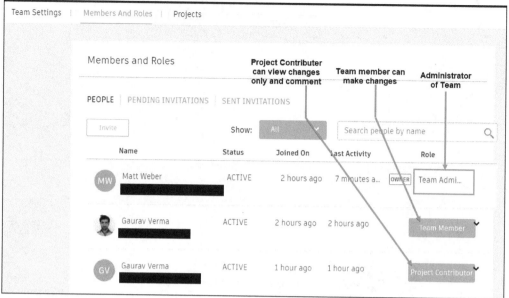

Figure-116. Members and roles page

- To invite a new member, click on the Invite button on this page. The **Invite Team Members** dialog box will be displayed; refer to Figure-117.

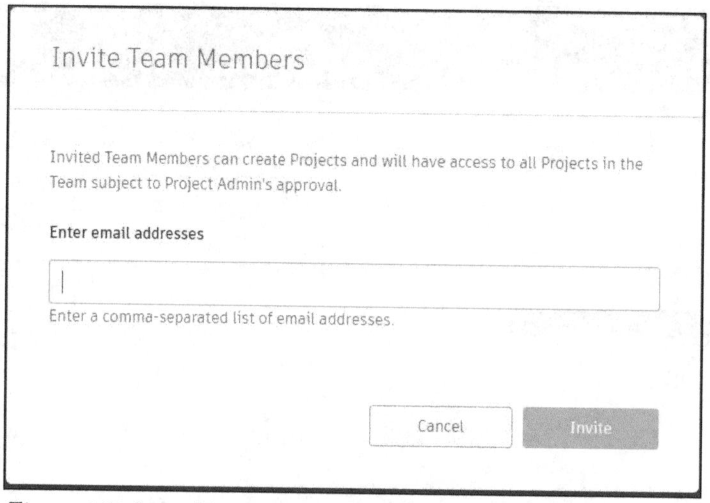

Figure-117. Invite Team Members dialog box

- Specify the email id's of team members separated by comma (,) in the **Enter email addresses** edit box and click on the **Invite** button. Once they accept invitation on their email ids, they will become members of the team. You can later assign the roles to members as desired using the Administrator page. You can now logout and close the browser to exit the admin panel of Fusion team.

After creating a team, make sure you are log-in by team name in the **Data Panel** and create a new project using **New Project** button in the **Data Panel**. This new project will be shared with all the team members for modifications based on roles assigned to them; refer to Figure-118.

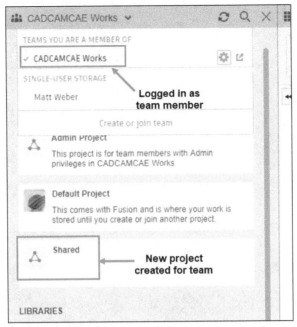

Figure-118. Sharing project with team

Cloud Account of User

To open the cloud account of user, you need to click on the **Open on the Web** button next to the name of user in **Data Panel**; refer to Figure-119. On selecting this button, details of user's cloud storage will be displayed in web browser. You can check and modify the files as desired. Note that you can delete any file available in Cloud account using the tools in web browser will is not possible in **Data Panel**; refer to Figure-120.

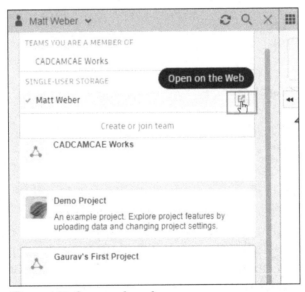

Figure-119. Open on the web

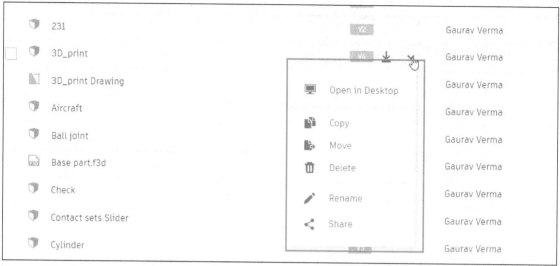

Figure-120. Shortcut menu for modifying files in clound account using web browser

BROWSER

The **BROWSER** presents an organized view of your design steps in a tree like structure on the left of the Fusion 360 screen; refer to Figure-121. When you select a feature or component in the **BROWSER**, it is also highlighted in the graphics window.

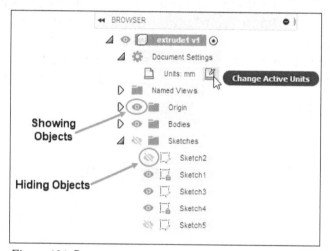

Figure-121. Browser

- Click on the **Eye** button to toggle the visibility of respective object.
- If you select any object in **BROWSER** then it is highlighted in blue color which indicates that the respective object is active for various operations.
- To change the units of the design or sketch, select the **Change Active Units** button as highlighted in the above figure. The **CHANGE ACTIVE UNITS** dialog box will be displayed. Select desired unit and click on the **OK** button.

NAVIGATION BAR

The **Navigation Bar** is available at the bottom of graphics window of Fusion 360. It provides access of navigation commands for design. To start a navigation command, click on any tool from the **Navigation Bar**; refer to Figure-122.

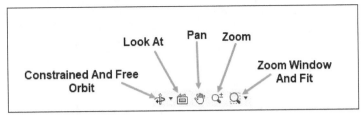

Figure-122. Navigation Bar

There are various tool in **Navigation Bar** to navigate the model which are discussed next.

- **Orbit**- The **Orbit** tools are used for rotating the current model. There are two types of orbit tools i.e. **Free Orbit** and **Constrained Orbit**. **Free Orbit** is used to rotate the design view freely and **Constrained Orbit** is used to rotate the view in constrained motion relative to current origin point. The **Constrained Orbit** tool is displayed on clicking down arrow next to **Free Orbit** tool. After selecting this tool, click at desired location on model and drag the cursor to rotate model view. You can do the same function by using **SHIFT** + Middle Mouse Button dragging.
- **Look At**- This tool is used to view the selected face parallel to screen. To use this tool, select desired face of model, and then click on the **Look At** tool from **Navigation Bar**.

- **Pan**- It is used to move the design parallel to the screen. You can do the same function by middle mouse button drag.

- **Zoom-** It is used to increase or decrease the magnification of the current view. After selecting this tool, click on the model, hold, and drag mouse upward/downward to zoom in/out. You can do the same function by rolling the mouse wheel up/down.

- **Zoom Window** and **Fit** - **Zoom Window** tools are used for magnification of selected area. **Fit** tool is used to position the entire design on the screen. You can do the same function as **Fit** tool does by pressing **F6** from keyboard.

DISPLAY BAR

The tools of **Display Bar** are used to visualize the design in different viewports; refer to Figure-123.

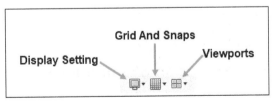

Figure-123. Display Bar

Display Settings

The options in the **Display Setting** flyout are used to enable or disable various commands related to **Visual Style**, **Visibility of objects**, **Camera settings**, and so on; refer to Figure-124.

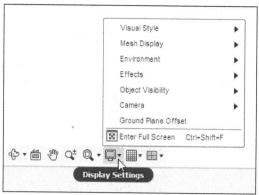

Figure-124. Display Settings flyout

- Select the desired visual style of 3D model from the **Visual Style** cascading menu. Various styles available are Shaded, Shaded with Hidden Edges, Shaded with Visible Edges only, Wireframe, Wireframe with Hidden Edges, Wireframe with Visible Edges only. (You can open a sample file of 3D model and check the effects of different visual styles.)

- The **Face Groups** check box available in the **Mesh Display** cascading menu is used to display or hide mesh group faces. You will learn more about this option while working on mesh.

- Select desired option from the **Environment** cascading menu to define color scheme of the canvas.

- Select desired check boxes for effects that you want to added in display style from the **Effects** cascading menu. Similarly, you can display or hide objects from the **Object Visibility** cascading menu.

- The options in **Camera** cascading menu are used to change the display to Orthographic, Perspective, or Perspective with Ortho Faces.

- Click on the **Ground Plane Offset** tool from the **Display Settings** flyout. The **GROUND PLANE OFFSET** dialog box will be displayed; refer to Figure-125. Clear the Adaptive check box and set desired offset value if you want to change the location of ground plane. Note that position of ground plane is generally changed for rendering which will be discussed later.

Figure-125. GROUND PLANE OFFSET dialog box

- If you want to run Autodesk Fusion 360 in full screen then press **CTRL+SHIFT+F** from keyboard.

Grids and Snaps

- The **Grid and Snaps** flyout is used to activate or deactivate various options like **Layout Grid**, **Layout Grid Lock**, **Snap to Grid**, and so on. You can also define the size of grid and snapping increments by using **Grid Settings** and **Set Increments** tools in this flyout.

Viewports

- The options in **Viewports** flyout are used to switch between four viewports and single viewport in the Fusion 360 screen. Click on the **Multiple Views** option from the **Viewports** flyout to display the current model in different views in four viewports. You can click in each of the viewport and reorient the model. If you want to set these views to synchronous again then select the **Synchronize Views** check box from **Viewports** flyout. If you want to move back to single viewport then select the **Single View** tool in **Viewports** flyout. By pressing **SHIFT+!** keys, you can toggle between single viewport and multiple viewports modes.

ViewCube

The **ViewCube** tool is used to rotate the view of model to access different faces of model. Click at desired location on model and drag the **ViewCube** to perform a free orbit. There are six faces in **ViewCube**; **Top**, **Bottom**, **Front**, **Back**, **Right**, and **Left** to access different orientations of model. To access the standard orthographic and isometric views, click on the faces and corners of the cube; refer to Figure-126. Right-click on the **ViewCube** to access options related to perspective mode and other view modes.

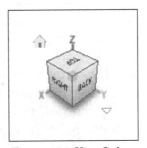

Figure-126. ViewCube

Mouse Functions

You can use the mouse shortcuts to zoom in/out, pan the view, and orbit the view; refer to Figure-127.

- Scroll the Middle Mouse Button downward/upward to zoom in or zoom out, respectively. Note that this function can be reversed by using options in **Preferences** dialog box.
- Click and hold the Middle Mouse Button to pan the view.
- Use the **Shift** + Middle Mouse Button to orbit the view.
- Press the left mouse button to select any object or tool.
- Press right mouse button (right-click) to access shortcut menus in the software.

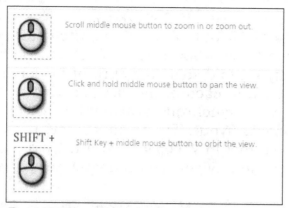

Figure-127. Mouse shourtcut keys

Timeline

The **Timeline** bar is very useful for saving time while editing. It records the design feature in chronological order; refer to Figure-128. Note that the **Capture Design History (Parametric Modeling)** option must be active in the **Design** node of **Preferences** dialog box to access **Timeline** bar.

Figure-128. Timeline

If you want to edit any feature of model, then double-click on the respective feature from **Timeline** bar. The respective dialog box/interface element will be displayed. Make desired changes and apply the parameters. The final design will be generated automatically with the applied changes; refer to Figure-129.

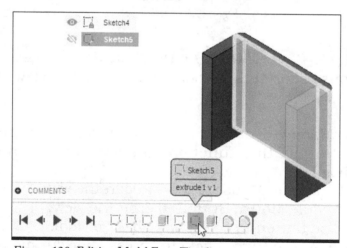

Figure-129. Editing Model From Timeline

CUSTOMIZING TOOLBAR AND MARKING MENU

In Autodesk Fusion 360, you can add or remove any tool in the displaying panels of **Toolbar**. You can also add or remove tool from right-click marking menu. These operations are discussed next.

Adding Tools to Panel

• Click on the **Three dot** button next to tool to be added in the panel from the drop-down of **Toolbar**. The shortcut menu to customize toolbar will be displayed; refer to Figure-130.

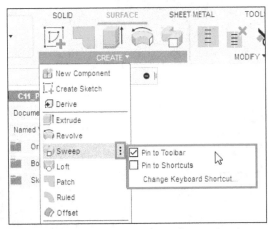

Figure-130. Shortcut menu for tool customization

- Select the **Pin to Toolbar** check box from the shortcut menu. The tool will be added in the panel.

Assigning Shortcut Key

- Click on the **Change Keyboard Shortcut** option from the shortcut menu as shown in Figure-130. The **Change Keyboard Shortcut** dialog box will be displayed; refer to Figure-131.

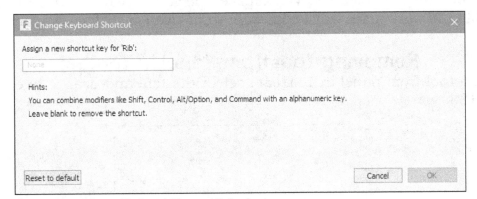

Figure-131. Change Keyboard Shortcut dialog box

- Use desired combine modifier like **SHIFT**, **CTRL**, or **ALT** and then press desired alphanumeric key to create a shortcut button.
- If you want to remove shortcut key then use **Backspace** to delete shortcut key from edit box. Click on the **OK** button from the dialog box to apply the settings.

Adding or Removing tool from Shortcut menu/Marking menu

- Select the **Pin to Shortcuts** check box from the shortcut menu for tool that you want to add in right-click shortcut menu/marking menu; refer to Figure-132.

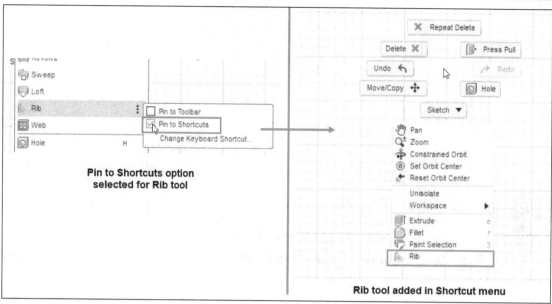

Figure-132. Adding tool to shortcut menu

- Clear the **Pin to Shortcuts** check box from the shortcut menu for tool that you want to remove from right-click shortcut menu/marking menu.

Note that the shortcut menu is Workspace specific so what you find in shortcut menu for **DESIGN** workspace will be different from **MANUFACTURE** workspace and so on.

Removing Tool from Panel in Toolbar

To remove a tool from panel in **Toolbar**, select the tool and drag it to canvas; refer to Figure-133.

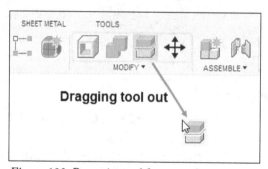

Figure-133. Removing tool from panel

Similarly, you can change position of tool by dragging it within the panel.

Resetting Panel and Toolbar Customization

Right-click on any of the tool in **Toolbar**. A shortcut menu will be displayed as shown in Figure-134.

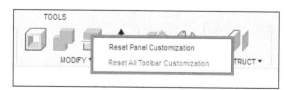

Figure-134. Shortcut menu for Resetting Toolbar

Click on the **Reset Panel Customization** option to reset current panel to original state. Note that this option will be available only when you have added or removed a tool from the panel.

Click on the **Reset All Toolbar Customization** option from the shortcut menu if you want to reset the complete toolbar including all the panels to its original state.

SELF ASSESSMENT

Q1. ....is used to manage the fusion 360 project.

Q2. Which tool is used to share the files to GrabCAD?

Q3. The...............button is used for producing CNC codes.

Q4. Which of the following workspace is not available in Autodesk Fusion 360 application?

a. Patch Workspace
b. Model Workspace
c. Assemble Workspace
d. Drawing Workspace

Q5. Discuss the use of **Save** tool with example.

Q6. Discuss the process of exporting a project in computer with example.

Q7. Discuss the use of **3D Print** tool with example.

Q8. Which tool is used to generate **Toolpaths** for model?

Q9. How to reset the default layout?

Q10. Explain the process of creating **Script** with example.

Q11. Mark the following statements as True or False:

a. Autodesk Fusion 360 software can be used to perform CAD operations.
b. Autodesk Fusion 360 software can be used to perform CAM operations.
c. Autodesk Fusion 360 software can be used to perform CAE operations.
d. Autodesk Fusion 360 software can be used to perform PLM operations.
e. Autodesk Fusion 360 software can be used to perform direct 3D printing on 3D Printer.

Q12. Which of the following tool is used to create a new part model in Autodesk Fusion?

a. New Design
b. New Drawing->From Animation
c. New Drawing->From Design
d. New Drawing->New Drawing Template

Q13. Which of the following workspace is not available in educational version of Autodesk Fusion 360?

a. Animation
b. Manufacturing
c. Generative Design
d. Simulation

Q14. Which of the following tool is used to save the file in local drive in Autodesk Fusion 360?

a. Save
b. Save As
c. Export
d. Upload

Q15. STL format is used as standard for 3D printing which abbreviates Standard Triangle Language. (T/F)

Q16. Which parameter of an STL file can decrease the angular distance between two edges of triangle as shown in next figure (from left to right)?

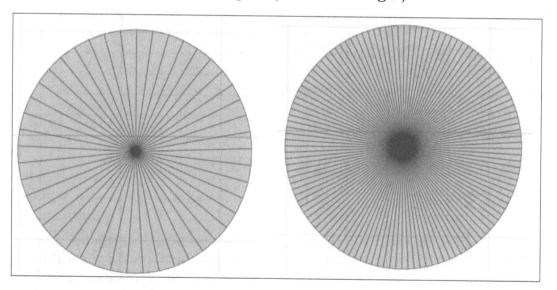

a. Surface Deviation
b. Normal Deviation
c. Aspect Ratio
d. Maximum Edge Length

Chapter 2

Sketching

Topics Covered

The major topics covered in this chapter are:

- *Introduction*
- *Starting Sketch*
- *Sketch Creation Tools*
- *Sketch Editing Tools*
- *Sketch Palette*
- *Constraints*
- *Sketching Practical and Practice*

INTRODUCTION

In Engineering, sketches are based on dimensions of real world objects. These sketches work as building blocks for various 3D operations. In this chapter, we will be working with sketch entities like; line, circle, arc, polygon, ellipse and so on to create base feature for various 3D operations. Note that the sketching environment is the base of 3D Models so you should be proficient in sketching.

There is no separate workspace for Sketching and tools to create sketches are available in **Design** workspace. So, we will be working in **Design** workspace in this chapter. We will learn about various tools of toolbar used in sketching and 3D Sketching.

STARTING SKETCH

* To start a new sketch, click on the **Create Sketch** tool from **CREATE** drop-down in **SOLID** tab of **Toolbar**; refer to Figure-1. The three primary planes will be displayed on canvas screen; refer to Figure-2.

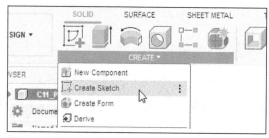

Figure-1. Create Sketch tool

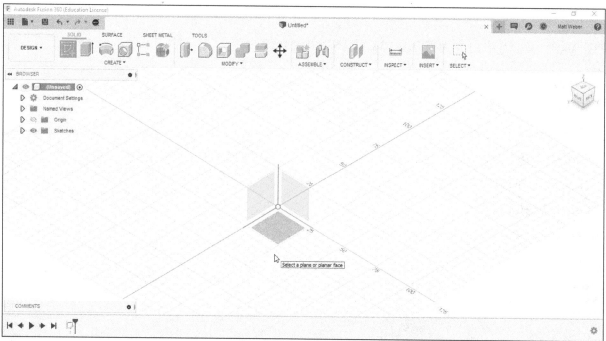

Figure-2. Select sketch plane

* Click on desired plane from the canvas screen. The selected plane will become parallel to the screen and act as current sketching plane. Also, the tools to create sketch will be displayed in **SKETCH** contextual tab in the **Toolbar**; refer to Figure-3. Now, we are ready to draw sketch on the selected plane.

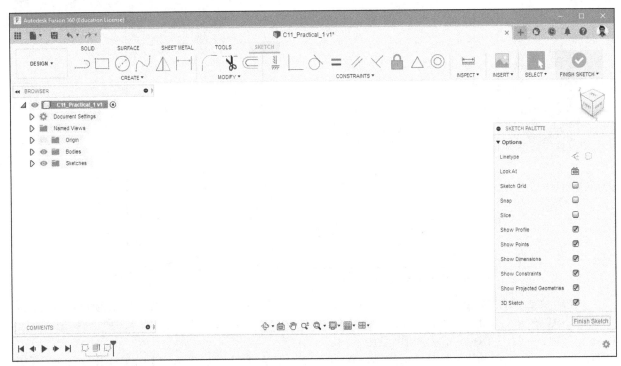

Figure-3. Sketching mode

First, we will start with the sketch creation tools and later, we will discuss the other tools.

SKETCH CREATION TOOLS

There are various tools in **CREATE** drop-down for creating sketch entities. These tools are discussed next.

Line

In Autodesk Fusion 360, the **Line** tool is used to create a line or arc. The procedure to use this tool is discussed next.

- Click on **Line** tool of **CREATE** drop-down from **Toolbar**; refer to Figure-4. You can also press **L** key to select the **Line** tool. You will be asked to specify start point of the line.

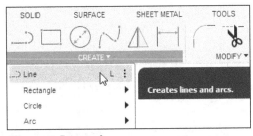

Figure-4. Line tool

- Click at desired location to specify start point and move the cursor away from selected point in desired direction in which you want to create the line. **Angle** and **Length** edit boxes will be displayed with the preview of line.
- Enter desired value of length in the **Length** edit box. Press **TAB** key to toggle to **Angle** edit box; refer to Figure-5.

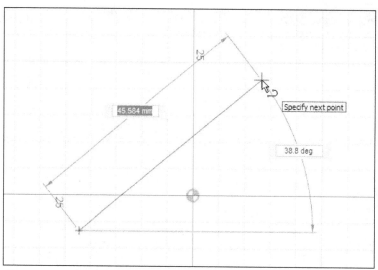

Figure-5. Creating the line

- If you are creating a freestyle line then click on the screen at desired location without specifying any value in **Length** and **Angle** edit boxes. Press **ESC** after specifying end points to exit the tool.
- To create a line for construction purpose, click on the **Construction** button from **SKETCH PALETTE** displayed at the right in the application window after activating **Line** tool; refer to Figure-6. (You will learn more about **SKETCH PALETTE** later in this chapter.) You can also right-click on the line after creating it and then click on **Normal/Construction** button from **Marking menu** to do the same; refer to Figure-7.

Figure-6. Construction button in SKETCH PALETTE

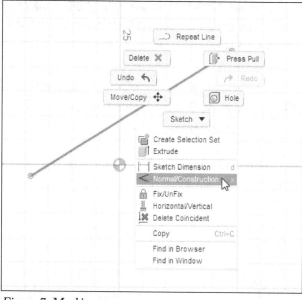

Figure-7. Marking menu

Marking menu

It is the radical display of most frequently used tool and tools related to selected entities. This menu also provides a quick access to the tools of the toolbar. This menu is the fastest way to activate the tool. To access this menu, right-click anywhere on the screen within the graphics window. To select any tool of this menu, move the cursor towards the tool and the tool will be highlighted. Click on it to activate the tool.

Creating Arc using Line Tool

You can create an arc using the **Line** tool which has one of its end point shared to another entity. In simple words, the starting point of arc created by **Line** tool will lie on end point of another line or arc. The procedure to create arc is given next.

- Click on the end point of a line or curve to define start point of arc, hold the LMB (Left - Mouse Button), and drag to the point where you want to specify the end point of arc; refer to Figure-8.

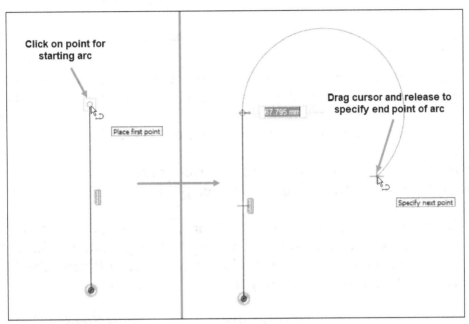

Figure-8. Creating line arc

Rectangle

The tools in **Rectangle** cascading menu are used to create rectangles. There are three tools in **Rectangle** cascading menu; **2-Point Rectangle**, **3-Point Rectangle**, and **Center Rectangle**; refer to Figure-9.

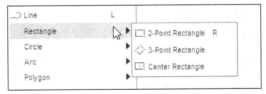

Figure-9. Rectangle cascading menu

You can also activate the **Rectangle** tool by pressing **R** key while in **Sketch** mode. The procedures to use these tools are discussed next.

2-Point Rectangle

- Click on the **2-Point Rectangle** tool of the **Rectangle** cascading menu from **CREATE** drop-down in **SKETCH** contextual tab. You will be asked to specify the first point.
- Click on the sketch canvas to specify first point of rectangle. You will be asked to specify the other corner point; refer to Figure-10.

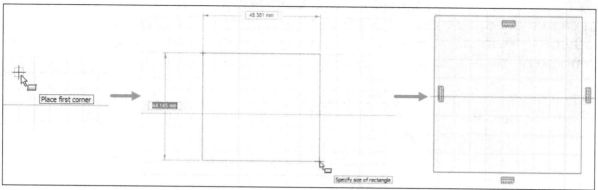

Figure-10. Creation of 2-point rectangle

- Specify the length and width of rectangle in the dynamic input boxes. To switch between the length dynamic input box and width dynamic input box, press **TAB** key.
- After specifying the parameters, press **ENTER** to create the rectangle. To exit the tool, press **ESC** key.
- If you want to create a freestyle rectangle then click on the screen to select the first corner point and other diagonal corner point of the rectangle on the sketch canvas.
- To add dimensions, right-click on the line of rectangle and click on **Sketch Dimension** tool in **Sketch** drop-down from the **Marking menu**. You can also activate the **Dimension** tool by pressing **d** from keyboard after selecting the line. You will learn more about dimensioning later in this chapter.

3-Point Rectangle

- Click on the **3-Point Rectangle** tool of **Rectangle** cascading menu from **CREATE** drop-down. You will be asked to specify location of first corner point of rectangle.
- Click at desired location to specify first corner point of the rectangle on the sketch canvas; refer to Figure-11.

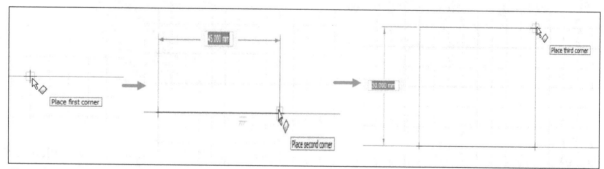

Figure-11. Creation of 3-point retangle

- Next, click to specify the second corner point (end point of base line of rectangle) on the canvas or specify the dimension in dimension box.
- Next, click to specify the third point of the rectangle to define the length of rectangle and complete the rectangle creation.

Center rectangle

- Click on the **Center Rectangle** tool of the **Rectangle** cascading menu from **CREATE** drop-down. You will be asked to specify center point of the rectangle.

- Click on the canvas to specify the center point of the rectangle; refer to Figure-12. You will be asked to specify corner point of rectangle.

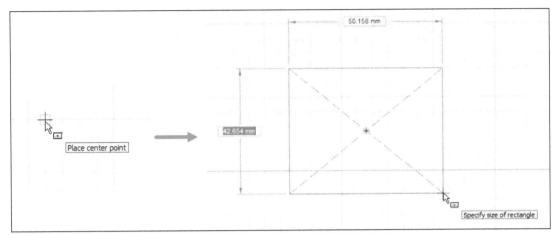

Figure-12. Creation of center rectangle

- Enter the dimension for rectangle in the edit boxes and press **ENTER** to create the rectangle.
- If you want to create freestyle rectangle then click on the screen to select the center point and corner point of rectangle.

Note that you can switch between different types of rectangles from the **Rectangle** section of **SKETCH PALETTE** after activating the **Rectangle** tool.

Circle

There are five tools in **Circle** cascading menu to create circles; **Center Diameter Circle**, **2-Point Circle**, **3-Point Circle**, **2-Tangent Circle**, and **3-Tangent Circle**; refer to Figure-13.

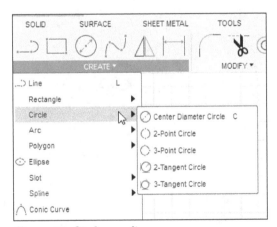

Figure-13. Circle cascading menu

The shortcut key to use the **Center Diameter Circle** tool is by pressing **C** key from keyboard. The tools to create circle are discussed next.

Center Diameter Circle

The **Center Diameter Circle** tool is used to create circle specifying center location and diameter value. The procedure to use this tool is given next.

- Click on the **Center Diameter Circle** tool from **Circle** cascading menu in the **CREATE** drop-down of **SKETCH** contextual tab. You will be asked to specify center point of circle.
- Click to specify the center point for the circle; refer to Figure-14. You will be asked to specify circumferential point of circle to define diameter.

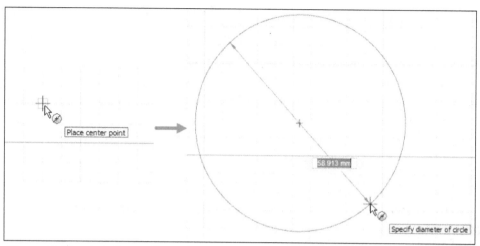

Figure-14. Creation of center diameter circle

- Click to specify the circumferential point or enter desired value of diameter in the dynamic input box.

2-Point Circle

The **2-Point Circle** tool is used to create circle by specifying two circumferential points. The procedure to use this tool is discussed next.

- Click on the **2-Point Circle** tool of **Circle** cascading menu from **CREATE** drop-down and specify the first point for circle; refer to Figure-15.

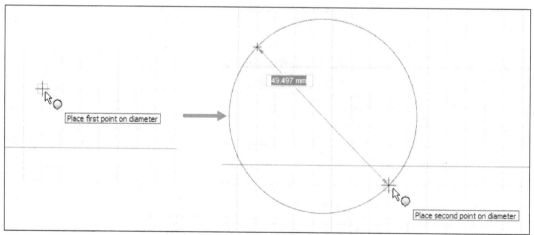

Figure-15. Creation of 2-point circle

- Specify the second point on the canvas screen or enter the diameter dimension in the dynamic input box and press **ENTER** key.

3-Point Circle

The **3-Point Circle** tool is used to create circles using 3 circumferential points as references. The procedure to use this tool is discussed next.

- Click on **3-Point Circle** tool of **Circle** cascading menu from **CREATE** drop-down.
- Click on the screen to specify the first point, second point, and third point of circle; refer to Figure-16. Note that you can also select on other entities to define circumferential points for circle.

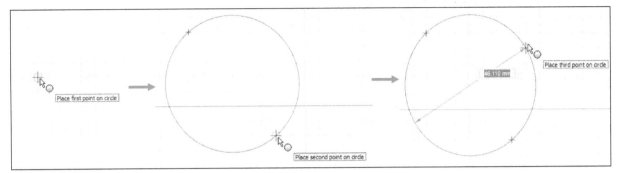

Figure-16. Creation of 3-point circle

- After specifying points, right-click on the screen and click **OK** button from **Marking menu** to complete the process. You can also press **ESC** key to exit the circle tool.
- If you want to add dimensions then right-click on the selected circle and click on the **Sketch Dimension** tool from **Marking menu**. You can also select **Sketch Dimension** tool from **SKETCH** drop-down.

2-Tangent Circle

The **2-Tangent Circle** tool is used to create a circle which is tangent to two selected lines. The procedure to use this tool is discussed next.

- Select **2-Tangent Circle** tool of **Circle** cascading menu from **CREATE** drop-down.
- Click at desired locations on lines that should be tangent to the circle; refer to Figure-17. Preview of circle will be displayed.

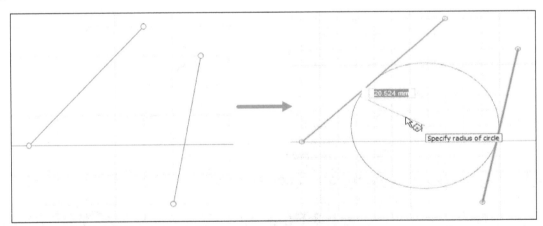

Figure-17. Creation of 2-tangent circle

- Enter desired radius value of the circle in the dynamic input box and press **ENTER** to create a circle or click at desired location to define diameter of the circle.
- Press **ESC** to exit the tool.

3-Tangent Circle

The **3-Tangent Circle** tool is used to create a circle using three tangent lines. The procedure to use this tool is discussed next.

- Click on **3-Tangent Circle** tool of **Circle** cascading menu from **CREATE** drop-down.
- Select three lines to be tangent to the circle. You can also use window selection to select the lines; refer to Figure-18. The preview of circle will be displayed.
- Right-click on the screen and click **OK** button from **Marking menu** to create the circle. You can also press **ESC** key to exit the tool.

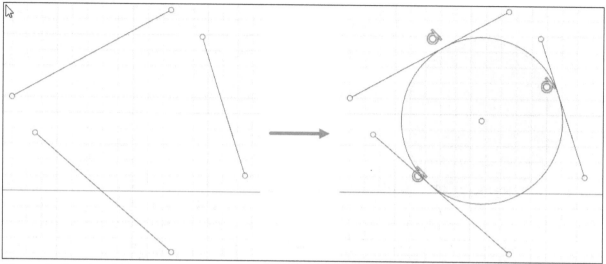

Figure-18. Creation of 3-tangent circle

Arc

The **Arc** tool is used to create arcs. There are three tools in the **Arc** cascading menu; **3-Point Arc**, **Center Point Arc**, and **Tangent Arc**; refer to Figure-19.

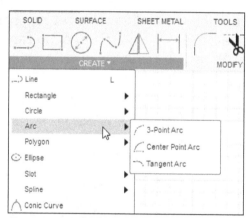

Figure-19. Arc cascading menu

The tools of **Arc** cascading menu are discussed next.

3-Point Arc

The **3-Point Arc** tool is used to create an arc using three points (start point, end point, and circumferential point). To use this tool, you need to specify two end points and one circumferential point of the arc. The procedure to use this tool is discussed next.

- Click on **3-Point Arc** tool of **Arc** cascading menu from **CREATE** drop-down. You will be asked to specify points for arc.
- Click on desired location to specify the start point of the arc; refer to Figure-20.

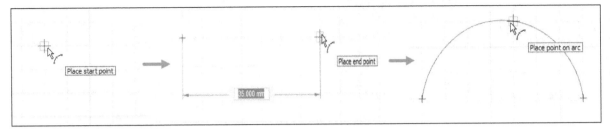

Figure-20. Creation of 3-point arc

- Click on the screen to specify the end point of the arc.
- Click at desired location to define the radius of arc. The arc will be created.
- Press **ESC** to exit the tool.

Center Point Arc

The **Center Point Arc** tool is used to create arc by defining center point, start point, and end point of the arc. The procedure is given next.

- Click **Center Point Arc** tool of **Arc** cascading menu from **CREATE** drop-down. You will be asked to specify center point of the arc.
- Click on the screen canvas to specify the center point of arc.
- Click to specify the start point of the arc or enter desired value of radius in dynamic input box.
- Enter desired angle value in dynamic input box to define length of arc or click on the screen to specify the end point of the arc; refer to Figure-21.

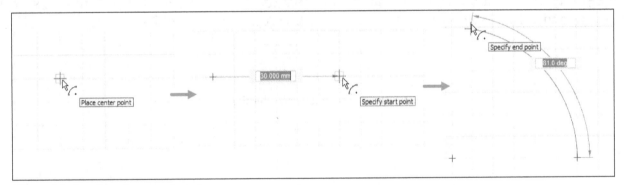

Figure-21. Creation of center point arc

- The arc will be displayed. Right-click on the screen and click **OK** button from **Marking menu** to exit the tool. You can also press **ESC** key to exit the tool.

Tangent Arc

The **Tangent Arc** tool is used to create arc tangent to selected entity.

- Click on the **Tangent Arc** tool of **Arc** cascading menu from **CREATE** drop-down.
- Click near the end point of an entity (line, arc, or curve) from screen to specify the start point of the arc; refer to Figure-22. A rubber-band arc will be attached to cursor starting from selected point.
- Click at desired location to specify the end point of the arc.

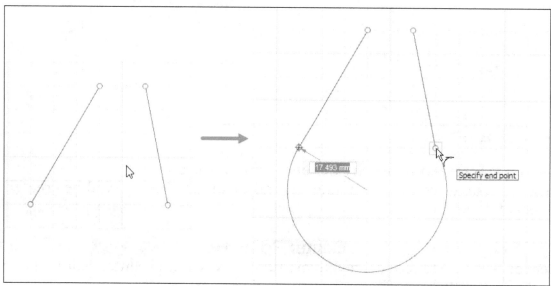

Figure-22. Creation of tangent arc

- Right-click on the screen and click **OK** button from **Marking menu** to complete the process. You can also press **ESC** key to exit the arc tool.

What happens when you try to create an arc at corner location where two entities meet and selected point is common to both the entities; refer to Figure-23. Will the arc be created tangent to line 1 or line 2? What if the arc is being created tangent to line 1 but we want to create it tangent to line 2. In such conditions, after selecting end point, move the cursor in line with entity to be tangent for few millimeters and then specify the end point at desired location; refer to Figure-24.

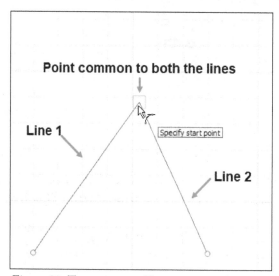

Figure-23. Tangent arc problem at corner point

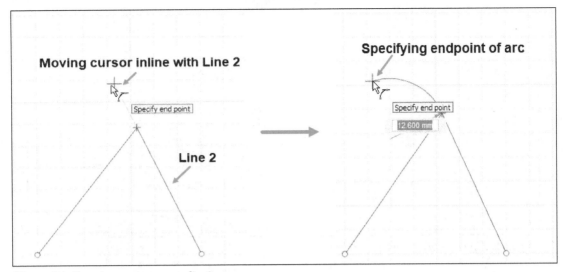

Figure-24. Creating arc tangent to line2

Polygon

The **Polygon** tool is used to create polygon of desired number of sides. There are three tools to create polygon in **Polygon** cascading menu; **Circumscribed Polygon**, **Inscribed Polygon**, and **Edge Polygon**; refer to Figure-25. The tools to create the polygons are discussed next.

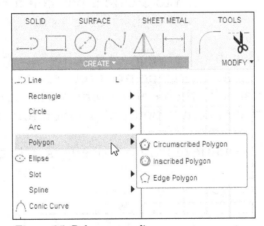

Figure-25. Polygon cascading menu

Circumscribed Polygon

Use the **Circumscribed Polygon** tool to create a polygon formed outside the reference circle. Note that midpoints of polygon lines will lie on the reference circle. The procedure to use this tool is discussed next.

- Click on the **Circumscribed Polygon** tool of **Polygon** cascading menu from **CREATE** drop-down; refer to Figure-25. You will be asked to specify center point of the reference circle.
- Click on the screen to specify the center point of the polygon; refer to Figure-26. You will be asked to specify radius of reference circle.

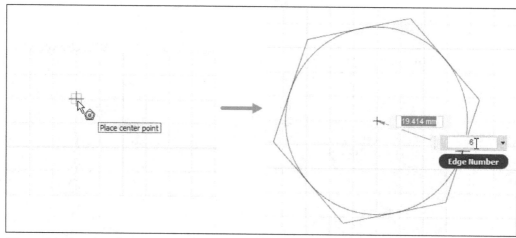

Figure-26. Creation of circumscribed polygon

- Enter the value of radius in the edit box and press **TAB** to activate **Edge Number** edit box.
- Specify desired number of edges of the polygon in **Edge Number** edit box.
- After specifying the parameters, press **ENTER** to create polygon.

Inscribed Polygon

The **Inscribed Polygon** tool is used to create a polygon formed inside the reference circle. The procedure to use this tool is discussed next.

- Click on **Inscribed Polygon** of **Polygon** cascading menu from **CREATE** drop-down. You will be asked to specify center point of reference circle.
- Click on the screen to specify the center point of the circle; refer to Figure-27.
- Enter the distance between center point of polygon and vertex of polygon edge in dynamic input box.
- Press **TAB** and specify the value of number of edges in **Edge Number** dynamic input box.
- Press **ENTER** to create polygon.

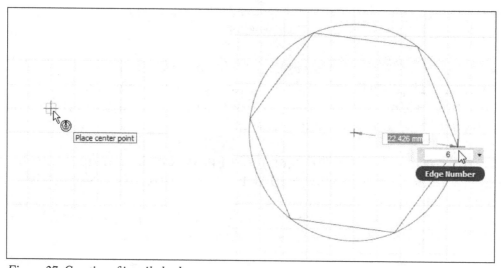

Figure-27. Creation of inscribed polygon

Edge Polygon

The **Edge Polygon** tool is used to create a polygon by defining single edge and number of edges of the polygon. The procedure to use this tool is discussed next.

- Click on the **Edge Polygon** tool of **Polygon** cascading menu from **CREATE** drop-down. You will be asked to specify start and end points of the edge line.
- Click on the screen to select the start point and end point of edge. You can also enter the value of angle and length of the edge in dynamic input boxes; refer to Figure-28. After entering the value in dynamic input boxes, move the cursor to desired side of edge where you want to create the polygon.
- Specify the value of number of edges of polygon in **Edge Number** input box.

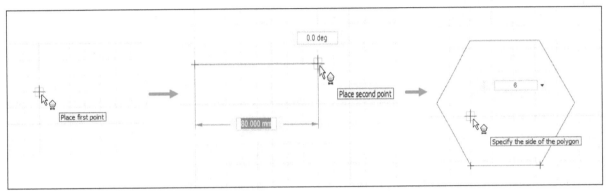

Figure-28. Creation of edge polygon

- Press **ENTER** or click in desired side to create the polygon.

Ellipse

The **Ellipse** tool is used to create ellipses in the sketch. The procedure to create an ellipse is discussed next.

- Click on **Ellipse** tool of **CREATE** drop-down from **Toolbar**; refer to Figure-29. You will be asked to specify center point for ellipse.

Figure-29. Ellipse tool

- Click on the screen to specify the center point for ellipse.
- Move the cursor in desired direction and specify the value of major diameter in dynamic input box. You can also specify the value of angle in **Angle** dynamic input box if you want major axis to be inclined.

- Click on the screen and move the cursor upward/downward. You will be asked to specify length of minor axis. Enter desired value of minor diameter in the input box; refer to Figure-30. The ellipse will be created.

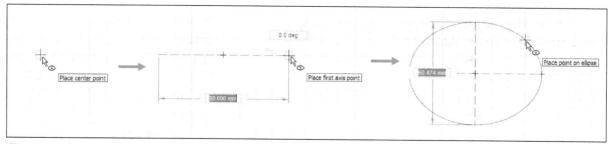

Figure-30. Creation of ellipse

- Right-click on the screen and click **OK** button to exit this tool.

Slot

The **Slot** tool is used to create slots. There are five tools in **Slot** cascading menu; **Center to Center Slot, Overall Slot, Center Point Slot, Three Point Arc Slot**, and **Center Point Arc Slot**; refer to Figure-31.

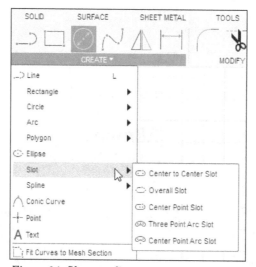

Figure-31. Slot cascading menu

The tools of **Slot** cascading menu are discussed next.

Creating Center To Center Slot

The **Center To Center Slot** tool is used to create the linear slot defined by the distance of slot arc centers and width of slot. The procedure to use this tool is discussed next.

- Click on **Center To Center Slot** tool of **Slot** cascading menu from **CREATE** drop-down.
- Click on the screen to specify the center point of first arc of slot. You will be asked to specify center point of other arc of slot.
- Click to specify the point or enter desired parameters in the input boxes.
- Move the cursor upward/downward. You will be asked to specify width of slot/ diameter of slot arc.
- Enter the value in dynamic input box or click on the screen; refer to Figure-32. The slot will be created.

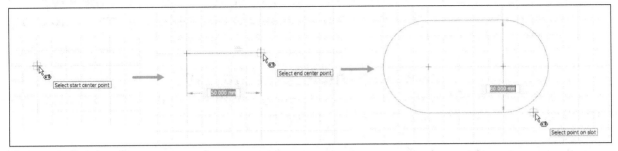

Figure-32. Creation of center to center slot

Creating Overall Slot

The **Overall Slot** tool creates a linear slot defined by total length and total width. The procedure to use this tool is discussed next.

- Click on **Overall Slot** tool of **Slot** cascading menu from **CREATE** drop-down.
- Click on the screen to specify the start point of the slot; refer to Figure-33. You will be asked to specify end point of slot.

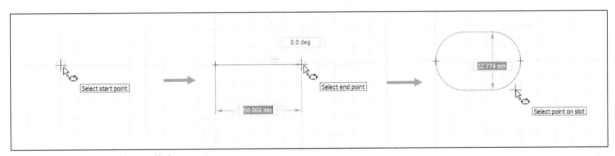

Figure-33. Creation of overall slot

- Click on the screen to specify the end point of the slot or enter the distance between start point and end point in the dynamic input box.
- Move the cursor upward/downward and enter the width of slot in dynamic input box or click at desired location.
- Press **ENTER** to create the slot and exit the tool.

Creating Center Point Slot

The **Center Point Slot** tool is used to create a linear slot which is defined by center point of slot, location of arc center, and width of slot. The procedure to use this tool is discussed next.

- Click on **Center Point Slot** tool of **Slot** cascading menu from **CREATE** drop-down.
- Click on the screen to specify the center of slot; refer to Figure-34.

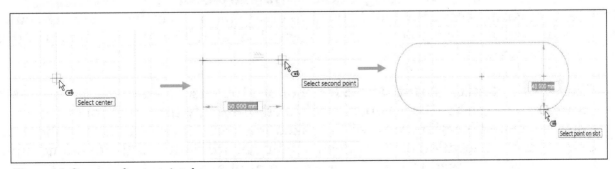

Figure-34. Creation of center point slot

- Click to specify the half length of the center to center distance of the slot.
- Specify the value of width for slot in the dynamic input box and press **ENTER** key. The slot will be created.

Creating Three Point Arc Slot

The **Three Point Arc Slot** tool is used to create a slot along specified arc with defined width. The procedure to use this tool is given next.

- Click on the **Three Point Arc Slot** tool of **Slot** cascading menu from **CREATE** drop-down. You will be asked to specify the start point of center arc.
- Click at desired location. You will be asked to specify end point of center arc.
- Click to specify the end point of arc. You will be asked to specify a circumferential point of the arc.
- Click at desired location and move the cursor upward/downward. Preview of arc will be displayed with dynamic width value.
- Enter desired value of slot width in dynamic input box and the press **ENTER** to create the slot; refer to Figure-35.

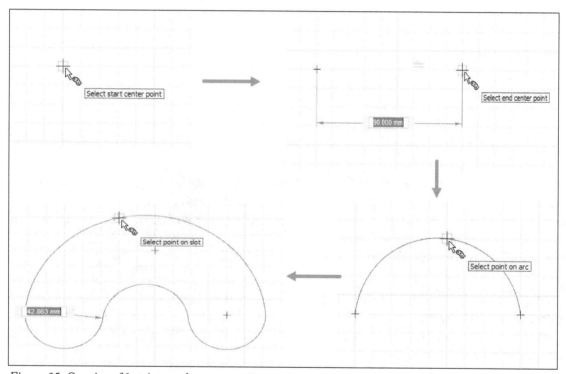

Figure-35. Creation of 3-point arc slot

Creating Center Point Arc Slot

The **Center Point Arc Slot** tool is used to create an arc slot by specifying center point of arc, start & end point of arc, and width of slot. The procedure to use this tool is given next.

- Click on the **Center Point Arc Slot** tool of **Slot** cascading menu from **CREATE** drop-down in the **Toolbar**. You will be asked to specify the center of arc.
- Click at desired location. You will be asked to specify the start point of the arc.
- Click at desired location or enter desired value of arc radius in the dynamic input box. You will be asked to specify the end point of arc.

- Click at desired location to specify end point or enter desired value of angular span of arc in the dynamic input box. You will be asked to specify the width of slot.
- Click to specify the slot width or enter desired value in dynamic input box to create the slot; refer to Figure-36.

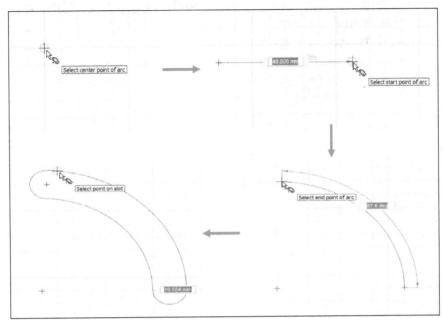

Figure-36. Creation of center point arc slot

Spline

The **Spline** tools are used to create spline curves. There are two tools in **Spline** cascading menu; **Fit Point Spline** and **Control Point Spline**; refer to Figure-37.

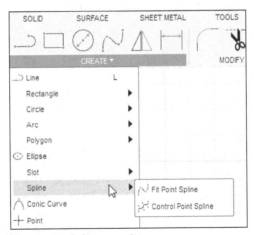

Figure-37. Spline cascading menu

The tools of **Spline** cascading menu are discussed next.

Fit Point Spline

The **Fit Point Spline** tool is used to create spline curve through the selected points. It is a freeform curve creation tool. The procedure to use this tool is discussed next.

- Click **Fit Point Spline** tool of **Spline** cascading menu from **CREATE** drop-down in the **Toolbar**; refer to Figure-37.

- Click on the screen to specify the start point for the spline. You will be asked to specify next point of spline.
- Specify the other points at desired locations to form spline of desired shape; refer to Figure-38.
- After specifying desired number of points, click on the **Tick** (**Create and Continue**) button to end the spline path.
- Press **ENTER** or **ESC** to exit the tool.

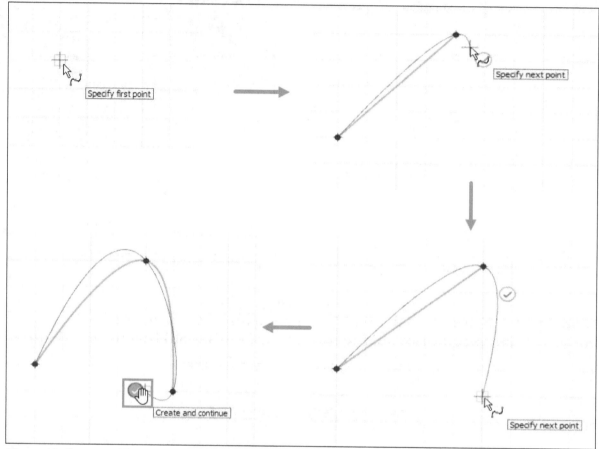

Figure-38. Creation of fit point spline

Control Point Spline

The **Control Point Spline** tool is used to create the spline curve driven by the selected control points. The procedure to use this tool is discussed next.

- Click on **Control Point Spline** tool of **Spline** cascading menu from **CREATE** drop-down in the **Toolbar**; refer to Figure-37.
- Click on the screen to specify the start point for the spline. You will be asked to specify next point of spline.
- Specify the other points at desired locations to form spline of desired shape; refer to Figure-39.
- After specifying desired number of points, click on the **Tick** (**Create and Continue**) button to end the spline path.
- Press **ENTER** or **ESC** to exit the tool.

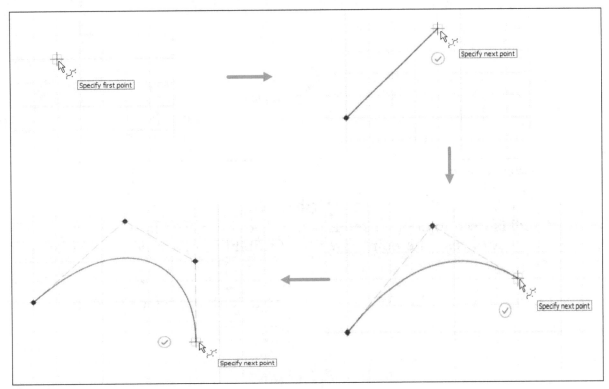

Figure-39. Creation of control point spline

Conic Curve

The **Conic Curve** tool creates a curve driven by points and rho value. The shape of the curve can be elliptical, parabolic, or hyperbolic depending on the value of Rho. The procedure to use this tool is discussed next.

*   Click on **Conic Curve** tool of **CREATE** drop-down from **Toolbar**; refer to Figure-40.

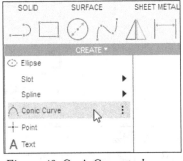

Figure-40. Conic Curve tool

*   Click on the screen to specify the start point of the conic curve.
*   Click on the screen to specify the end point of the conic curve.
*   Click on the screen to specify control point.
*   Enter the **Rho** value in **Rho** dynamic input box for the conic curve; refer to Figure-41. The curve will be created. The value of rho lies between **0** to **0.99**.

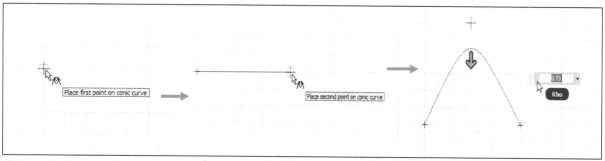

Figure–41. Creation of conic curve

Point

The **Point** tool is used to create sketch points. Sketch points find their major usage when creating surfaces. The point gives the flexibility to parametrically change the surface design.

• Click on **Point** tool of **CREATE** drop-down from **Toolbar**; refer to Figure-42.

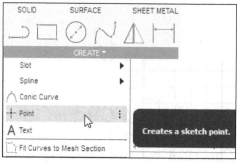

Figure–42. Point tool

• Click at desired locations in the sketch canvas to create points; refer to Figure-43.

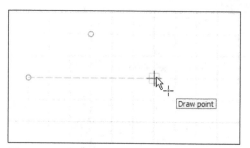

Figure–43. Creation of point

• Press **ESC** key to exit this tool or click **OK** button from **Marking menu**.

Text

The **Text** tool is used to create text which can be used for embossing/engraving on a solid model. The **Text** tool is also used to create notes and other information for the model.

• Click on **Text** tool of **CREATE** drop-down from **Toolbar**; refer to Figure-44. The **TEXT** dialog box will be displayed along with the symbol of text attached to the cursor asking you to specify first corner of the text frame; refer to Figure-45.

Note that **Text** option is selected by default in the **Type** section of the dialog box. Select **Text** option to create the text inside a rectangular frame; refer to Figure-45 and select **Text On Path** option to create the text along selected path; refer to Figure-46.

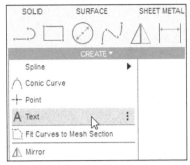

Figure-44. Text tool

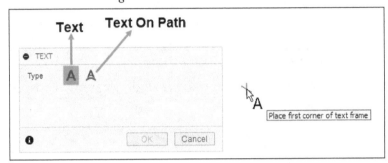

Figure-45. TEXT dialog box with text symbol attached to cursor

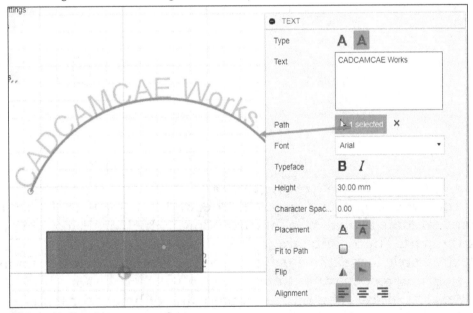

Figure-46. Creating text on path

- Click at desired location to place first corner of text frame. You will be asked to specify second corner of text frame.
- Move the cursor away and click at desired location to place second corner of text frame. The text frame will be created and the **TEXT** dialog box will be updated; refer to Figure-47.

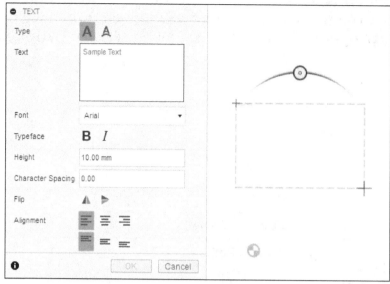

Figure-47. Updated TEXT dialog box with frame created

- Click in the **Text** edit box and enter desired text. Preview of text will be displayed in real-time; refer to Figure-48.

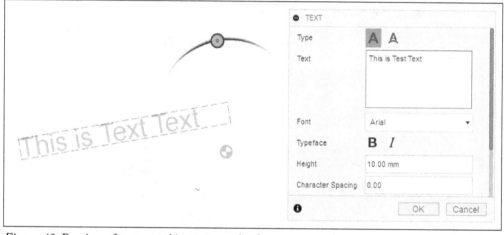

Figure-48. Preview of text created in a rectangular frame

- Click on the **Font** drop-down and select desired font for text. There are various single line text fonts available in this drop-down which are used for CAM to name or mark the part. These fonts are of **.shx** format like **cdm.shx**.
- Select desired style for text in **Typeface** section. You can set the text as bold and italic by selecting respective buttons.
- Click in **Height** edit box and enter desired value of height for text.
- Click in the **Character Spacing** edit box and enter desired value to specify change in default spacing between characters.
- If you want to flip the text then select desired button from the **Flip** section.
- Select **Align Left, Align Center**, or **Align Right** option from Alignment section to align the text horizontally and select **Align Top, Align Middle**, or **Align Bottom** option from **Alignment** section to align the text vertically.
- Click **OK** button of the **TEXT** dialog box to complete the process.
- If you want to create the text on path, then create the sketch for path and select **Text On Path** option from **Type** section of the dialog box; refer to Figure-45. The **TEXT** dialog box will be updated; refer to Figure-49.

Figure-49. TEXT dialog box with Text On Path options

- The **Path** selection button is active by default and you are asked to select the path.
- Select the path created for the text to follow. The **TEXT** dialog box will be updated and the selected path will be displayed in the **Path** section of the dialog box; refer to Figure-50.

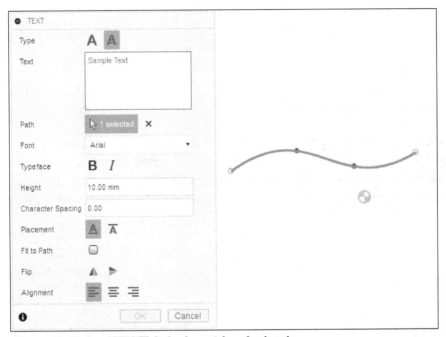

Figure-50. Updated TEXT dialog box with path selected

- Click in the **Text** edit box and enter desired text. Preview of text will be displayed in real-time; refer to Figure-51.

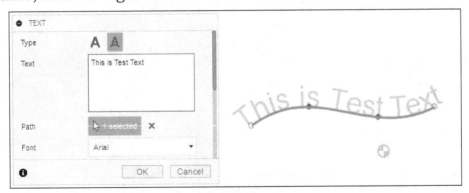

Figure-51. Preview of text created on selected path

- Select the placement of the text above or below the path by selecting respective button from **Placement** section of the dialog box.
- Select **Fit to Path** check box to distribute the text evenly along the entire path.
- Specify the other parameters in the dialog box as discussed earlier. Click **OK** button from the dialog box to complete the process.

After creating the text, you can edit it anytime by double-clicking on it in sketch mode.

Mirror

The **Mirror** tool is used to create mirror copy of the selected entities with respect to a reference called mirror line. The procedure to create the mirror entities is given next.

- Click on **Mirror** tool of **CREATE** drop-down from **Toolbar**; refer to Figure-52. The **MIRROR** dialog box will be displayed; refer to Figure-53.

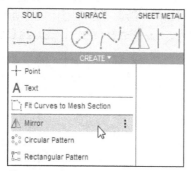

Figure-52. Mirror tool

Figure-53. MIRROR dialog box

- The **Objects** selection button is active by default and you are asked to select the objects. Select the entities or sketch whose mirror copy is to be created.
- Click on **Select** button of **Mirror Line** section and select the reference line for creating mirror copy. The preview of mirror will be displayed; refer to Figure-54.

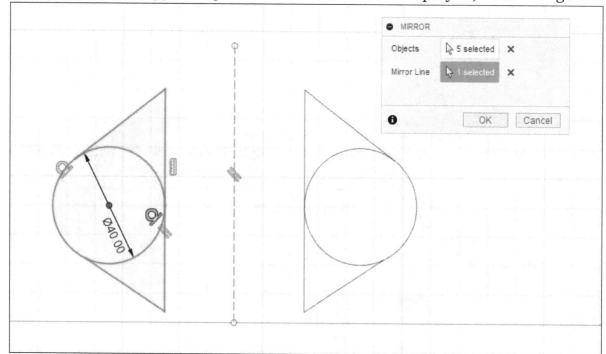

Figure-54. Procedure of mirror

- If preview of mirror is as per your requirement then click on **OK** button from **MIRROR** dialog box to complete the process.

Sometimes, we need to create multiple copies of the sketch entities like in sketch of keyboard or piano. For such cases, Autodesk Fusion 360 provides two tools, **Circular Pattern** and **Rectangular Pattern**. These tools are discussed next.

Circular Pattern

The **Circular pattern** tool is used to create multiple copies of an entity in circular fashion. The procedure to use this tool is discussed next.

- Click on **Circular Pattern** tool of **CREATE** drop-down from **Toolbar**; refer to Figure-55. The **CIRCULAR PATTERN** dialog box will be displayed; refer to Figure-56.

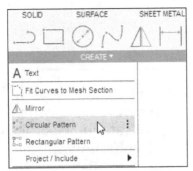

Figure-55. Circular Pattern tool

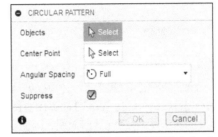

Figure-56. CIRCULAR PATTERN dialog box

- The **Objects** selection button is active by default and you are asked to select the objects. Select the entities or sketch to apply the circular pattern.
- Click on **Select** button for **Center Point** section and select the reference point to be used as center point for circular pattern. The preview of circular pattern will be displayed; refer to Figure-57.

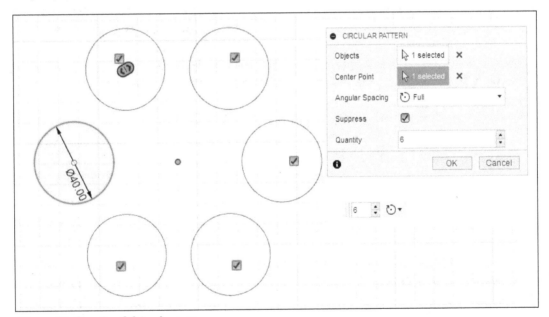

Figure-57. Creation Of circular pattern

- Select **Full** option from **Angular Spacing** drop-down if you want to create a pattern over full 360 round. Select the **Angle** option from **Angular Spacing** drop-down if you want to define the span of pattern in angle value. Select the **Symmetric** option from **Angular Spacing** drop-down if you want to create a circular pattern symmetric to the center line; refer to Figure-58.

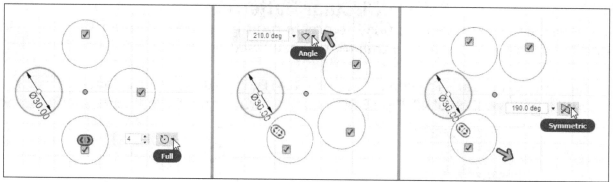

Figure-58. Tools of Type drop down

- Select the **Suppress** check box if you want to suppress any copy in pattern and then clear the check box for that entity in the preview to be suppressed. Clear the **Suppress** check box from the dialog box to generate all instances of pattern.
- Click in the **Quantity** edit box and enter desired value for number of copies of object to be created in the circular pattern. You can also set the quantity by moving drag handle.
- After specifying the parameters, click on **OK** button from **CIRCULAR PATTERN** dialog box to complete the process.

Rectangular Pattern

The **Rectangular Pattern** tool is used to create multiple copies of an entity in horizontal and vertical direction. You can create pattern in two linear directions at the same time. The procedure to create rectangular pattern is given next.

- Click on the **Rectangular Pattern** tool of **CREATE** drop-down from **Toolbar**; refer to Figure-59. A **RECTANGULAR PATTERN** dialog box will be displayed; refer to Figure-60.

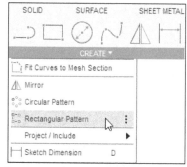

Figure-59. Rectangular Pattern tool

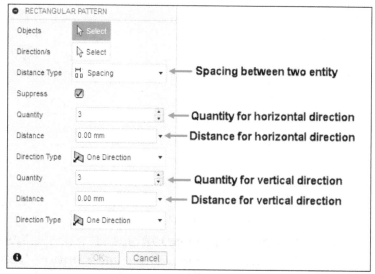

Figure-60. RECTANGULAR PATTERN dialog box

- The **Objects** section button is active by default and you are asked to select the objects. Select the sketch or entities for rectangular pattern.
- Click on **Select** button of **Direction/s** section and select the reference lines for directions of rectangular pattern. If you do not select reference lines then patterns are automatically created along horizontal and vertical axes; refer to Figure-61.

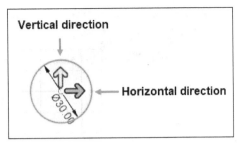

Figure-61. Rectangular pattern direction

- Select **Extent** option from **Distance Type** drop-down if you want to specify the total span in which object copies will be created. Select **Spacing** option from **Distance Type** drop-down to set distance between two consecutive copies of object. In our case, we are selecting **Extent** option.
- Select the **Suppress** check box if you want to suppress any entity in the pattern.
- Click in the **Quantity** edit box and enter desired value for number of copies.
- Click in the **Distance** edit box and enter desired distance within which copies will be created; refer to Figure-62.

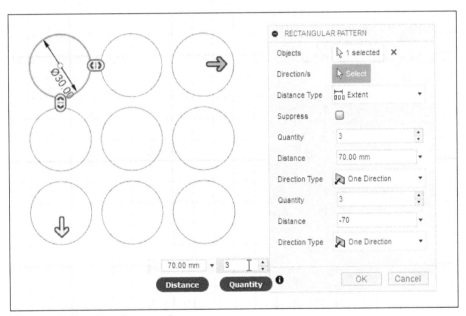

Figure-62. Creation of rectangular pattern

- Select **One Direction** option from **Direction Type** drop-down if you want to create copies in only one direction. Select **Symmetric** option from **Direction Type** drop-down if you want to create copies in both sides of selected direction.
- Similarly, specify the parameters for other direction.
- After specifying the parameters, click **OK** button from **RECTANGULAR PATTERN** dialog box to complete the process.

Project/Include

The tools in this cascading menu are used to project edges/curves of selected solids or surfaces on the sketch. These tools will be discussed later in this book while working with 3D models. Hover on the **Project/Include** option of **CREATE** drop-down from **Toolbar** to display **Project/Include** cascading menu; refer to Figure-63.

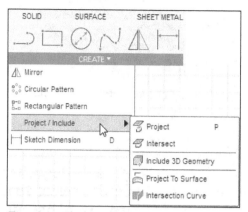

Figure-63. Project or Include cascading menu

Sketch Dimension

The **Sketch Dimension** tool is used to dimension the entities automatically based on selections. This tool can create different type of dimension like horizontal, vertical, diameter, radius, or inclined dimensions. The procedure to use this tool is discussed next.

- Click on the **Sketch Dimension** tool of **CREATE** drop-down from **Toolbar**; refer to Figure-64.

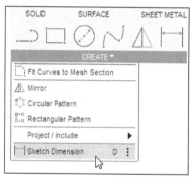

Figure-64. Sketch Dimension tool

- Select the entity you want to dimension and then click at desired distance to place dimension; refer to Figure-65.

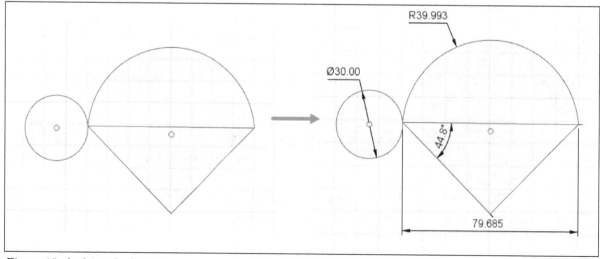

Figure-65. Applying the dimension

Based on how you move the cursor, the dimension will be horizontal, vertical, or inclined. So, after selecting an inclined line, you need to move the cursor in different ways starting from the object to get different type of dimensions. If you want to create inclined dimension for a line then move the cursor perpendicular to the line after selecting it. If you want to create horizontal dimension of an inclined line then move the cursor straight along horizontal axis. In the same way, move the cursor straight vertically to create vertical dimension of an inclined line. This technique can also be applied while dimensioning two points.

SKETCH EDITING TOOL

The standard tools to edit sketch entities are categorized in this section. The tools in this section are discussed next.

Fillet

The **Fillet** tool is used to create fillet at the corners created by intersection of two entities. Fillet is sometimes also referred to as round. The procedure to use this tool is discussed next.

- Click on **Fillet** tool of **MODIFY** drop-down from **Toolbar**; refer to Figure-66. You will be asked to select the entities.

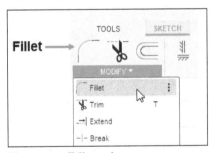

Figure-66. Fillet tool

- Select the first entity and then second entity.
- Enter the value of radius in floating window; refer to Figure-67. The fillet will be created.

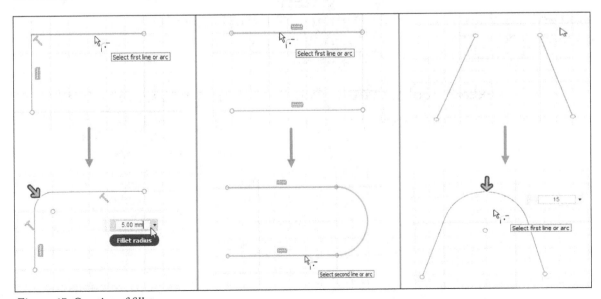

Figure-67. Creation of fillet

Trim

The **Trim** tool is used to remove unwanted portion of a sketch entity. While removing the segments, the tool considers intersection point as the reference for trimming. The procedure to use this tool is discussed next.

- Click on **Trim** tool of **MODIFY** drop-down from **Toolbar**; refer to Figure-68.

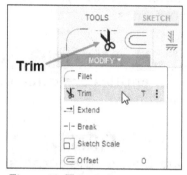

Figure-68. Trim tool

- Click on the entity to trim. You can also trim multiple entities by clicking on them.
- The entity will be trimmed. Figure-69 shows the procedure of trimming.

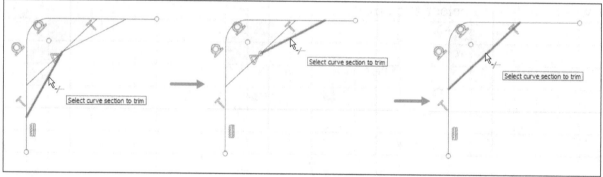

Figure-69. Procedure of trim tool

- Press **ESC** key to exit this tool.

Extend

The **Extend** tool does the reverse of **Trim** tool. This tool is available below the **Trim** tool. The **Extend** tool extends the sketch entities up to the nearest intersecting entities. The procedure to use this tool is discussed next.

- Click on the **Extend** tool of **MODIFY** drop-down from **Toolbar**; refer to Figure-70.

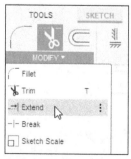

Figure-70. Extend tool

- Hover the cursor on the entity that you want to extend. Preview of the extension will display.
- Click on the entity if the preview is as per your requirement. Figure-71 shows the process of extending entities.

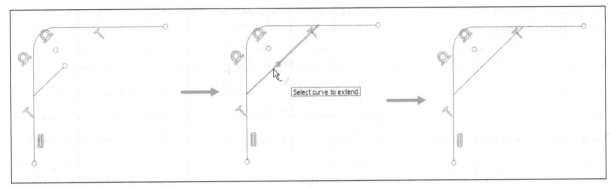

Figure-71. Procedure of extend tool

- Press **ESC** key to exit this tool.

Break

The **Break** tool is used to break the entity at selected reference points. The procedure to use this tool is discussed next.

- Click on **Break** tool of **MODIFY** drop-down from **Toolbar**; refer to Figure-72.

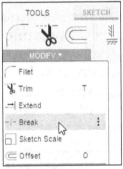

Figure-72. Break tool

- Hover the cursor on the entity. The preview of break will be displayed. Note that there must be any intersecting curve on selected entity to apply break.
- If preview is as per your requirement then click on the entity to break; refer to Figure-73.

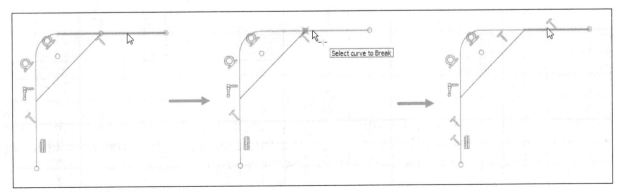

Figure-73. Creation of break

* Press **ESC** key to exit this tool.

Sketch Scale

The **Sketch Scale** tool is used to enlarge or reduce the size of selected entities with respect to the selected point based on specified scale factor. The procedure to use this tool is discussed next.

* Click on the **Sketch Scale** tool of **MODIFY** drop-down from **Toolbar**; refer to Figure-74. The **SKETCH SCALE** dialog box will be displayed; refer to Figure-75.

Figure-74. Sketch Scale tool

Figure-75. SKETCH SCALE dialog box

* The **Entities** selection box is active by default. Click on the entities to be scaled.
* Click on **Select** button of **Point** section and select the reference point about which selected entities will be scaled up/down. The updated **SKETCH SCALE** dialog box will be displayed; refer to Figure-76. Make sure the point to be selected as reference has not been selected during entities selection.
* Click in the **Scale Factor** edit box or dynamic input box and enter desired value for scale. The preview will be displayed. You can also set the scaling value by moving the drag handle from screen. A value between **0** to **1** will reduce the size and value higher than **1** will increase the size of selected entities.
* If preview is as per your requirement then click on the **OK** button of **SKETCH SCALE** dialog box to complete the process.

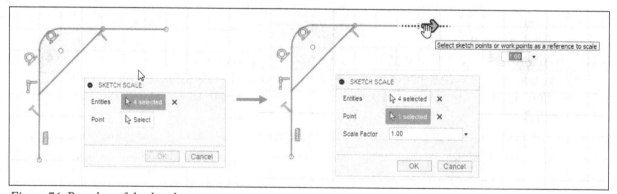

Figure-76. Procedure of sketch scale

Offset

The **Offset** tool is used to create copies of the selected entities at a specified distance. If you are the user of **AutoCAD** then this is the most common tool used for creating layouts. The procedure to use this tool is discussed next.

- Click on **Offset** tool of **MODIFY** drop-down from **Toolbar**; refer to Figure-77. The **OFFSET** dialog box will be displayed; refer to Figure-78.

Figure-77. Offset tool

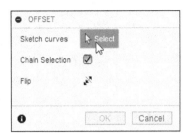

Figure-78. OFFSET dialog box

- The **Sketch curves** selection button is active by default. Select the entities for offset. The updated **OFFSET** dialog box will be displayed.
- Select the **Chain Selection** check box if you want to select all the entities in current chain. Otherwise, clear the check box if you want to select only segments of chain for offset.
- Click in the **Offset position** edit box and enter desired distance for offset; refer to Figure-79. You can also set the offset distance by moving the drag-handle from screen.

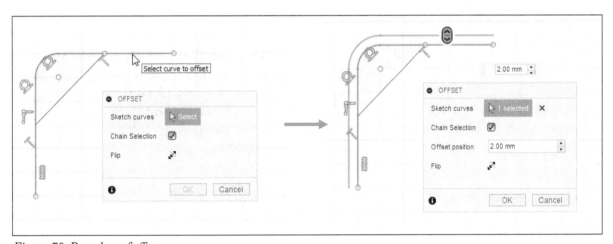

Figure-79. Procedure of offset

- Click **OK** button of **OFFSET** dialog box to finish the process.

SKETCH PALETTE

When sketch mode is active, the **SKETCH PALETTE** dialog box is displayed at bottom right corner of Fusion 360 screen. This dialog box contain commonly used sketch options and sketch constraints; refer to Figure-80.

Figure-80. SKETCH PALETTE dialog box

The various tools of Sketch Palette are discussed next.

Options

In **Options** section of **SKETCH PALETTE** dialog box, there are various options to activate or deactivate sketch functions. These options are discussed next.

*   **Linetype-** Select **Construction** button from **Linetype** section while creating any sketch object like line, circle, rectangle, etc. The object will be created as construction geometry and select **Centerline** button from **Linetype** section while creating sketch of only line and rectangle object to create them as centerline. The object will be created as centerline geometry; refer to Figure-81. Click again on these buttons to return to normal sketching mode.

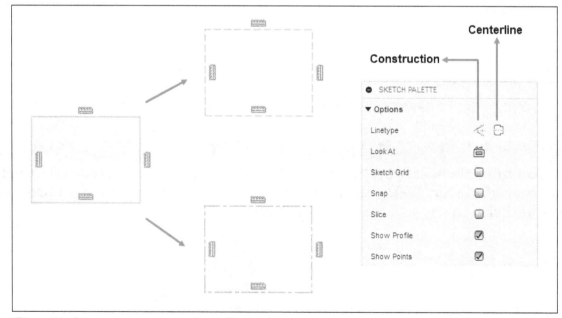

Figure-81. Creating construction and centerline geometry

- **Look At**- If due to any reason, you have rotated the model while sketching and want to make the sketching plane parallel to screen again then click on the **Look At** button once.

- **Sketch Grid**- This option controls the display of sketch grid. To view grid, select the **Grid** check box otherwise clear the check box to hide.

- **Snap**- This option controls the grid Snap. To activate grid snapping, select **Snap** check box and clear the **Snap** check box to disable. Note that the snapping of key points like end point, intersection point, mid point, etc. will still be active even if you clear the **Snap** check box.

- **Slice**- This option cuts the solid/surface part using **Sketch Plane** if it is coming before sketching plane while creating sketch. This tool does not change the geometry because it is only a display option. To enable, select the **Slice** check box and clear the **Slice** check box to disable.

- **Show Profile**- This option is used to display closed profiles in the sketch. This tool is useful to check geometry when you are going to apply protrusion tools like **Extrude** or **Revolve** later on the sketch. If your sketch is not a closed profile then you cannot create solid protrusions using that sketch. To enable this tool, select the **Show Profile** check box and to disable it clear the check box.

- **Show Points**- The **Show Points** option is used to display points at the end of different open sketch entities. Select the check box to enable this option.

- **Show Dimensions**- This option is used to display the dimensions of entities. Select the check box to enable this option.

- **Show Constraints**- This option is used to display or hide the constraints in the sketch. Constrains are used to restrict the movement of object in defined fashion. To display constraints in the sketch, select the **Show Constraints** check box and clear the check box to hide constraints.

- **Show Projected Geometries**- This tool is used to display the projected geometries on the sketch. Select the check box to enable and clear the check box to disable this option. Note that projected geometries are generally displayed in dark pink color.

- **3D Sketch**- This option enables to pull sketch objects out of the sketch plane and create a 3D sketch. To enable this tool, select the **3D Sketch** check box and to disable it clear the check box.

Feature Options

The **Feature Options** section of **SKETCH PALETTE** is displayed when sketch creation tools like rectangle, circle, slot, polygon, and so on are active which have variation in defining their parameters; refer to Figure-82. Select desired option from the **Feature Options** section to switch to different input style for same tool.

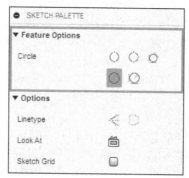

Figure-82. SKETCH PALETTE when Circle tool is active

CONSTRAINTS

The tools in **CONSTRAINTS** panel of **SKETCH** contextual tab are used to apply constraints to the sketch; refer to Figure-83. Constraints are used to restrict the shapes/position of sketch entities with respect to other entities.

Figure-83. CONSTRAINTS panel

• Make sure the **Show Constraints** check box is selected from **SKETCH PALETTE** dialog box before working with constraints; refer to Figure-84.

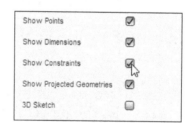

Figure-84. Show Constraints check box

Various constraints available in Autodesk Fusion are discussed next.

The **Horizontal/Vertical Constraint** makes one or more selected lines to become horizontal/Vertical. To apply this constraint, click on the **Horizontal/Vertical** button and select the entities. Press **ESC** to exit the tool. You can also make two points horizontally or vertically in-line by using this constraint.

What if you were trying to make a line vertical constrained but it is becoming horizontal constrained. In such cases, you should drag the end point of line towards vertical reference and then apply the Horizontal/Vertical constraint.

The **Coincident Constraint** makes selected point to be coincident with a line, arc, circle, or ellipse. To apply this constraint, select the **Coincident** button from **CONSTRAINTS** panel in **SKETCH** contextual tab and select the entities for coincident. Press **ESC** to exit the tool.

The **Tangent Constraint** makes selected arc, circle, spline, or ellipse to become tangent to other arc, circle, spline, ellipse, line, or edge. To apply this constraint, click on the **Tangent** button and select required entity. Press **ESC** to exit the tool; refer to Figure-85.

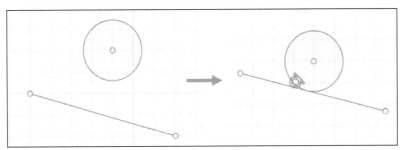

Figure-85. Applying tangent constraint

The **Equal Constraint** makes the selected lines to have equal length and the selected arcs or circles to have equal radii/diameter. To apply this constraint, click on the **Equal** button and select required entity to apply the equal constraint; refer to Figure-86. Press **ESC** to exit the tool.

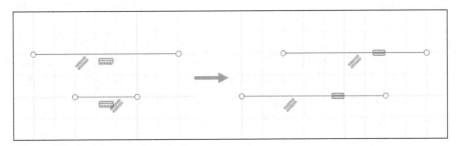

Figure-86. Applying equal constraint

The **Parallel Constraint** makes the selected lines parallel to each other. To apply this constraint, click on the **Parallel** button and then select desired entities; refer to Figure-87. Press **ESC** to exit the tool.

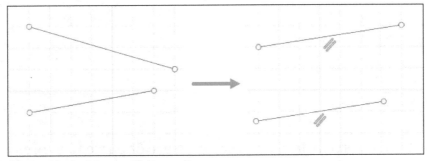

Figure-87. Applying parallel constraint

The **Perpendicular Constraint** makes the selected lines perpendicular to each other. To apply this constraint, click on the **Perpendicular** button and select the entities; refer to Figure-88. Press **ESC** to exit the tool.

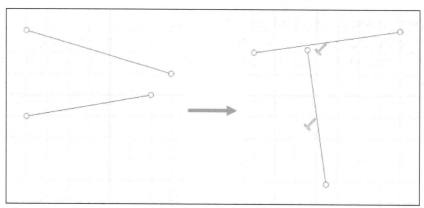

Figure-88. Applying perpendicular constraint

🔒 The **Fix/Unfix Constraint** makes the selected entity to be fixed at current position. If you apply this constraint to a line or an arc, its location will be fixed but you can change its size by dragging the endpoints or specifying value. To apply this constraint, click on the **Fix/Unfix** button and select the entities. Press **ESC** to exit the tool.

Fix option is useful when you want to restrict the movement of an entity while applying other constraints. Like in Figure-89, if you want the arc to move in place of line then you can fix the line by applying **Fix** constraint before applying **Midpoint** constraint.

△ The **Midpoint Constraint** makes selected point of an entity to be placed on the midpoint of another sketch entity. To apply this constraint, click on the **Midpoint** button. Select the point and then select the entity whose midpoint is to be used for constraining; refer to Figure-89. Press **ESC** to exit the tool.

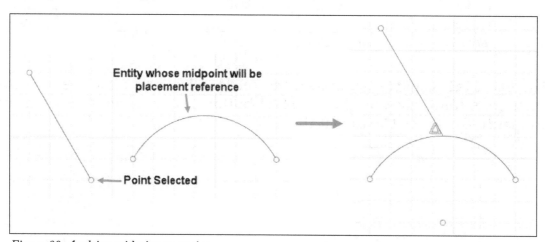

Figure-89. Applying midpoint constraint

◎ The **Concentric Constraint** makes selected arc or circle to share the same center point with other arc, circle, point, vertex, or circular edge. To apply this constraint, click the **Concentric** button and then select the two entities to apply the concentric constraint; refer to Figure-90. Press **ESC** to exit the tool.

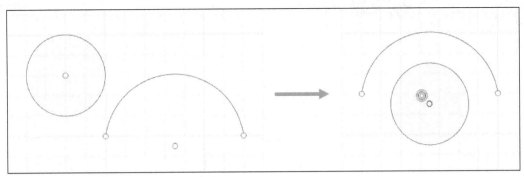

Figure-90. Applying concentric constraint

The **Collinear Constraint** makes the selected lines to lie on the same infinite line; refer to Figure-91. To use this constraint, select the **Collinear constraint** button and select the entities. Press **ESC** to exit the tool.

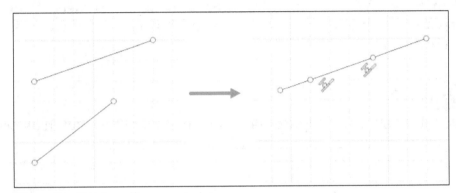

Figure-91. Applying collinear constraint

The **Symmetry Constraint** makes two selected lines, arcs, points, and ellipses to remain symmetric about a reference line. To apply this constraint, click on the **Symmetric** button and select the two objects to be symmetric about the reference line. You will be asked to select reference object. Select the line to be used as symmetry reference. The constraint will be applied; refer to Figure-92. Press **ESC** to exit the tool.

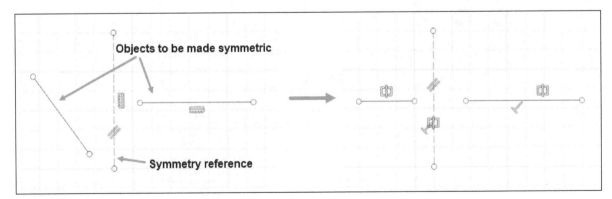

Figure-92. Applying symmetry constraint

The **Curvature Constraint** is used to create a curvature continuous (G2) condition between selected sketch entities which meet at sharp vertex. To use this tool, click on **Smooth** button and select desired entities near sharp vertex; refer to Figure-93. Press **ESC** to exit the tool.

While applying curvature constraint to splines, make sure you select the splines not the control handles.

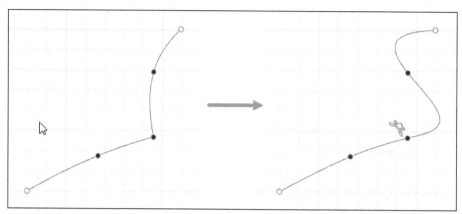

Figure-93. Applying curvature constraint

Except for Horizontal/Vertical and Fix constraint, you can also select an external entity such as an edge, plane, axis, or sketch curve of an external sketch as reference to apply the constraint.

PRACTICAL 1

Create the sketch as shown in Figure-94. Also, dimension the sketch as per the figure.

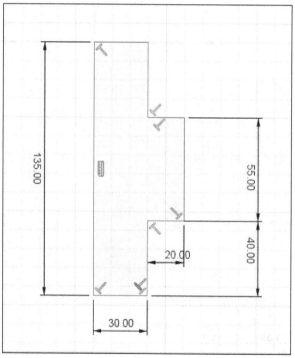

Figure-94. Practical 1

Steps to be Performed:

Below is the step by step procedure of creating the sketch shown in the Figure-94.

Starting Sketching Environment

*   Start **Autodesk Fusion 360** if not started already.
*   Select the **New Design** tool from **File** menu to start a new design; refer to Figure-95.

- Select the **Create Sketch** tool from **CREATE** drop-down; refer to Figure-96. You will be asked to select a sketching plane; refer to Figure-97.

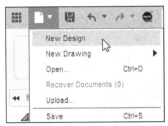

Figure-95. New Design tool

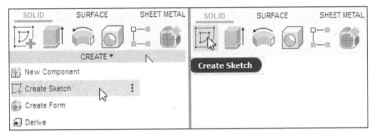

Figure-96. Create Sketch tool

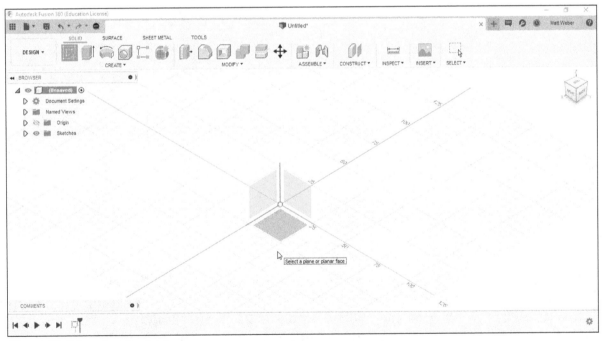

Figure-97. Select plane for sketch

- Select the **"Top"**(XY) plane for sketching.

Creating Lines

- Click on the **Line** button from **CREATE** drop-down or press **L** key from keyboard to select line. The **Line** tool will become active and you will be asked to select the start point.
- Click on the origin and move the cursor towards right; refer to Figure-98.

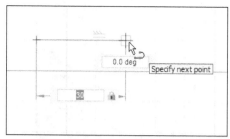

Figure-98. Starting creation of line

- Type **30** in the dimension box displayed near line and press **TAB** key to lock the dimension. Click on the screen to create a line.
- Move the cursor vertically upwards and type the value **40** in the dimension box and press **TAB** key to lock the dimension and then click on the screen to complete the line.
- Move the cursor towards right and type **20** in the dimension box. Lock the dimension as discussed earlier and click on screen; refer to Figure-99.

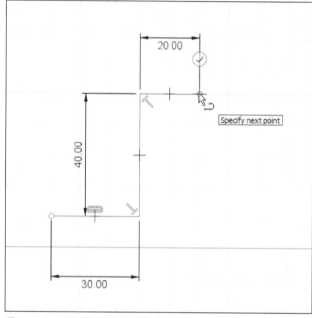

Figure-99. Sketch after specifying 20 value

- Move the cursor upward and specify the value as **55** in the dimension box. Press **TAB** key to lock the dimension and then click on the screen to complete the line.
- Move the cursor left and specify the value as **20** in the dimension box. Press **TAB** key to lock the dimension and then click on the screen to complete the line.
- Move the cursor upwards and specify the value as **40** in the dimension box. Press **TAB** key to lock the dimension and then click on the screen to complete the line.
- Move the cursor towards left and specify **30** in the dimension box press **TAB** key to lock the dimension and then click on the screen to complete the line.
- Move the cursor downward and click on the origin to close the sketch. Press **ESC** to exit the line tool.

The sketch after performing the above steps is displayed as shown in Figure-100.

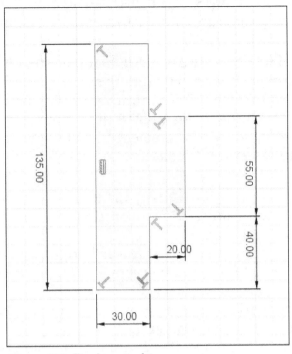

Figure-100. Final practical 1

PRACTICAL 2

In this practical, we will create the sketch as shown in Figure-101.

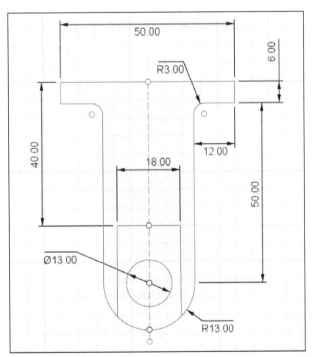

Figure-101. Practical 2

Steps to be performed:

Below is the step by step procedure of creating the sketch shown in the Figure-101.

Starting Sketching Environment

- Start **Autodesk Fusion 360** if not started already.
- Select the **New Design** tool from **File** menu to start a new design.
- Select the **Create Sketch** tool from **CREATE** drop-down.
- After selecting this tool, you need to select the **plane** for creating sketch.
- Select the **Top Plane** to create the sketch.

Creating Lines

- Click on the **Line** button from **CREATE** drop-down or press **L** key. You are asked to specify the start point of the line.
- Click on the origin and move the cursor towards right. Specify **25** in the dimension box and press **TAB** key to lock the dimension. Click on the screen to complete the line.
- Move the cursor downward and specify **6** in the dimension box. Lock the dimension as discussed earlier and click on screen.
- Move the cursor to the left and specify **12** in the dimension box. Lock the dimension as discussed earlier and click on screen
- Move the cursor downwards and enter **50** in the dimension box. Till this point, our sketch should display like Figure-102.

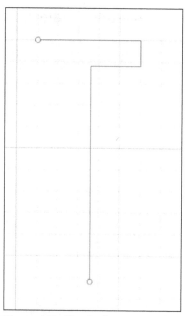

Figure-102. Sketch after creating lines

Creating Arcs

- Select the **Tangent Arc** tool from **Arc** cascading menu in **CREATE** drop-down.
- Click on the end point of the vertical line recently created and move the cursor downwards in such a way that the end point of arc is in-line with the origin.
- Specify the dimension value **13** in the dimension box and press **ENTER** to complete the arc; refer to Figure-103.

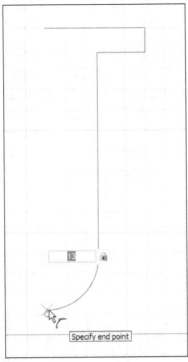

Figure-103. Arc creation

Note that you can also create tangent arc by dragging the cursor while creating line using **Line** tool. To specify dimension of tangent arc, you need to drag the cursor while creating line, hold the left click, specify the radius value for tangent arc, and then release the left click.

Creating Fillet

- Click on the **Fillet** tool from **MODIFY** drop-down/panel in the **SKETCH** contextual tab in the **Toolbar**.
- Select the lines as shown in Figure-104.
- Enter the value as **3** in the radius box.

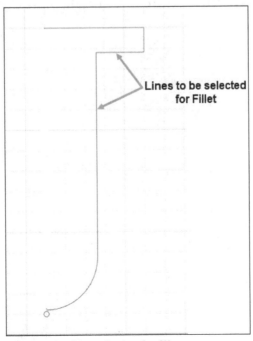

Figure-104. Line selection for fillet

Creating Mirror Copy

- Press **L** key and create a line from origin to end point of the Arc.
- Press **ESC** key from keyboard to exit the tool.
- Now, select the newly created line, right-click on it and select **Normal Construction** from **Marking menu** or select the newly created line and press **x** from keyboard to convert line into construction geometry. Press **ESC** to exit selection.
- Select the **Mirror** tool from **CREATE** drop-down and select all the entities we have sketched except construction line; refer to Figure-105.

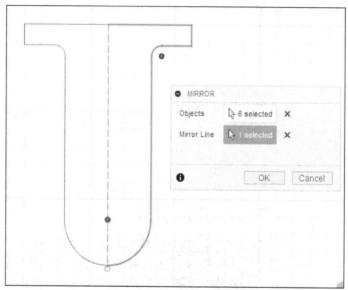

Figure-105. Entities selected for creating mirror

- Click on the **Mirror Line** selection button in the dialog box and select the construction line from sketch. Preview of mirror feature will be displayed.
- Click **OK** button from the **MIRROR** dialog box.

Creating Circle and Lines to complete.

- Click on the **Center Diameter Circle** tool from **Circle** cascading menu.
- Click at the center point of the bottom arc and move the cursor; refer to Figure-106.

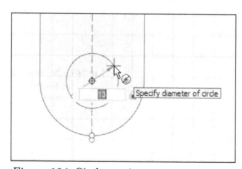

Figure-106. Circle creation

- Enter the dimension as **13** in the input box.
- Click on the **Line** tool from **CREATE** drop-down and click anywhere on the center line (except mid point and end points) to specify the start point of the line; refer to Figure-107.

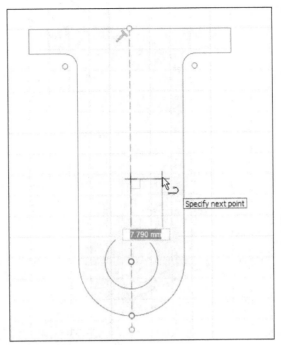

Figure-107. Point selected on centerline

- Move the cursor horizontally towards right and specify the value as **9**. Press **TAB** and click on the screen.
- Move the cursor vertically downwards and click when line passes the arc; refer to Figure-108.

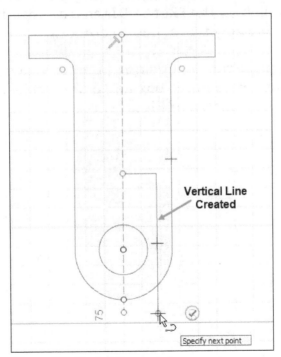

Figure-108. Vertical line created

- Press **ESC** to exit the tool. Click on the **Trim** tool from **MODIFY** drop-down and remove the extra portion outside the arc. Press **ESC** to exit the **Trim** tool.
- Mirror both the lines as we did earlier. The sketch should display as shown in Figure-109.

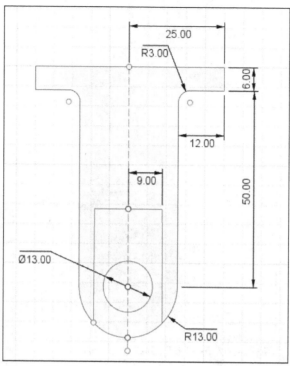

Figure-109. Sketch after all sketching operation

Dimensioning Sketch

Some of the dimensions are applied automatically while creating sketch. To create rest of the dimensions, click on the **Sketch Dimension** tool from **CREATE** panel/drop-down. Dimension symbol will get attached to cursor.

- Click on the arc and place the dimension at proper spacing if not created already. Enter the radius in dynamic input box as 13 and press **ENTER.**
- Click on the two lines as shown in Figure-110.

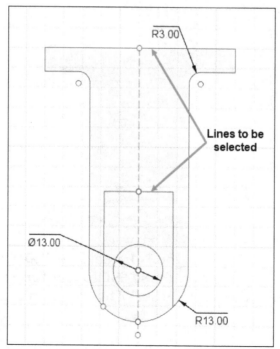

Figure-110. Lines to be selected for dimensioning

• Click to place the dimension at its proper place. In the dynamic input box, enter the value as 40.

Similarly, dimension all the entity in the sketch until the sketch is fully defined. The final sketch after dimensioning will be displayed as shown in Figure-111. You can delete the extra dimensions before applying real dimensions.

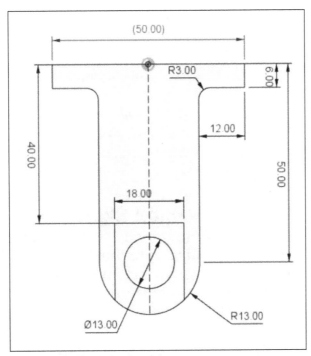

Figure-111. Final sketch

FULLY DEFINED, UNDER-DEFINED, AND OVER-DEFINED SKETCHES

Till this point, you have worked on various sketching tools and you have created sketch for two practical sessions. You might have noticed that some entities are displayed in blue color while some are displayed in bold black color. Also, there are dimensions which are enclosed in bracket. Here, we will discuss why these colors and different dimensions appear.

Fully Defined Sketches

A fully defined sketch is the one whose specifications can not be changed unintentionally. In some complex sketches, when you change dimension of one entity, the dimension of other entity gets changed automatically. If you have the sketch fully defined then the dimension of the entities will not change unintentionally. In technical terms, a fully defined sketch is the one in which entities have zero degree of freedom. In Autodesk Fusion, the sketch that is fully defined will be displayed in bold black color; refer to Figure-112. In a fully defined sketch, if you try to drag any entity of the sketch then it will not move because its degree of freedom is zero. A fully defined sketch is always recommended for creating features in CAD.

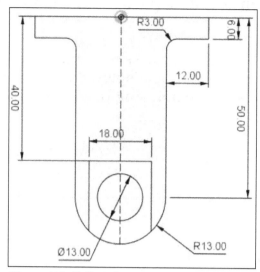

Figure-112. Fully defined sketch

Under Defined Sketches

An under-defined sketch is the one in which all the required dimensions are not applied. An under-defined sketch is displayed in blue color in Autodesk Fusion 360. When you change value of a dimension or drag an entity in the under-defined sketch then other entities may change shape/size accordingly. An under-defined sketch is not recommended for CAD modeling.

Over Defined Sketches

An over defined sketch is the one where more dimensions are applied than those required to fully define the sketch. In over-defined sketches, extra dimensions are bound by parenthesis like the **50** dimension as shown in Figure-111.

PRACTICAL 3

Create the sketch as shown in Figure-113. Also, dimension the sketch as per the figure.

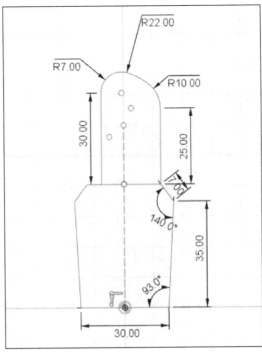

Figure-113. Practical 3

Steps to be performed:

Below is the step by step procedure of creating the sketch shown in Figure-113.

Starting Sketching Environment

- Start **Autodesk Fusion 360** if not started already.
- Select the **New Design** tool from **File** menu to start a new design.
- Select the **Create Sketch** tool from **CREATE** drop-down.
- On selecting this tool, you will be asked to select a plane for creating sketch.
- Select the **Top Plane** to create the sketch.

Creating Lines

- Select **Line** tool from **CREATE** drop-down and create a vertical line starting from origin and moving upward. (No need of dimensions, just make it a long one as it is going to be centerline.)
- Select the line and right-click on this line. Marking menu will be displayed. Select **Normal/Construction** tool to make this line as a center line; refer to Figure-114. Press **ESC** to exit the tool.

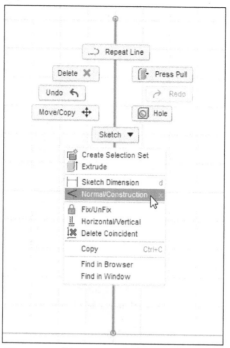

Figure-114. Creating centerline

- Press **L** key to start **Line** tool again and click at the end point of center line to create a line.
- Move the cursor towards right and specify the value as **15** in the dimension box. Press **TAB** key to lock the dimension and click on the screen.
- Move the cursor upward with some outward angle. Specify length as **35** in dimension box. Press **TAB** key to lock the dimension and switch to angle dimension. Specify the angle value as **93** degree and click on the screen.
- Move the cursor to the left, specify the value as **7** in the length input box and **140** degree in angle input box; refer to Figure-115.

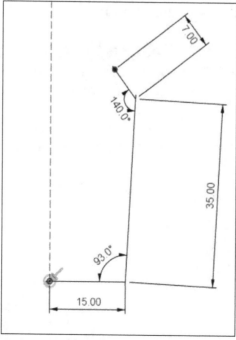

Figure-115. Creating lines with dimensions

Creating Mirror copy

- Select the **Mirror** tool from **CREATE** drop-down and select the entities we have sketched except center line for mirror.
- Click on the **Mirror Line** selection button and select the center line of sketch as mirror line. Preview of mirror will be displayed; Figure-116.

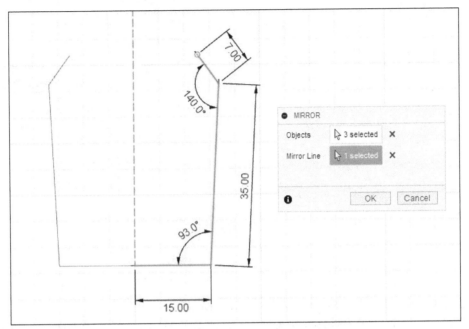

Figure-116. Preview of mirror

- Click **OK** button from **MIRROR** dialog box to complete the mirror copy.
- Now, select **Line** tool and create a line joining the open ends of sketch; refer to Figure-117.

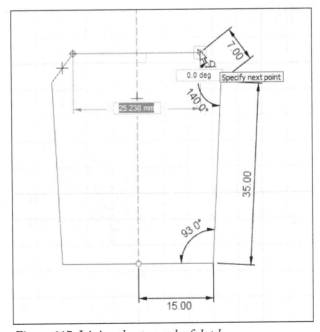

Figure-117. Joining the open ends of sketch

- Create a vertical line from one end of the sketch of length **25** mm as shown in Figure-118.

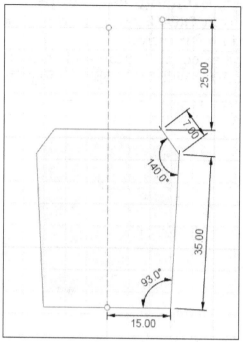

Figure-118. Creating line from right side of sketch

Creating Arcs

- Click **Tangent Arc** tool from **Arc** cascading menu in **CREATE** drop-down.
- Click on the end point of vertical line recently created and move the cursor upwards. Create three consecutive arc joining each other; as shown in Figure-119. Press **ESC** to exit the tool.

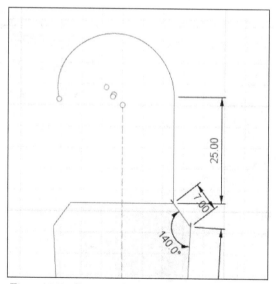

Figure-119. Creating three consecutive arcs

- Select **Sketch Dimension** tool from **CREATE** drop-down or press **D** key to select this tool.
- Apply radius dimensions to three arcs as **10**, **22**, and **7**, respectively; refer to Figure-120. Make sure horizontal and vertical constraints are applied to the straight lines in the sketch.

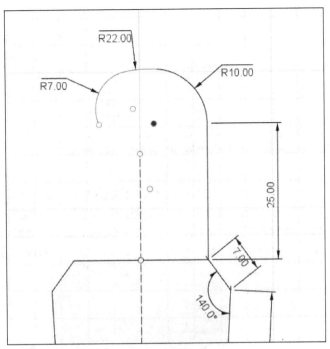

Figure-120. After sketch dimension on arc

- Select the **Line** tool and create a line joining end of recently created arc to horizontal line in the sketch; refer to Figure-121. Press **ESC** tool to exit the command.

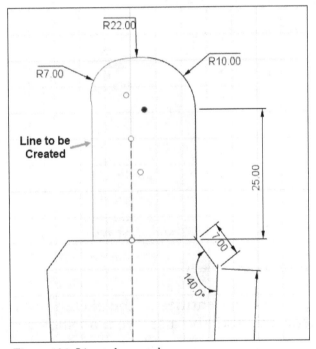

Figure-121. Line to be created

You may find that there is not a smooth transition between newly created line and arc as in our case. In such cases, apply the tangent constraint between line and arc; refer to Figure-122.

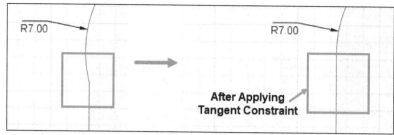

Figure-122. Applying tangent constraint for smooth transition

Dimensioning Sketch

- Select the unwanted dimensions and delete them by pressing **DELETE** key.
- Click on the **Sketch Dimension** tool or press **D** key to select the tool.
- Select the entity and apply the dimension; refer to Figure-123.

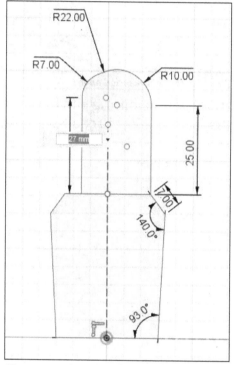

Figure-123. Select entities for dimension

- Drag the dimension to proper place.

Similarly, dimension all the other entities in the sketch until it is fully defined. The final sketch after dimensioning will be displayed; as shown in Figure-124.

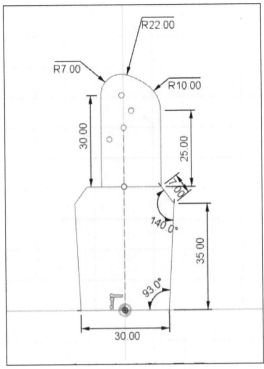

Figure-124. Final sketch of practical 3

Following are some sketches for practice.

PRACTICE 1

Create the sketch as shown in Figure-125.

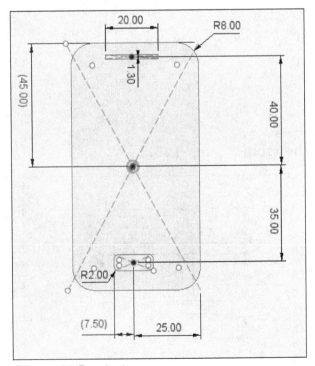

Figure-125. Practice 1

PRACTICE 2

Create the sketch as shown in Figure-126.

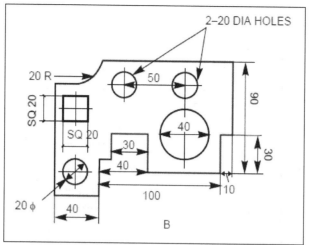

Figure-126. Practice 2

PRACTICE 3

Create the sketch as shown in Figure-127.

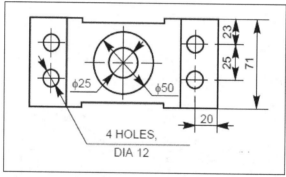

Figure-127. Practice 3

PRACTICE 4

Create the sketch as shown in Figure-128.

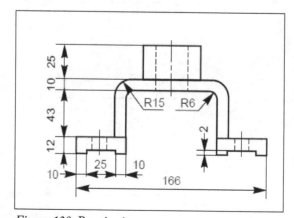

Figure-128. Practice 4

SELF ASSESSMENT

Q1. In Autodesk Fusion 360, which of the following workspace has tool to start a new sketch?

a. RENDER
b. DESIGN
c. DRAWING
d. SIMULATION

Q2. Which of the following shortcut key can be used to convert selected line into a construction line?

a. L
b. c
c. y
d. x

Q3. While the **Line** tool is active in sketch, if you drag the cursor then a tangent arc is created. (T/F)

Q4. Which of the following tool is not available in Autodesk Fusion Sketch mode to create circle?

a. Center Radius Circle
b. 3-Point Circle
c. 3-Tangent Circle
d. 2-Point Circle

Q5. Which of the following **Polygon** tool should be used to create polygon if distance between two parallel edges of polygon and number of edges are given?

a. Circumscribed Polygon
b. Inscribed Polygon
c. Edge Polygon
d. All of the above.

FOR STUDENT NOTES

Chapter 3

3D Sketch and
Solid Modeling

Topics Covered

The major topics covered in this chapter are:

- *3D Sketches Introduction*
- *Selection Tools*
- *Solid Modeling*
- *Create drop-down*
- *Construction Geometry*
- *Project/Include Tools*

3D SKETCHES

3D Sketches are those sketches which are not confined to single plane. Assume a line connecting end points of two lines in YZ and ZX planes; refer to Figure-1. Such a sketch can be called 3D Sketch; refer to Figure-2.

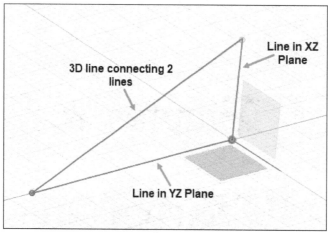

Figure-1. 3D line

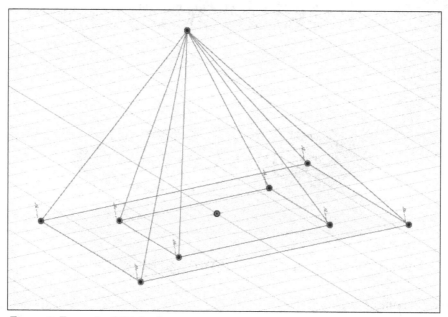

Figure-2. Example of 3D sketch

In Autodesk Fusion 360, there are no specific tools for creating 3D sketches except a check box to enable or disable 3D sketching in **SKETCH PALETTE**. So, we are back to basics of plane orientations for creating 3D sketches. The process of creating a 3D sketch is explained next with the help of an example.

Creating 3D Sketch

Create the sketch of frame as shown in Figure-3.

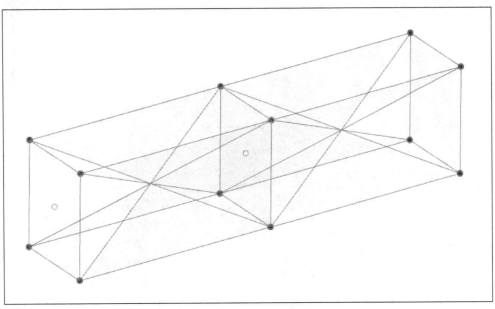

Figure-3. Load bearing frame sketch

Creating Base Sketches and Planes

- After starting a new design file in Autodesk Fusion, click on the **Create Sketch** button from the **CREATE** drop-down. You will be asked to select a sketching plane.
- Select the XY plane as sketching plane and create a center point rectangle as shown in Figure-4.

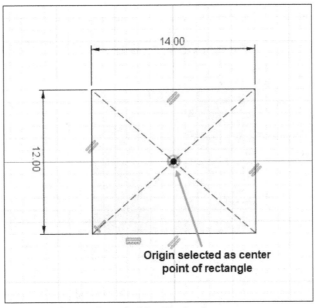

Figure-4. Rectangle created for 3D sketch

- Nope! Don't exit the sketching environment yet!! Select the whole sketch by window selection; refer to Figure-5 and press **CTRL+C** from keyboard. This will keep a copy of current sketch in clipboard memory of software.
- Click on the **Finish Sketch** button from **FINISH SKETCH** panel of the **Toolbar** to exit sketching environment.

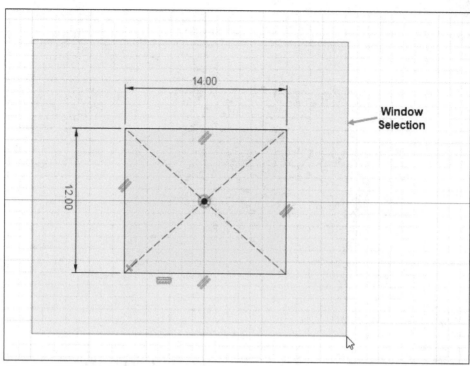

Figure-5. Window selection

- Click on the **Offset Plane** tool from the **CONSTRUCT** drop-down of the **SOLID** tab in the **Toolbar** and select **XY Plane** from **Origin** node of **BROWSER**; refer to Figure-6. (You will learn more about the plane and construction geometries later in this chapter.)

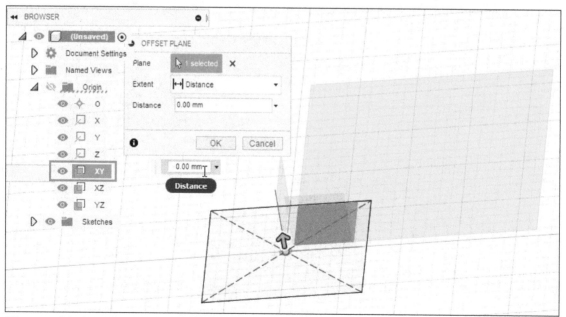

Figure-6. Creating offset plane

- Specify the distance value as **300** in **OFFSET PLANE** dialog box and click on the **OK** button. A new plane will be created at a distance of **300** from XY plane.
- Click again on the **Offset Plane** tool from the **CONSTRUCT** drop-down in the **Toolbar** and create another plane at an offset distance of **300** from earlier created plane; refer to Figure-7.

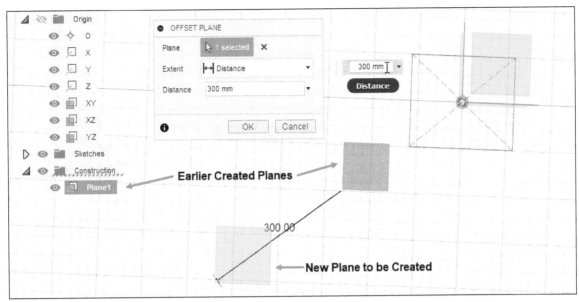

Figure-7. New plane to be created

- Select **Plane1** from the **Construction** node in **BROWSER** and click on the **Create Sketch** button from **CREATE** drop-down in the **Toolbar**. The sketching environment will be activated.
- Press **CTRL+V** from keyboard to paste earlier copied sketch. The **MOVE/COPY** dialog box will be displayed with sketch pasted; refer to Figure-8.

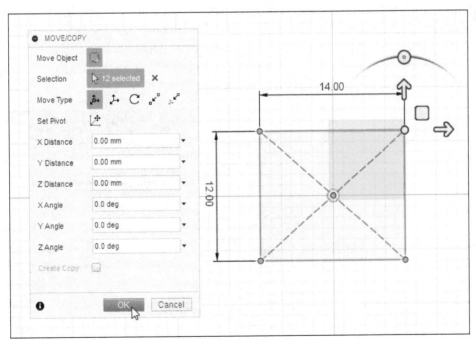

Figure-8. Sketch pasted with MOVE COPY dialog box

- There is no need to change any parameter because our sketch is well constrained and same as we wanted. Click on the **OK** button from the dialog box and exit the sketching environment.
- Create the same sketch on other offset plane. The sketches will be displayed as shown in Figure-9.

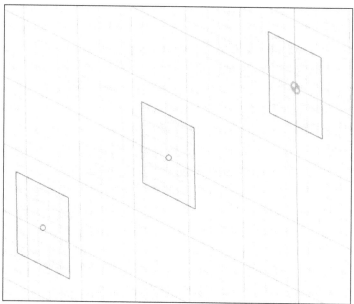

Figure-9. Sketches after copy paste

Creating 3D Lines

(Till the time, this book is authored, we are able to use only lines and splines for creating 3D sketches.)

- Click on the **Preferences** option from the **User Account** drop-down at the top-right corner of the application window. The **Preferences** dialog box will be displayed. (This dialog box has been discussed earlier in Chapter 1.)
- Select the **Design** option from **General** node at the left in the dialog box. The options will be displayed as shown in Figure-10.

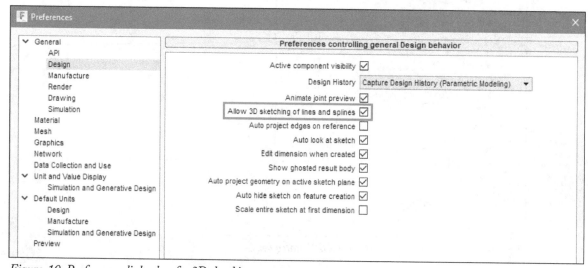

Figure-10. Preferences dialog box for 3D sketching

- Select the **Allow 3D sketching of lines and splines** check box as highlighted in above figure and click on the **OK** button from the dialog box.

Now, we are ready to create 3-dimensional lines.

- Click on the **Create Sketch** tool from the **CREATE** drop-down in the **Toolbar** and select any of the sketching plane. The sketching plane will become parallel to screen.
- Rotate the model by using **SHIFT+Middle Mouse Button (MMB)**.
- Click on the **Line** tool from **CREATE** drop-down and hover the cursor on earlier created sketches. You will find that the snap points of all the sketches get highlighted when you hover the cursor over them.
- Now what left is just clicks and clicks. Join various points of sketches to form frame sketch as shown in Figure-11.

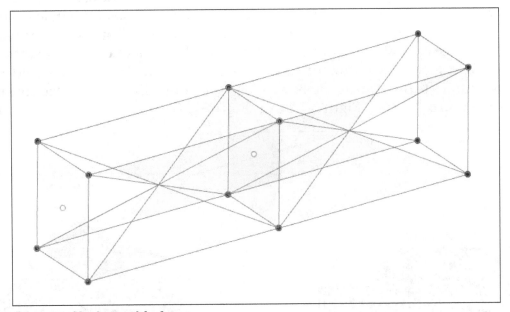

Figure-11. Sketch created for frame

- Click on the **Finish Sketch** tool from the **FINISH SKETCH** panel of the **Toolbar** to exit the sketching environment. So, it is the property of snap points for various elements that allow creating 3D sketch entities.

SELECTION TOOLS

You have earlier selected the objects by clicking on them or by making a window to select them. In Autodesk Fusion, sometimes you will reach in situations where only features are to be selected or only faces are to be selected. In such cases, we apply selection filters. You can also define the pattern by which we will be selecting objects using the options in **SELECT** drop-down of the **Toolbar**; refer to Figure-12. These tools are discussed next.

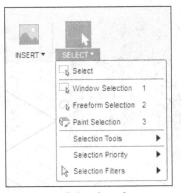

Figure-12. Select drop down

Select Tool

This tool is always active while working in Autodesk Fusion. Because of this tool, you are able to select any tool from toolbar or object from canvas.

Window Selection

This tool is active by default when you start working on Autodesk Fusion. This tool is used to activate window selection mode. You can activate this tool by selecting it from **SELECT** drop-down or by pressing **1** key from keyboard (Note that 1 is not to be pressed from **NUMPAD** on your keyboard). After activating window selection mode, there are two ways in which you can create window to select objects; **Window selection** and **Cross Window selection** (sometimes called Cross selection). Click in the canvas and drag the cursor towards right to perform window selection. Release the button when you have formed window of desired size. Click in the canvas and drag the cursor towards left to perform cross window selection. Both the selection methods and their uses are given in Figure-13 and Figure-14.

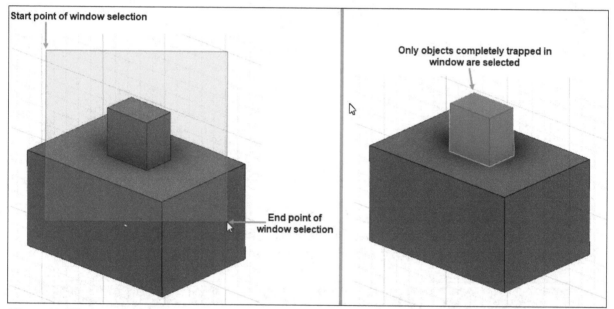

Figure-13. Window selection mode

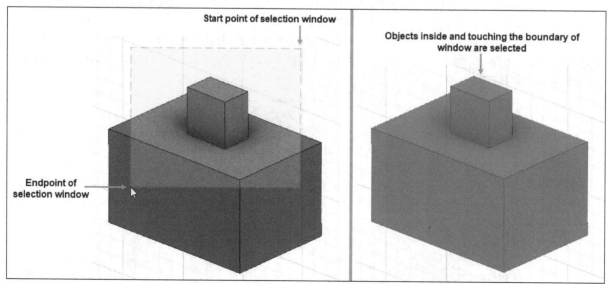

Figure-14. Cross window selection mode

Freeform Selection

Click on the **Freeform Selection** tool from **SELECT** drop-down or press **2** from keyboard to activate this selection mode. Click and drag the cursor to create a freeform boundary for selection. Dragging cursor towards right will make objects completely inside the boundaries selected and dragging cursor towards left will make objects inside or touching the boundaries selected.

Paint Selection

Click on the **Paint Selection** tool from **SELECT** drop-down or press **3** from keyboard to activate this selection mode. In this selection mode, all the objects over which cursor passes while dragging will be selected.

Selection Tools

The selection tools discussed earlier are conventional selection tools. You can also select the objects by theirs names, types, and sizes. These tools are available in the **Selection Tools** cascading menu of **SELECT** drop-down; refer to Figure-15.

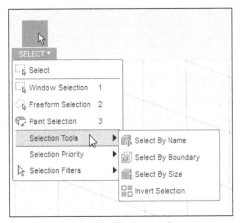

Figure-15. Selection Tools cascading menu

Select By Name Method

• Click on the **Select By Name** tool from the **Selection Tools** cascading menu of **SELECT** drop-down in the **Toolbar**. The **SELECT BY NAME** dialog box will be displayed; refer to Figure-16.

Figure-16. SELECT BY NAME dialog Box

• Type the name of object to be selected in the **Name** edit box of the dialog box.
• Select the **Bodies** check box to allow selection of bodies and select the **Components** check box to allow selection of components.

- Click on the **Find** button in the dialog box. All the components and bodies matching the specified name will be selected; refer to Figure-17.

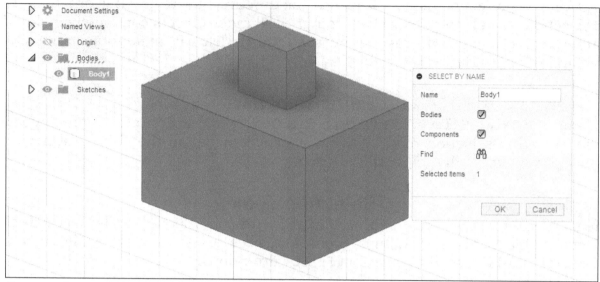

Figure-17. Bodies selected by name

- Click on the **OK** button from the dialog box to complete the process.

Select By Boundary Method

The **Select By Boundary** method is used to select objects which fall under define shape of cylinder, box, or sphere. The procedure to use this method is given next.

- Click on the **Select By Boundary** tool from the **Selection Tools** cascading menu of **SELECT** drop-down. The **SELECT BY BOUNDARY** dialog box will be displayed; refer to Figure-18.

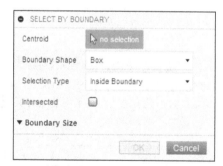

Figure-18. SELECT BY BOUNDARY dialog box

- Select desired option from the **Boundary Shape** drop-down to define the shape of selection boundary.
- Click on desired point in the model to place centroid of boundary shape. Preview of boundary will be displayed; refer to Figure-19.
- Specify desired parameters in the **Boundary Size** section to define size of boundary shape.
- Select the **Inside Boundary** option from the **Selection Type** drop-down if you want to select the objects inside the boundary. Select the **Outside Boundary** option from the **Selection Type** drop-down if you want to select the objects outside the boundary shape.
- Select the **Intersected** check box if you want to select all the objects that are intersecting with created boundary shape.

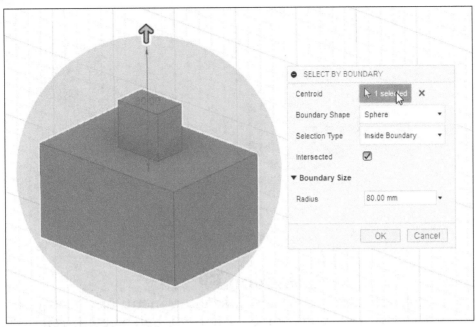

Figure-19. Sphere boundary preview

- Click on the **OK** button from the dialog box to make selection.

Select By Size Method

The select by size method is used when you want to select the entities based on their sizes. The procedure for using this method is given next.

- Click on the **Select By Size** tool from the **Selection Tools** cascading menu of the **SELECT** drop-down in **Toolbar**. The **SELECT BY SIZE** dialog box will be displayed; refer to Figure-20.

Figure-20. SELECT BY SIZE dialog box

- Specify the range of size within which entities will be select and then click on the **OK** button.

Selection Priority

The options in **Selection Priority** cascading menu of **SELECT** drop-down in **Toolbar** are used to define which entities will be given preference while selecting objects; refer to Figure-21. Select the **Select Body Priority** option if you want to select bodies in place of faces/edges. Similarly, select the other options in the cascading menu to select respective entities.

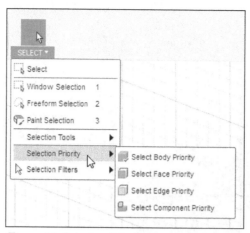

Figure-21. Selection Priority cascading menu

Selection Filters

The check boxes in **Selection Filters** cascading menu are used to define which objects can be selected and which objects cannot be selected; refer to Figure-22. Select desired check boxes to activate selections of respective objects. In default condition, the **Select All** check box is selected.

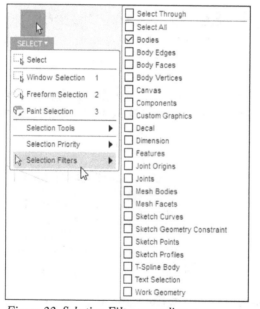

Figure-22. Selection Filters cascading menu

SOLID MODELING

Till this point, we have worked on 2D sketches and have created on Top or Front plane. We have also created 3D sketches. Now, we will learn the tools for solid modeling. **Solid modeling** is the most advanced method of geometric modeling in three dimensions. Solid modeling is the representation of the solid objects in your computer in digital form. Various tools to perform solid modeling are available in the **SOLID** tab of **Toolbar**. We will learn about various tools and commands used to create solid models in next topic.

CREATE DROP-DOWN

In **Create** drop-down, there are various 3D modeling tools used to create and manage solid objects; refer to Figure-23. These tools are discussed next.

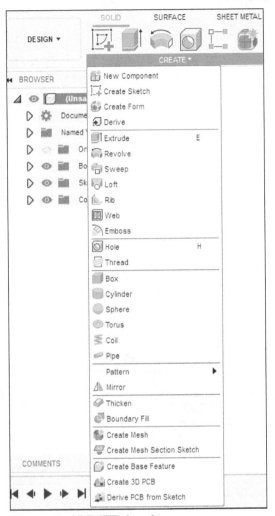

Figure-23. CREATE drop down

New Component

Click on **New Component** tool from **Create** drop-down; refer to Figure-24. A new component will start. You will find use of this tool later in Assembly chapter of the book. (No! we are not skipping this tool, we will discuss it later :-)

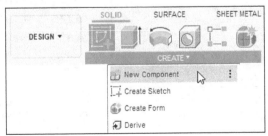

Figure-24. New Component

Extrude

The **Extrude** tool is used to create a solid volume by adding height to the selected sketch. In other words, this tool adds material in the direction perpendicular to the plane of sketch by using the boundaries of sketch. The procedure to use this tool is discussed next.

- Create a closed profile sketch as discussed in previous chapter.
- After creating sketch, exit the sketch environment by selecting the **Finish Sketch** tool from **Toolbar** or **Right-click** outside the sketch and select **Finish Sketch** button from the **Sketch** drop-down in **Marking menu**; refer to Figure-25.

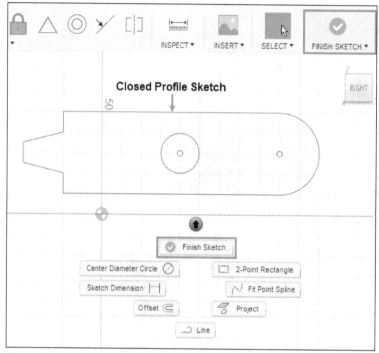

Figure-25. Sketch for extrusion

- Click on **Extrude** tool of **Create** drop-down from **Toolbar**; refer to Figure-26. The **EXTRUDE** dialog box will be displayed; refer to Figure-27.

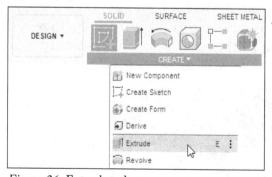

Figure-26. Extrude tool

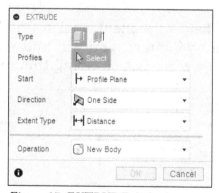

Figure-27. EXTRUDE dialog box

- Select the type of extrusion you want to create from **Type** section of the dialog box. Select **Extrude** button to extrude entire area within selected closed profiles as solid. Select **Thin Extrude** button to extrude a thin wall along selected closed profiles.
- The **Extrude** option is selected by default in the **Type** section of the dialog box and the profile selection is active by default. Select the sketch to be extruded by clicking on the shaded area in the sketch. If your sketch has multiple closed profiles then make sure to select correct shaded area. For example, in Figure-25, we can select inside the circle, outside the circle, or both to create extrusion. The extruded body will be of the shape selected here.

Start drop-down

- There are three tools in **Start** drop-down; refer to Figure-28.

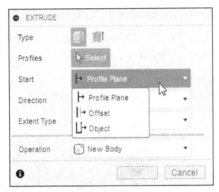

Figure-28. Start drop down

- Select **Profile Plane** tool, if you want to extrude the sketch starting from current sketching plane; refer to Figure-29.

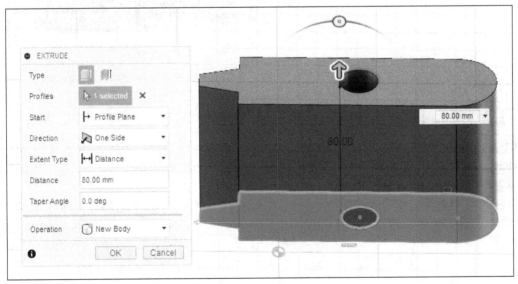

Figure-29. Extrude starting from profile plane

- Click on **Offset** tool if you want to extrude the sketch starting from some offset distance. The updated **EXTRUDE** dialog box will be displayed; refer to Figure-30.

Figure-30. Offset tool in EXTRUDE dialog box

- Click in **Offset** edit box and enter desired value of distance between starting of extrusion and sketch plane; refer to Figure-31.

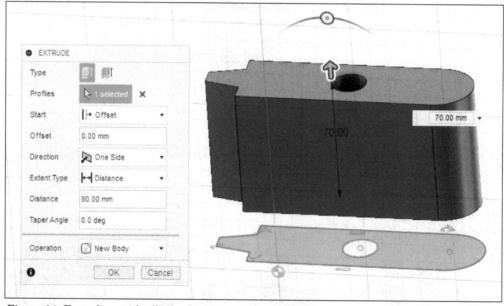

Figure-31. Extruding with offset

- Select the **Object** tool from **Start** drop-down in the dialog box, if you want to extrude the profile starting from selected surface/face (or at a distance from selected face) of another body.
- After selecting **Object** tool, click on **Select** button of **Object** section and select the object from where you are going to start the extrude; refer to Figure-32.

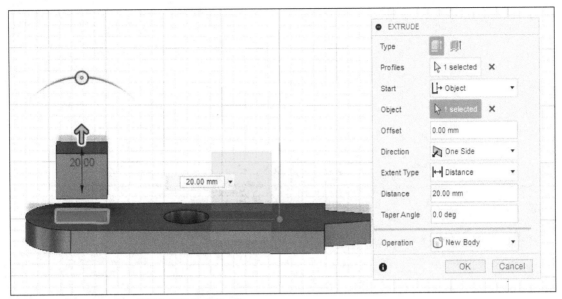

Figure-32. Object tool in EXTRUDE dialog box

- Click in **Distance** edit box and specify the height of extrusion. You can also drag the Arrow handle displayed on the sketch upward to define extrude height.
- If you want to create tapered walls of extrude feature then specify desired angle value in the **Taper Angle** edit box. You can specify both positive and negative values in this edit box. You can also use the rotation handle displayed in the preview to specify taper value. Applying taper angle is also called draft in some software.

Direction drop-down

- There are three tools in **Direction** drop-down of **EXTRUDE** dialog box; refer to Figure-33.

Figure-33. Direction drop down

- Click on **One Side** option from **Direction** drop-down if you want to specify the height of extrusion in one direction.
- Click on **Two Sides** option from **Direction** drop-down if you want to specify the height of extrusion in both upward and downward directions of sketching plane; refer to Figure-34. The options to specify parameters for side 1 and side 2 will be displayed in the dialog box.

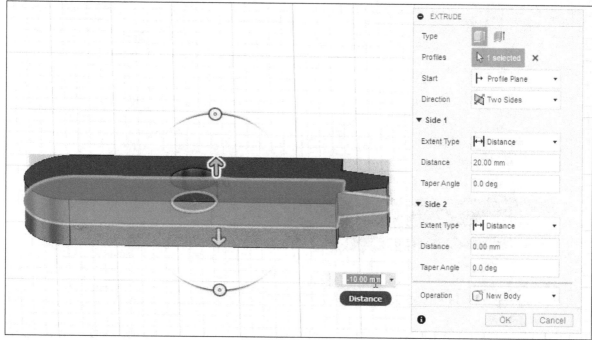

Figure-34. Extrude with two sides

- Click on **Symmetric** option from **Direction** drop-down if you want to specify the height of extrusion symmetric in both upward and downward directions of sketching plane. You will be asked to specify height of extrusion for one side and the profile will be extruded symmetrically on the other side with same value. Select the **Half length** or **Whole length** button from **Measurement** section as required; refer to Figure-35.

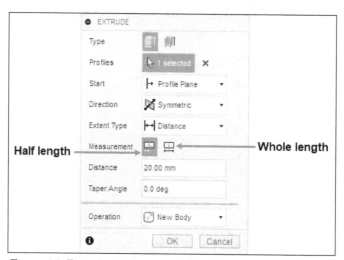

Figure-35. Extrude with symmetric direction

Extent Type drop-down

- Select **Distance** option in **Extent Type** drop-down from **EXTRUDE** dialog box if you want to set the height of extrusion manually. Click in the **Distance** edit box and specify desired height.
- Select **To Object** option from **Extent Type** drop-down if you want to define a surface/face at which extrusion will end; refer to Figure-36. After selecting this option, click on the face or surface up to which extrusion is to be created.

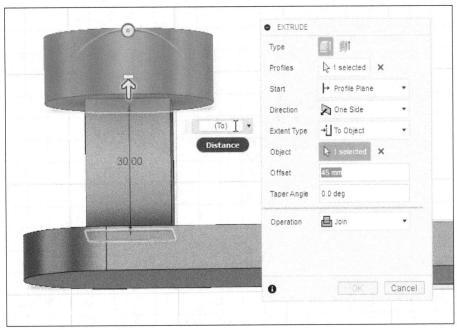

Figure-36. To Object extrude

- Click on **All** tool from **Extent Type** drop-down, if you want to cut through the model or part from extrusion. The extrusion will goes through from the body of model; refer to Figure-37. If you want to add material in place of removing then select the **Join** option from the **Operation** drop-down in the dialog box.

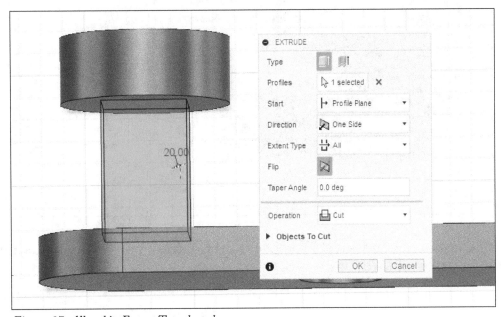

Figure-37. All tool in Extent Type drop down

- Click on the **Flip** button to reverse direction of extrusion.

Operation drop-down

- There are five options in **Operation** drop-down of **EXTRUDE** dialog box; refer to Figure-38. These options are discussed next.

Figure-38. Operation drop down

- Select **Join** option of **Operation** drop-down if you want to combine the extrude feature with intersecting bodies; refer to Figure-39.

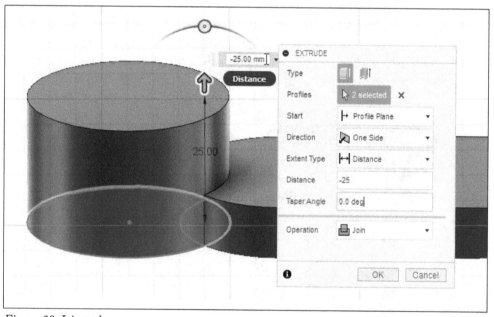

Figure-39. Join tool

- Select **Cut** option if you want to remove material from base body by using current extrude feature; refer to Figure-40.

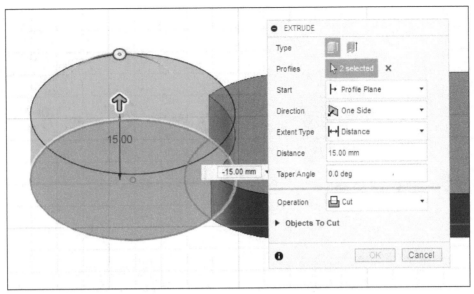

Figure-40. Cut tool

- Select **Intersect** option from **Operation** drop-down if you want to create the region commonly bounded by extrude feature and intersecting solid bodies; refer to Figure-41.

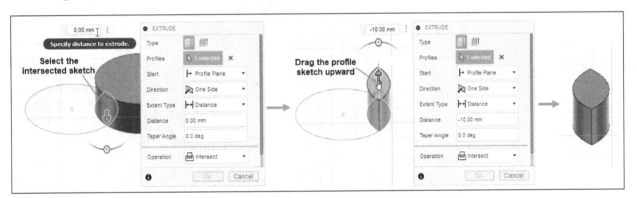

Figure-41. Intersect tool

- Select **New Body** option from **Operation** drop-down if you want to create a new body. A new body creation becomes important when you are creating molds and want all the tool bodies to be individual objects.
- Select the **New Component** option from **Operation** drop-down if you want to create a new component of the current extrude feature. New components are created for assembly. You will learn more about this option later in the book.

Extruding Thin Wall

If you want to extrude a thin wall along selected closed profiles, then select **Thin Extrude** button from **Type** section of the dialog box. The options in the **EXTRUDE** dialog box will be updated; refer to Figure-42.

Figure-42. Thin Extrude options in EXTRUDE dialog box

Wall Location drop-down

There are three options in **Wall Location** drop-down of **EXTRUDE** dialog box; refer to Figure-43. These options are discussed next.

Figure-43. Wall Location drop down

- Select **Inside** option from the drop-down if you want to offset/move the extrusion to the inside of the profile; refer to Figure-44.

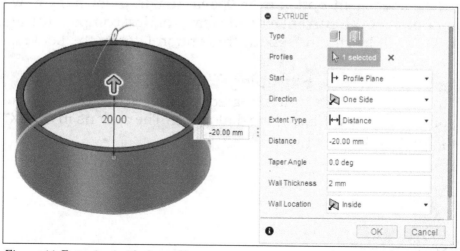

Figure-44. Extrusion inside the profile

- Specify desired thickness of extrusion in the **Wall Thickness** edit box of the dialog box.
- Specify the parameters in the dialog box as discussed earlier and click on **OK** button. The extrusion inside the profile will be created.
- Similarly, select **Outside** option from the drop-down to offset/move the extrusion to the outside of the profile and select **Center** option from the drop-down to center the extrusion on the profile.

Revolve

Revolve tool is used to create a solid volume by revolving a sketch about selected axis. In other words, if you revolve a sketch about an axis then the volume covered by revolving sketch boundary is called revolve feature. The procedure to use this tool is discussed next.

- Create a closed profile sketch with a center line as discussed in previous chapter.
- Click on **Revolve** tool from **CREATE** drop-down of **Toolbar**; refer to Figure-45. The **REVOLVE** dialog box will be displayed; refer to Figure-46.

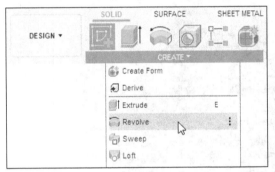

Figure-45. Revolve tool

Figure-46. REVOLVE dialog box

- Click on **Select** button for **Profile** selection from **REVOLVE** dialog box and then select the sketch for profile from the screen canvas.
- Click on **Select** button for **Axis** selection from **REVOLVE** dialog box and then select the sketch edge, line, or center line from the screen canvas; refer to Figure-47.

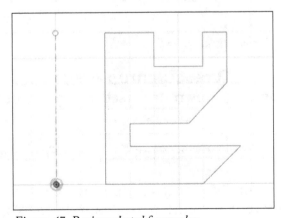

Figure-47. Region selected for revolve

- After selecting the entities, the preview of revolve feature will be displayed; refer to Figure-48.

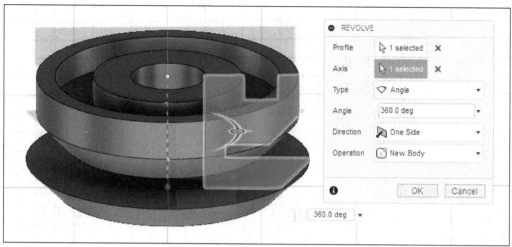

Figure-48. Preview of revolve feature

- Specify desired value of revolve angle in the **Angle** dynamic input box. You can also set the value of angle by moving the drag handle; refer to Figure-49.

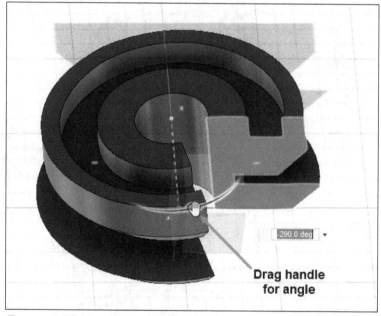

Figure-49. Moving the drag handle

Direction drop-down

The options in **Direction** drop-down are used to define in which direction profile will be revolved.

- Select **One Side** option from **Direction** drop-down to revolve the profile in one direction.
- Select **Two Side** option from **Direction** drop-down if you want to revolve the profile in both sides with different revolution angles; refer to Figure-50.

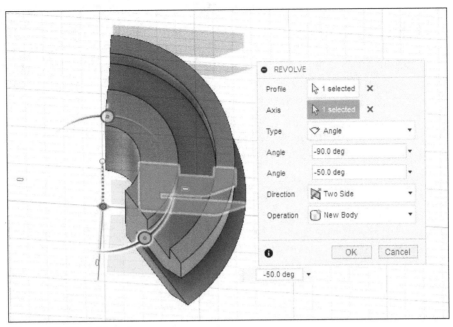

Figure-50. Two side tool

- Select the **Symmetric** option from **Direction** drop-down if you want to revolve the model in both direction with same angle i.e. symmetrically.

Type

- Select the **Angle** option from **Type** drop-down if you want to revolve the profile by specified angle.
- Select the **To** option from **Type** drop-down if you want to revolve the model up to a plane, face, surface, or vertex; refer to Figure-51.

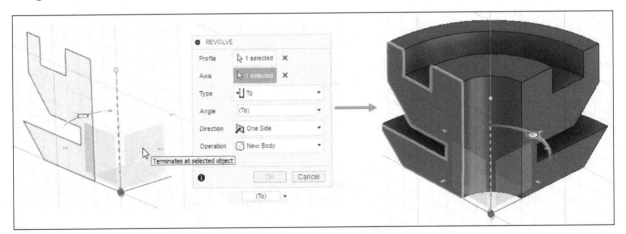

Figure-51. Preview of To option

- Select the **Full** option from **Type** drop-down if you want to revolve the profile to 360 degree angle.
- Set the other options as discussed earlier.
- Click **OK** button from **REVOLVE** dialog box to create the feature.

Sweep

The **Sweep** tool is used to create a solid volume by moving a sketch along the selected path. In other words, if you move a sketch along a path then the volume that is covered by moving sketch boundary is called sweep feature. Note that to create a solid sweep feature, you will need two sketches, one for the profile and another for the path already created in the canvas. The steps to use this tool are discussed next.

- Click on **Sweep** tool from **CREATE** drop-down; refer to Figure-52. The **SWEEP** dialog box will be displayed; refer to Figure-53.

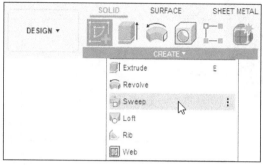

Figure-52. Sweep tool

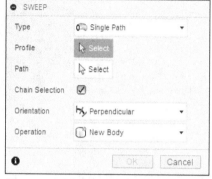

Figure-53. SWEEP dialog box

In this dialog box, there are three options in **Type** drop-down to create sweep feature; **Single Path**, **Path + Guide Rail**, and **Path + Guide Surface**. The procedures to create sweep feature by using these options are discussed next.

Single Path

The **Single Path** tool sweeps the selected profile along a single path.

- Click on the **Select** button of **Profile** section from the **SWEEP** dialog box and select the sketch for profile from the canvas screen.
- Click on the **Select** button of **Path** section from **SWEEP** dialog box and select the path from the canvas screen; refer to Figure-54. Note that few more options are now available in the dialog box.

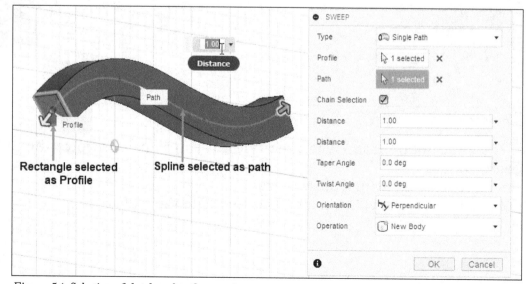

Figure-54. Selection of sketch and path

- Select the **Chain Selection** check box if you want to select chain of entities connected with path curve being selected; refer to Figure-55.

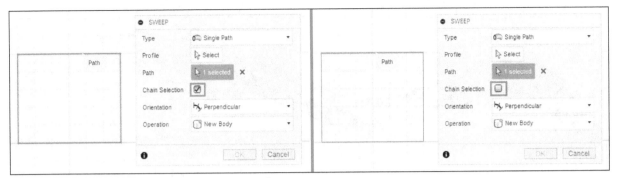

Figure-55. Selecting path with and without chain selection

- The distance in the **Distance** edit boxes are not the actual distance value rather it is the percentage of total distance along the path. **1** is the maximum value which is the total path distance of sweep. Enter **0.5** value in the edit box to create the sweep along only half length of the path. You can also adjust the length of the sweep by entering value in dynamic input box or by moving the drag arrow handles displayed on the model.
- Click in the **Taper Angle** edit box and specify desired angle value to apply draft to the faces of feature.
- Click in the **Twist Angle** edit box and specify the angle value by which you want to twist the profile along path while creating sweep feature; refer to Figure-56.

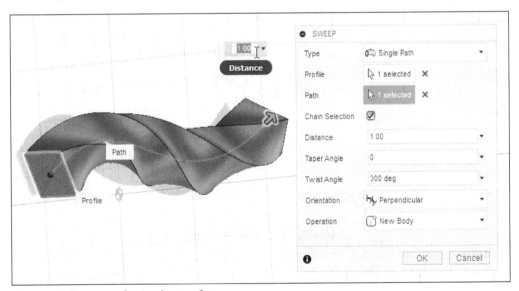

Figure-56. Preview of twisted sweep feature

- The options in the **Orientation** drop-down of **SWEEP** dialog box are used to define how the profile is pulled along the sweep path.
- **Perpendicular** option is the default selection of **Orientation** drop-down. This option creates the sweep feature in such that profile section is always perpendicular to the path. This allow more curvature on the path; refer to Figure-57.

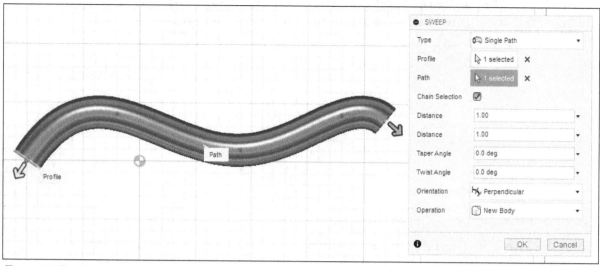

Figure-57. Perpendicular orientation

- **Parallel** option of **Orientation** drop-down creates sweep feature in such a way that the profile section is always parallel to its original orientation; refer to Figure-58.

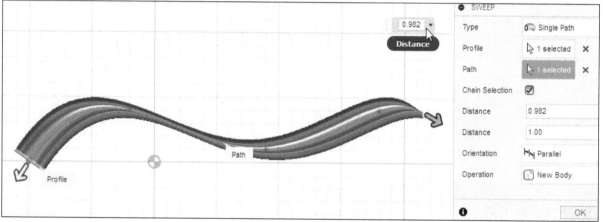

Figure-58. Parallel orientation

- After specifying the parameters, click on **OK** button to create the sweep feature.

Path + Guide Rail

Selecting this option sweeps the selected sketch along the path and uses the guide rail to control the shape. The procedure to use this tool is discussed next.

- Select the **Path + Guide Rail** option of **Type** drop-down from **SWEEP** dialog box. The options in the dialog box will be displayed as shown in Figure-59.

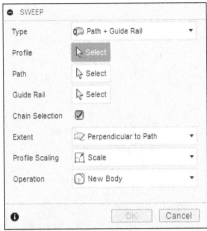

Figure-59. Path + Guide Rail option

- Click on the **Select** button for **Profile** from the **SWEEP** dialog box and then select the sketch created for profile from the canvas screen.
- Click on the **Select** button for **Path** from **SWEEP** dialog box and then select the sketch/edge for path from the canvas screen.

Note that you can also select edges of solids/surfaces as path or guide rails and faces of solids as profile section.

- Click on the **Select** button for **Guide Rail** from the **SWEEP** dialog box and then select the sketch for guide from the canvas screen; refer to Figure-60. Preview of sweep feature will be displayed and additional options will be available in the **SWEEP** dialog box.

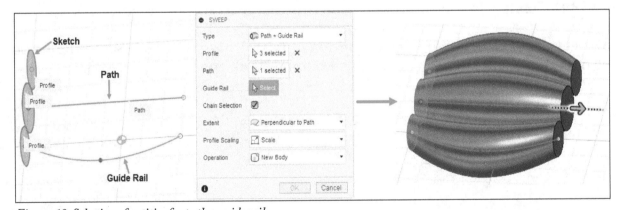

Figure-60. Selection of entities for path + guide rail

- Click in the **Extent** drop-down and select the **Perpendicular to Path** option if you want to create the sweep feature till the profile is perpendicular to path. Select the **Full Extents** option from the drop-down if you want to create sweep till the end point of path.
- Specify desired values for distance in **Path Distance** and **Guide Rail Distance** edit boxes as discussed earlier. Note that the **Guide Rail Distance** edit box will be available only when **Full Extents** option is selected in the **Extents** drop-down.
- Select the **Scale** option from **Profile Scaling** drop-down in **SWEEP** dialog box, if you want to equally scale the profile in both X and Y direction; refer to Figure-61. **The profile will be scaled up depending on the distance between path and guide rail.**

- Select the **Stretch** option from **Profile Scaling** drop-down if you want to scale the profile in X direction only. In this way, profile will keep following guide rail while moving along the path; refer to Figure-61.
- Click on **None** option from **Profile Scaling** drop-down if you don't want to scale the profile along sweep. On selecting this option, guide rail is used as an orientation guide only; refer to Figure-61.

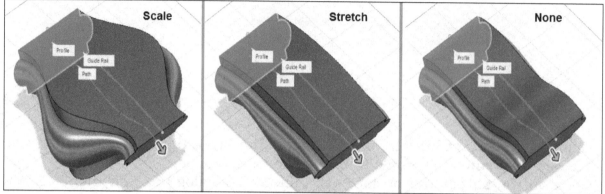

Figure-61. Preview of profile orientation

Note that the profile, path, and guide rail are same in all three cases.

- The tools of **Operation** drop-down are same as discussed earlier in this book. Click on the **OK** button from the dialog box to create the features.

Path + Guide Surface

This option sweeps the selected sketch along the path and uses the guide surface to control the shape. The procedure to use this tool is discussed next.

- Select the **Path + Guide Surface** option of **Type** drop-down from **SWEEP** dialog box. The options in the dialog box will be displayed as shown in Figure-62.

Figure-62. Path + Guide Surface option

- Select the profile for sweep and click on the **Select** button for **Path**.
- Select the curve/edge for path and click on the **Select** button for **Guide Surface**.
- Select the surface by which you want to control the shape of sweep feature; refer to Figure-63. Preview of feature will be displayed.

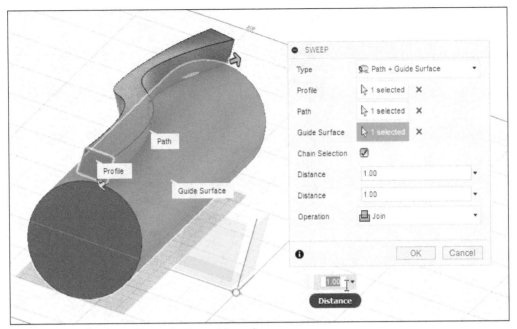

Figure-63. Sweep feature created using guide surface

- Set the parameters as discussed earlier and click on the **OK** button to create the feature.

Part file of this example is available in resources folder of this book. you can try different combinations of sweep for this example.

CONSTRUCTION GEOMETRY

Till now, we have created all the features on default planes but sometimes, we need to create features that cannot be created on default planes. In such cases, we create some construction geometries. Construction geometries are used as references for other features. Some of the well known entities that come under construction geometries are:

- Planes
- Axes
- Points
- Coordinate Systems
- Curves
- Sketches

You have learned about sketches earlier in the book. Now, we will work on other features. All the tools for construction geometry features are available in the **Construct** drop-down; refer to Figure-64. These tools are discussed next.

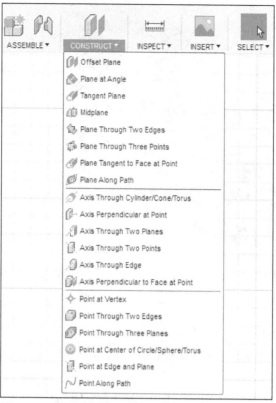

Figure-64. CONSTRUCT panel drop down

Offset Plane

The **Offset Plane** tool is used to create a plane at specified distance from a face/ plane reference. The procedure to use this tool is discussed next.

- Click on the **Offset Plane** tool of **CONSTRUCT** drop-down from **Toolbar**; refer to Figure-65. The **OFFSET PLANE** dialog box will be displayed; refer to Figure-66.

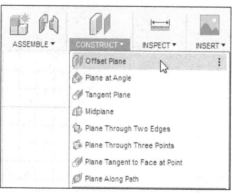

Figure-65. Offset Plane tool

Figure-66. OFFSET PLANE dialog box

- Plane selection is active by default. Click on the plane/face of model to be used as reference for offset plane; refer to Figure-67. The preview for offset plane will be displayed.

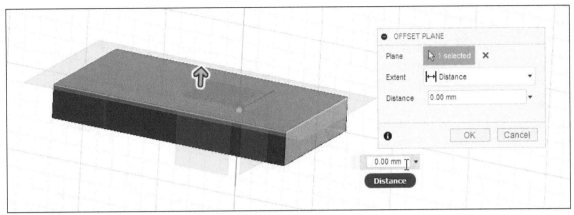

Figure-67. Selection of plane for creating offset plane

- Click in the **Distance** edit box of **OFFSET PLANE** dialog box and enter the distance at which you want to create the plane. You can also set the distance by moving the drag handle.
- After specifying the distance for offset plane, click on the **OK** button from **OFFSET PLANE** dialog box. The plane will be created; refer to Figure-68.

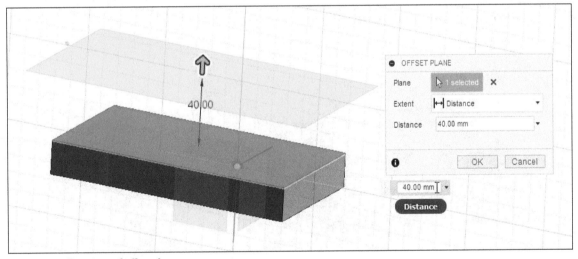

Figure-68. Preview of offset plane

Plane at Angle

The **Plane at Angle** tool is used to create a plane at specified angle. The procedure to use this tool is discussed next.

- Click on the **Plane at Angle** tool of **CONSTRUCT** drop-down from **Toolbar**; refer to Figure-69. The **PLANE AT ANGLE** dialog box will be displayed; refer to Figure-70.

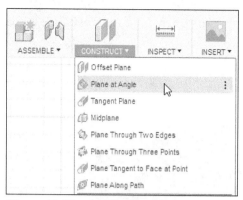
Figure-69. Plane at Angle tool

Figure-70. PLANE AT ANGLE dialog box

- **Line** selection of **PLANE AT ANGLE** dialog box is active by default. Click on the line/ edge/axis from the model as reference for creating a plane. The updated **PLANE AT ANGLE** dialog box will be displayed; refer to Figure-71.

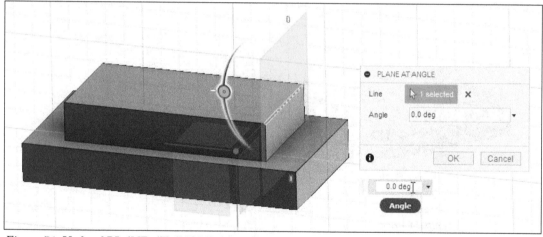
Figure-71. Updated PLANE AT ANGLE dialog box

- Specify desired angle value in **Angle** edit box or drag the handle.
- After specifying the angle for plane, click on the **OK** button from **PLANE AT ANGLE** dialog box. The plane will be created; refer to Figure-72.

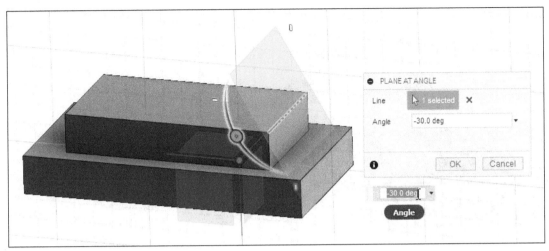

Figure-72. Preview of plane at angle

Tangent Plane

The **Tangent Plane** tool is used to create a plane tangent to the selected face/surface. The procedure to use this tool is discussed next.

- Click on the **Tangent Plane** tool of **CONSTRUCT** drop-down from **Toolbar**; refer to Figure-73. The **TANGENT PLANE** dialog box will be displayed; refer to Figure-74.

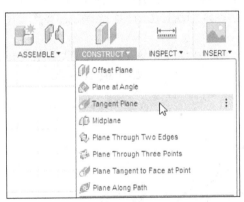

Figure-73. Tangent Plane tool

Figure-74. TANGENT PLANE dialog box

- The **Face** selection of **TANGENT PLANE** dialog box is active by default. Click on the round face of model to which plane should be tangent. The updated dialog box will be displayed; refer to Figure-75.

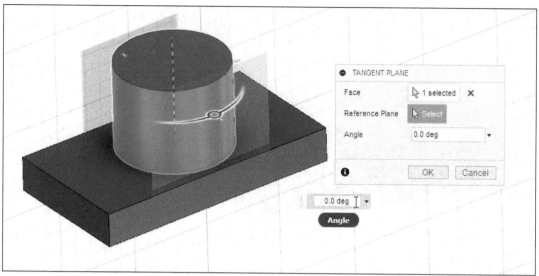

Figure-75. Updated TANGENT PLANE dialog box

- Selection of **Reference Plane** is optional. If you want to select the reference plane to define orientation of the tangent plane then click on the **Select** button of **Reference Plane** section and click on plane/face. The plane will orient accordingly to selected plane.
- Click in the **Angle** edit box and specify the angle of plane with respect to the reference plane. If you have not specified reference plane then angle will be calculated from original position. You can also set the angle by moving the drag handle.
- After specifying the parameters, click on the **OK** button from **TANGENT PLANE** dialog box to create a tangent plane; refer to Figure-76.

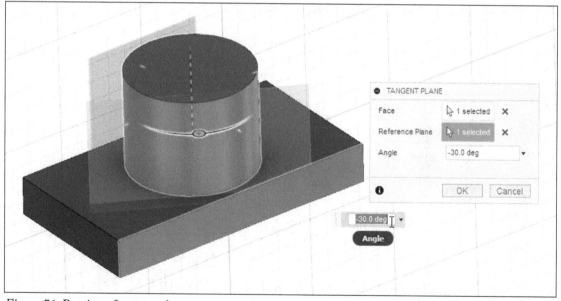

Figure-76. Preview of tangent plane

Midplane

The **Midplane** tool is used to create a plane at the midpoint of two faces/planes. The procedure to use this tool is discussed next.

- Click on the **Midplane** tool of **CONSTRUCT** drop-down from **Toolbar**; refer to Figure-77. The **MIDPLANE** dialog box will be displayed; refer to Figure-78.

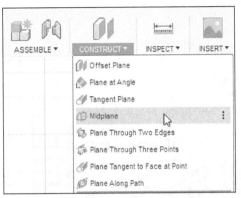

Figure-77. Midplane tool

Figure-78. MIDPLANE dialog box

- Now, you need to select two plane or faces to create a mid plane; refer to Figure-79.

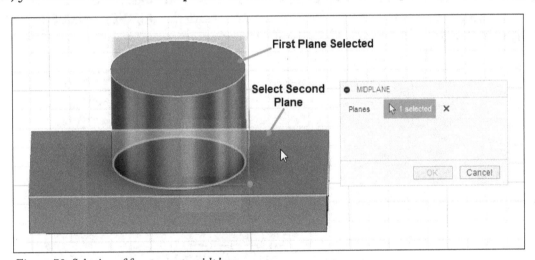

Figure-79. Selection of face to create midplane

- After selecting the second plane or face, preview of plane will be displayed; refer to Figure-80.

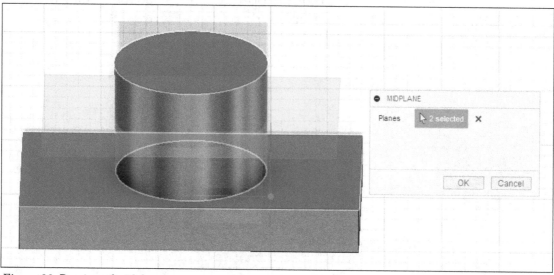

Figure-80. Preview of midplane

• Click on the **OK** button from the dialog box. The plane will be created.

Plane Through Two Edges

The **Plane Through Two Edges** tool is used to create a plane with the reference as two linear edges. The procedure to use this tool is discussed next.

• Click on the **Plane Through Two Edges** tool of **CONSTRUCT** drop-down from **Toolbar**; refer to Figure-81. The tool will be activated and the **PLANE THROUGH TWO EDGES** dialog box will be displayed; refer to Figure-82.

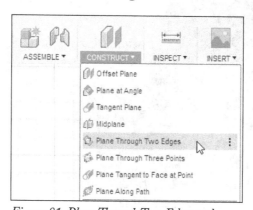

Figure-81. Plane Through Two Edges tool

Figure-82. PLANE THROUGH TWO EDGES dialog box

• Click on two straight edges of model to create plane; refer to Figure-83.

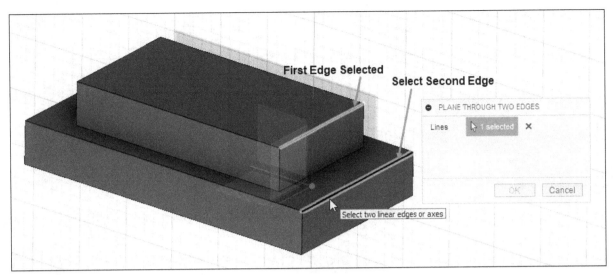

Figure-83. Selection of edges

- On selecting the second edge, the preview of plane will be displayed; refer to Figure-84.

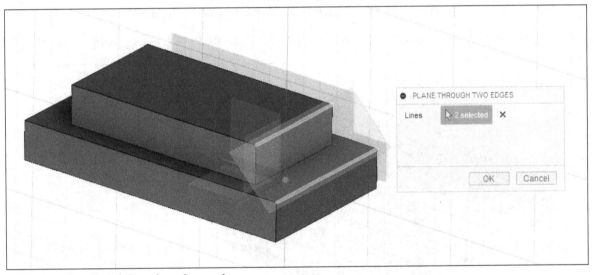

Figure-84. Preview of plane through two edges

- Click on the **OK** button from the dialog box to create the plane.

Plane Through Three Points

The **Plane Through Three Points** tool is used to create a plane with the help of three points as reference. The procedure to use this tool is discussed next.

- Click on the **Plane Through Three Points** tool of **CONSTRUCT** drop-down from **Toolbar**; refer to Figure-85. The **PLANE THROUGH THREE POINTS** dialog box will be displayed; refer to Figure-86.

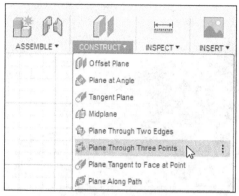

Figure-85. Plane Through Three Points tool

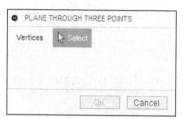

Figure-86. PLANE THROUGH THREE POINTS dialog box

- Select three points from model as reference to create a plane; refer to Figure-87.

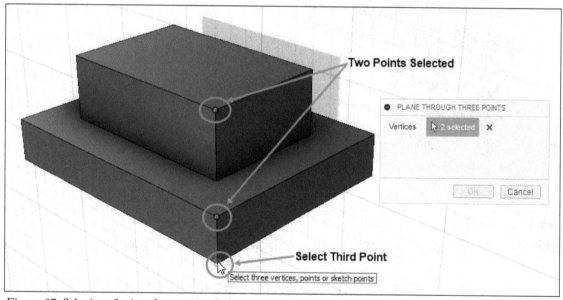

Figure-87. Selection of points for creating plane

- After selecting the third point, the preview of plane will be displayed along with model; refer to Figure-88. Click on the **OK** button from the dialog box.

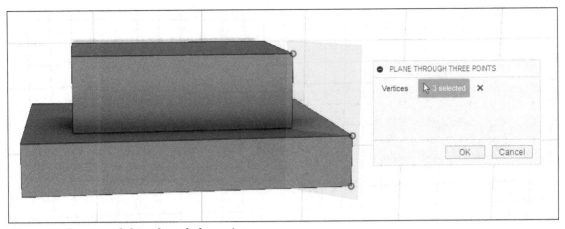

Figure-88. Preview of plane through three points

Plane Tangent to Face at Point

The **Plane Tangent to Face at Point** tool is used to create a plane which is tangent to selected face and aligned to selected point. The procedure to use this tool is discussed next.

- Click on the **Plane Tangent to Face at Point** tool of **CONSTRUCT** drop-down from **Toolbar**; refer to Figure-89. The **PLANE TANGENT TO FACE AT POINT** dialog box will be displayed; refer to Figure-90.

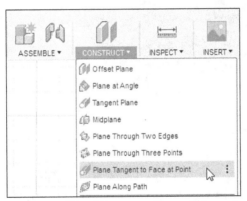

Figure-89. Plane Tangent to Face at Point tool

Figure-90. PLANE TANGENT TO FACE AT POINT dialog box

- Click on the model to select a point and face; refer to Figure-91. You can select round as well as flat faces in this case.

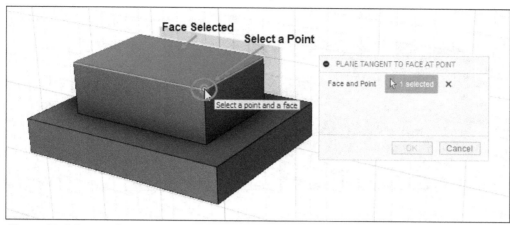

Figure-91. Selection of point and face

- On selecting the second face or point, preview of plane will be displayed; refer to Figure-92. Click on the **OK** button from the dialog box to create the plane.

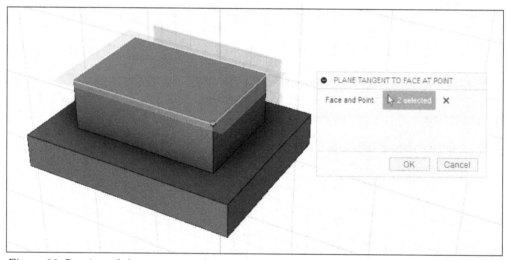

Figure-92. Preview of plane tangent to face at point

Plane Along Path

The **Plane Along Path** tool is used to create a plane along selected path. The plane will be perpendicular to the path at selected point. The procedure to use this tool is discussed next.

- Click on the **Plane Along Path** tool of **CONSTRUCT** drop-down from **Toolbar**; refer to Figure-93. The **PLANE ALONG PATH** dialog box will be displayed; refer to Figure-94.

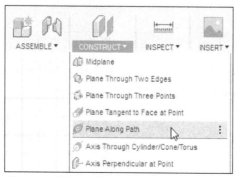

Figure-93. Plane Along Path tool

Figure-94. PLANE ALONG PATH dialog box

- **Path** selection of **PLANE ALONG PATH** dialog box is active by default. Click on the edge/curve from model as reference for plane; refer to Figure-95.

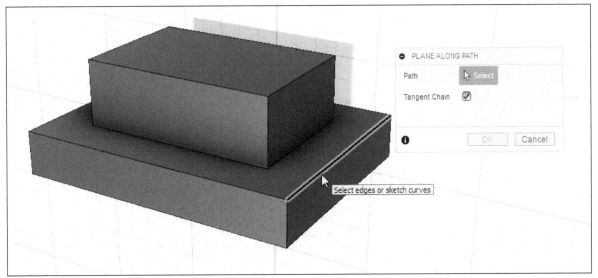

Figure-95. Selection of path

- On selecting the path, the updated **PLANE ALONG PATH** dialog box will be displayed along with the preview of plane; refer to Figure-96.

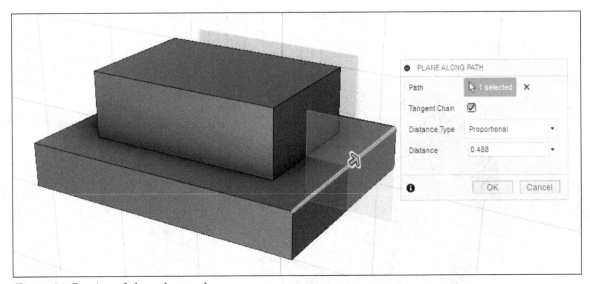

Figure-96. Preview of plane along path

- Click in the **Distance** edit box and specify the distance of plane in percentage between 0 to 1. You can also adjust the distance by moving the drag handle.
- After specifying the parameters, click on the **OK** button from **PLANE ALONG PATH** dialog box. The plane will be created.

Till now, we have created plane through different references. In next section, we will create axis through different references.

Axis Through Cylinder/Cone/Torus

The **Axis Through Cylinder/Cone/Torus** tool is used to create axis through center of Cylinder, Cone, or Torus body. The procedure to use this tool is discussed next.

*   Click on the **Axis Through Cylinder/Cone/Torus** tool of **CONSTRUCT** drop-down from **Toolbar**; refer to Figure-97. The **AXIS THROUGH CYLINDER/CONE/TORUS** dialog box will be displayed; refer to Figure-98.

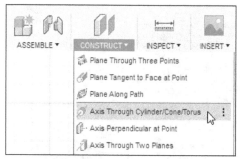

Figure-97. Axis Through Cylinder Cone Torus tool

Figure-98. AXIS THROUGH CYLINDER CONE TORUS dialog box

*   Click on the round face of Cylinder, Cone, or Torus from model, preview of the axis will be displayed; refer to Figure-99. Click on the **OK** button from the dialog box to create the axis.

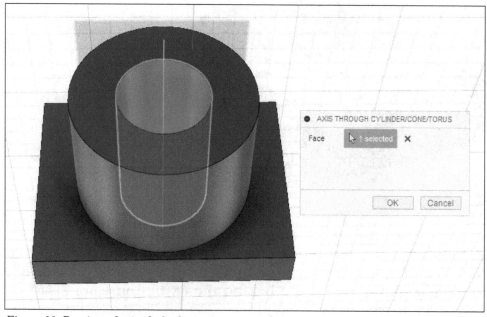

Figure-99. Preview of axis of cylinder

Axis Perpendicular at Point

The **Axis Perpendicular at Point** tool is used to create an axis perpendicular to the selected face at selected point. The procedure to use this tool is discussed next.

- Click on the **Axis Perpendicular at Point** tool of **Construct** drop-down from **Toolbar**; refer to Figure-100. The tool will be activated and **AXIS PERPENDICULAR AT POINT** dialog box will be displayed; refer to Figure-101.

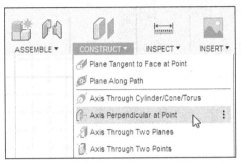

Figure-100. Axis Perpendicular at Point tool

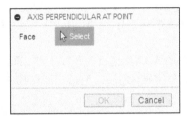

Figure-101. AXIS PERPENDICULAR AT POINT dialog box

- Click at desired location on the face of model; refer to Figure-102.

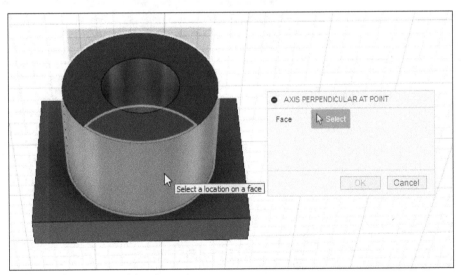

Figure-102. Selection of face for axis

- After selecting the point on any face, preview of axis will be displayed; refer to Figure-103. Click on the **OK** button from the dialog box to create axis.

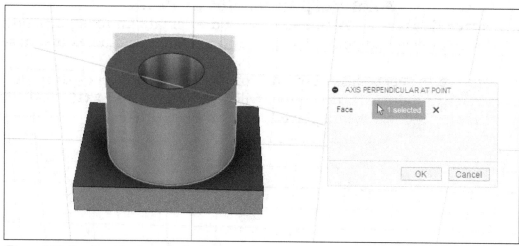

Figure-103. Preview of axis perpendicular at point

Axis Through Two Planes

The **Axis Through Two Planes** tool is used to create an axis at the intersection of two planes or planer faces. The procedure to use this tool is discussed next.

• Click on the **Axis Through Two Planes** tool of **CONSTRUCT** drop-down from **Toolbar**; refer to Figure-104. The **AXIS THROUGH TWO PLANES** dialog box will be displayed; refer to Figure-105.

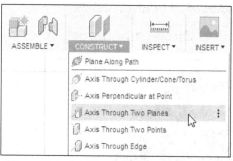

Figure-104. Axis Through Two Planes tool

Figure-105. AXIS THROUGH TWO PLANES dialog box

• Planes selection of **AXIS THROUGH TWO PLANES** dialog box is active by default. You need to select two planes to create an axis; refer to Figure-106.

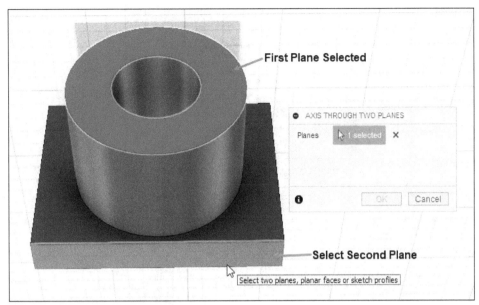

Figure-106. Selection of planes for creating axis

- On selecting the second plane, preview of axis will be displayed; refer to Figure-107.

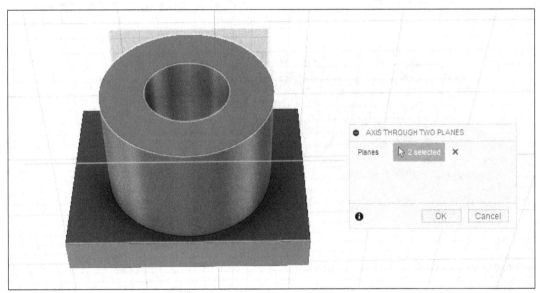

Figure-107. Preview of axis through two planes

- Click on the **OK** button from the dialog box to create the axis.

Axis Through Two Points

The **Axis Through Two Points** tool is used to create an axis passing through two reference points. The procedure to use this tool is discussed next.

- Click on the **Axis Through Two Points** tool of **CONSTRUCT** drop-down from **Toolbar**; refer to Figure-108. The tool will be activated and **AXIS THROUGH TWO POINTS** dialog box will be displayed; refer to Figure-109.

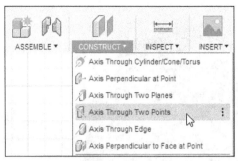

Figure-108. Axis Through Two Points tool

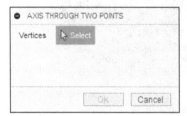

Figure-109. AXIS THROUGH TWO POINTS dialog box

- Select the two points from model; refer to Figure-110. Preview of axis will be displayed. Click on the **OK** button from the dialog box to create the axis.

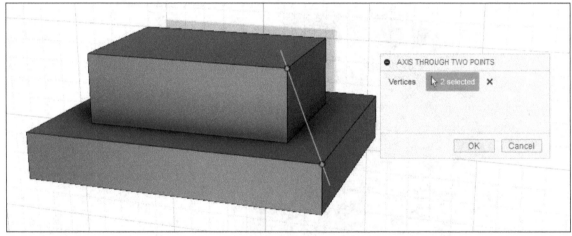

Figure-110. Preview of axis through two points

Axis Through Edge

The **Axis Through Edge** tool is used to create an axis on the selected edge/axis. The procedure to use this tool is discussed next.

- Click on the **Axis Through Edge** tool of **CONSTRUCT** drop-down from **Toolbar**; refer to Figure-111. The tool will be activated and **AXIS THROUGH EDGE** dialog box will be displayed; refer to Figure-112.

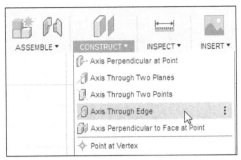

Figure-111. Axis Through Edge tool

Figure-112. AXIS THROUGH EDGE dialog box

- Click on desired edge of model to create the axis; refer to Figure-113. Preview of axis will be displayed.

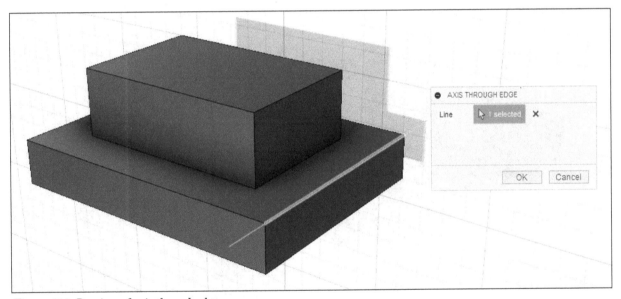

Figure-113. Preview of axis through edge

- Click on the **OK** button from the dialog box to create the axis.

Axis Perpendicular to Face at Point

The **Axis Perpendicular to Face at Point** tool is used to create an axis by using a face and a point as reference. The procedure to use this point is discussed next.

- Click on the **Axis Perpendicular to Face at Point** tool of **CONSTRUCT** drop-down from **Toolbar**; refer to Figure-114. The tool will be activated and **AXIS PERPENDICULAR TO FACE AT POINT** dialog box will be displayed; refer to Figure-115.

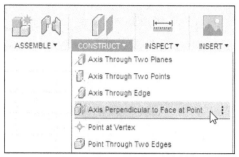

Figure-114. Axis Perpendicular to Face at Point tool

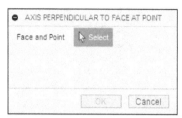

Figure-115. AXIS PERPENDICULAR TO FACE AT POINT dialog box

- Click on the model to select the point and face as a reference for creating plane; refer to Figure-116.

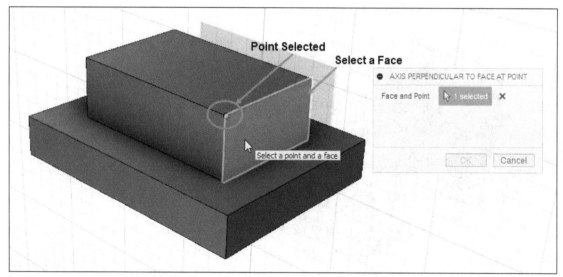

Figure-116. Selection of point and face for creating axis

- On selecting the face, preview of axis will be displayed; refer to Figure-117.

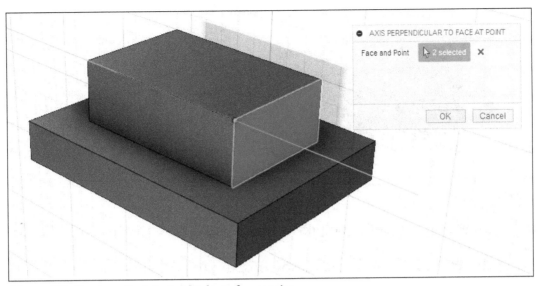

Figure-117. Preview of axis perpendicular to face at point

Note that axis will always be created passing through the point.

Till now, we have learned to create plane and axis. In next topics, we will learn the procedure to create points. Points can be used as origin, coordinate system location, or reference for other features. The tools to create points are discussed next.

Point at Vertex

The **Point at Vertex** tool is used to create a point by selecting any vertex of the model. The procedure to use this tool is discussed next.

- Click on the **Point at Vertex** tool of **CONSTRUCT** drop-down from **Toolbar**; refer to Figure-118. The tool will be activated and **POINT AT VERTEX** dialog box will be displayed; refer to Figure-119.

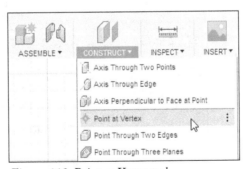

Figure-118. Point at Vertex tool

Figure-119. POINT AT VERTEX dialog box

- Click on the vertex of model to create a constructional point; refer to Figure-120.

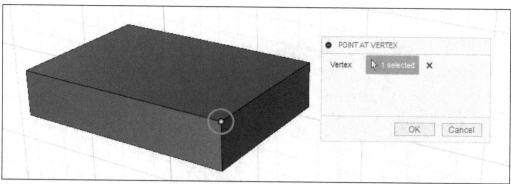

Figure-120. Preview of point at vertex

Note that you can create only one point at a time using this tool.

- Click on the **OK** button from the dialog box to create point.

Point Through Two Edges

The **Point Through Two Edges** tool is used to create a point at intersection of two edges/axes intersecting each other. The procedure to use this tool is discussed next.

- Click on the **Point Through Two Edges** tool of **CONSTRUCT** drop-down from **Toolbar**; refer to Figure-121. The tool is activated and **POINT THROUGH TWO EDGES** dialog box will be displayed; refer to Figure-122.

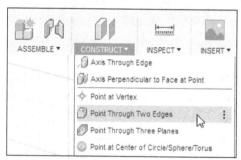

Figure-121. Point Through Two Edges tool

Figure-122. POINT THROUGH TWO EDGES dialog box

- Select the two intersecting lines on the model to create point; refer to Figure-123.

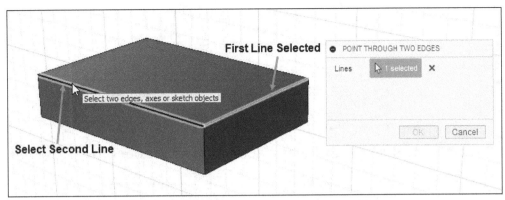
Figure-123. Selection of lines to create point

• On selecting the second line, preview of point will be displayed. Click on the **OK** button from the dialog box to create the point; refer to Figure-124.

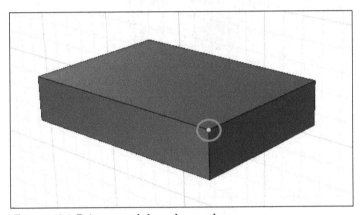
Figure-124. Point created through two edges

Point Through Three Planes

The **Point Through Three Planes** tool is used to create a point at the intersection of three planes/faces. The procedure to use this tool is discussed next.

• Click on the **Point Through Three Planes** tool of **CONSTRUCT** drop-down from **Toolbar**; refer to Figure-125. The tool will be activated and **POINT THROUGH THREE PLANES** dialog box will be displayed; refer to Figure-126.

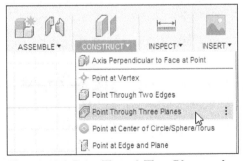
Figure-125. Point Through Three Planes tool

Figure-126. POINT THROUGH THREE PLANES dialog box

- Select the three intersecting planes/faces from the model; refer to Figure-127.

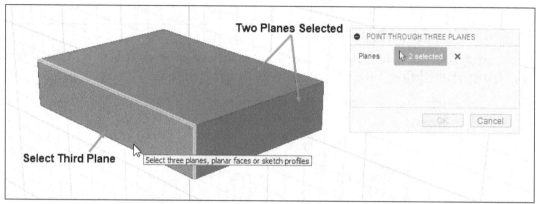

Figure-127. Selection of three planes to create point

- On selecting the third plane, preview of the constructional point will be displayed; refer to Figure-128.

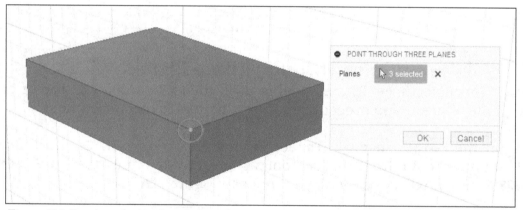

Figure-128. Preview of point through three planes

- Click on the **OK** button from the dialog box to create the point.

Point at Center of Circle/Sphere/Torus

The **Point at Center of Circle/Sphere/Torus** tool is used to create a point at the center of selected circle/sphere/torus. The procedure to use this tool is discussed next.

- Click on the **Point at Center of Circle/Sphere/Torus** tool of **Construct** drop-down from **Toolbar**; refer to Figure-129. The tool will be activated and respective dialog box will be displayed; refer to Figure-130.

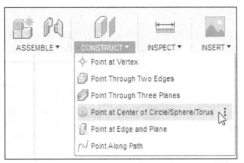

Figure-129. Point at Center of Circle Sphere Torus tool

Figure-130. POINT AT CENTER OF CIRCLE SPHERE TORUS dialog box

- Select the circle, or round face of sphere/torus to create the center point. Preview of point will be displayed; refer to Figure-131.

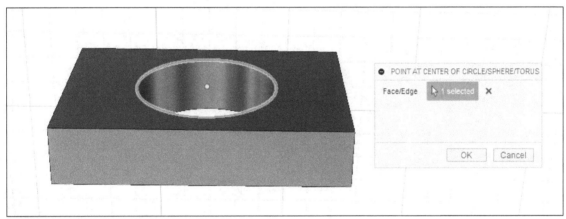

Figure-131. Preview of point at center

- Click on the **OK** button from the dialog box to create the point.

Point at Edge and Plane

The **Point at Edge and Plane** tool is used to create a point with the help of edge and plane as references. The procedure to use this tool is discussed next.

- Click on the **Point at Edge and Plane** tool of **CONSTRUCT** drop-down from **Toolbar**; refer to Figure-132. The tool will be activated and respective dialog box will be displayed; refer to Figure-133.

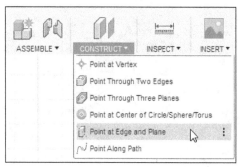

Figure-132. Point at Edge and Plane tool

Figure-133. POINT AT EDGE AND PLANE dialog box

- Select the edge and plane/face to create a point at the intersection; refer to Figure-134.

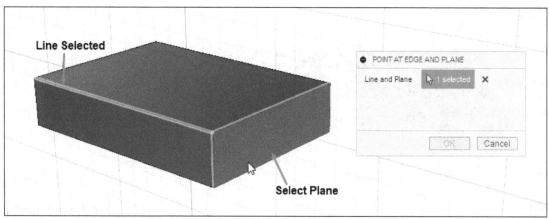

Figure-134. Selection of edge and plane

- On selecting the plane, preview of point will be displayed. Click on the **OK** button from the dialog box to create the point; refer to Figure-135.

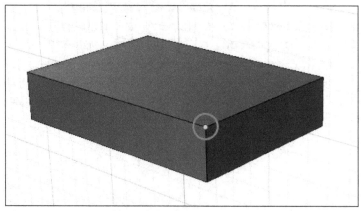

Figure-135. Point at edge and plane created

Point Along Path

The **Point Along Path** tool is used to create a point along selected path. The procedure to use this tool is discussed next.

- Click on the **Point Along Path** tool of **CONSTRUCT** drop-down from **Toolbar**; refer to Figure-136. The **POINT ALONG PATH** dialog box will be displayed; refer to Figure-137.

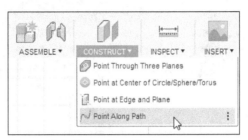

Figure-136. Point Along Path tool

Figure-137. POINT ALONG PATH dialog box

- **Path** selection of **POINT ALONG PATH** dialog box is active by default. Click on the edge/curve from model as reference for point. On selecting the path, the updated **POINT ALONG PATH** dialog box will be displayed; refer to Figure-138.

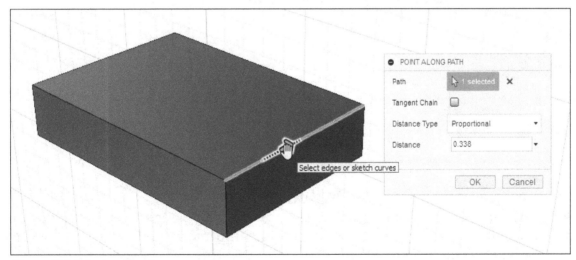

Figure-138. Selection of path to create point

- Click in the **Distance** edit box and specify the distance of point in percentage between 0 to 1. You can also adjust the distance by moving the drag handle.
- After specifying the parameters, click on the **OK** button from **POINT ALONG PATH** dialog box. The point will be created; refer to Figure-139.

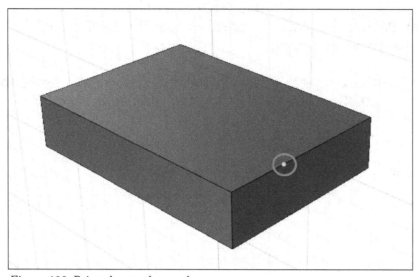

Figure-139. Point along path created

PROJECT/INCLUDE TOOLS

There are various tools in **Project/Include** cascading menu of **CREATE** drop-down in **Toolbar** of **Sketch** mode, which are used to generate sketch entities on current sketching plane or surfaces by using edges, faces, and other geometries of selected objects; refer to Figure-140. You can also generate intersection curves using 3D objects. The tools in this cascading menu are given next.

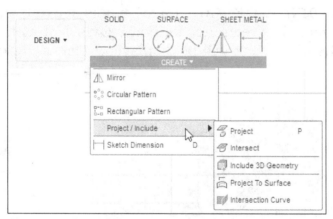

Figure-140. Project Include cascading menu

Project Tool

The **Project** tool is used to project selected surface/face/edge/point/sketch of other features on selected sketching plane. Like, you can project the boundaries of selected face on the sketching plane. The procedure to use this tool is given next.

- Click on the **Project** tool from **Project/Include** cascading menu of **SKETCH** drop-down in the **Toolbar** after activating sketch on desired sketching plane. The **PROJECT** dialog box will be displayed; refer to Figure-141.

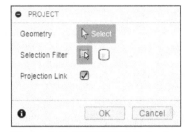

Figure-141. PROJECT dialog box

- Hover the cursor at desired geometry. Preview of project feature will be displayed; refer to Figure-142.

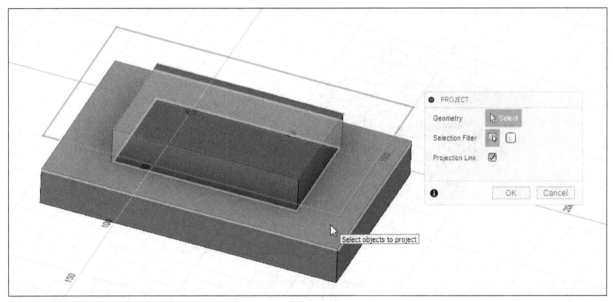

Figure-142. Preview of projected geometry

- Click at desired face/edge/point/curve to project it on sketching plane. You can select multiple objects to project them on sketching plane.
- Click on the **OK** button from the dialog box to apply projection.

Intersect Tool

The **Intersect** tool is used when you want to create sketch geometries at the intersection of selected object and intersecting active sketching plane. The procedure to use this tool is given next.

- Start a sketch on desired plane.
- Click on the **Intersect** tool from **Project/Include** cascading menu of the **CREATE** drop-down in the **Toolbar**. The **INTERSECT** dialog box will be displayed; refer to Figure-143 and you will be asked to select the objects that are intersecting with current sketching plane.

Figure-143. INTERSECT dialog box

- Select the objects that are intersecting with current sketching plane and their intersection curves are to be generated; refer to Figure-144.

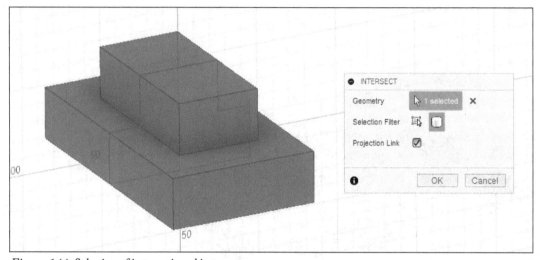

Figure-144. Selection of intersecting objects

- On selecting the objects that are intersecting with current sketching plane, a preview of the intersection curve will be displayed. Click on the **OK** button to create the generated curves; refer to Figure-145.

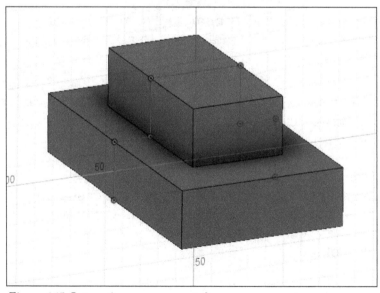

Figure-145. Intersection curves generated

Include 3D Geometry Tool

The **Include 3D Geometry** tool is used to include the boundaries of selected objects in current sketch. Note that this tool will not project the curves on selected sketching plane but you can use the curves for other purposes like as a reference. The procedure to use this tool is given next.

- Start a sketch on desired plane.
- Click on the **Include 3D Geometry** tool from the **Project/Include** cascading menu of the **CREATE** drop-down in **Toolbar**. You will be asked to select the 3D geometry to be included in current sketch.
- Select desired object. Boundary curves of selected object will be created on itself; refer to Figure-146.

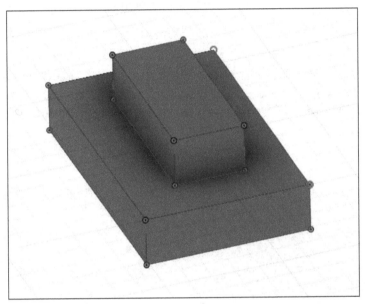

Figure-146. Boundary curves generated

- Press **ESC** to exit the tool.

Project to Surface Tool

The **Project to Surface** tool is used to project selected curves on the selected faces/ surfaces. The procedure to use this tool is given next.

- Click on the **Project to Surface** tool from the **Project/Include** cascading menu of the **CREATE** drop-down in the **Toolbar**. The **PROJECT TO SURFACE** dialog box will be displayed; refer to Figure-147. You will be asked to select a sketching plane.

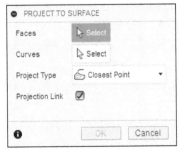

Figure-147. PROJECT TO SURFACE dialog box

- Select the plane which is parallel to face/surface on which sketch is to be projected. The sketching environment will become active. Click on the **Project to Surface** tool, the **PROJECT TO SURFACE** dialog box will be displayed; refer to Figure-148.

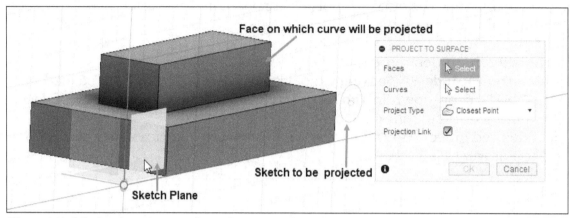

Figure-148. Selection for project to surface tool

- Select the face(s) on which you want to project selected curves.
- Click on the **Select** button for Curves from the dialog box and select the curves to be projected. Preview of projection will be displayed; refer to Figure-149.

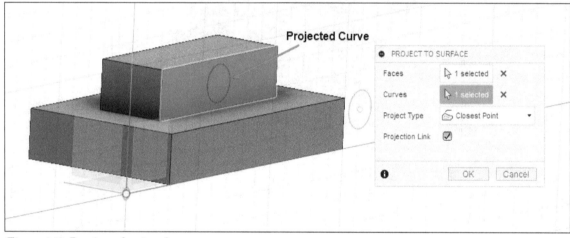

Figure-149. Preview of projected curve

- Select the **Closest Point** option from the **Project Type** drop-down if your object is well within the boundaries of selected face or you want to project along the surface vector to the closest point on the faces; refer to Figure-149. If some of the part of curve is outside the nearest surface then rest of the curve will not be projected; refer to Figure-150. Select the **Along Vector** option from the **Project Type** drop-down if you want to define the direction along which the curve will be projected and select a direction references like edge, axis, face, or plane. Preview of the projection will be displayed; refer to Figure-151.

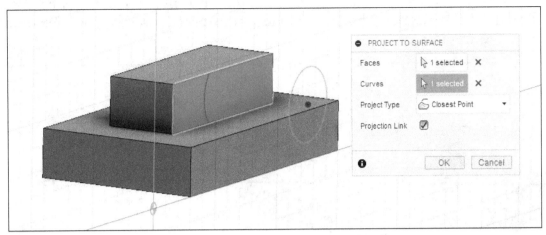

Figure-150. Closest point projection

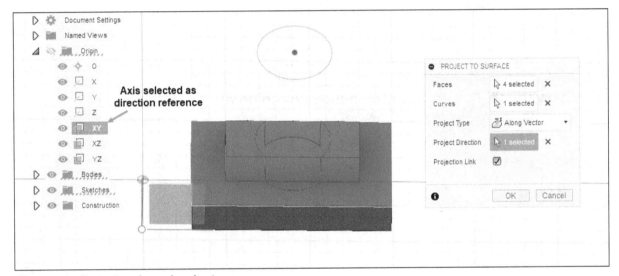

Figure-151. Projection along selected axis

- Click on the **OK** button from the dialog box to create projection.

Intersection Curve Tool

The **Intersection Curve** tool is used to create projection curves generated by selected curve and selected surfaces/faces. The procedure to use this tool is given next.

- Start a new sketch on the plane perpendicular to projection direction.
- Click on the **Intersection Curve** tool from the **Project/Include** cascading menu of the **CREATE** drop-down in **Toolbar**. The **INTERSECTION CURVE** dialog box will be displayed; refer to Figure-152.

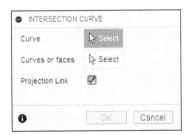

Figure-152. INTERSECTION CURVE dialog box

- Select the curve that you want to be projected and then select the faces/surfaces on which curve is to be projected. Preview of intersection curve will be displayed; refer to Figure-153.

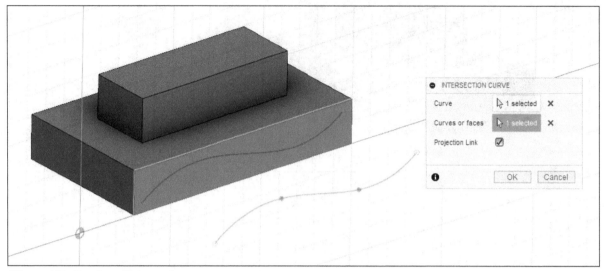

Figure-153. Preview of intersection curve

- Click on the **OK** button from the dialog box to create curves.

SELF ASSESSMENT

Q1. Discuss the procedure of cutting material from a body.

Q2. Which tool is used to combine two bodies?

Q3. Discuss the use of **Sweep** tool with example?

Q4. Discuss the procedure of creating plane using **Offset Plane** tool with example.

Q5. Create a plane tangent to the face of model.

Q6. Create a plane using three points as reference.

Q7. Create a plane using edge of a model as reference.

Q8. Create a point at the center of cylinder.

FOR STUDENT NOTES

Chapter 4

Advanced 3D Modeling

Topics Covered

The major topics covered in this chapter are:

- *Introduction*
- *Loft Tool*
- *Rib Tool*
- *Web Tool*
- *Hole Tool*
- *Thread Tool*
- *Box Tool*
- *Cylinder Tool*
- *Sphere Tool*
- *Torus Tool*
- *Coil Tool*

- *Pipe Tool*
- *Pattern Tool*
- *Mirror Tool*
- *Thicken Tool*
- *Boundary Fill Tool*
- *Create Form Tool*
- *Create Base Feature Tool*
- *Create Mesh Tool*
- *Create Mesh Section Sketch Tool*
- *Create 3D PCB Feature Tool*

INTRODUCTION

In the last chapter, we have learned the procedure to use some 3D modeling tools and creating the constructional geometry. In this chapter, we will use the constructional geometry tools with some advanced 3D modeling tools. The tools are available in panel of MODEL workspace.

LOFT

The **Loft** tool is used to create a solid volume by joining two or more sketches created on different planes; refer to Figure-1. The procedure to use this tool is given next.

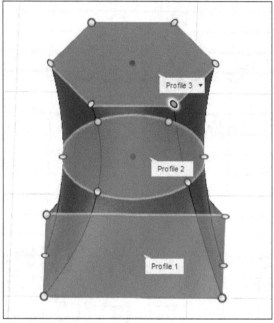

Figure-1. Lofted feature

- Click on the **Loft** tool from **CREATE** drop-down in the **Toolbar**; refer to Figure-2. The **LOFT** dialog box will be displayed as shown in Figure-3.

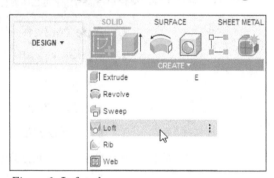

Figure-2. Loft tool

- By default, **Profiles** selection box is active and you are asked to select sketches for loft feature. You can also select faces/surfaces as profile for loft.
- Click one by one on the sketches created at different planes. Note that you need to select the sketches in order by which they can be joined to each other successively. The preview of loft will be displayed; refer to Figure-4.

Figure-3. LOFT dialog box

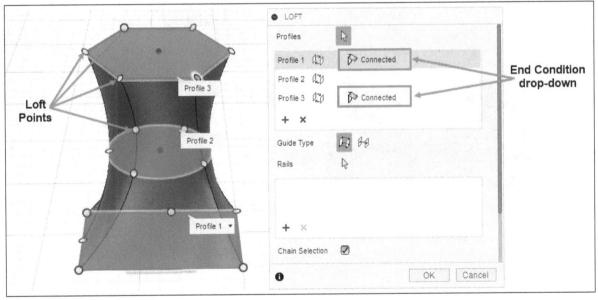

Figure-4. Preview of loft

- To make adjustments to the loft, you can move the loft points along the profile edge.
- Click the **End condition** drop-down of **Profile 1** from **Profiles** selection box. The options will be displayed as per the geometry selected.

If you have selected sketches only as profiles then **Connected** and **Direction** options are available in the drop-down; refer to Figure-5. If you have selected sketches and faces of model or only faces of the model as profiles then **Connected (G0)**, **Tangent(G1)**, and **Curvature(G2)** options are available in the **End condition** drop-down; refer to Figure-6. If you have selected a point as profile at the end then **Sharp** and **Point Tangent** options will be available in the **End condition** drop-down; refer to Figure-7.

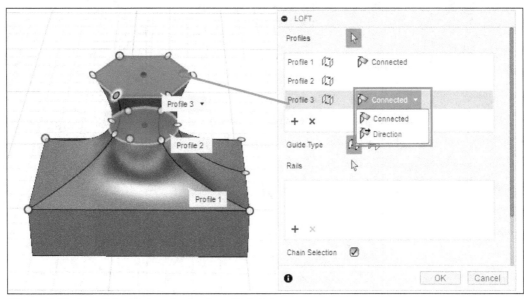

Figure-5. End condition options for sketch profile in loft

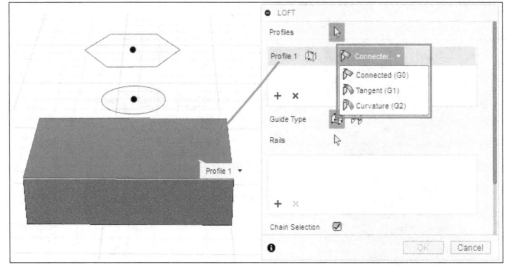

Figure-6. End condition options for face profile in loft

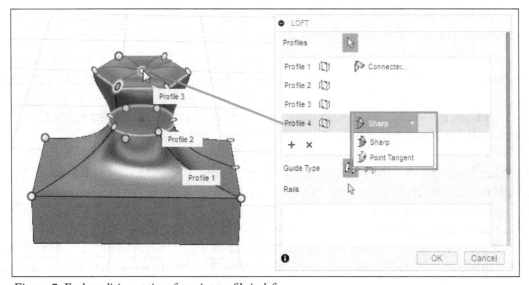

Figure-7. End condition options for point profile in loft

- Select desired end condition option from the **End condition** drop-down. One line explanation of each end condition option is given in Figure-8.

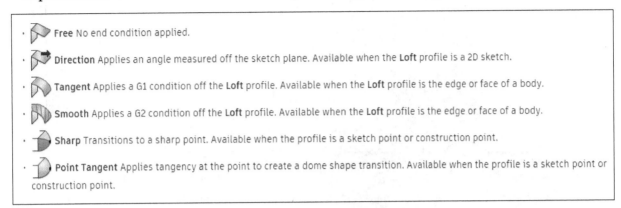

Figure-8. End condition

- Select the **Direction** option from the **End Condition** drop-down. The updated **LOFT** dialog box will be displayed with options to specify **Takeoff Weight** and **Takeoff Angle**; refer to Figure-9. As the name suggests, these options are used to specify scale and angle at which loft will start or end at selected profile.

Figure-9. Updated LOFT dialog box

- If you have selected **Tangent**, **Curvature**, or **Point Tangent** option from the **End Condition** drop-down then **Tangency Weight** edit box will be displayed in the dialog box to specify scale factor of tangency.
- Enter the value as **4** in **Takeoff Weight** edit box and specify the outward angle in **Takeoff Angle** edit box by **20** degree. You will find a bulge in the loft feature; refer to Figure-10.

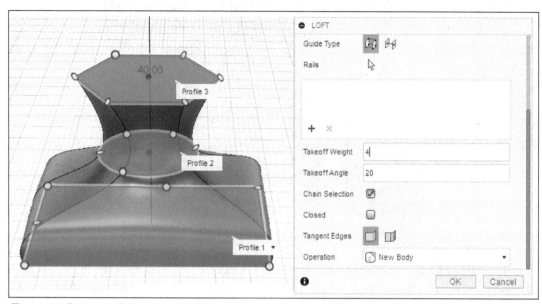

Figure-10. Preview after entering the value

- Select the **Chain Selection** check box if you also want to select the adjacent edges and included as one profile. Clear the check box if you want to select the each edge of profile individually. This check box hardly has any effect in **Model** workspace but you will find its uses in **Patch** and **Sculpt** workspaces while creating loft feature.
- After setting desired parameters, click **OK** button from **LOFT** dialog box to create the loft feature.

Loft feature can be very useful in joining two faces or surfaces to create solid volume; refer to Figure-11. You can adjust the shape of loft by using the drag points on the edges.

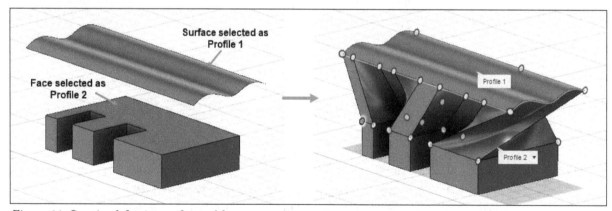

Figure-11. Creating loft using surface and face

Guide Type

There are two ways to control transition between profiles of loft; using **Rails** and using **Centerline**. **Rails** are 2D or 3D curves that affect the loft shape between sections at boundaries. To refine the shape of loft, you need to add number of rails. These rails must intersect each section and terminate on or beyond the first and last sections. **Rails** must be tangentially continuous. A **Centerline** tool is a type of rail to which the loft sections are held normal. It behaves like a sweep path. Centerline lofts maintain a more consistent transition between the cross-sectional areas of selected loft sections. Center lines follow the same criteria as rails, except they need not intersect the sections and only one centerline can be selected. We have discussed two examples next for tent and ring which use rails and centerline, respectively.

Using Rails

- To select rails for the loft, click the **Rails** button in **Guide Type** section; refer to Figure-12.

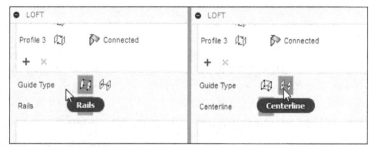

Figure-12. Rails and Centerline tool

- In this case, we have created a sketch of tent like structure using a rectangle and four arcs (to be used as rail) on different planes; refer to Figure-13.

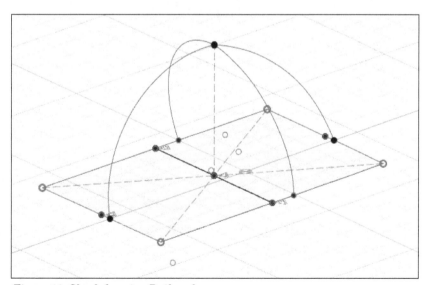

Figure-13. Sketch for using Rails tool

- Select the **Loft** tool from **CREATE** drop-down. The **LOFT** dialog box will be displayed.
- Select the rectangle and point of the sketch in **Profiles** section of **LOFT** dialog box. The preview of model will be shown.
- Click on the **Arrow** selection button of **Rails** from **Guide Type** section box, to select the rails for the loft.
- Select the four arcs of the model; refer to Figure-14.

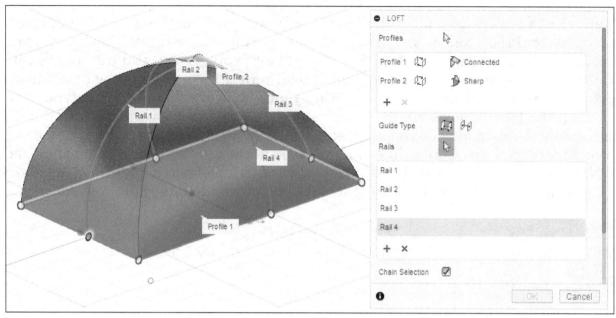

Figure-14. Preview of loft on selecting Rails

- Click **OK** button of **LOFT** dialog box to complete the process of making loft. The shape of loft will be modified according to rails selected.

Using Centerline

- To use the **Centerline** guide type, we have created a sketch of ring like structure with the help of circle and rectangle on different planes; refer to Figure-15.

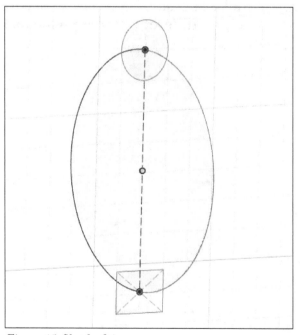

Figure-15. Sketch of ring

- Select the **Loft** tool from **CREATE** drop-down. The **LOFT** dialog box will be displayed.
- Select the circle and rectangle as profile in **Profiles** section of **LOFT** dialog box.
- Since these two profiles are coplanar, you may not get any preview and an error might welcome you.

- Click on **Centerline** button from **Guide Type** section and select the circle to be used as center line. The preview will be displayed; refer to Figure-16.

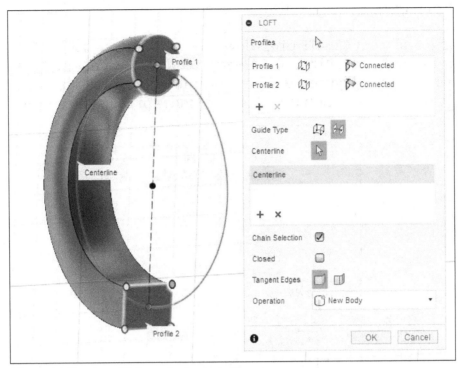

Figure-16. Preview of loft on selecting centerline

- By default, **Closed** check box is clear so you will be getting half of the ring. Select the **Closed** check box to complete the loft. Preview of the loft feature will be modified; refer to Figure-17.

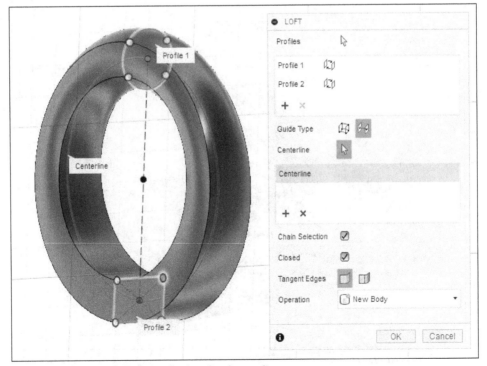

Figure-17. Preview of loft on selecting closed centerline

- Click **OK** button of **LOFT** dialog box to complete the process.
- The tools of **Operation** drop-down are same as discussed earlier.

RIB

The **Rib** tool is used to create supporting features in the model by using curve(s). The procedure to use this tool is discussed next.

- Click on **Rib** tool from **CREATE** drop-down; refer to Figure-18 or right-click on the screen canvas and select **Rib** from **Marking menu** displayed. The **RIB** dialog box will be displayed; refer to Figure-19. Note that the curve for rib should be created in such a way that its projection is within solid faces of the model; refer to Figure-20.

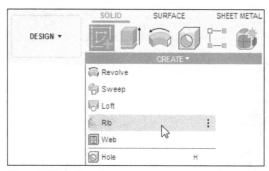

Figure-18. Rib tool

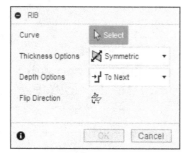

Figure-19. RIB dialog box

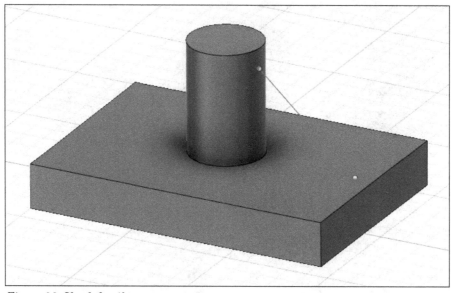

Figure-20. Sketch for rib

- The **Curve** section of **RIB** dialog box is active by default. Select the sketch curve joining edges of solid model to create rib. The updated **RIB** dialog box will be displayed; refer to Figure-21.

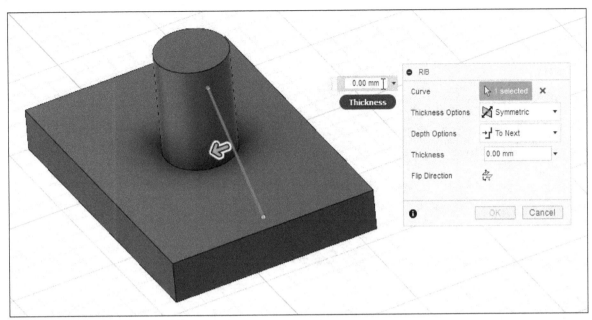

Figure-21. Updated RIB dialog box

- Select the **Symmetric** option from **Thickness Options** drop-down to set the thickness of rib on both side of curve. Select **One Direction** option from **Thickness Options** drop-down to apply the thickness of rib in one direction or one side.
- Enter desired value of thickness in **Thickness** edit box or floating window.
- Select the **To Next** option in **Depth Options** drop-down from **RIB** dialog box to create the rib up to next surface or model; refer to Figure-22.

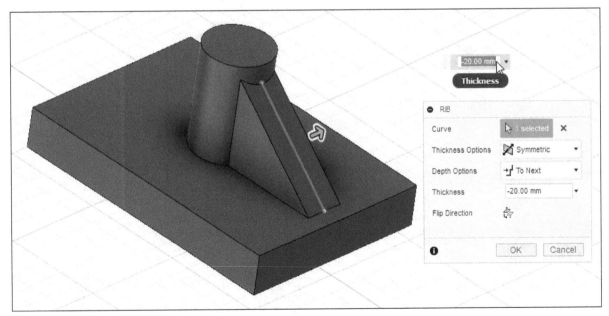

Figure-22. To Next tool in Rib

- Select the **Depth** option from **Depth Options** drop-down to specify the depth of rib. Type the numerical value of depth in **Depth** edit box or adjust the depth by moving the **Drag arrow**.
- You can also enter the value of depth in minus(-) to reverse the direction of depth in **Depth** edit box; refer to Figure-23.

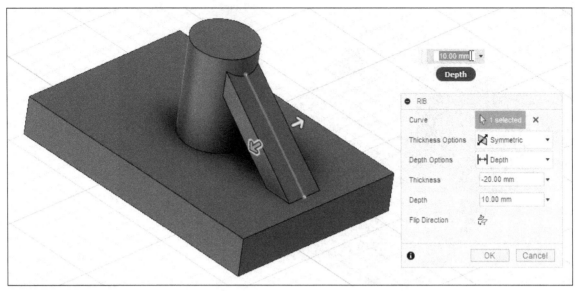

Figure-23. Depth tool in Rib

- Click on the **Flip Direction** button to flip the direction of rib.
- After specifying the parameters, click **OK** button of **RIB** dialog box to complete the process.

WEB

The **Web** tool works like **Rib** tool which is used to create geometry from sketch curves that intersect with pre-existing bodies. The **Web** tool uses multiple curves to create several merged ribs at once. The procedure to use this tool is discussed next.

- Click on **Web** tool from **CREATE** drop-down of **SOLID** tab in the **Toolbar**; refer to Figure-24. The **WEB** dialog box will be displayed; refer to Figure-25.

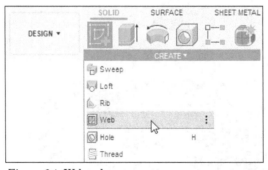

Figure-24. Web tool

Figure-25. WEB dialog box

- The **Curve** selection is active by default. Select the sketch to create web; refer to Figure-26.

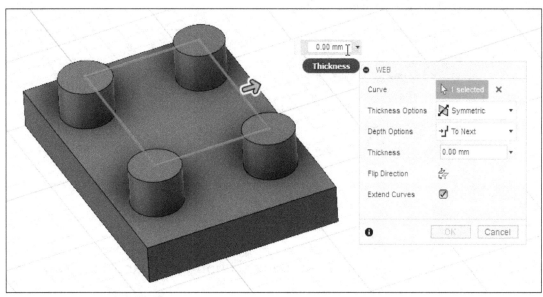

Figure-26. Selecting the sketch

- Enter desired value of thickness in the **Thickness** edit box or floating window. You can also set the thickness by moving the drag handle.
- The options for **Thickness Options** and **Depth Options** drop-downs are same as discussed earlier. Preview of rib will be displayed according to parameters specified; refer to Figure-27.
- Click **OK** button from **WEB** dialog box to complete the process.

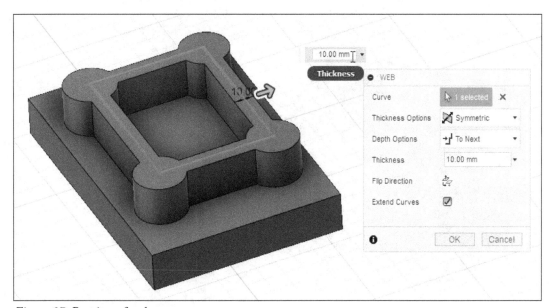

Figure-27. Preview of web

HOLE

The **Hole** tool is used to create holes that comply with the real machining tools. With the help of **Hole** tool, three types of holes are created. The procedure to use this tool is discussed next.

• Click on **Hole** tool from **CREATE** drop-down or right-click on the screen canvas and select **Hole** from **Marking menu** displayed; refer to Figure-28. The **HOLE** dialog box will be displayed; refer to Figure-29.

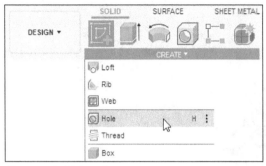

Figure-28. Hole tool

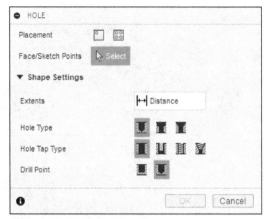

Figure-29. HOLE dialog box

• There are two types of tools in **Placement** section which are discussed next.

Single Hole

• Click on **Single Hole** button from **Placement** section to create one hole. The **Select** button for **Face** section will be activated.
• Select the face/plane on which you want to create hole. The updated **HOLE** dialog box will be displayed with preview of hole; refer to Figure-30.

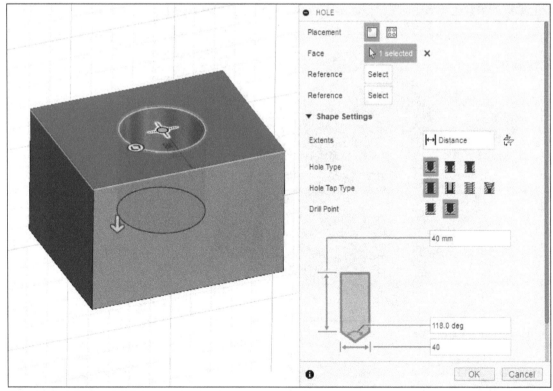

Figure-30. Updated HOLE dialog box

• Click on **Select** button of first **Reference** section from **HOLE** dialog box, click at the center of hole and then click at the edge from where you want to specify the distance of hole; refer to Figure-31.

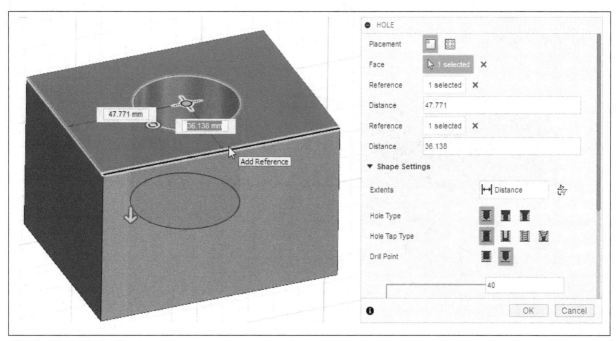

Figure-31. Adding references

- Enter desired value of distance in the dynamic input box or **Distance** edit box in dialog box. Click on another edge to define other reference and set desired distance value.
- Click on **Simple** button in **Hole Type** section to create simple hole of specified dimensions.
- Click on **Counterbore** button in **Hole Type** section to create a counterbore hole and specify the dimension in their respective edit box at the bottom in the dialog box; refer to Figure-32.

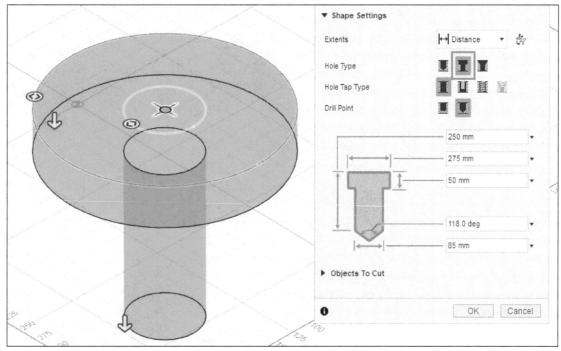

Figure-32. Counterbore hole

- Click on **Countersink** button in **Hole Type** drop-down to create a countersink hole and specify the dimension in their respective edit box at the bottom in the dialog box; refer to Figure-33.

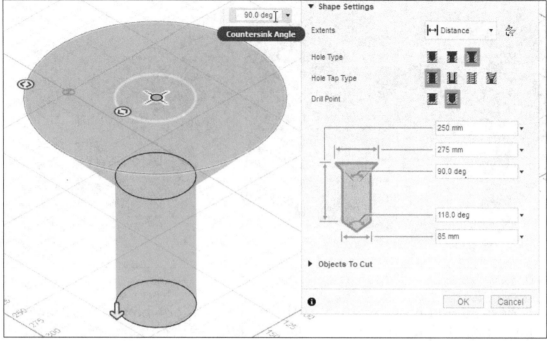

Figure-33. Countersink hole

- Click on desired button in the **Hole Tap Type** section to define the tapping inside the hole. Select the **Simple** button to create straight holes without tapping.
- Select the **Clearance** button to create standard holes with defined fit. The options will be displayed accordingly; refer to Figure-34. Select desired options to define size of hole.

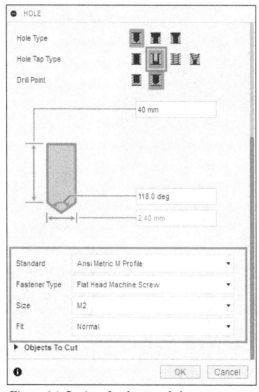

Figure-34. Options for clearance hole tap type

- Select the **Tapped** button to create holes as per their class and designation in standard hole profile database; refer to Figure-35.

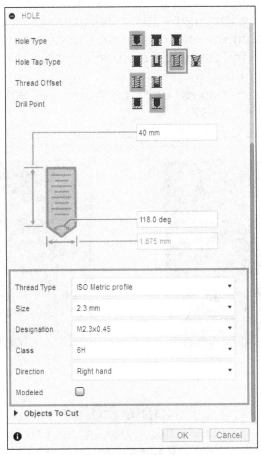

Figure-35. Options for tapped hole tap type

- Select desired options to define shape and size of hole. If you want to model threads created by tapping in the hole then select the **Modeled** check box. Preview of hole will be displayed with selected thread type; refer to Figure-36.

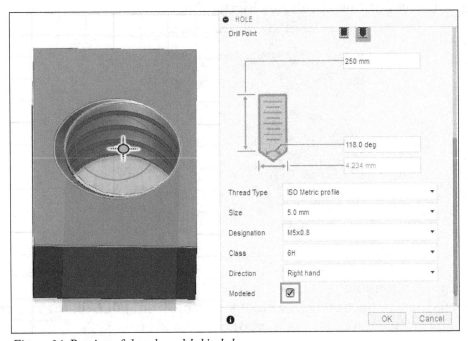

Figure-36. Preview of threads modeled in hole

- Select the **Distance** option from **Extents** drop-down to specify the distance of hole in the model. Select the **To** option from **Extents** drop-down to set the length of hole up to next model or face. Select the **All** option from **Extents** drop-down to create a hole passing through all the features in its path.
- Click on **Flip Direction** button to reverse the direction of hole.
- Expand the **Objects To Cut** node and select check boxes for bodies that you want to be cut by the hole.
- After specifying the parameters, click **OK** button of **HOLE** dialog box to complete the process; refer to Figure-37.

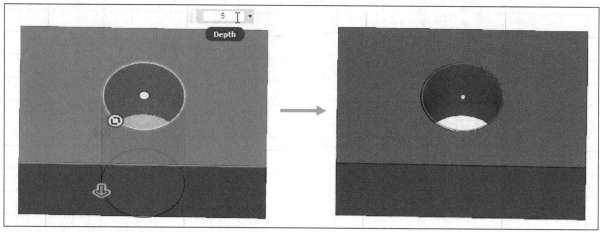

Figure-37. Preview of hole

Multiple Holes

The **Multiple Holes** button in **HOLE** dialog box is used to create multiple similar hole at a time. To use this tool, you must have sketch points already created on the model where you want to cut holes. The procedure to use this tool is discussed next.

- Click on the **Multiple Holes** button in **Placement** section from **HOLE** dialog box. The **Sketch Point** selection tool will be activated. Select the sketch points for defining positions of holes; refer to Figure-38.

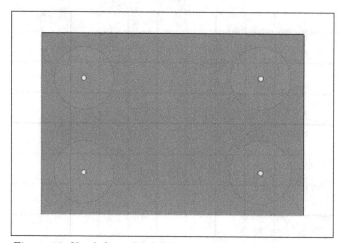

Figure-38. Sketch for multiple holes

- Set the other options as discussed earlier.
- After specifying the parameters, click **OK** button from **HOLE** dialog box to finish the process. The preview will be displayed; refer to Figure-39.

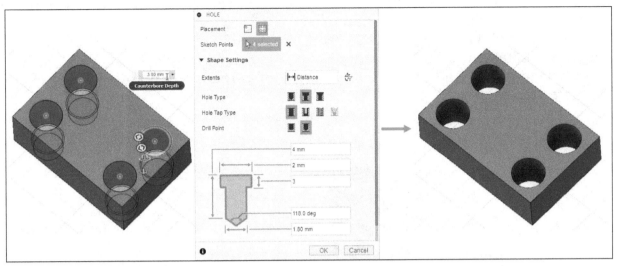

Figure-39. Preview of multiple holes

THREAD

The **Thread** tool is used to cut helical thread on cylindrical faces. Using this tool, you can save the custom threads in library. The procedure to use this tool is discussed next.

• Click on the **Thread** tool from the **CREATE** drop-down in the **Toolbar**; refer to Figure-40. The **THREAD** dialog box will be displayed; refer to Figure-41.

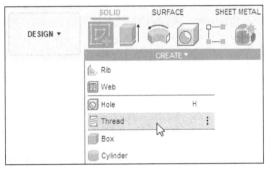

Figure-40. Thread tool

Figure-41. THREAD dialog box

- The **Faces** selection of **THREAD** dialog box is active by default. Select the round faces to which threads are to be applied.
- Select the **Modeled** check box to create the model feature of thread otherwise it will be created as cosmetic feature.
- On selecting **Modeled** check box, preview of thread will be displayed; refer to Figure-42.

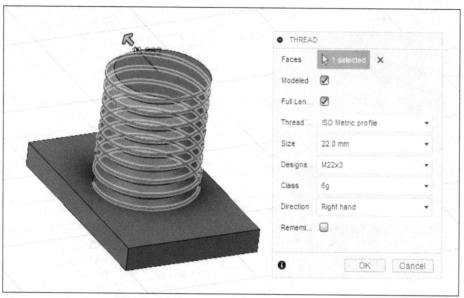

Figure–42. Preview of thread on selecting Modeled check box

- Clear the **Full Length** check box to specify the length of thread on the selected face. Otherwise, thread will be created up to the entire length of selected face.
- Click on the **Thread Type** drop-down to select the type of thread profile; refer to Figure-43. Most of the engineering thread profiles are available in this drop-down.

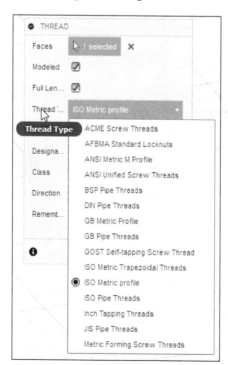

Figure–43. Thread Type drop down

- Click on the **Size** drop-down and select desired size of thread. You can also adjust the size by using the **Drag Handle**.
- Click on the **Designation** drop-down and select the designation of thread; refer to Figure-44.
- Click on the **Class** drop-down and select the class of thread.
- Click on **Direction** drop-down and select the direction for thread; refer to Figure-45.

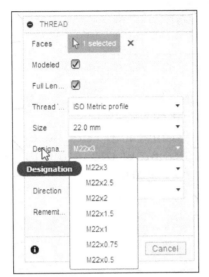

Figure-44. Thread Designation drop down

Figure-45. Thread Direction drop down

- Select the **Remember Size** check box to remember these thread setting for the next time, when the **Thread** tool will be invoked.
- Click **OK** button from **THREAD** dialog box to create the thread; refer to Figure-46.

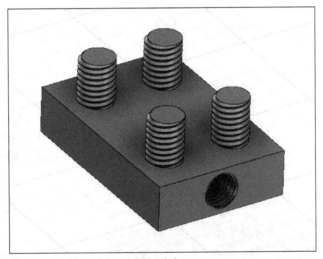

Figure-46. Thread created in solid

BOX

The **Box** tool is used to create a rectangular box on selected work plane. The procedure to use this tool is discussed next.

- Click on **Box** tool in **CREATE** drop-down from **Toolbar**; refer to Figure-47. You will be asked to select the plane for creating the sketch for box.

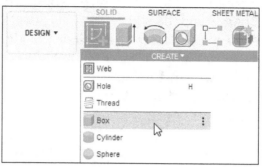

Figure-47. Box tool

- Select **Top** plane or any other desired plane/face.
- Click on the screen to specify the first point of sketch.
- Move the cursor to display preview of rectangle and click at desired location to specify other corner of rectangle. You can also enter the numerical value of **Length** and **Width** in their specific edit box; refer to Figure-48.

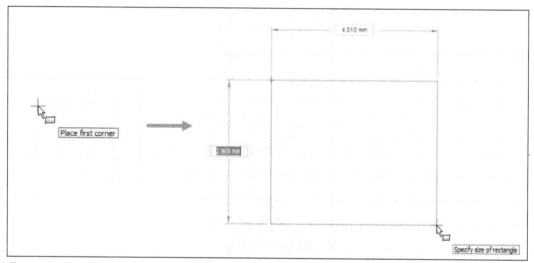

Figure-48. Entering dimension for rectangle

- You can toggle between these two edit boxes by pressing **TAB** key. The **Box** dialog box will be displayed along with the preview of box; refer to Figure-49.

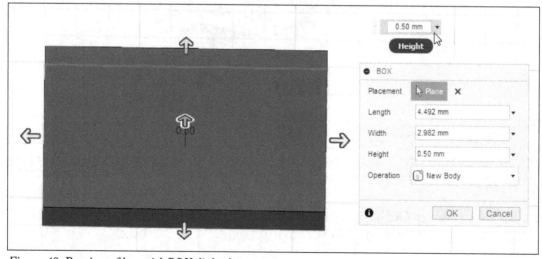

Figure-49. Preview of box with BOX dialog box

- Specify desired value of height in the **Height** edit box of the **BOX** dialog box.
- The options of **Operation** section were discussed earlier.
- Click **OK** button from the **BOX** dialog box to create the box; refer to Figure-50.

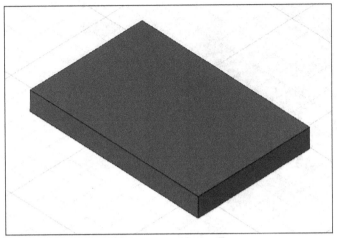

Figure-50. Box created

CYLINDER

This tool is used to create a cylindrical body by adding depth to a circular region. The procedure to use this tool is discussed next.

- Click on **Cylinder** tool in **CREATE** drop-down from **Toolbar**; refer to Figure-51. You are asked to select plane for creating base sketch. Select the desired plane/face.

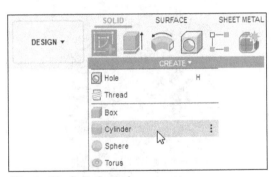

Figure-51. Cylinder tool

- Click on the screen to specify the center point of base circle.
- Enter desired value of diameter in the floating window and press **ENTER** to create a sketch of circle; refer to Figure-52.

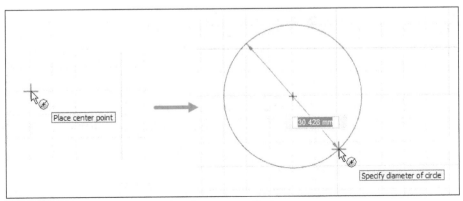

Figure-52. Specifying diameter of circle to create cylinder

- The **CYLINDER** dialog box will be displayed along with preview of cylinder; refer to Figure-53.

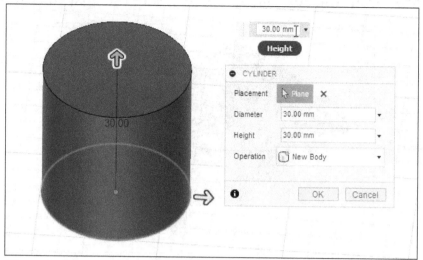

Figure-53. Preview of cylinder with CYLINDER dialog box

- If you want to edit the diameter of cylinder then click in the **Diameter** edit box of **CYLINDER** dialog box and specify desired value.
- Click in the **Height** edit box and specify the height of cylinder.
- Click **OK** button from the **CYLINDER** dialog box to create the cylinder; refer to Figure-54.

Figure-54. Cylinder created

SPHERE

The **Sphere** tool is used to create a T-spline, solid, or surface sphere on a work plane or face. The procedure to use this tool is discussed next.

- Click on the **Sphere** tool in **CREATE** drop-down from **Toolbar**; refer to Figure-55. You are asked to select the plane for creating the sketch for circle. Select desired plane/face.

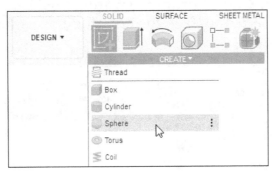

Figure-55. Sphere tool

- Click on the screen at desired location to specify the center point of sphere. The **SPHERE** dialog box will be displayed along with the preview of sphere; refer to Figure-56.

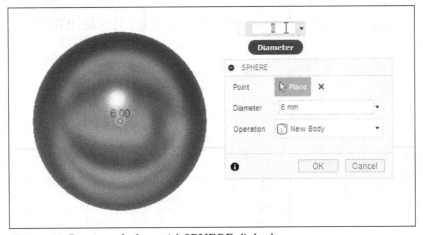

Figure-56. Preview of sphere with SPHERE dialog box

- Click in the **Diameter** edit box and specify desired value of diameter for sphere.
- The options of **Operation** drop-down are same as discussed earlier.
- Click **OK** button of **SPHERE** dialog box to finish the process; refer to Figure-57.

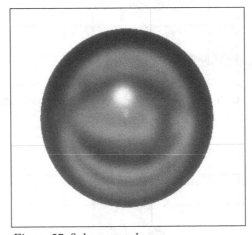

Figure-57. Sphere created

TORUS

The **Torus** tool is used to create T-spline, solid, or surface torus on a work plane or face. The procedure to use this tool is discussed next.

- Click on **Torus** tool in **CREATE** drop-down from **Toolbar**; refer to Figure-58. You are asked to select the plane/face for creating the center circle of torus. Select desired face/plane.

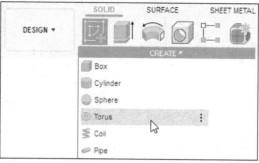

Figure-58. Torus tool

- Click on the screen to specify the center point of sketch. Specify the diameter of torus in dynamic input box and press **ENTER**; refer to Figure-59. The **TORUS** dialog box will be displayed along with the preview of Torus; refer to Figure-60.

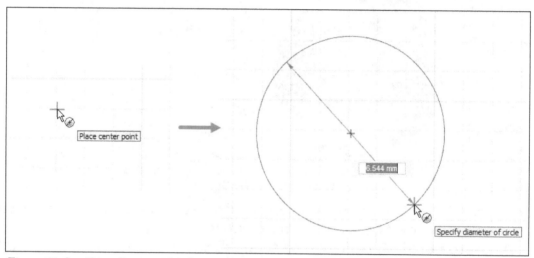

Figure-59. Specifying diameter of circle to create torus

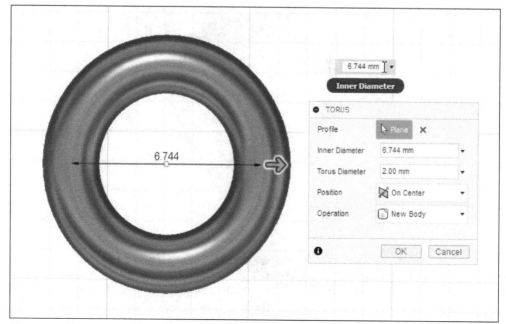

Figure-60. Preview of torus with TORUS dialog box

- If you want to edit the diameter of circle earlier created then click in the **Inner Diameter** edit box and specify desired value.
- Click in the **Torus Diameter** edit box and enter the diameter of torus. You can also adjust the diameter of torus by moving the drag handle.
- Select the **On Center** option of **Position** drop-down from **TORUS** dialog box, if you want to use inner circle as center of the torus. Select the **Inside** option of **Position** drop-down if you want to position the torus inside of the inner circle. Select the **Outside** option of **Position** drop-down if you want to position the torus section outside of the inner circle.
- The options of **Operation** drop-down are same as discussed earlier.
- Click **OK** button of **TORUS** dialog box to finish the process; refer to Figure-61.

Figure-61. Torus created

COIL

The **Coil** tool is used for creating helix-based features like springs. The procedure to use this tool is discussed next.

- Click on the **Coil** tool of **CREATE** drop-down from **Toolbar**; refer to Figure-62. You will be asked to select a face/plane to create base sketch.

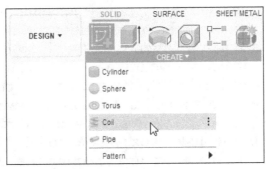

Figure-62. Coil tool

- Click on desired face/plane. You will be asked to specify center of the base circle.
- Click on the screen to specify the center point of sketch. Type desired value of diameter of coil in dynamic input box and press **ENTER**; refer to Figure-63.

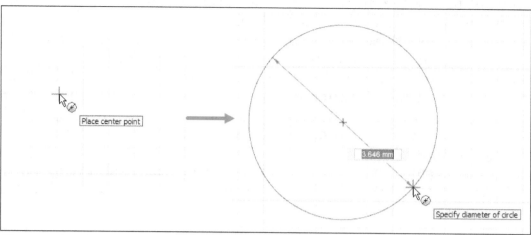

Figure-63. Specifying diameter of circle to create coil

- The **COIL** dialog box will be displayed along with the preview of coil; refer to Figure-64.

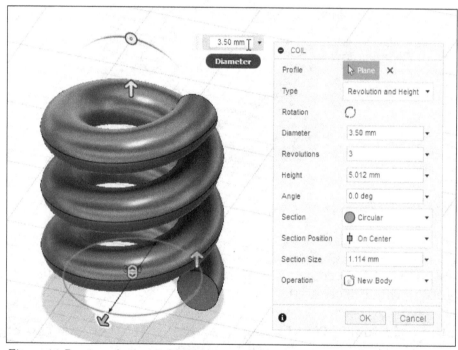

Figure-64. Preview of coil with COIL dialog box

- Select the **Revolution and Height** option from **Type** drop-down if you want to create the coil by specifying number of revolutions in the coil and height of coil. Select the **Revolution and Pitch** option from **Type** drop-down if you want to create the coil by specifying number of revolutions in the coil and distance between two successive turns of the coil. Select the **Height and Pitch** option from **Type** drop-down to specify total height of the coil and distance between two successive turns of the coil. Select the **Spiral** option from **Type** drop-down to create a spiral coil. Note that the options in the dialog box will be modified based on your selection in **Type** drop-down. In our case, we are selecting the **Revolution and Height** option; refer to Figure-65.

- Click on the **Rotation** button to flip the rotation side of coil to clockwise or counterclockwise direction; refer to Figure-66.

Figure-65. Coil Type drop down

Figure-66. Rotation button

- Click in the **Diameter** edit box and enter desired value to change the diameter of coil.
- Click in the **Revolutions** edit box and enter desired value for number of revolutions of coil.
- Click in the **Height** edit box and specify the total height of coil. You can also set the height of coil by moving the drag handle.
- Click in the **Angle** edit box and specify desired angle value to create taper coil.
- Select **Circular** option from **Section** drop-down of **COIL** dialog box to set the circular profile of coil. Select the **Square** option from **Section** drop-down to set square profile of coil. Select **Triangular(External)** option from **Section** drop-down to set the external triangular profile of coil. Select **Triangular(Internal)** option of **Section** drop-down to set internal triangular profile of coil; refer to Figure-67.

Figure-67. Coil Section drop down

- Select **On Center** option of **Section Position** drop-down from **COIL** dialog box to position the coil in such a way that the base circle is at center of the coil. Click on the **Inside** button of **Section Position** drop-down if you want to position the coil inside the base circle. Click on the **Outside** section of **Section Position** drop-down if you want to position the coil outside the base circle.

- Click in the **Section Size** edit box and specify the size of coil section. If section is circle then this will be diameter and if section is triangle or square then it will be length of one edge of section.
- The options of **Operations** drop-down were discussed earlier.
- Click **OK** button of **COIL** dialog box to finish the process of creating a coil; refer to Figure-68.

Figure-68. Coil created

PIPE

The **Pipe** tool is used to create pipes of different sections that follow a selected path. The procedure to use this tool is discussed next.

- Click on **Pipe** tool of **CREATE** drop-down from **Toolbar**; refer to Figure-69. The **PIPE** dialog box will be displayed and you will be asked to select a path for pipe; refer to Figure-70.

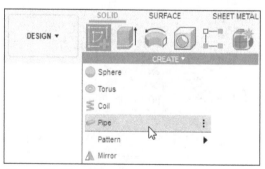

Figure-69. Pipe tool

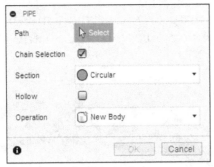

Figure-70. PIPE dialog box

- Select the sketch or edge of a model for path of the pipe; refer to Figure-71.

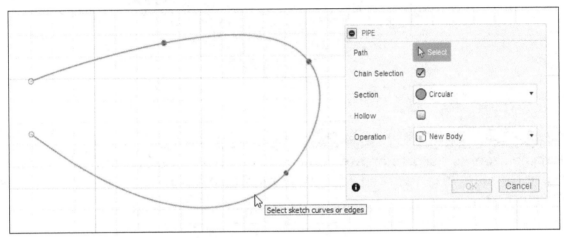

Figure-71. Selection of path to create pipe

- On selecting the path, the updated **PIPE** dialog box will be displayed. Click in the **Distance** edit box and enter desired value. The value of distance lies between 0 to 1. You can also set the distance (length) of pipe on the selected path by moving the drag handle.
- The options of **Section** drop-down were discussed earlier in **Coil** tool.
- Click in the **Section Size** edit box and specify the diameter of pipe.
- Select the **Hollow** check box to create a hollow pipe. The updated **PIPE** dialog box will be displayed along with the preview of pipe; refer to Figure-72.

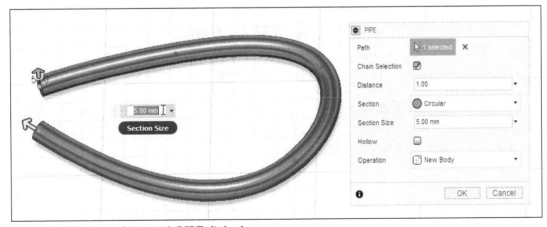

Figure-72. Preview of pipe with PIPE dialog box

- Click in the **Section Thickness** edit box and specify thickness of pipe.
- The options of **Operation** drop-down were discussed earlier in this book.
- Click **OK** button of **PIPE** dialog box to complete the process of creating the pipe; refer to Figure-73.

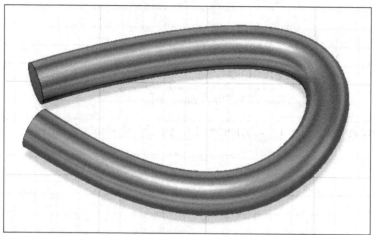

Figure-73. Pipe created

PATTERN

The **Pattern** tool is used to create copies of features or bodies at regular intervals. There are three tools in **Pattern** cascading menu; refer to Figure-74. These tools are discussed next.

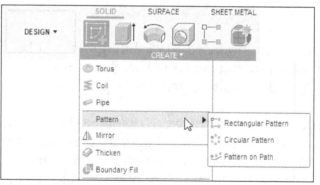

Figure-74. Pattern cascading menu

Rectangular Pattern

The **Rectangular Pattern** tool is used to create copies of objects in one or two directions. The procedure to use this tool are discussed next.

- Click on **Rectangular Pattern** tool of **Pattern** cascading menu from **CREATE** drop-down in the **Toolbar**; refer to Figure-75. The **RECTANGULAR PATTERN** dialog box will be displayed; refer to Figure-76.

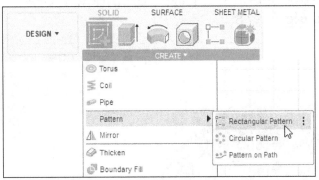
Figure-75. Rectangular Pattern tool in solid

Figure-76. RECTANGULAR PATTERN dialog box in solid

- Select **Faces** option from **Type** drop-down if you want to select faces for pattern. Select **Bodies** option from **Type** drop-down if you want to select bodies for creating copies. Select **Features** option from **Type** drop-down if you want to select features for creating copies. Select **Components** option from **Type** drop-down if you want to select component for creating copies; refer to Figure-77.

Figure-77. Pattern Type drop down

- We are using **Bodies** option from **Type** drop-down. Select the body for creating its copies; refer to Figure-78.

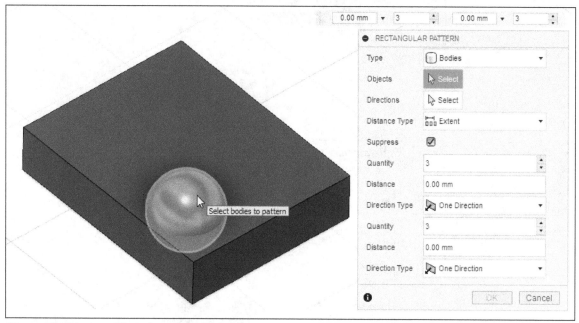

Figure-78. Selection of body for rectangular pattern

- Click on **Select** button of **Directions** option and select the default axes or edges of the model to define direction in which copies will be created; refer to Figure-79.

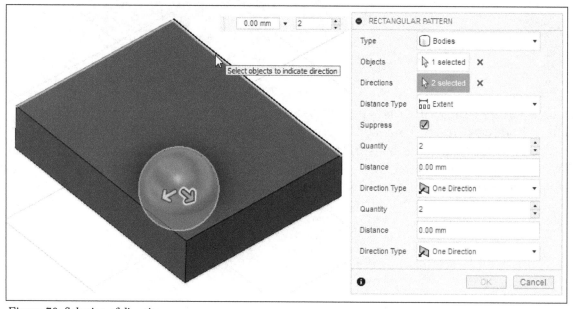

Figure-79. Selection of direction

- Select the **Extent** option from **Distance Type** drop-down if you want to specify total length within which copies will be created. Select the **Spacing** option from **Distance Type** drop-down to set the distance between two consecutive instances of object. In our case, we have selected **Extent** option.
- Select the **Suppress** check box if you want to suppress any instance of pattern.
- Click in the **Quantity** edit box and specify the number of instances to be created in the pattern along selected direction.

- Click in **Distance** edit box and specify the distance value within which pattern instances will be created. You can also set the distance by moving the drag handle.
- Select the **One Direction** option from **Direction Type** drop-down if you want to create the copies in one side of selected direction. Click on **Symmetric** option from **Direction Type** drop-down to create the copies in both sides of selected direction.
- Similarly, specify the quantity, distance, and direction type for other direction.
- After specifying the parameter, click **OK** button from **RECTANGULAR PATTERN** dialog box to finish the process; refer to Figure-80.

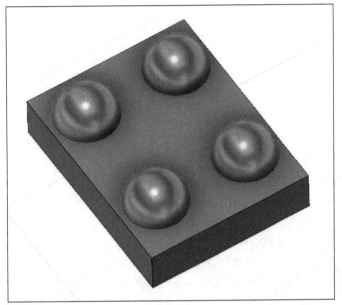

Figure-80. Solid model after applying rectangular pattern tool

Circular Pattern

The **Circular Pattern** tool is used to create copies of selected objects around selected axis/edge. The procedure to use this tool is discussed next.

- Click on **Circular Pattern** tool from **Pattern** cascading menu; refer to Figure-81. The **CIRCULAR PATTERN** dialog box will be displayed; refer to Figure-82.

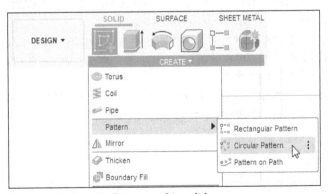

Figure-81. Circular Pattern tool in solid

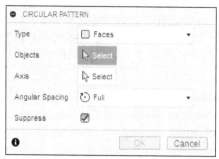

Figure-82. CIRCULAR PATTERN dialog box in solid

- Select desired option from **Type** drop-down. The options of **Type** drop-down are same as discussed in **Rectangular Pattern** tool.
- **Objects** selection is active by default. Select the component to be patterned; refer to Figure-83.

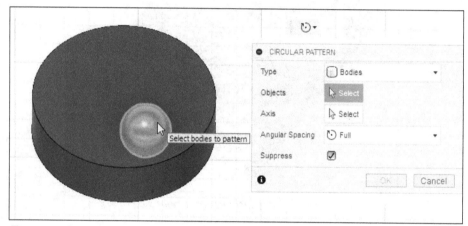

Figure-83. Selection of body for circular pattern

- Click on **Select** button of **Axis** section and select the edge/axis about which the circular pattern will be created.
- Select the **Full** option from **Angular Spacing** drop-down if you want to create the copies of object around 360 degree path. Select **Angle** option from **Angular Spacing** drop-down if you want to create the copies within specified angle range. Click on **Symmetric** option from **Angular Spacing** drop-down if you want to create copies symmetrically in specified angle range.
- Select the **Suppress** check box if you want to suppress the any instance of pattern.
- Click in the **Quantity** edit box and enter the number of copies to be created in pattern.
- After specifying the parameters, click on the **OK** button from **CIRCULAR PATTERN** dialog box to finish the process; refer to Figure-84.

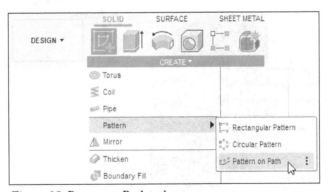

Figure-84. Solid model after applying circular pattern tool

Pattern on Path

The **Pattern on Path** tool is used to create copies of the selected object along selected path. The procedure to use this tool is discussed next.

• Click on the **Pattern on Path** tool of **Pattern** cascading menu from **CREATE** drop-down; refer to Figure-85. The **PATTERN ON PATH** dialog box will be displayed; refer to Figure-86.

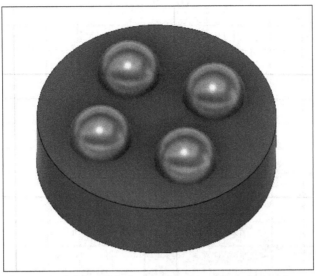

Figure-85. Pattern on Path tool

Figure-86. PATTERN ON PATH dialog box

- Click on **Faces** option from **Type** drop-down. Other options of **Type** drop-down are same as discussed earlier.
- **Objects** selection is selected by default. Select the faces for creating the copies; refer to Figure-87.

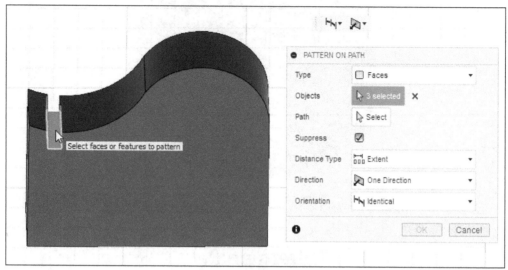

Figure-87. Selection of faces

- Click on **Select** button for **Path** option and select the path for creating copies.
- On selecting path, the updated **PATTERN ON PATH** dialog box will be displayed; refer to Figure-88.

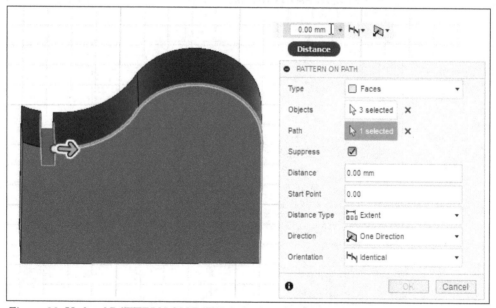

Figure-88. Updated PATTERN ON PATH dialog box on selection of path

- Select the **Suppress** check box if you want to suppress any instance of pattern.
- Click in the **Distance** edit box and specify the distance value along the path.
- Click in the **Quantity** edit box and specify the number of copies to be created.
- The **Start Point** option is used to specify the distance from the start point to place where the first pattern instance will be created. It is shown in the screen canvas as a **white point** and determines the starting point of the selected path. It can be set to any value between zero and one using the **Start Point** edit box, but it is easier to adjust it manually by dragging the start point; refer to Figure-89.

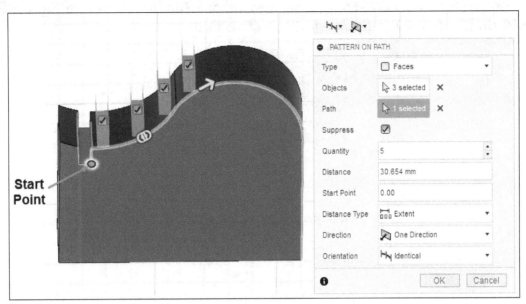

Figure-89. Pattern creation with start point

- The options of **Distance Type** and **Direction** drop-down were discussed earlier in this chapter.
- Select **Path Direction** option from **Orientation** drop-down if you want to rotate each instance according to the pattern path. Select **Identical** option from **Orientation** drop-down to keep the instances in the same orientation as the original object.
- Click **OK** button from **PATTERN ON PATH** dialog box to complete the process of creating the copies; refer to Figure-90.

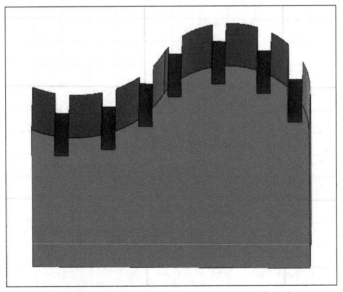

Figure-90. Solid model after applying pattern on path tool

MIRROR

The **Mirror** tool is used to create mirror copy of selected objects to the opposite side of a selected face or plane. The procedure to use this tool is discussed next.

- Click on **Mirror** tool of **CREATE** drop-down from **Toolbar**; refer to Figure-91. The **MIRROR** dialog box will be displayed; refer to Figure-92.

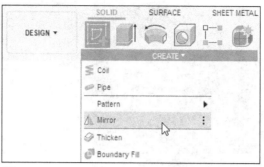

Figure-91. Mirror tool in solid

Figure-92. MIRROR dialog box in solid

- Select desired option from **Type** drop-down to select respective objects for mirroring. The other options of **Type** drop-down have already been discussed earlier.
- Select the bodies to be mirrored.
- Click on **Select** button of **Mirror Plane** section and select the plane/face to be used as mirror plane. Preview of mirror copy will be displayed; refer to Figure-93.

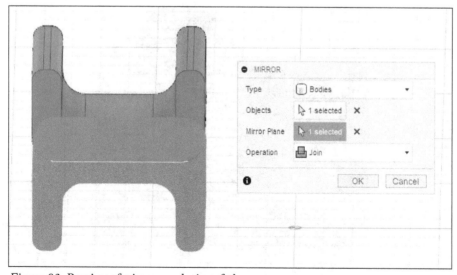

Figure-93. Preview of mirror on selection of plane

- Select desired option from **Operation** down which have been discussed earlier in this chapter.

• If you have selected **Features** in **Type** drop-down then **Compute Option** drop-down will also be displayed in the dialog box; refer to Figure-94. Select the **Optimized** option if you want to create copy of feature with minimum processing time and less flexibility to individually change features later. Select **Identical** option if you want to create copy of feature using same background programming which was used while creating the base feature. Select the **Adjust** option if you want to recompute each instance of the feature as a separate feature.

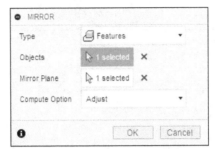

Figure-94. Compute Option in MIRROR dialog box

• Click **OK** button of **MIRROR** dialog box to finish the process of creating mirror copy; refer to Figure-95.

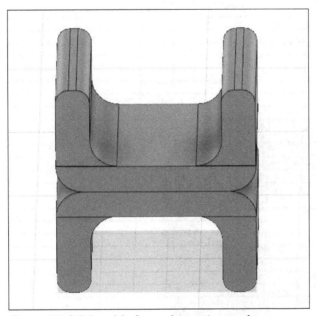

Figure-95. Solid model after applying mirror tool

THICKEN

The **Thicken** tool is used to add thickness to surfaces to make them solid. (You will learn about surfaces in Chapter 9 Surface Modeling of this book.) The procedure to use this tool is discussed next.

• Open any file that has surfaces or create a surface by following basic tools of Chapter 9 of this book.
• Click on **Thicken** tool of **CREATE** drop-down from **Toolbar**; refer to Figure-96. A **THICKEN** dialog box will be displayed; refer to Figure-97.

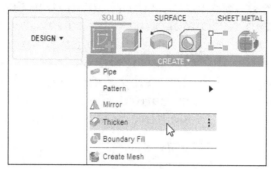

Figure-96. Thicken tool

Figure-97. THICKEN dialog box

- **Faces** selection is active by default. Select the surface to add thickness; refer to Figure-98.

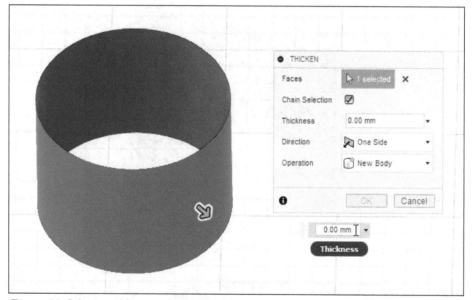

Figure-98. Selection of faces to add thickness

- Click in the **Thickness** edit box of dialog box and specify desired value of thickness.
- Click on **One Side** option from **Direction** drop-down to apply the thickness to one side of surface. Click on **Symmetric** option from **Direction** drop-down to apply the thickness to both sides of surface equally. In our case, we have selected **Symmetric** option.
- The options of **Operation** drop-down were discussed earlier in this chapter.
- Click **OK** button of **THICKEN** dialog box to finish the process; refer to Figure-99.

Figure-99. Solid model after applying thicken tool

BOUNDARY FILL

The **Boundary Fill** tool is used to create a volume from bounded by selected planes, bodies, and surfaces. This volume is called cell and can be used to cut, add, or create new body in the model. The procedure to use this tool is discussed next.

• Click on **Boundary Fill** tool of **CREATE** drop-down from **Toolbar**; refer to Figure-100. The **BOUNDARY FILL** dialog box will be displayed; refer to Figure-101.

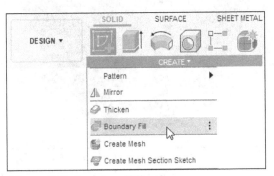

Figure-100. Boundary Fill tool

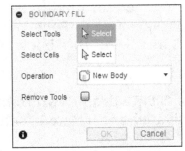

Figure-101. BOUNDARY FILL dialog box

- The **Select Tools** selection is active by default. Select the bodies/planes/surfaces for boundary fill; refer to Figure-102.

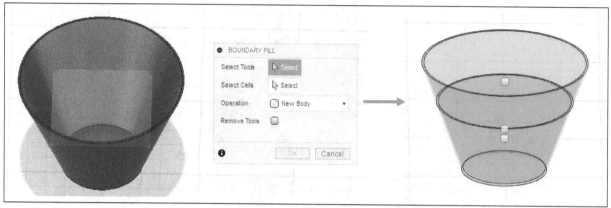

Figure-102. Selection for boundary fill

- Click on **Select** button of **Select Cells** section and select the check boxes for cells to be created.
- The tools of **Operation** section were discussed earlier in this chapter.
- Click **OK** button of **BOUNDARY FILL** dialog box to finish the process; refer to Figure-103. If the parent body is overlapping with new boundary fill bodies then hide the parent body using eye icon in **BROWSER** to check the result.

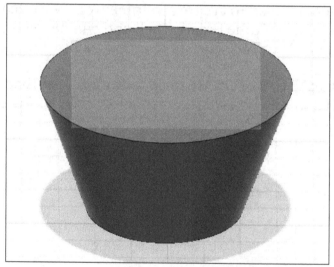

Figure-103. Result of boundary fill after hiding parent body

CREATE FORM

The **Create Form** tool is used to create or modify the form of model using freestyle editing tools. Clicking this tool takes you to **Form** mode. You will learn more about **Form** mode later in Chapter 12 and 13 of this book. The procedure to use this tool is discussed next.

- Click on **Create Form** tool of **CREATE** drop-down from **Toolbar**; refer to Figure-104. The **Form mode** will be activated and related toolbar will be displayed; refer to Figure-105.

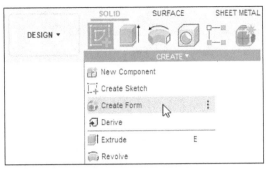

Figure-104. Create Form tool

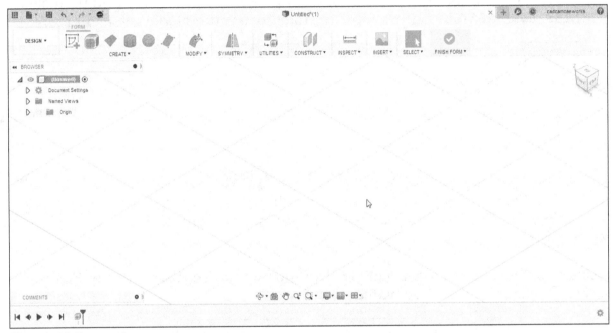

Figure-105. Form mode

- Now, create or modify the model as desired.
- Click **Finish Form** button from **Toolbar** to exit the **Form mode** and return to **Model Workspace**; refer to Figure-106.

Figure-106. Finish Form button

CREATE BASE FEATURE

The **Create Base Feature** tool inserts a base feature operation in the **Timeline**. Any operations performed on the design while in a base feature are not recorded in the **Timeline**. This tool is used when you want to safeguard some of the operations while sharing the file.

- Click on **Create Base Feature** tool of **CREATE** drop-down from **Toolbar**; refer to Figure-107. The tools in the **Toolbar** will be same as for **Model** workspace.

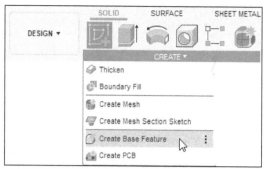

Figure-107. Create Base Feature tool

- Do desired operation that you do not want to record in **Timeline** and then click on **FINISH BASE FEATURE** button from **Toolbar** to exit the mode; refer to Figure-108.

Figure-108. Finish Base Feature button

CREATE MESH

The **Create Mesh** tool is used to create a mesh of the current model. (You will learn about mesh modeling in Chapter 14 of this book.) The procedure to use this tool is discussed next.

- Click on **Create Mesh** tool of **CREATE** drop-down from **Toolbar**; refer to Figure-109. **Mesh** workspace will be activated.

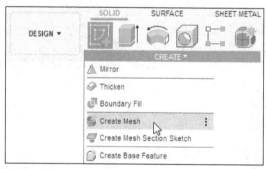

Figure-109. Create Mesh tool

- Make sure that you have selected the **Mesh Workspace** check box in **Preview** section of **Preferences** dialog box. You can access **Preferences** dialog box by clicking on **Preferences** tool in **User Account** drop-down; refer to Figure-110.

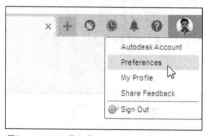

Figure-110. Preferences tool

• Create and modify desired model as desired in **Mesh Workspace.**
• Click on **FINISH MESH** button from **Toolbar** to exit **Mesh Workspace** and return to **Model Workspace**; refer to Figure-111.

Figure-111. Finish Mesh button

CREATING MESH SECTION SKETCH

The Mesh section sketch is created at the intersection of mesh with selected plane. The procedure to create mesh section sketch is discussed next.

• Click on the **Create Mesh Section Sketch** tool from the **CREATE** drop-down in the **SOLID** tab of **Toolbar**. The **CREATE MESH SECTION SKETCH** dialog box will be displayed; refer to Figure-112.

Figure-112. CREATE MESH SECTION SKETCH dialog box

• Select desired mesh body. The updated **CREATE MESH SECTION SKETCH** dialog box will be displayed; refer to Figure-113.

Figure-113. Updated CREATE MESH SECTION SKETCH dialog box

• You can select desired plane for section or you can use dynamic handles to set location for section plane; refer to Figure-114.

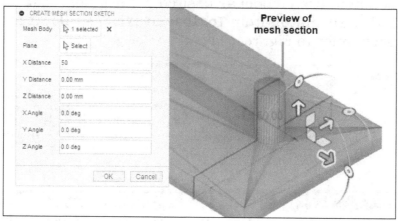

Figure-114. Preview of mesh section

- After setting desired parameters, click on the **OK** button from the dialog box to create the section sketch.

CREATE 3D PCB FEATURE

The **Create 3D PCB** tool in **CREATE** drop-down is used to create and modify PCB design in collaboration with **EAGLE** users. This tool is available only if the **PCB Feature** check box in **Preview** section of **Preferences** dialog box is selected. You can access **Preferences** dialog box by clicking on **Preferences** tool in **User Account** drop-down. The procedure to use this tool is given next.

- Click on **Create 3D PCB** tool from **CREATE** drop-down in **Toolbar**; refer to Figure-115. The **CREATE 3D PCB** dialog box will be displayed; refer to Figure-116 and you will be asked to select sketch profile for PCB base.

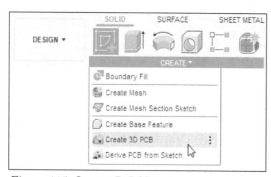

Figure-115. Create 3D PCB tool

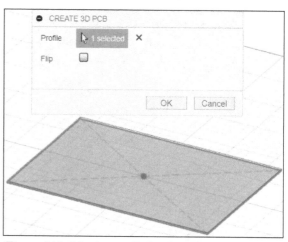

Figure-116. Selecting profile for 3D PCB

- Select desired sketch profile from canvas. The **3D PCB Workspace** will be activated and related toolbar will be displayed; refer to Figure-117.

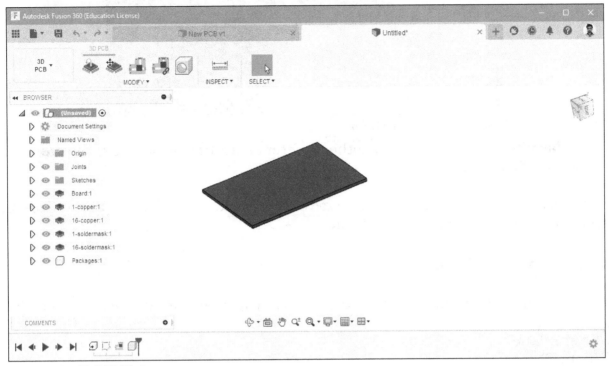

Figure-117. 3D PCB Workspace

Adding Holes to PCB

The **PCBHole** tool is used to create holes for adding components to the PCB. The procedure to use this tool is given next.

- Click on the **PCBHole** tool from the **MODIFY** drop-down in the **3D PCB** tab of the **Ribbon**. The **PCBHOLE** dialog box will be displayed.
- Select the face of PCB where you want to place the hole. Preview of hole will be displayed; refer to Figure-118.

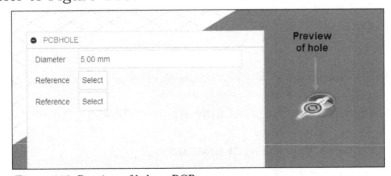

Figure-118. Preview of hole on PCB

- Click at the center of hole and select desired edge of PCB to define first reference. The options to define distance from selected reference will be displayed; refer to Figure-119.

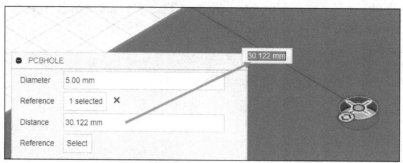

Figure-119. Specifying distance

- Similarly, specify distance for other reference and click on the **OK** button. The hole will be created.

The tools of 3D PCB are beyond the scope of this book. You will learn about PCB and Electronic Design in our another book.

SELF ASSESSMENT

Q1. Which of the following is not the default plane available in Autodesk Fusion 360?

a. Front Plane
b. Right Plane
c. Top Plane
d. Left Plane

Q2. Discuss the procedure of creating multiple holes with example.

Q3. Which of the following tool is not available in **CREATE** drop-down of Autodesk Fusion 360?

a. Thread
b. Box
C. Torus
d. Curve

Q4. **Thicken** tool is used to add thickness and length. (T/F)

Q5. The **Pattern On Path** tool is available in the **CREATE** drop-down. (T/F)

Q6. Discuss the use of **Pipe** tool with examples.

Q7. Discuss the use of **Web** tool with examples.

Q8. Which tool is used to create a solid body connecting two sketch from different planes?

Q9. Discuss the use of **Pattern On Path** tool with examples.

Q10. The --------- tool is used to cut helical thread on cylindrical faces.

Chapter 5

Practical and Practice

The major topics covered in this chapter are:

- *Practical*
- *Practice*

PRACTICAL 1

Create the model as shown in Figure-1. The views of the model with dimension are given in Figure-2.

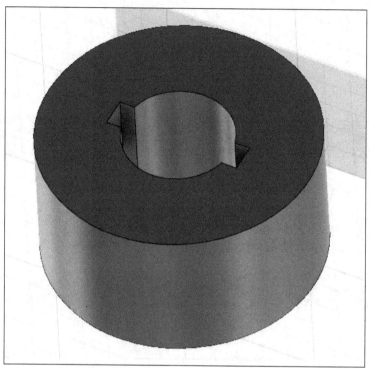

Figure-1. Model for Practical 1

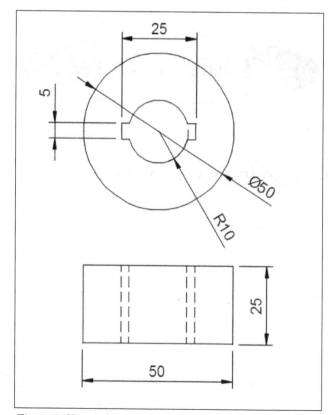

Figure-2. Views of Practical 1

Before we start working on Practical 1, it is important to understand two terms; first angle projection and third angle projection. These are the standards of placing views in the engineering drawings. The views placed in the above figure are using third angle projection. In first angle projection, the top view of model is placed below the front view and right side view is placed at the left of the front view. You will learn more about projection in chapter related to drafting.

Creating the Sketch

- Click on **Create Sketch** tool from **CREATE** drop-down in **Toolbar**. You will be asked to select a plane for creating sketch.
- Select **Top Plane** from the canvas screen.
- Click on **Center Diameter Circle** tool of **Circle** cascading menu from **CREATE** drop-down in **Toolbar**. You will be asked to place the center point of circle.
- Click at the **Origin** to place the center point of circle. Enter **50** in floating window as diameter of circle and press **ENTER** key.
- Sketch another circle of radius **10** (diameter **20**) taking center point as origin.

How to show radius in place of diameter of circle? Create the circle with diameter value. Right-click on diameter dimension and select **Toggle Radius** button from Marking menu; refer to Figure-3.

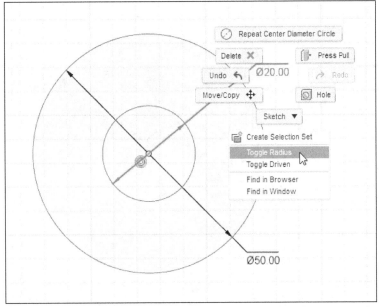

Figure-3. Toggle Radius Option

- Select **Center Rectangle** tool of **Rectangle** cascading menu from **CREATE** drop-down in **Toolbar**.
- Click at **Origin** as center point for rectangle and enter **5** as width & **25** as length in floating window to create a rectangle; refer to Figure-4.

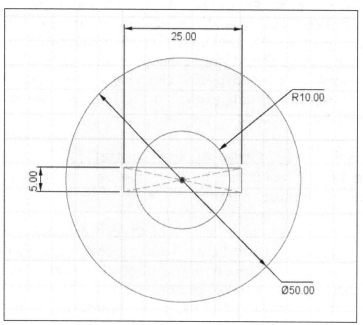

Figure-4. Sketch after creating circle and rectangle

- Select **Trim** tool of **MODIFY** drop-down from **Toolbar** and trim the entities in such a way that the sketch is displayed as shown in Figure-5.

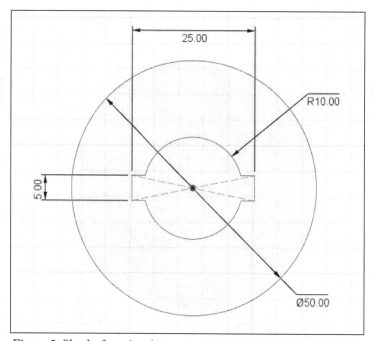

Figure-5. Sketch after trimming

- Click on **Finish Sketch** button from **Toolbar** to exit the sketch mode.

Creating Extrude Feature

- Click on **Extrude** tool of **CREATE** drop-down from **Toolbar** or press **E** key from keyboard to select **Extrude** tool. The **EXTRUDE** dialog box will be displayed.
- Profile selection is active by default. Select the recently created sketch profile. The updated **EXTRUDE** dialog box will be displayed.

- Specify the height of extrusion as **25** in **Distance** edit box or floating window. The preview of extrusion will be displayed; refer to Figure-6. (You need to rotate the model by using **SHIFT+Middle Mouse Button** to check the preview.)

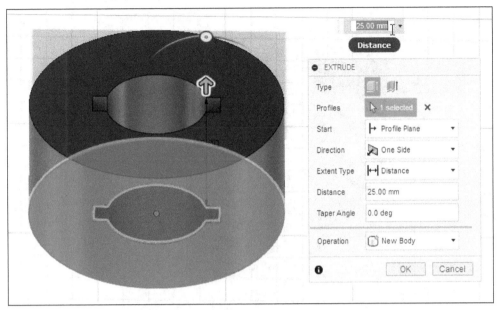

Figure-6. Preview of extrusion

- Click **OK** button from **EXTRUDE** dialog box to complete the process of extrusion. The model will be displayed as shown in Figure-7.
- Save the file and close it.

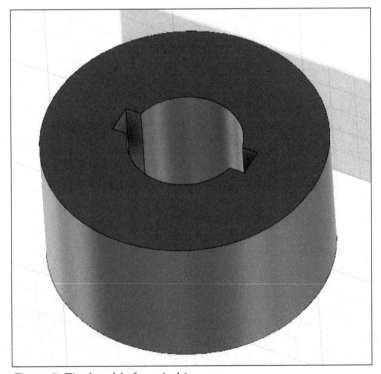

Figure-7. Final model of practical 1

PRACTICAL 2

Create the model as shown in Figure-8. The dimensions are given in Figure-9. Assume the missing dimensions.

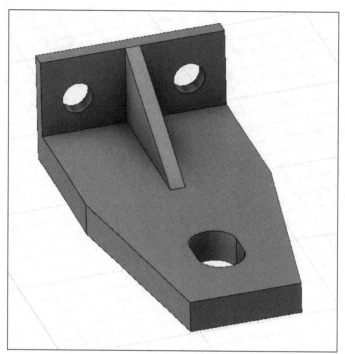

Figure-8. Model For Practical 2

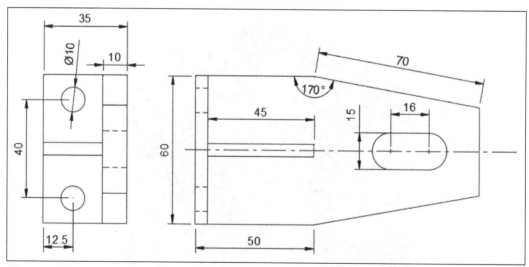

Figure-9. Practical 2 drawing

Creating sketch for model

- Start a new sketch using **Top Plane.**
- Select **Centre Rectangle** tool of **Rectangle** cascading menu from **CREATE** drop-down.
- Click on the origin to specify the centre point of rectangle and enter **60** and **50** in floating window as dimensions of rectangle; refer to Figure-10. Press **TAB** to toggle between edit boxes.

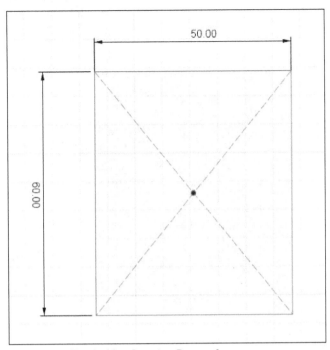

Figure-10. Sketch After Creating Rectangle

- Select **Line** tool from **CREATE** drop-down. Create a line of length as **70** and angle of **80** degree as shown in Figure-11.

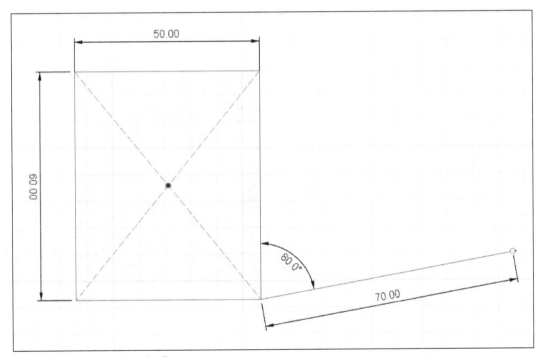

Figure-11. Creating Angular Line

- Select **Mirror** tool of **CREATE** drop-down from **Toolbar** and create a mirror copy of recently created line about centerline as shown in Figure-12.

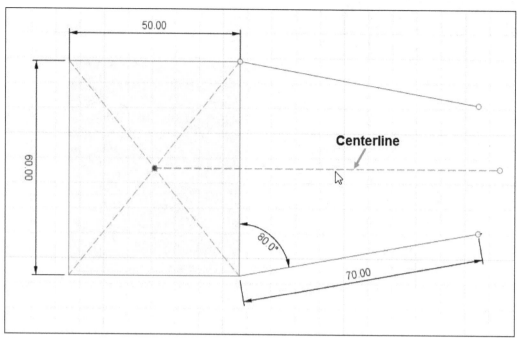

Figure-12. Creating Mirror Copy

• Create a line joining end points of two earlier created lines as shown in Figure-13.

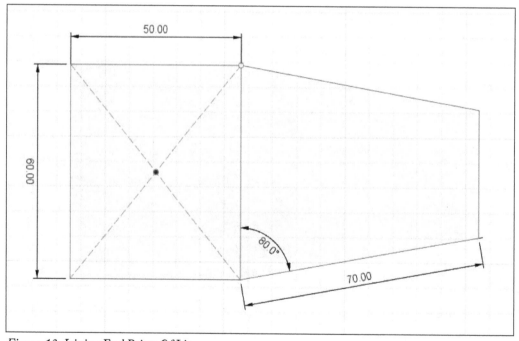

Figure-13. Joining End Points Of Lines

• Select **Center Point Slot** tool of **Slot** cascading menu from **CREATE** drop-down and create a slot of diameter as **15** and length as **8** at a distance of **65** from origin on the centerline; refer to Figure-14.

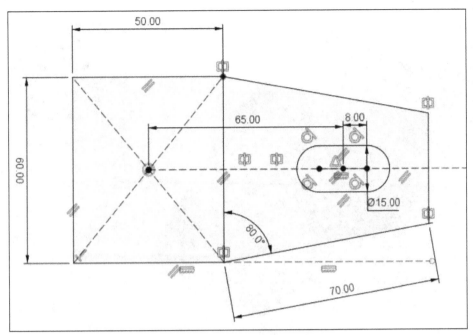

Figure-14. Creating Slot

- Press **T** from keyboard or click on the **Trim** tool from **MODIFY** drop-down in **Toolbar** to activate trimming tool. Select the inner lines of rectangle to remove them; refer to Figure-15.

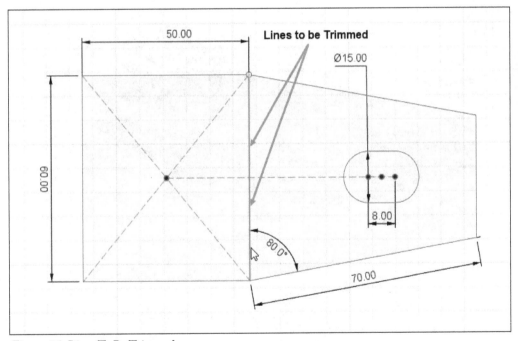

Figure-15. Lines To Be Trimmed

- Click on the **Finish Sketch** button to exit sketch.

Creating Extrude Feature

- Select **Extrude** tool from **CREATE** drop-down or press **E** key. The **EXTRUDE** dialog box will be displayed.
- Profile selection is active by default. Click in the sketch outside the slot.
- Specify **10** as height of extrusion in **Distance** edit box of **EXTRUDE** dialog box; refer to Figure-16. (Rotate the model to get clear view.)

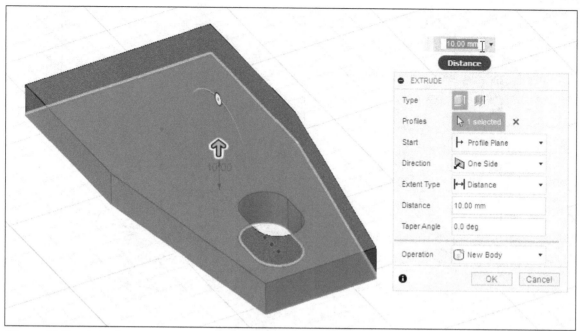

Figure-16. Preview of extrusion

- Click **OK** button from **EXTRUDE** dialog box to complete the process of extrusion.
- Click on the **Create Sketch** tool from **CREATE** drop-down. You are asked to select the plane to create sketch. Select the side face of model as shown in Figure-17.

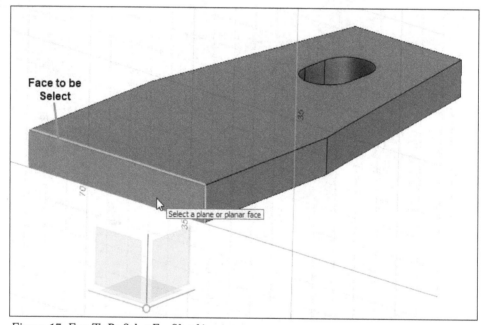

Figure-17. Face To Be Select For Sketching

- Select the **2-Point Rectangle** tool from **Rectangle** cascading menu in the **CREATE** drop-down of **Toolbar**. Create a rectangle of dimensions **25** and **60** with two holes of diameter **10** as shown in Figure-18.
- Click on the **Finish Sketch** button to exit sketching.

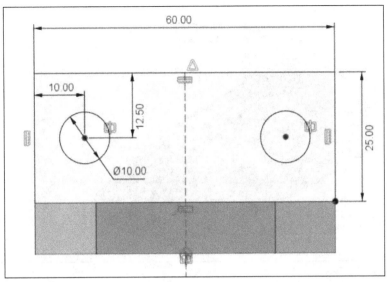

Figure-18. Creating the rectangle with holes

- Select **Extrude** tool from **CREATE** drop-down. The **EXTRUDE** dialog box will be displayed.
- Select the upper portion of recently created sketch in **Profiles** selection and enter **-5** as height of extrusion.
- Select **Join** from **Operation** drop-down of **EXTRUDE** dialog box to join the two extrusions.
- Click **OK** button from **EXTRUDE** dialog box to finish this process; refer to Figure-19.

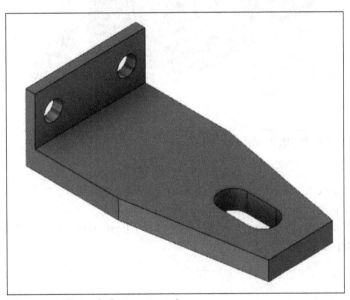

Figure-19. Extrude feature created

Creating Midplane

- Select **Midplane** tool of **CONSTRUCT** drop-down from **Toolbar.** The **MIDPLANE** dialog box will be displayed. You are asked to select the faces between which you want to create a midplane.
- Select both the faces as shown in Figure-20.
- The preview of midplane will be displayed. Click **OK** button from the **MIDPLANE** dialog box to complete the process; refer to Figure-21.

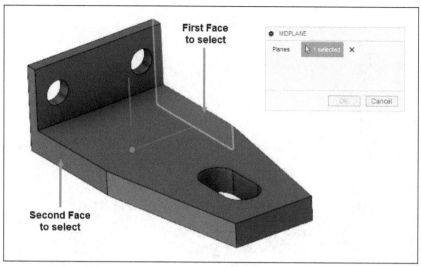

Figure-20. Face selection

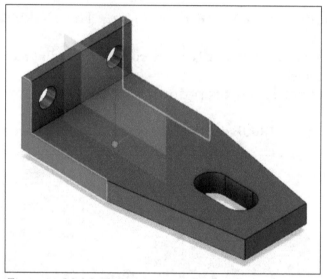

Figure-21. Midplane created

Creating Rib

- Select **Create Sketch** tool from **CREATE** drop-down. You will be asked to select the plane for creating sketch. Select the newly created midplane from **BROWSER**.
- Create a line as shown in Figure-22 with end points coincident to the vertices of model and finish the sketch.

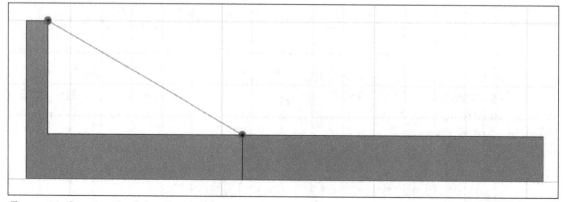

Figure-22. Creating sketch for rib

- Select **Rib** tool from **CREATE** drop-down. The **RIB** dialog box will be displayed.
- Select the recently created line for **Curve** section.
- Select **Symmetric** option from **Thickness Options** drop-down.
- Select **To Next** option from **Depth Options** drop-down.
- Click in **Thickness** edit box and enter **5** as thickness of rib; refer to Figure-23. (You might need to click on the **Flip Direction** button from the dialog box to get desired preview.)

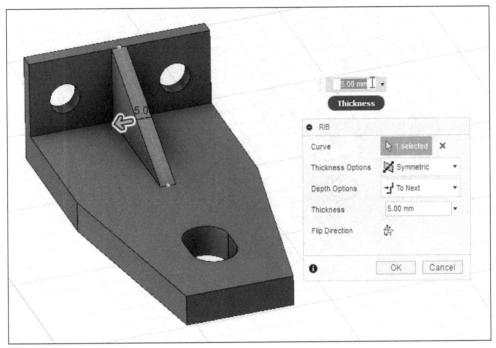

Figure-23. Preview of rib

- Click **OK** button of **RIB** dialog box to finish the process. The model will be displayed as shown in Figure-24.

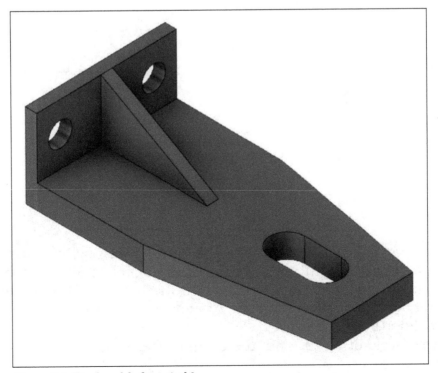

Figure-24. Final model of practical 2

PRACTICAL 3

Create the model as shown in Figure-25. The dimensions of the model are given in Figure-26.

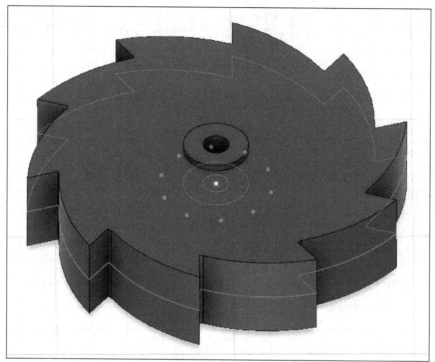

Figure–25. Model for Practical 3

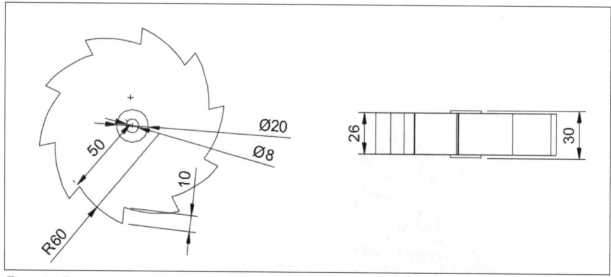

Figure-26. Practical 3 drawing views

Creating sketch for model

- Start a new design file in Autodesk Fusion. Select **Create Sketch** tool from **CREATE** drop-down in the **Toolbar**. You are asked to select sketching plane. Select **Top Plane** to create sketch. Select **Center Diameter Circle** tool of **Circle** cascading menu from **CREATE** drop-down.
- Create three circle of **8, 20,** and **100** as diameter dimension of circles; refer to Figure-27.

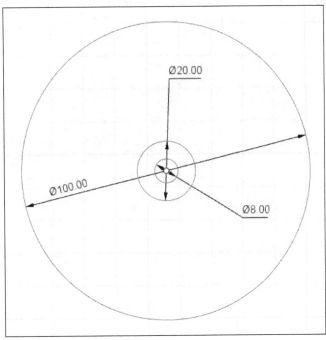

Figure-27. Creating circles

- Press **L** key to select **Line** tool and create a line of length **10**.
- Select **3-Point Arc** tool from **Arc** cascading menu and create an arc of **60** radius and **35** length as shown in Figure-28.

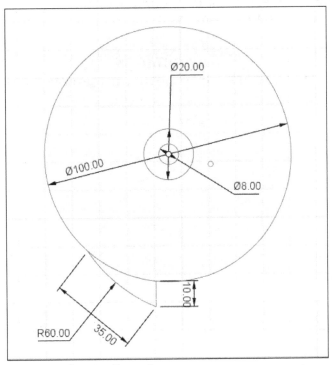

Figure-28. Creating line and arc

Creating Circular Pattern

- Select **Circular Pattern** tool from **CREATE** drop-down. The **CIRCULAR PATTERN** dialog box will be displayed.
- Objects selection is active by default. Select the recently created line and arc for circular pattern.

- Click on **Select** button of **Center Point** section and select the major circle to define its center as center point for pattern; refer to Figure-29.

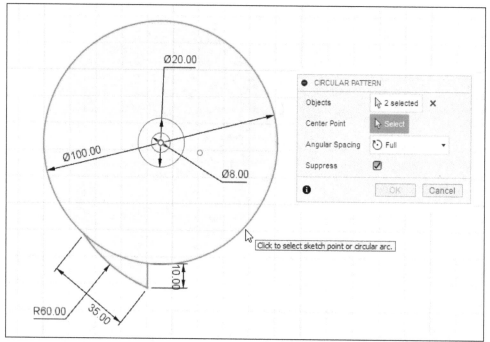

Figure-29. Creating circular pattern

- The updated **CIRCULAR PATTERN** dialog box will be displayed. Click in **Quantity** edit box and specify **10** as quantity of circular pattern.
- Click **OK** button from **CIRCULAR PATTERN** dialog box to complete the process.

Trimming

- Select the **Trim** tool from **MODIFY** drop-down and remove all the extra entities from sketch as shown in Figure-30.

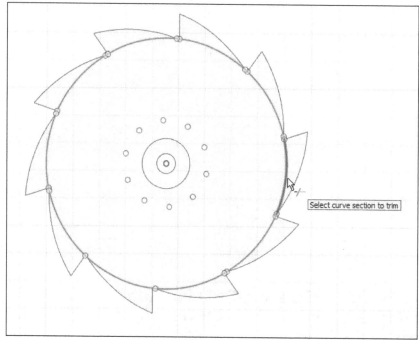

Figure-30. Trimming the extra entities

- After removing the extra entities from sketch, the sketch will be displayed as shown in Figure-31.

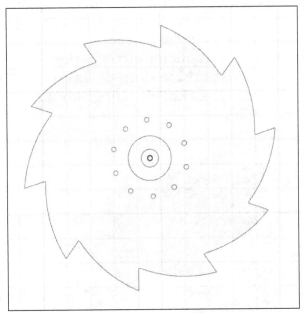

Figure-31. After removing extra entities

Creating Extrude

- Select **Extrude** tool from **CREATE** drop-down or press **E** key. The **EXTRUDE** dialog box will be displayed.
- Profile selection is active by default. Click between **20** diameter and **8** diameter circle to extrude the middle section.
- Select **Symmetric** option of **Direction** drop-down from **EXTRUDE** dialog box.
- Click in **Distance** edit box of **EXTRUDE** dialog box and specify **15** as height of extrusion. Rotate the model to check preview; refer to Figure-32.

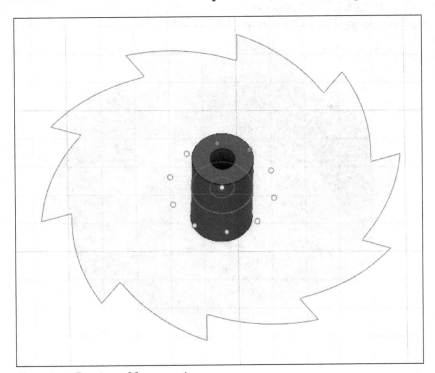

Figure-32. Preview of first extrusion

- Click **OK** button of **EXTRUDE** dialog box to finish this extrusion. The sketch of extrude feature will hide automatically. Click on the crossed eye icon before **Sketch 1** in the **Sketches** node of **BROWSER** to display the sketch again.
- Right-click on the screen and select **Repeat Extrude** button from **Marking Menu**. The **EXTRUDE** dialog box will be displayed.
- Select the area outside the **20** diameter circle earlier created.
- Select **Symmetric** button of **Direction** drop-down from **EXTRUDE** dialog box.
- Click in **Distance** edit box of **EXTRUDE** dialog box and specify **13** as height of extrusion; refer to Figure-33.

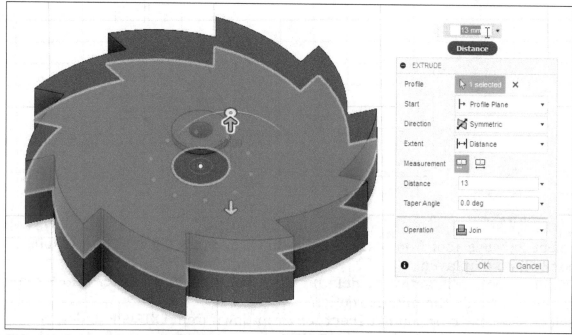

Figure-33. Preview of second extrusion

- Click **OK** button from **EXTRUDE** dialog box to complete the process of extrusion. The model will be displayed as shown in Figure-34.

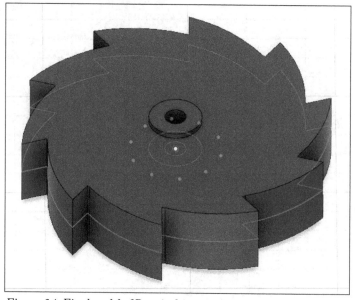

Figure-34. Final model of Practical 3

You can now hide the sketch and save the file.

PRACTICAL 4

Create the model as shown in Figure-35. The dimensions of the model are given in Figure-36.

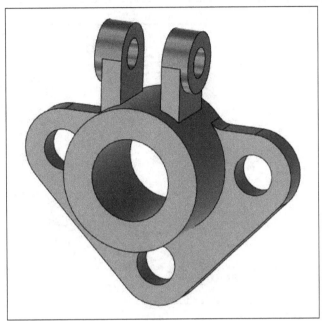

Figure-35. Model for Practical 4

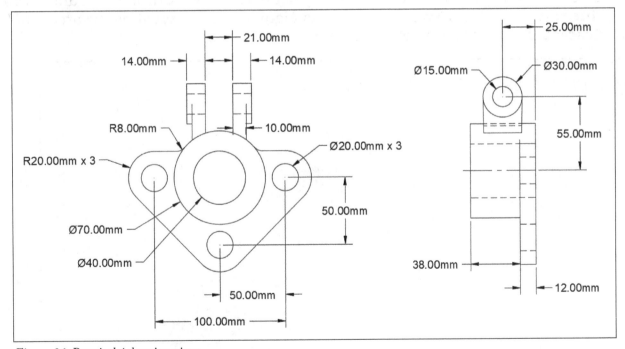

Figure-36. Practical 4 drawing views

Creating sketch for model

- Start a new design file in Autodesk Fusion. Click on the **Create Sketch** tool from **CREATE** drop-down in the **Toolbar**. You are asked to select sketching plane.
- Select **Top Plane** to create sketch. Select **Center Diameter Circle** tool of **Circle** cascading menu from **CREATE** drop-down.

- Create two circles of diameter **70** mm and **40** mm and create three circles of diameter **20** mm each; refer to Figure-37. After creating the circles, apply **Horizontal/Vertical** constraint to the circles.

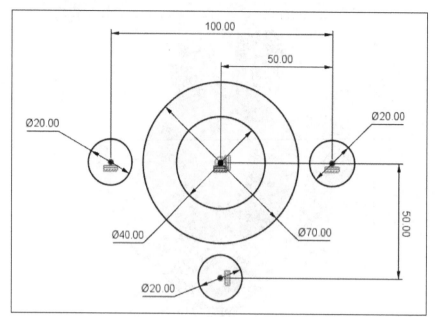

Figure-37. Creating circles

- Select **3-Point Arc** tool of **Arc** cascading menu from **CREATE** drop-down and create three arcs of radius **20** mm each; refer to Figure-38. After creating the arcs, apply **Concentric** constraint to all the arcs.

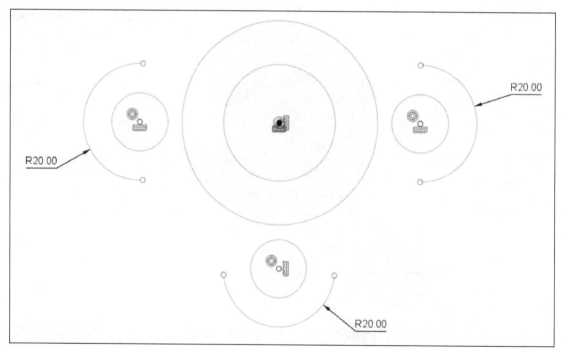

Figure-38. Creating arcs

- Select **Line** tool from **CREATE** drop-down and create three lines joining the arcs; refer to Figure-39. After creating the lines, apply **Tangent** constraint to all the lines.

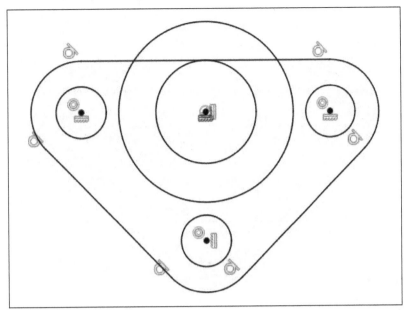

Figure-39. Creating lines joining the arcs

- Select **3-Point Arc** tool of **Arc** cascading menu and create two arcs of radius **8** mm each; refer to Figure-40. After creating the arcs, apply **Tangent** constraint to the arcs.

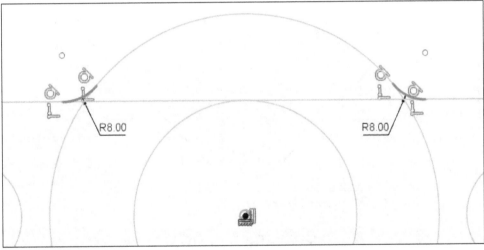

Figure–40. Creating arcs

Trimming

- Select the **Trim** tool from **MODIFY** drop-down and remove extra entities from sketch as shown in Figure-41.

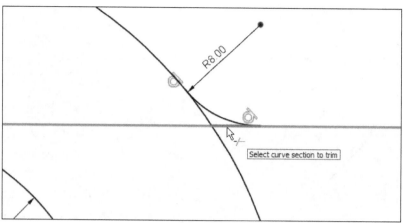

Figure–41. Removing extra entities

*   After removing extra entities from sketch, the sketch will be displayed as shown in Figure-42.

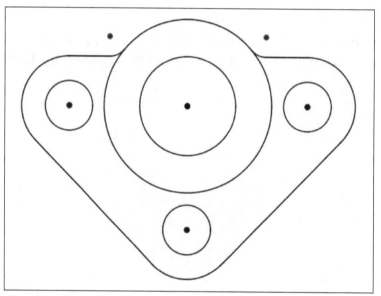

Figure–42. Sketch after trimming

Creating Extrude

*   Select **Extrude** tool from **CREATE** drop-down or press **E** key. The **EXTRUDE** dialog box will be displayed.
*   Profile selection is active by default. Select the profile from the sketch as shown in Figure-43.

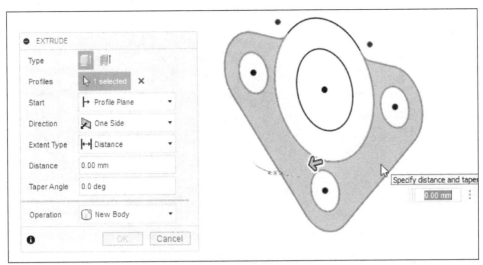

Figure-43. Extruding the sketch

- Select **One Side** option of **Direction** drop-down from **EXTRUDE** dialog box.
- Click in **Distance** edit box of **EXTRUDE** dialog box and specify **12** as distance of extrusion.
- Click **OK** button of **EXTRUDE** dialog box to finish this extrusion. The first sketch will be extruded and another sketch to be extruded will hide automatically; refer to Figure-44.

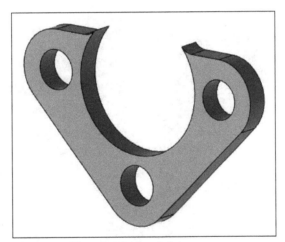

Figure-44. First sketch extruded

- Click on the crossed eye icon before **Sketch 1** in the **Sketches** node of **BROWSER** to display the sketch again.
- Right-click on the screen and select **Repeat Extrude** button from **Marking Menu**. The **EXTRUDE** dialog box will be displayed.
- Select the profile to extrude as shown in Figure-45.

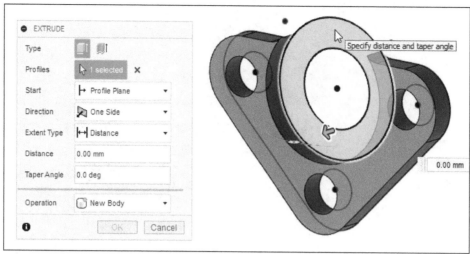

Figure–45. Extruding the second sketch

- Select **One Side** option of **Direction** drop-down from **EXTRUDE** dialog box.
- Click in **Distance** edit box of **EXTRUDE** dialog box and specify **50** as distance of extrusion.
- Click on the **OK** button from **Extrude** dialog box to finish the extrusion. The second sketch will be extruded; refer to Figure-46.

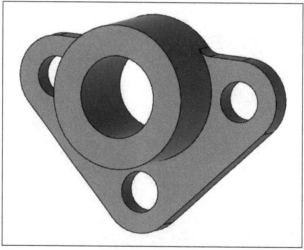

Figure–46. Second sketch extruded

Creating Offset Plane

- Select **Offset Plane** tool from **CONSTRUCT** drop-down. The **OFFSET PLANE** dialog box will be displayed and you are asked to select the plane.
- The **Select** button of **Plane** section is active by default. Select **YZ** plane from **BROWSER**.
- Click in the **Distance** edit box of the dialog box and enter **24.5** as distance value; refer to Figure-47.
- Click on **OK** button from the dialog box to complete the process.

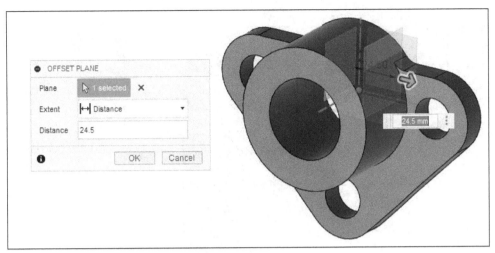

Figure–47. Creating the offset plane

Creating the Sketch

- Select **Create Sketch** tool from **CREATE** drop-down. You are asked to select the plane.
- Select newly created offset plane from **BROWSER** as sketching plane. The sketching environment will be displayed.
- Select **Center Diameter Circle** tool of **Circle** cascading menu from **CREATE** drop-down and create two circles of diameter **15** mm and **30** mm; refer to Figure-48.

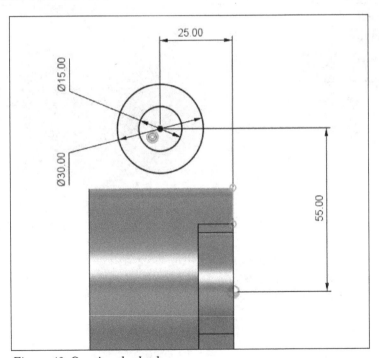

Figure–48. Creating the sketch

Intersecting the Sketch

- Select **Slice** check box from **SKETCH PALETTE** dialog box. The slice of the model will be created; refer to Figure-49.

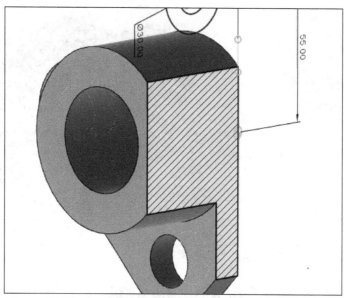

Figure-49. Slice created

- Select **Intersect** tool of **Project/Include** cascading menu from **CREATE** drop-down. The **INTERSECT** dialog box will be displayed. You are asked to select the geometries.
- The **Select** button of **Geometry** section is active by default. Select sliced part of the model as geometry to be intersect; refer to Figure-50.

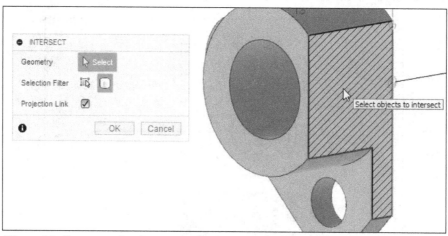

Figure-50. Geometry selected

- Click on **OK** button from **INTERSECT** dialog box to complete the process.
- Hide the body by clicking on eye icon before **Body1** in the **Bodies** node of **BROWSER**. The intersected geometry will be displayed; refer to Figure-51.

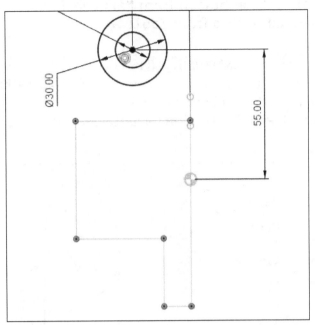

Figure-51. Intersected geometry

- Select **Line** tool from **CREATE** drop-down and create the lines as shown in Figure-52.

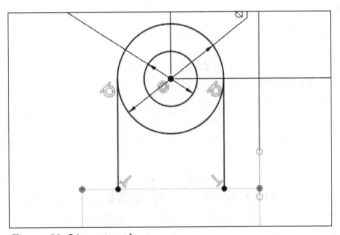

Figure-52. Lines created

- Select the intersected geometry by pressing **CTRL** key and right-click; refer to Figure-53. A **Marking Menu** will be displayed.

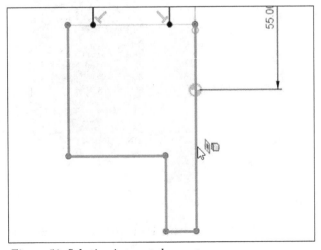

Figure-53. Selecting intersected geometry

- Select **Normal/Construction** button from **Marking Menu**.
- Click on **Finish Sketch** button from **Toolbar**.

Extruding the sketch

- Select **Extrude** tool from **CREATE** drop-down. The **EXTRUDE** dialog box will be displayed. You are asked to select the profile.
- The **Select** button of **Profile** section is active by default. Select the sketch to extrude as shown in Figure-54.

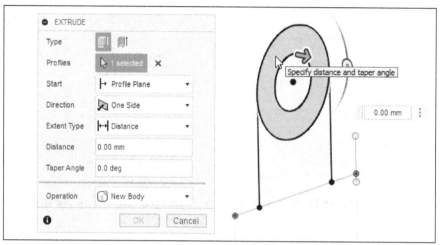

Figure-54. Selecting the sketch to extrude

- Specify the parameters as shown in the Figure-54. Click in the **Distance** edit box and enter **-14** mm as distance of extrusion.
- Click on **OK** button from **EXTRUDE** dialog box to complete the process. The another sketch to be extruded will hide automatically.
- Click on the crossed eye icon before **Sketch 2** in the **Sketches** node of **BROWSER** to display the sketch again.
- Select the sketch and click **RMB**. A **Marking Menu** will be displayed. Click **Repeat Extrude** button from **Marking Menu**. The **EXTRUDE** dialog box will be displayed again along with profile selected for extrusion.
- Specify the parameters as specified for previous extrusion; refer to Figure-55.

Figure-55. Extruding another sketch

- Click on **OK** button from the dialog box to complete the extrusion.

- Select recently extruded sketch and click **RMB**. A **Marking Menu** will be displayed; refer to Figure-56.

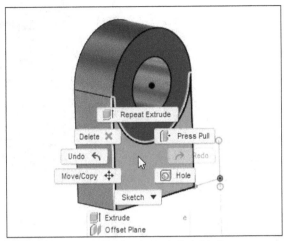

Figure-56. Selecting recently extruded sketch

- Click **Repeat Extrude** button from **Marking Menu**. The **EXTRUDE** dialog box will be displayed along with profile selected for extrusion.
- Click in the **Distance** edit box and enter **-4** as the distance for extrusion.
- Select **Cut** option from **Operation** drop-down. The preview of extrusion will be displayed; refer to Figure-57.

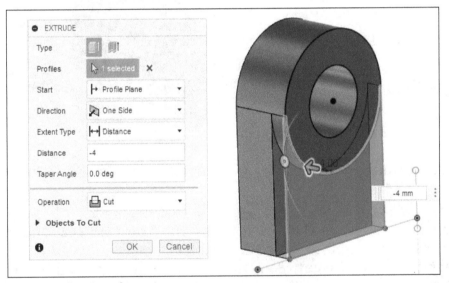

Figure-57. Preview of extrusion

- Click on **OK** button from **EXTRUDE** dialog box to complete the process.

Mirroring the body

- Click on the crossed eye icon before **Body1** in the **Bodies** node of **BROWSER** to display the body.
- Select **Mirror** tool from **CREATE** drop-down. The **MIRROR** dialog box will be displayed. You are asked to select the object.
- Select **Bodies** option from **Type** drop-down of dialog box.
- The **Select** button of **Objects** section is active by default. Select recently created body to be mirror; refer to Figure-58. You are asked to select the mirror plane.

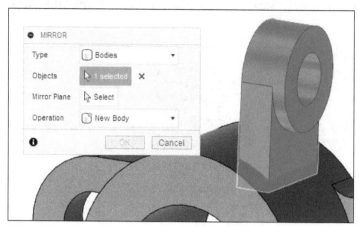

Figure-58. Body selected to be mirror

- Click on the **Select** button of **Mirror Plane** section and select **YZ** plane as mirror plane. The preview of mirror will be displayed; refer to Figure-59.

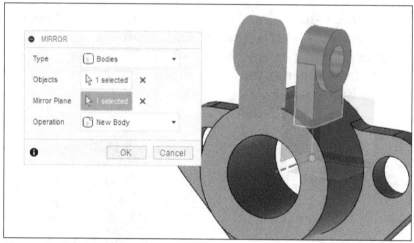

Figure-59. Preview of mirror

- Click on **OK** button from dialog box to complete the process.

Combining the bodies

- Select **Combine** tool from **MODIFY** drop-down. The **COMBINE** dialog box will be displayed. You are asked to select target body.
- The **Select** button of **Target Body** section is active by default. Select the body from the model as shown in Figure-60. You are asked to select the tool body.

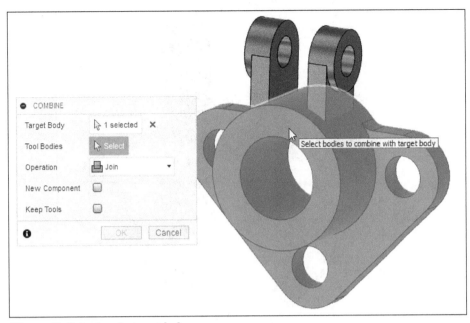

Figure-60. Selecting the target body

- The **Select** button of **Tool Bodies** section is active by default. Select the bodies to combine with target bodies as shown in Figure-61.

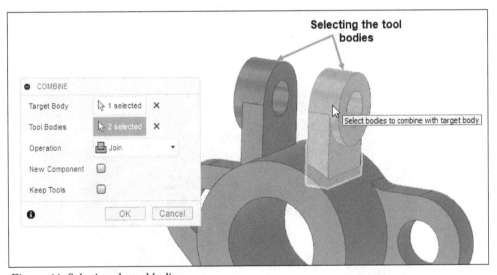

Figure-61. Selecting the tool bodies

- Click on **OK** button from the dialog box to complete the process of combining the bodies. The model will be displayed as shown in Figure-62.

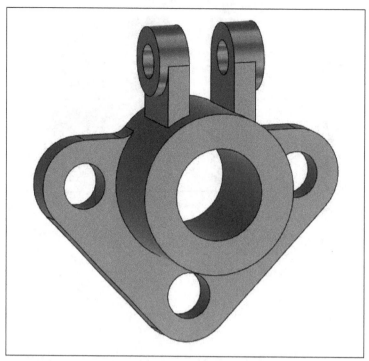

Figure-62. Final model of Practical 4

PRACTICE 1

Create the model as shown in Figure-63. The dimensions are given in Figure-64.

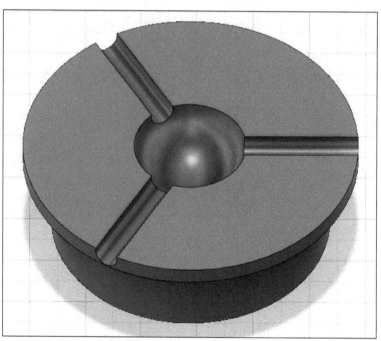

Figure-63. Model for Practice 1

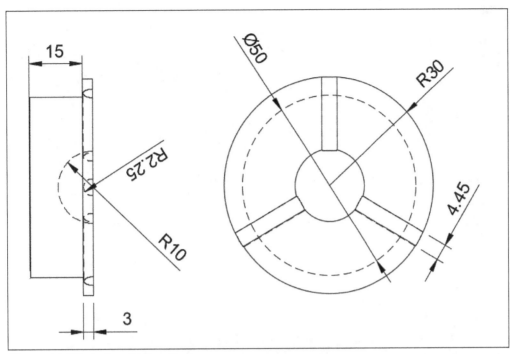

Figure-64. Dimensions for Practice 1

PRACTICE 2

Create the model as shown in Figure-65. The dimensions are given in Figure-66.

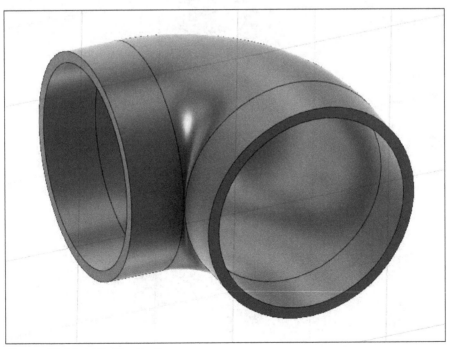

Figure-65. Model for Practice 2

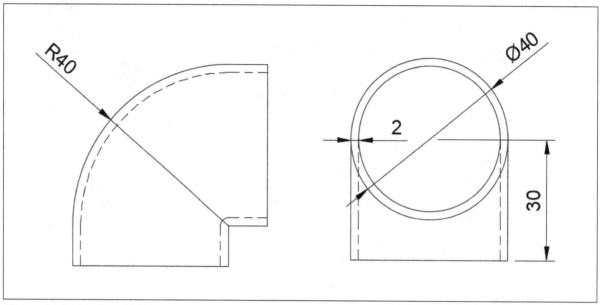

Figure-66. Dimensions for Practice 2

PRACTICE 3

Create the model as shown in Figure-67. The dimensions are given in Figure-68.

Figure-67. Model for Practice 3

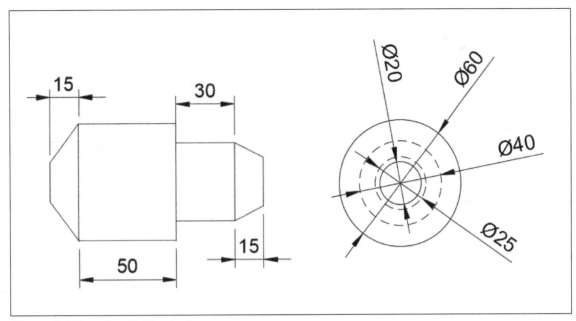

Figure-68. Dimensions for Practice 3

PRACTICE 4

Create the model as shown in Figure-69. The dimensions are given in Figure-70

Figure-69. Model for Practice 4

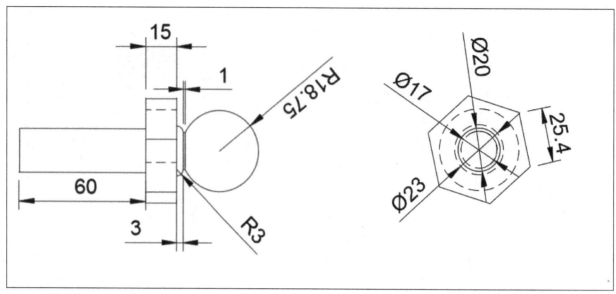

Figure-70. Dimensions for Practice 4

PRACTICE 5

Create the model as shown in Figure-71. The dimensions are given in Figure-72.
(Hint: You will need **Rectangular Pattern** tool and **Move/Copy** tool in this model.)

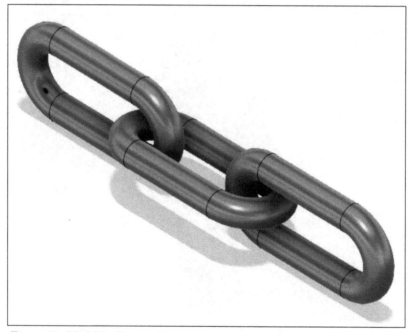

Figure-71. Model for Practice 5

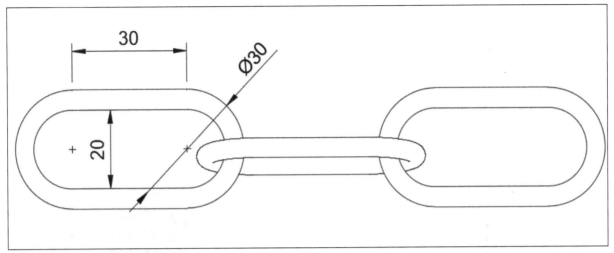

Figure-72. Dimensions for Practice 5

PRACTICE 6

Create the model as shown in Figure-73. The dimensions are given in Figure-74.

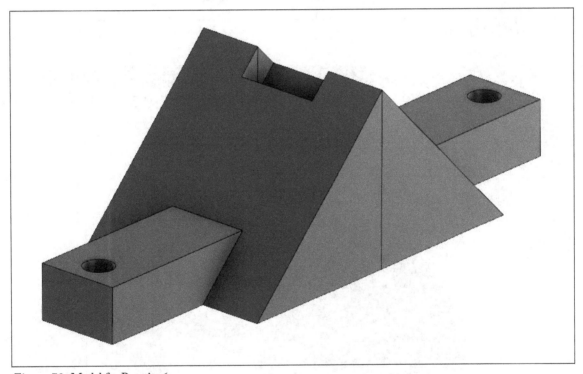

Figure-73. Model for Practice 6

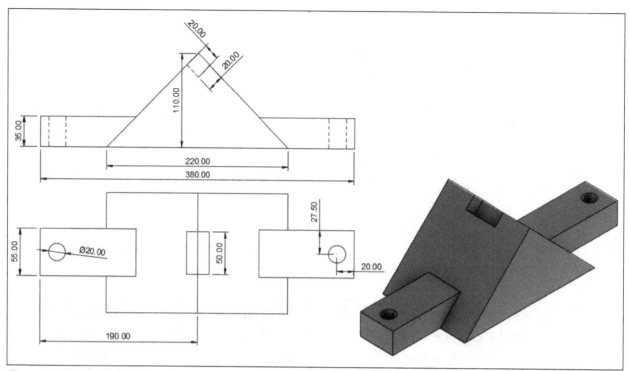

Figure-74. Dimensions for Practice 6

PRACTICE 7

Create the model as shown in Figure-75. The dimensions are given in Figure-76.

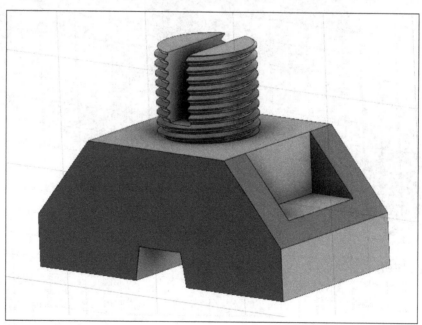

Figure-75. Model for Practice 7

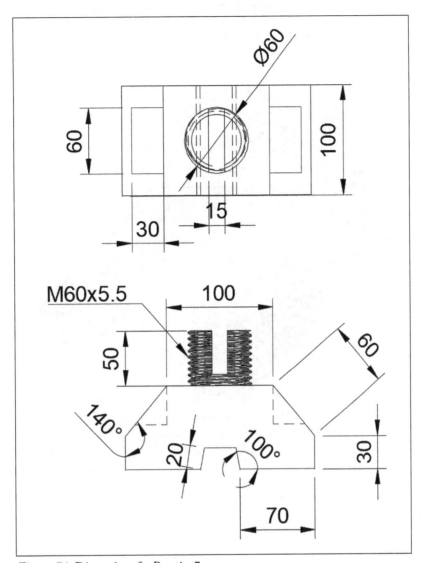

Figure-76. Dimensions for Practice 7

PRACTICE 8

Create the model as shown in Figure-77. The dimensions are given in Figure-78.

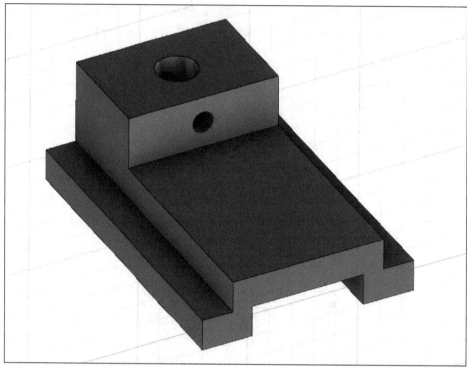

Figure-77. Model for Practice 8

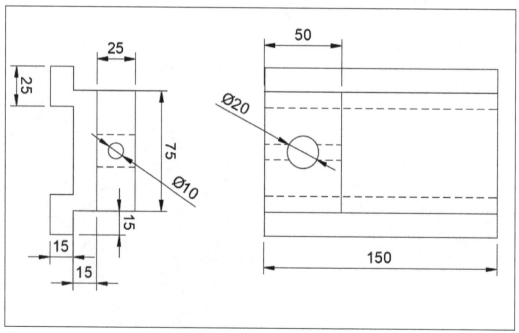

Figure-78. Dimensions for Practice 8

PRACTICE 9

Create the model as shown in Figure-79. The dimensions are given in Figure-80.

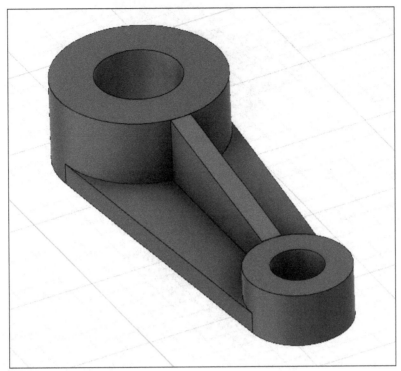

Figure-79. Model for Practice 9

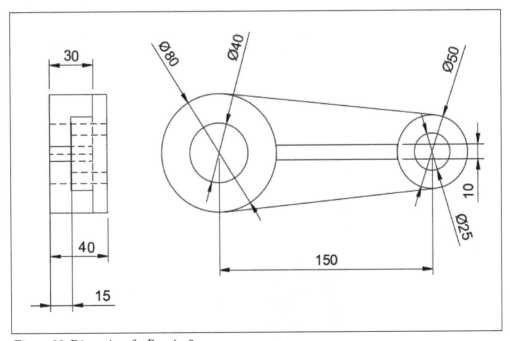

Figure-80. Dimensions for Practice 9

PRACTICE 10

Create the model as shown in Figure-81. The dimensions are given in Figure-82.

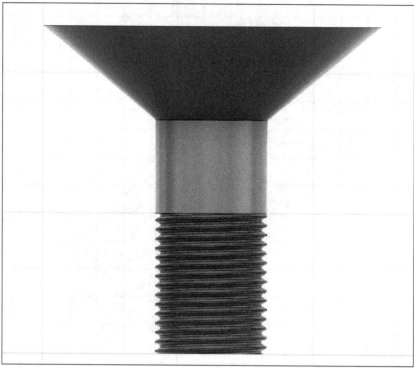

Figure-81. Model for Practice 10

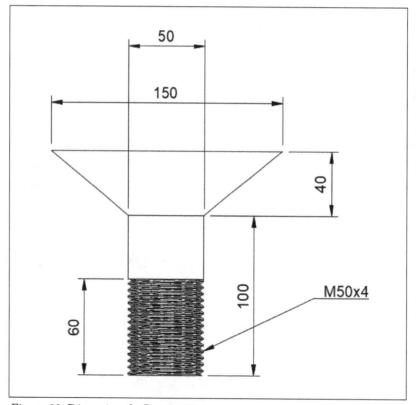

Figure-82. Dimensions for Practice 10

Chapter 6

Solid Editing

Topics Covered

The major topics covered in this chapter are:

- **Solid Editing Tools**
- **Press Pull Tool**
- **Fillet Tool**
- **Chamfer Tool**
- **Shell Tool**
- **Draft Tool**
- **Scale Tool**
- **Combine Tool**
- **Replace Face Tool**
- **Split Face Tool**
- **Split Body Tool**
- **Silhouette Split Tool**
- **Move/Copy Tool**
- **Align Tool**
- **Delete Tool**
- **Physical Material Tool**
- **Appearance Tool**
- **Manage Materials Tool**
- **Change Parameters Tool**

SOLID EDITING TOOLS

In the previous chapters, we have learned to create solid models and remove materials from them. In this chapter, we will learn to edit the models. Like the other tools, Autodesk Fusion 360 has packed all the editing tools into one drop-down. These tools are available in **MODIFY** drop-down from **Toolbar**; refer to Figure-1. The tools in this drop-down are discussed next.

Figure-1. Modify drop down

PRESS PULL

The **Press Pull** tool is used to modify the body by drag and drop methods. The options displayed to perform press pull operations depends on the selected geometry. The procedure to use this tool is discussed next.

• Select **Press Pull** tool of **MODIFY** drop-down from **Toolbar**; refer to Figure-2. The **PRESS PULL** dialog box will be displayed; refer to Figure-3.

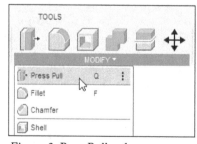

Figure-2. Press Pull tool

Figure-3. PRESS PULL dialog box

- Select a face, if you want to offset or move the face. The **OFFSET FACE** dialog box will be displayed. The **Offset Face** tool will be discussed later. The **Offset Face** tool will be discussed later.
- Select an edge, if you want to fillet the edge. The **FILLET** dialog box will be displayed. The **Fillet** tool will be discussed later in this unit.
- Select a sketch, if you want to extrude the profile. The **EXTRUDE** dialog box will be displayed. This tool was discussed earlier in this book.

Offset Face

The **Offset Face** tool is used to offset or move the face. The procedure to use this tool is discussed next.

- Select **Press Pull** tool from **MODIFY** drop-down. The **PRESS PULL** dialog box will be displayed. You can also select **Press Pull** tool by pressing **Q** key.
- Click on the face of the model to offset. The **OFFSET FACE** dialog box will be displayed; refer to Figure-4.

Figure-4. OFFSET FACE dialog box

- Select **Automatic** button from **Offset Type** drop-down if you do not want to update the sketch or base feature on applying the offset. This action is not added to time line.
- Select **Modify Existing Feature** button from **Offset Type** drop-down if you want to update the sketch or base feature on applying offset. This action is not added to time line.
- Select **New Offset** button from **Offset Type** drop-down if you want to create a new face. This action is added to time line.
- Click in **Distance** edit box of **OFFSET FACE** dialog box or floating window from screen and enter desired value as the distance for offset. You can also move drag handle to set the distance. The preview of offset face will be displayed; refer to Figure-5.

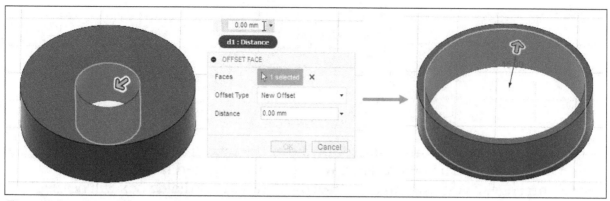

Figure-5. Preview on applying offset face

- Click **OK** button from **OFFSET FACE** dialog box to complete the process.

FILLET

The **Fillet** tool is used to apply round at the edges. This tool works in the same way as the **Sketch Fillet** do. It is recommended that, apply the fillets after creating all the features required in the model because fillets can increase the processing time during modifications. The procedure to use this tool is given next.

- Select **Fillet** tool of **MODIFY** drop-down from **Toolbar**; refer to Figure-6 or press **Q** and select an edge. The **FILLET** dialog box will be displayed; refer to Figure-7.

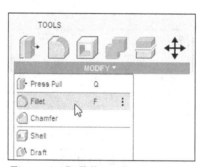

Figure-6. 3D Fillet tool

Figure-7. 3D FILLET dialog box

- **Edges** selection is active by default. Click on the edge(s) from the model to apply fillet. You can select multiple edges also without holding the **CTRL** key. The updated **FILLET** dialog box will be displayed; refer to Figure-8.

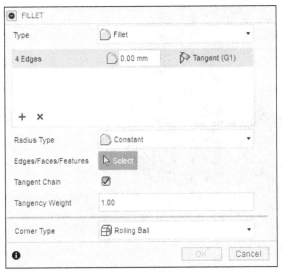

Figure-8. Updated 3D FILLET dialog box

- The options of the updated **FILLET** dialog box are discussed next.

Radius Type Drop-Down

There are three options in **Radius Type** drop-down; refer to Figure-9. These options are used to specify the type of round to be created. These options are discussed next.

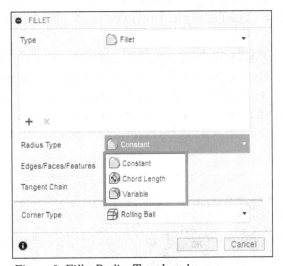

Figure-9. Fillet Radius Type drop down

Constant Radius Type

- Select **Constant** button from **Radius Type** drop-down to apply the fillet of fix size on the selected edge.
- Click in the **Radius** edit box of **FILLET** dialog box and enter **10** as the radius of fillet. You can also move the drag-handle from canvas screen to set the radius of fillet; refer to Figure-10.

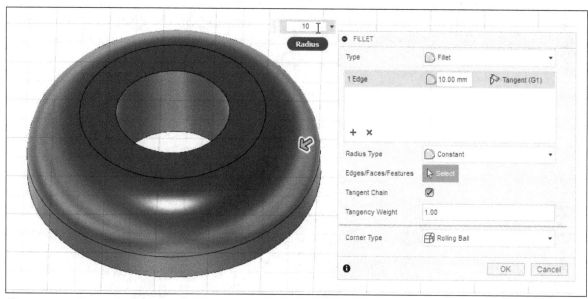

Figure-10. Preview of constant radius type

- Select the **Tangent Chain** check box to include tangentially connected edges in fillet.
- Specify desired value in **Tangency Weight** edit box to increase or decrease the scale of tangency weight.
- Select the **Curvature (G2)** option from **Continuity** drop-down if you want to apply the curvature continuity to the selected fillet.
- Select desired option from **Corner Type** section to define fillet shape at vertex formed by three edges meeting at single point.
- Click **OK** button from **FILLET** dialog box to apply the fillet.

Chord Length Radius Type

- Select **Chord Length** option of **Radius Type** drop-down from **FILLET** dialog box if you want to specify the chord length to control the size for the fillet.
- Click in the **Chord Length** edit box and enter desired value for fillet. You can also set the value by moving the drag handle. The preview will be displayed; refer to Figure-11.

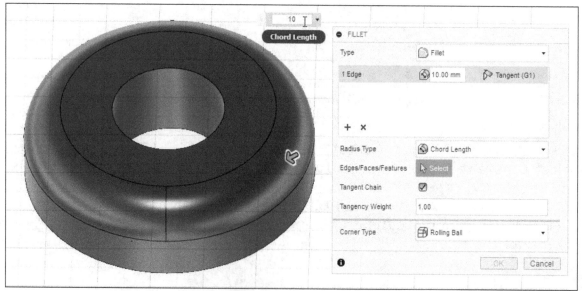

Figure-11. Preview of chord length fillet

- The other parameters are same as discussed in last topic.

Variable Radius Type

- Select **Variable** option from **Radius Type** drop-down of **FILLET** dialog box if you want to specify radii at selected points along the edge. The updated **FILLET** dialog box will be displayed.
- Click on the edge to create points for applying fillet and specify radius for each point in the dialog box or dynamic input box; refer to Figure-12.

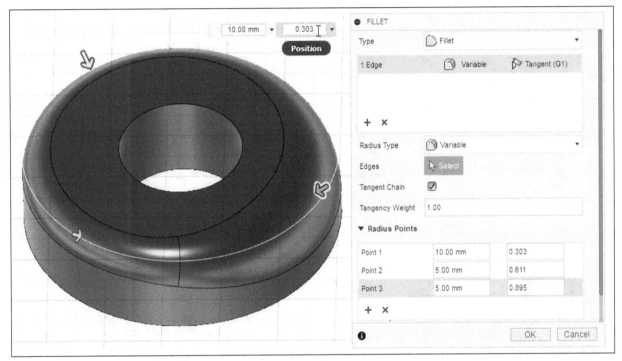

Figure-12. Preview of variable radius type

- To delete the created point, select the specific point from the edge and click on the **Remove selection set** button from the **FILLET** dialog box. Point earlier created to apply fillet will be displayed with arrows.
- The other parameters of this dialog box were discussed earlier.
- After specifying the parameters, click on the **OK** button from **FILLET** dialog box to create the fillet.

RULE FILLET

The **Rule Fillet** option is used to add fillets or rounds to a design based on specified rules. The edges are determined by specified rules rather than selection in the canvas. The procedure to use this tool is discussed next.

- Select **Rule Fillet** option from **Type** drop down in **FILLET** dialog box; refer to Figure-13. The **FILLET** dialog box with rule fillet options will be displayed; refer to Figure-14.

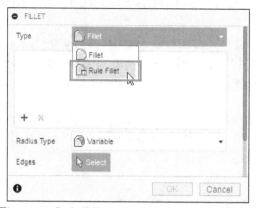

Figure-13. Rule Fillet tool in FILLET dialog box

Figure-14. FILLET dialog box with rule fillet options

- **Faces/Features** selection is active by default. Select desired faces/features from model to apply rule fillet.
- Click in the **Radius** edit box and enter desired value of radius for fillet. The preview of fillet will be displayed; refer to Figure-15. You can also set the radius by moving the drag handle.

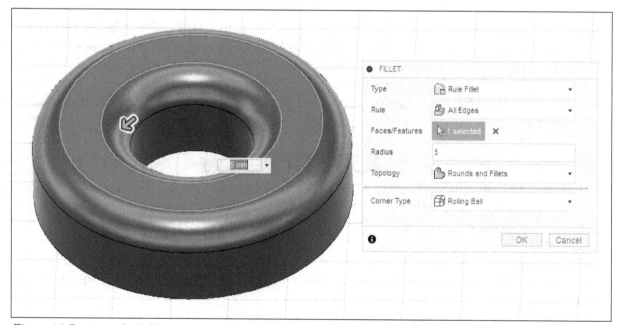

Figure-15. Preview of rule fillet

- Select **All Edges** option from **Rule** drop-down if you want to add the fillets to all edges of the input faces/features and select **Between Faces/Features** option if you want to add the fillets at the intersection of the input faces/features. Note that you can also select features from Time line.

- Select **Rounds Only** option from **Topology** drop-down if you want to create only rounds and select the **Fillets Only** option if you want to create only fillets. Selecting the **Rounds And Fillets** option in this drop-down will create both fillets and rounds.

- Click **OK** button from **FILLET** dialog box to finish this process.

FULL ROUND FILLET

The **Full Round Fillet** option is used to create full round fillet by selecting a face only and it will automatically find the two opposite edges. The procedure to use this tool is discussed next.

- Select **Full Round Fillet** option from **Type** drop-down in **Fillet** dialog box; refer to Figure-16. The **FILLET** dialog box with full round fillet options will be displayed; refer to Figure-17.

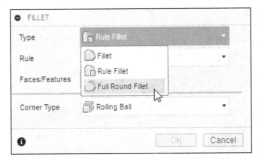

Figure-16. Full Round Fillet tool in FILLET dialog box

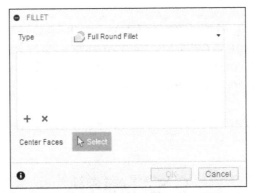

Figure-17. FILLET dialog box with full round fillet options

- The **Select** button of **Center Faces** section is active by default. Select desired center face of the model to create full round fillet; refer to Figure-18.

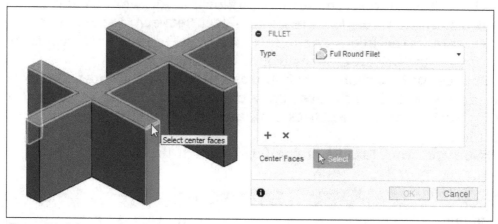

Figure-18. Selecting center face of the model

- On selecting center face of the model, full round fillet of the model will be created and the updated **FILLET** dialog box will be displayed; refer to Figure-19.

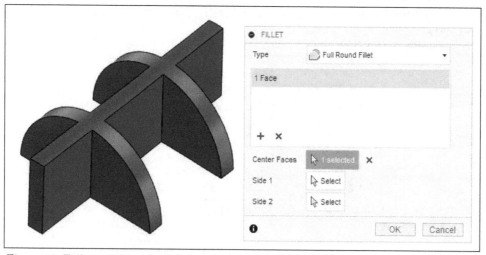

Figure-19. Full round fillet of model created with updated FILLET dialog box

- Click on **Select** button of **Side 1** section and press & hold **CTRL** key to select faces on first side of the full round fillet and click on **Select** button of **Side 2** section and press & hold **CTRL** key to select faces on second side of the full round fillet to modify the fillet.
- After specifying the parameters, click on **OK** button from the dialog box to finish the process.

CHAMFER

The **Chamfer** tool is used to bevel the sharp edges of the model. The procedure to create chamfer by using this tool is given next.

- Click on the **Chamfer** tool of **MODIFY** drop-down from **Toolbar**; refer to Figure-20. The **CHAMFER** dialog box will be displayed; refer to Figure-21.

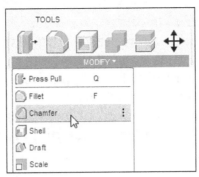

Figure-20. Chamfer tool

Figure-21. CHAMFER dialog box

- **Edges/Faces/Features** selection is active by default. Click on the edge of model to apply chamfer. The updated **CHAMFER** dialog box will be displayed; refer to Figure-22.

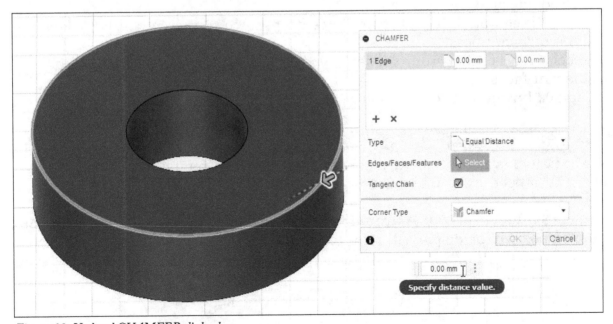

Figure-22. Updated CHAMFER dialog box

- Select the **Tangent Chain** check box to include tangentially connected edges in selection.
- Select **Equal Distance** option from **Type** drop-down of **CHAMFER** dialog box if you want to specify a equal distance for both sides of the chamfer.

- Select **Two Distance** option from **Type** drop-down of **CHAMFER** dialog box if you want to specify distance for each side of the chamfer.
- Enter the specific distance for each side in the respective edit box. You can also set the distance by moving the drag handle from canvas screen.
- Select **Distance and Angle** option from **Type** drop-down if you want to specify a distance and angle to create the chamfer.
- Click in the **Specify distance value** edit box and enter desired value for distance for chamfer. The preview of chamfer will be displayed; refer to Figure-23.

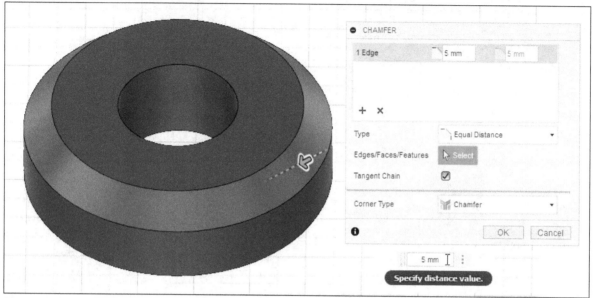

Figure-23. Preview of chamfer

- Select **Chamfer** option from **Corner Type** drop-down to create a chamfer to join beveled edges at the corner.
- Select **Miter** option from **Corner Type** drop-down to merge beveled edges into a mitered corner point.
- Select **Blend** option from **Corner Type** drop-down to blend beveled edges into adjacent faces.
- Click **OK** button of **CHAMFER** dialog box to finish the process.

SHELL

The **Shell** tool is used to make a solid part hollow and remove one or more selected faces. The procedure to use this tool is given next.

- Select **Shell** tool of **MODIFY** drop-down from **Toolbar**; refer to Figure-24. The **SHELL** dialog box will be displayed; refer to Figure-25.

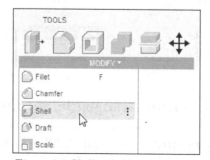

Figure-24. Shell tool

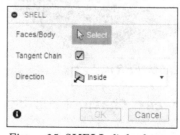

Figure-25. SHELL dialog box

- **Faces/Body** selection button is active by default. Click on the face/body of the solid to apply shell. The updated **SHELL** dialog box will be displayed; refer to Figure-26.

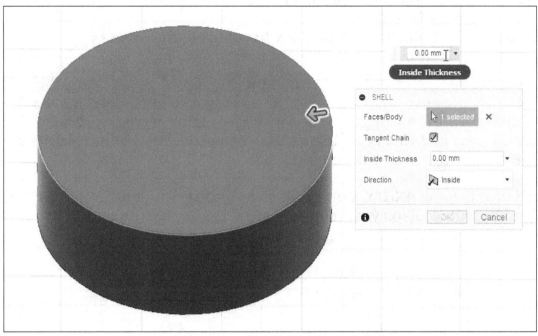

Figure-26. Updated SHELL dialog box

- Select the **Tangent Chain** check box to include tangentially connected edges in selection.
- Click in the **Inside Thickness** edit box and enter desired value of shell thickness. The preview of shell will be displayed; refer to Figure-27.

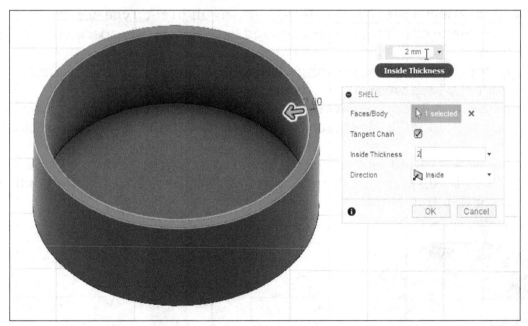

Figure-27. Preview of shell

- Select **Inside** option from **Direction** drop-down if you want to shell the faces towards the interior of the part.
- Select **Outside** option from **Direction** drop-down if you want to shell the faces towards the exterior of the part.

- Select **Both** option from **Direction** drop-down if you want to shell the faces towards the interior and the exterior of the part by equal amount. Click in the **Inside Thickness** and **Outside Thickness** edit boxes and specify desired value to specify thickness for shell.
- Click **OK** button from **SHELL** dialog box to complete the process.

DRAFT

The **Draft** tool is used to apply taper to the faces of a solid model. This tool is mainly useful when you are designing components for moulding or casting. **Draft** tool applies taper on the face and this taper allows easy & safe ejection of parts from dies. The procedure to use this tool is discussed next.

- Select **Draft** tool of **MODIFY** drop-down from **Toolbar**; refer to Figure-28. The **DRAFT** dialog box will be displayed; refer to Figure-29.

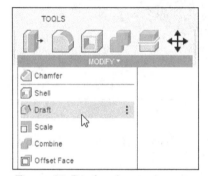

Figure-28. Draft tool

Figure-29. DRAFT dialog box

- The **Fixed Plane** draft type is selected by default in the **Type** section of the dialog box. Fixed Plane draft type creates tapered face using angle specified from selected fixed plane.
- **Pull Direction** selection button is active by default. Click on the face or plane of the model which you want to use as fixed reference.
- Click on **Select** button of **Faces** section and select the faces to be tapered.
- Click on the **Flip Pull Direction** button from **DRAFT** dialog box to flip the direction of draft.
- Click in the **Angle** edit box and enter desired value of angle for draft. The preview of draft will be displayed; refer to Figure-30. You can also set the angle of draft by moving the drag handle.

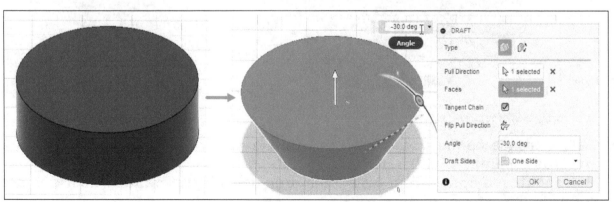

Figure-30. Preview of draft

- The options of **Draft Sides** drop-down were discussed earlier in this book.
- Click **OK** button from **DRAFT** dialog box to finish the process.
- Select **Parting Line** draft type from **Type** section of the dialog box to create draft along a parting line. The updated **DRAFT** dialog box will be displayed; refer to Figure-31.

Figure-31. Updated DRAFT dialog box

- Select desired face or plane of the model to define pull direction in **Pull Direction** section.
- Select desired plane, face, edge, or sketch curve on the model to use as parting tool in **Parting Tool** section.
- Select desired faces on the model which you want to draft in the **Faces** section of the dialog box.
- Specify desired value of angle for draft in the **Angle** edit box. The preview of parting line draft type will be displayed; refer to Figure-32.

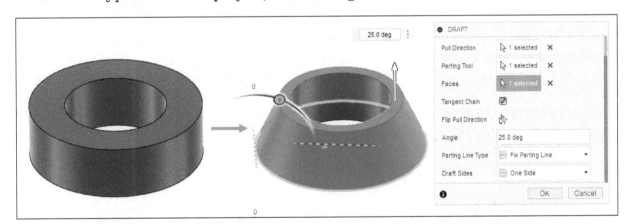

Figure-32. Preview of parting line draft type

- Select **Fix Parting Line** option from **Parting Line Type** drop-down to fix the parting line in place.
- Select desired option from **Draft Sides** drop-down as discussed earlier in this book.
- Select **Move Parting Line** option from **Parting Line Type** drop-down to move the parting line to apply draft.
- Select **Angle Above** option from **Direction** drop-down to maintain angle above parting line.
- Select **Angle Below** option from **Direction** drop-down to maintain angle below parting line.

- Select **Both** option from **Direction** drop-down to maintain angle on both sides of parting line.
- Click on the **Select** button from **Fixed Edges** section of the dialog box and select edges of the model to prevent deformation when parting line moves.
- The other options in the dialog box have been discussed earlier.
- Click **OK** button from **DRAFT** dialog box to finish the process.

SCALE

The **Scale** tool is used to enlarge or reduce selected bodies, sketches, or components based on specified scale factor. The procedure to use this tool is discussed next.

- Select **Scale** tool of **MODIFY** drop-down from **Toolbar**; refer to Figure-33. The **SCALE** dialog box will be displayed; refer to Figure-34.

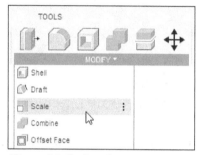

Figure-33. Scale tool

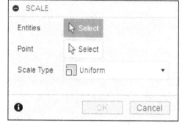

Figure-34. SCALE dialog box

- **Entities** selection is active by default. Click on the entities to be scaled up or down.
- On selecting the entity for scale, point is automatically selected in **Point** section. If you want to use a different point as scale reference then click on the **Select** button for **Point** option in the dialog box and select desired point.
- Select **Uniform** option from **Scale Type** drop-down if you want to scale the selected entities uniformly in all the directions.
- Select **Non-Uniform** option from **Scale Type** drop-down if you want to scale the entity non-uniformly in x, y, and z direction.
- Click in the **Scale Factor** edit box and specify desired value for scale. The preview will be displayed; refer to Figure-35.
- Click on **OK** button of **SCALE** dialog box to complete the process.

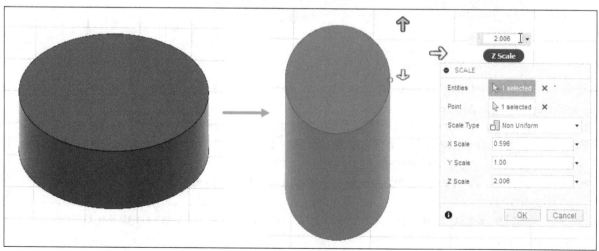

Figure-35. Preview of scale

COMBINE

The **Combine** tool is used to cut, join, and intersect the selected bodies. The procedure to use this tool is discussed next.

- Select **Combine** tool of **MODIFY** drop-down from **Toolbar**; refer to Figure-36. The **COMBINE** dialog box will be displayed; refer to Figure-37.

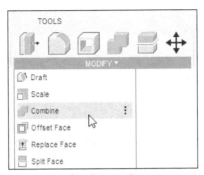

Figure-36. Combine tool

Figure-37. COMBINE dialog box

- The **Target Body** selection button is active by default. Click on the component from model to be used as a target body.
- Click on other components to select in **Tool Bodies** section. The preview of combine will be displayed; refer to Figure-38.

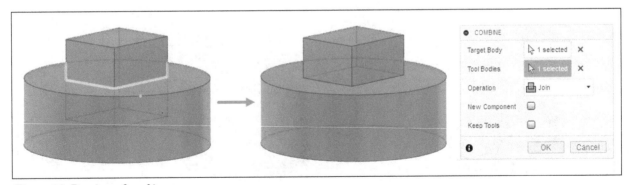

Figure-38. Preview of combine

- The options of **Operations** drop-down have been discussed earlier in this book.
- Select the **New Component** check box to create a new component with results.
- Select the **Keep Tools** check box to retain the tool bodies after the combine result.
- Click **OK** button of **COMBINE** dialog box to finish the process.

REPLACE FACE

The **Replace Face** tool is used to replace an existing face with a surface or work plane. The procedure to use this tool is discussed next.

*   Select **Replace Face** tool of **MODIFY** drop-down from **Toolbar**; refer to Figure-39. The **REPLACE FACE** dialog box will be displayed; refer to Figure-40.

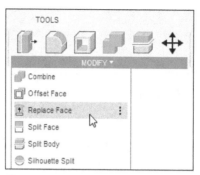

Figure-39. Replace Face tool

Figure-40. REPLACE FACE dialog box

*   **Source Faces** selection is active by default. Click on the component face to select the face.
*   Select the **Tangent Chain** check box to include tangentially connected faces in selection.
*   Click **Select** button of **Target Faces** options and click on the face or plane to apply this tool. The preview will be displayed; refer to Figure-41.

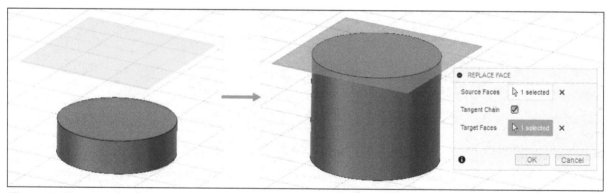

Figure-41. Preview of replace face

*   Click **OK** button of **REPLACE FACE** dialog box to finish the process.

SPLIT FACE

The **Split Face** tool is used to divide faces of a surface or solid to add draft, delete an area, or create features. The procedure to use this tool is discussed next.

- Select **Split Face** tool of **MODIFY** drop-down from **Toolbar**; refer to Figure-42. The **SPLIT FACE** dialog box will be displayed; refer to Figure-43.

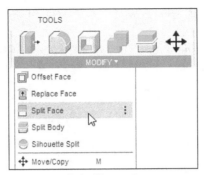

Figure-42. Split Face tool

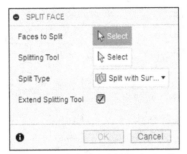

Figure-43. SPLIT FACE dialog box

- **Faces To Split** option is active by default. Click on the model face to be split. You can select multiple faces by holding the **CTRL** key.
- Click on the **Select** button of **Splitting Tool** option and select the face which will split the other faces.
- Select **Split With Surface** option from **Split Type** drop-down, if you want to project the splitting tool face which you had selected upon the splitting face.
- Select **Along Vector** option from **Split Type** drop-down, if you want to define the projection direction of splitting face by selecting a face, edge, or axis.
- Select **Closest Point** option from **Split Type** drop-down, if you want to define the direction of projection as the closest distance between the splitting body and target faces.
- Select the **Extend Splitting Tool** check box to extend the tool to completely intersect the target faces.
- After specifying the parameters, click on **OK** button from **SPLIT FACE** dialog box to finish the process; refer to Figure-44.

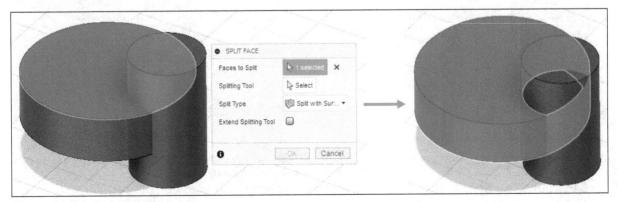

Figure-44. Preview of split face

SPLIT BODY

The **Split Body** tool works similarly to **Split Face** tool but in this tool, component or part is selected in place of faces. The procedure to use this tool is discussed next.

- Select **Split Body** tool of **MODIFY** drop-down from **Toolbar**; refer to Figure-45. The **SPLIT BODY** dialog box will be displayed; refer to Figure-46.

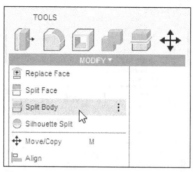

Figure-45. Split Body tool

Figure-46. SPLIT BODY dialog box

- **Body To Split** selection is active by default. Click on the component to be split.
- Click on **Select** button of **Splitting Tool** option and select the body of which splitting will be done.
- Select the **Extend Splitting Tool(s)** check box to extend the tool to completely intersect the target faces.
- After specifying the various parameters, click **OK** button of **SPLIT BODY** dialog box to complete the process; refer to Figure-47.

Figure-47. Preview of split body

SILHOUETTE SPLIT

The **Silhouette Split** tool is used to split selected body at the silhouette outline visible from the view direction. The procedure to use this tool is discussed next.

- Select **Silhouette Split** tool of **MODIFY** drop-down from **Toolbar**; refer to Figure-48. The **SILHOUETTE SPLIT** dialog box will be displayed; refer to Figure-49.

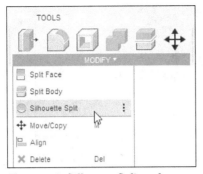

Figure-48. Silhouette Split tool

Figure-49. SILHOUETTE SPLIT dialog box

- **View Direction** selection option is active by default. Click on the plane or face to select the direction to split.
- Click on **Select** button of **Target Body** and select the body which you want to split.
- Select **Split Faces Only** option from **Operation** drop-down, if you want to split the faces of the selected body but keeps it as a single body.
- Select **Split Shelled Body** option from **Operation** drop-down, if you want to split the selected body into two bodies. To use this option , the selected body must be shelled first.
- Select **Split Solid Body** option from **Operation** drop-down, if you want to split the selected body at the parting line. The parting line of the selected body must be planar.
- After specifying the parameters, click **OK** button of **SILHOUETTE SPLIT** dialog box to finish the process; refer to Figure-50.

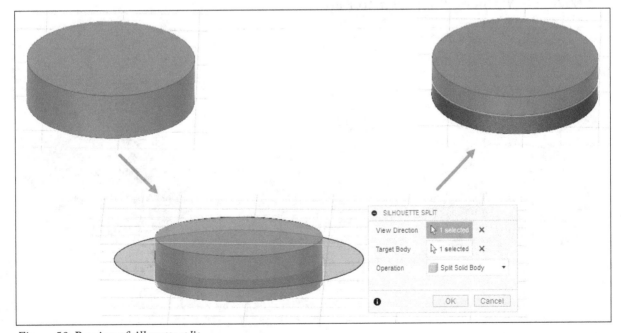

Figure-50. Preview of silhouette split

MOVE/COPY

The **Move/Copy** tool is used to move a face, body, sketch curve, components, or sketch geometry. Using this tool, you can also move the whole geometry. The procedure to use this tool is discussed next.

- Select **Move/Copy** tool of **MODIFY** drop-down from **Toolbar**; refer to Figure-51. The **MOVE/COPY** dialog box will be displayed; refer to Figure-52.

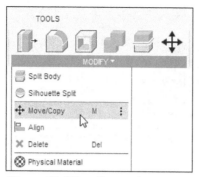

Figure-51. *Move Copy tool*

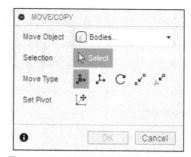

Figure-52. *MOVE COPY dialog box*

- Select **Component** option from **Move Object** drop-down, if you want to move a component.
- Select **Bodies** option from **Move Object** drop-down, if you want to move a body.
- Select **Faces** option from **Move Object** drop-down, if you want to move a face.
- Select **Sketch Objects** option from **Move Object** drop-down, if you want to move the selected sketch.
- In our case, we are selecting the **Bodies** option from **Move Object** drop-down.
- **Selection** section is active by default. Click on the body to be moved. The updated **MOVE/COPY** dialog box will be displayed.
- Select **Free Move** button from **Move Type** section, if you want to move selected body freely.
- On selecting the **Free Move** button, a Manipulator will be displayed on the selected body. Use the Manipulator to move the body or enter the value in respective field of **MOVE/COPY** dialog box; refer to Figure-53.

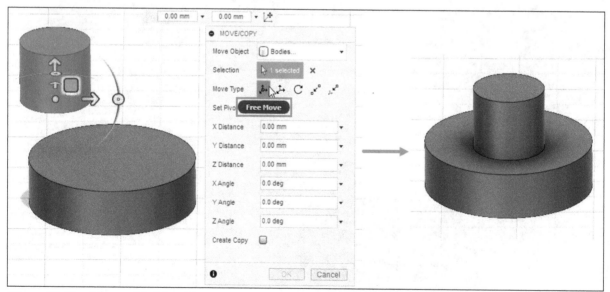

Figure-53. *Free Move tool in Move Type section*

- Select **Translate** button from **Move Type** section, if you want to move the body along X, Y, and Z directions.
- On selecting the **Translate** button, three arrows will be displayed on the entity. Drag desired arrow to move the body in respective direction. You can also enter the specific value of direction in the respective edit box from **MOVE/COPY** dialog box; refer to Figure-54.

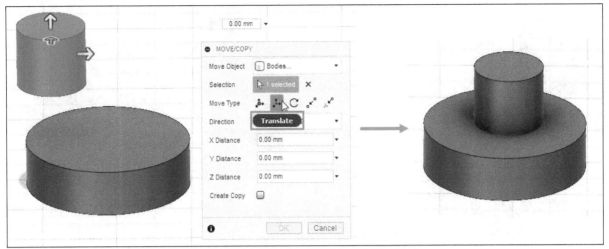

Figure-54. Translate tool in Move Type section

- Select **Rotate** button from **Move Type** section, if you want to rotate the entity.
- On selecting the **Rotate** option, the **Axis** option will be displayed in the **MOVE/ COPY** dialog box.
- Click on the **Select** button from **Axis** section of **MOVE/COPY** dialog box and select the axis/edge for rotation of entity. The drag handle will be displayed on the selected face.
- Move the drag handle as required or specify the value in the **Angle** edit box of **MOVE/COPY** dialog box; refer to Figure-55.

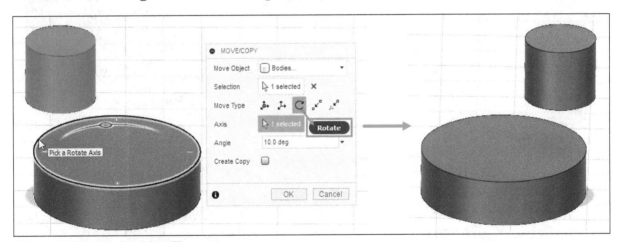

Figure-55. Rotate tool in Move Type section

- Select **Point to Point** button from **Move Type** section, if you want to move the body from one point to another point.
- Click on the **Select** button of **Origin Point** option from **MOVE/COPY** dialog box and select the start point of body from where you want to move the body.
- Click on the **Select** button of **Target Point** option from **MOVE/COPY** dialog box and select the finish point from model.
- After selecting the points, the preview will be displayed.
- If the preview is as required then click on **OK** button from **MOVE/COPY** dialog box to complete the process; refer to Figure-56.

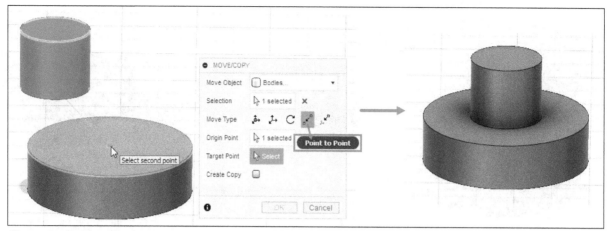

Figure-56. Point to Point tool in Move Type section

- Select **Point to Position** button from **Move Type** section, if you want to move the body by selecting a point and a position from model.
- Select a point of the body. The updated **MOVE/COPY** dialog box will be displayed; refer to Figure-57.

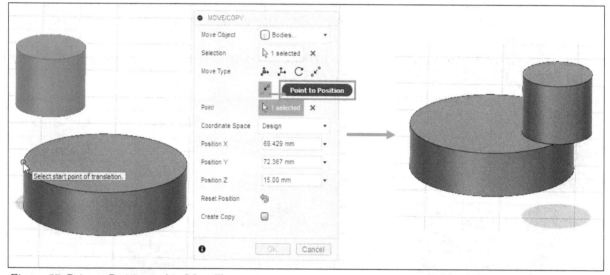

Figure-57. Point to Position tool in Move Type section

- Click in the **Position X** edit box and specify the value. Similarly, specify the value in **Position Y** and **Position Z** edit boxes as required.
- After specifying the parameters, click on the **OK** button from **MOVE/COPY** dialog box to complete the process.

ALIGN

The **Align** tool is used to align the selected object to another object. The procedure to use this tool is discussed next.

- Click on the **Align** tool of **MODIFY** drop-down from **Toolbar**; refer to Figure-58. The **ALIGN** dialog box will be displayed; refer to Figure-59.

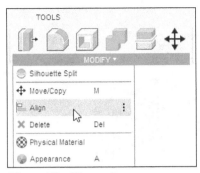

Figure-58. Align tool

Figure-59. ALIGN dialog box

- Select **Bodies** option from **Object** drop-down of **ALIGN** dialog box if you want to select bodies to align.
- Select **Components** option from **Object** drop-down of **ALIGN** dialog box if you want to select components to align.
- In our case, we are selecting **Bodies** option from **Object** drop-down.
- Click on **Select** button of **From** section of **ALIGN** dialog box and select point on the first body from which you want to align that body.
- Click on **Select** button of **To** section of **ALIGN** dialog box and select point on the second body to which the body to be aligned; refer to Figure-60.

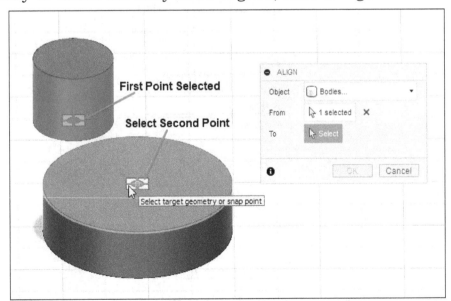

Figure-60. Selection of points to align body

- On selecting the point on second body, the two selected body will align to each other at selected location; refer to Figure-61.

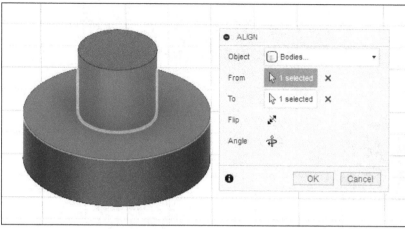

Figure-61. Preview of align

- Click on the **Flip** button, if you want to flip the aligned body.
- Click on **Angle** button, if you want to rotate the aligned body.
- After specifying the parameters, click on the **OK** button from **ALIGN** dialog box to finish the process.

DELETE

The **Delete** tool is used to delete the selected object or component. You can also press **DELETE** from keyboard to do the same. The procedure to use this tool is discussed next.

- Click on the **Delete** tool of **MODIFY** drop-down from **Toolbar**; refer to Figure-62. The **DELETE** dialog box will be displayed; refer to Figure-63.

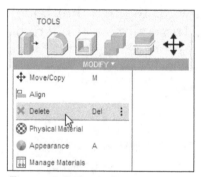

Figure-62. Delete tool

Figure-63. DELETE dialog box

- The **Entities** selection is active by default. Select the entity to be deleted from the existing model.
- After selecting the desired entities, click on the **OK** button from **DELETE** dialog box to delete the selected entity; refer to Figure-64.

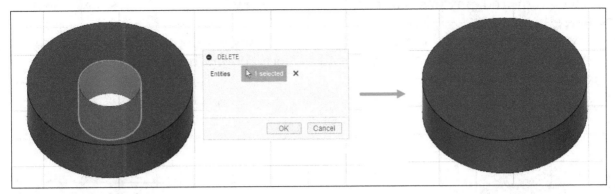

Figure-64. Entity deleted

PHYSICAL MATERIAL

The **Physical Material** tool is used to apply physical and visual material to the component or body. The procedure to use this tool is discussed next.

- Click on the **Physical Material** tool of **MODIFY** drop-down from **Toolbar**; refer to Figure-65. The **PHYSICAL MATERIAL** dialog box will be displayed; refer to Figure-66.

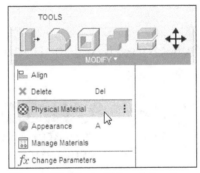

Figure-65. Physical Material tool

Figure-66. PHYSICAL MATERIAL dialog box

- Select required option from **Library** drop-down.

- Select required physical material from **Library** section of **PHYSICAL MATERIAL** dialog box.
- Now, drag the specific material from the **Library** section and drop it to the body or component. The physical material of body/component will be changed; refer to Figure-67.

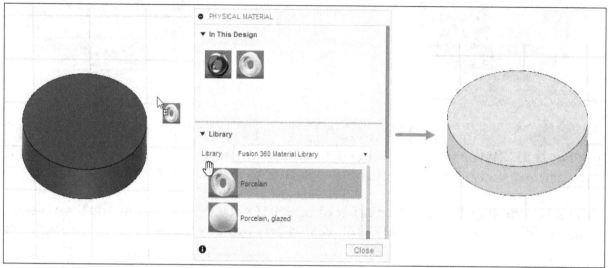

Figure-67. Applying physical material on the body

- After applying material, click on the **Close** button of **PHYSICAL MATERIAL** dialog box to apply changes.

APPEARANCE

The **Appearance** tool is used to change the appearance of body. This tool is generally used to change the color of the body. The procedure to use this tool is discussed next.

- Click on the **Appearance** tool of **MODIFY** drop-down from **Toolbar**; refer to Figure-68. The **APPEARANCE** dialog box will be displayed; refer to Figure-69.

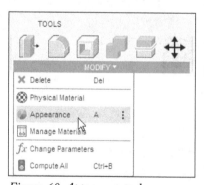

Figure-68. Appearance tool

Figure-69. APPEARANCE dialog box

- Select required radio button of **Apply To** section from **APPEARANCE** dialog box to select bodies/components or faces.
- Drag an appearance and drop it to the body or face; refer to Figure-70.

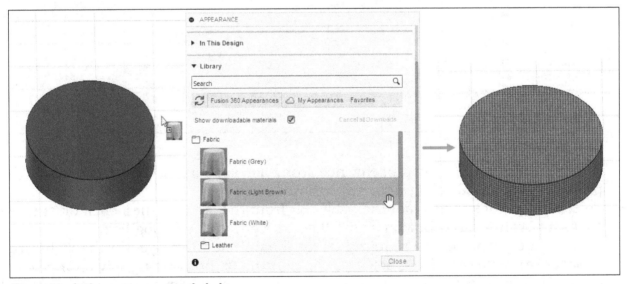

Figure-70. Applying appearance on the body

- After applying the appearance, click on the **Close** button from **APPEARANCE** dialog box to finish the process.

MANAGE MATERIALS

The **Manage Materials** tool is used to define the material properties shown in **PHYSICAL MATERIAL** and **APPEARANCE** dialog boxes discussed earlier. The procedure to use this tool is discussed next.

- Click on the **Manage Materials** tool of **MODIFY** drop-down from **Toolbar**; refer to Figure-71. The **Material Browser** dialog box will be displayed; refer to Figure-72.

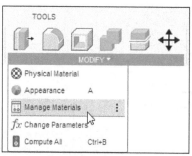

Figure-71. Manage Materials tool

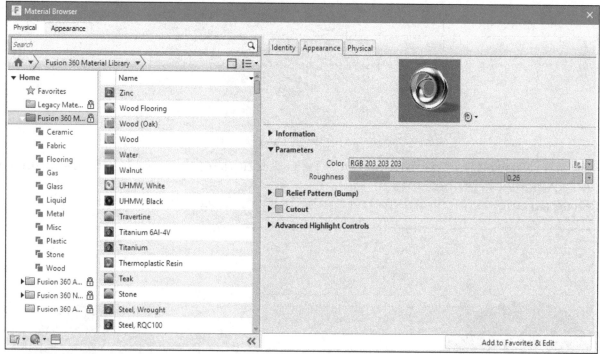

Figure-72. Material Browser dialog box

- Click on any material or appearance from list section to check or modify the property of selected material or appearance.
- To edit properties of any material, select it from the list and then click on the **Add to Favorites & Edit** button at the bottom-right in the dialog box.
- Change the properties as required using the options displayed in the right area of **Material Browser** dialog box and click on the **OK** button to apply changes.
- Close the dialog box once changes have been applied. Next time, when you will be applying material, make sure to select them from favorites list.

CHANGE PARAMETERS

The **Change Parameters** tool is used to change the parameters and apply equations to the model. The procedure to use this tool is discussed next.

- Click on the **Change Parameters** tool of **MODIFY** drop-down from **Toolbar**; refer to Figure-73. The **Parameters** dialog box will be displayed; refer to Figure-74.

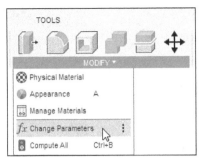

Figure-73. Change Parameters tool

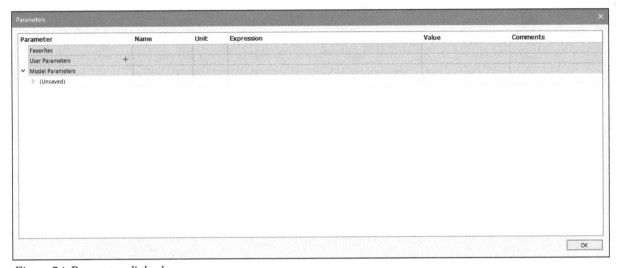

Figure-74. Parameters dialog box

- If you want to create some parameters which are to be used later in equation then click on the ⊞ plus button from **User Parameters** section of **Parameters** dialog box. The **Add User Parameter** dialog box will be displayed; refer to Figure-75.

Figure-75. Add User Parameter dialog box

- Click in the **Name** edit box and specify the name of parameter.
- Click on the **Unit** drop-down and select desired unit of the parameter.
- Click on the **Expression** edit box and enter desired expression like d1/4. The **Value** will be displayed.
- Click in the **Comment** edit box and enter required comment.
- After specifying the parameters, click on the **OK** button from **Add User Parameter** dialog box. The user defined parameter will be added in **Parameters** dialog box; refer to Figure-76.

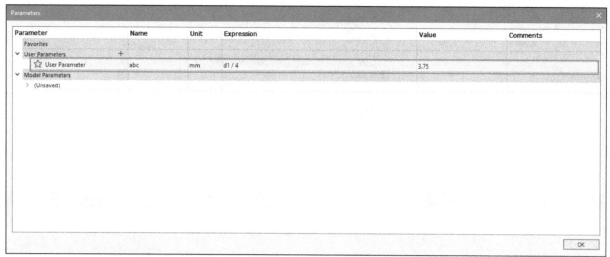

Figure-76. User Parameter added

- Click on the **Model Parameter** node from **Parameters** dialog box, all the model features will be displayed.
- Click on desired model feature node to check its parameters, the parameters applied in that feature will be displayed; refer to Figure-77.

Parameter	Name	Unit	Expression	Value
Favorites				
∨ User Parameters ✛				
☆ User Parameter	abc	mm	d1 / 4	3.75
∨ Model Parameters				
∨ Practice v1				
∨ Extrude1				
☆ AlongDistance	d1	mm	15 mm	15.00
☆ TaperAngle	d2	deg	0.0 deg	0.0
∨ Extrude3				
☆ AlongDistance	d6	mm	-50.00 mm	-50.00
☆ TaperAngle	d7	deg	0.0 deg	0.0
∨ Plane2				
☆ AlongDistance	d9	mm	0.00 mm	0.00
∨ Extrude4				
☆ AlongDistance	d10	mm	10.00 mm	10.00
☆ TaperAngle	d11	deg	0.0 deg	0.0

Figure-77. Parameters applied in the model

- If you want to apply equation in parameters then click on the dimension parameter and enter desired equation like d1*2; refer to Figure-78. Note that here d1 is dimension number applied automatically to the sketch dimension.

Parameter	Name	Unit	Expression	Value
Favorites				
∨ User Parameters	+			
☆ User Parameter	abc	mm	d1 / 4	3.75
∨ Model Parameters				
∨ Practice v1				
∨ Extrude1				
☆ AlongDistance	d1	mm	15 mm	15.00
☆ TaperAngle	d2	deg	0.0 deg	0.0
∨ Extrude3				
☆ AlongDistance	d6	mm	-50.00 mm	-50.00
☆ TaperAngle	d7	deg	0.0 deg	0.0
∨ Plane2				
☆ AlongDistance	d9	mm	d1 * 2	30.00
∨ Extrude4				
☆ AlongDistance	d10	mm	10.00 mm	10.00
☆ TaperAngle	d11	deg	0.0 deg	0.0

Figure-78. Entering equation

- After specifying the parameters, click on the **OK** button from **Parameters** dialog box
- The equation entered in the **Parameters** dialog box will be applied to the model; refer to Figure-79.

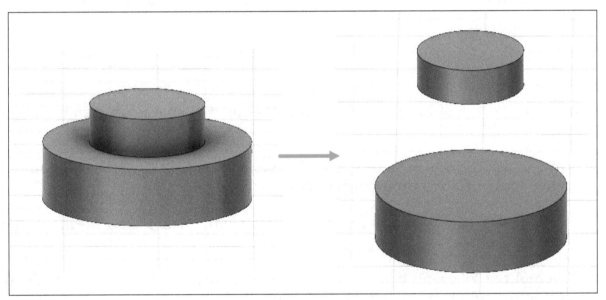

Figure-79. Parameter changed

Click on the **Compute All** tool from the **MODIFY** drop-down to re-calculate all the parameters.

SELF-ASSESSMENT

Q1. Which of the following tools can be used to dynamically use offset face or fillet function depending on the selection?

a. Fillet   b. Offset Face
c. Press Pull   d. Combine

Q2. Which of the following tool is used to apply round at the edges?

a. Fillet   b. Chamfer
c. Draft   d. Shell

Q3. Which of the following tool is used by casting mold designer to allow easy extraction of part from the die?

a. Fillet   b. Draft
c. Shell   d. Chamfer

Q4. Which of the following tool is used to create a body using intersection of two bodies?

a. Combine   b. Shell
c. Split Body   d. Split Face

Q5. The **Move/Copy** tool in Autodesk Fusion **Design** workspace can also be used to rotate the objects. (T/F)

Q6. What is the difference between applications of **Physical Material** tool and **Appearance** tool in **Modify** drop-down of **Design** workspace?

Q7. In Autodesk Fusion 360, you can create a fillet with changing radius at specified points. (T/F)

Q8. Which of the following tools is used to remove material from a solid body and convert it into a thin body of specified thickness?

a. Combine   b. Shell
c. Split Body   d. Split Face

Chapter 7

Assembly Design

Topics Covered

The major topics covered in this chapter are:

- *Introduction*
- *New Component Tool*
- *Joint Tool*
- *As-Built Joint Tool*
- *Joint Origin Tool*
- *Rigid Group Tool*
- *Drive Joints Tool*
- *Motion Link Tool*
- *Enable Contact Sets Tool*
- *Motion Study Tool*

INTRODUCTION

Most of the things you find around you in real-world are assembly of various components; the computer you are working on is an assembly, the automotive you may be driving or sitting in for travelling is an assembly, there are lots of examples. Till this point, we have learned to create solid parts. But most of the time, we need to assemble the parts to get some use of them. In this chapter, we will work on the assembly design of Autodesk Fusion 360 in which we will be assembling two or more components using some assembly constraints.

NEW COMPONENT

The **New Component** tool is used to create an empty component or component from the selected bodies. Note that we need to convert all parts to components before performing their assembly. The procedure to use this tool is discussed next.

- Click on the **New Component** tool of **ASSEMBLE** drop-down from **Toolbar**; refer to Figure-1. The **NEW COMPONENT** dialog box will be displayed; refer to Figure-2.

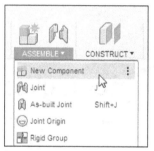

Figure-1. New Component tool

Figure-2. NEW COMPONENT dialog box

- Select **Standard** option from **Type** drop-down in the dialog box, if you want to create a standard component in the assembly structure.
- Select **Sheet Metal** option from **Type** drop-down, if you want to create a sheet metal component in the assembly structure.

Standard Component

In this topic, we will discuss the procedure of creating standard component.

- Select **Standard** option from **Type** drop-down of **NEW COMPONENT** dialog box if not selected by default.
- Select **External** radio button from dialog box to create an external component and referenced into the assembly in the current design. The options related to **External** radio button will be displayed in the **NEW COMPONENT** dialog box; refer to Figure-3.

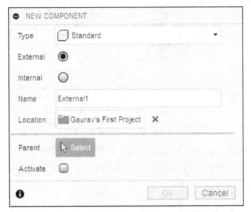

Figure-3. Options on selecting External radio button

- Click in the **Name** edit box of **NEW COMPONENT** dialog box and enter the name of component.
- Specify desired location where you want to save the new design that will contain the new component by clicking on the button from **Location** section of the dialog box.
- Click on **Select** button of **Parent** option and select the parent body under which new component will be created.
- Select the **Activate** check box from **NEW COMPONENT** dialog box to activate the new component. Note that all the modeling operations performed after activating the component will become child feature of that component.
- After specifying the parameters for standard component, click on the **OK** button from **NEW COMPONENT** dialog box. A standard component will be created; refer to Figure-4.

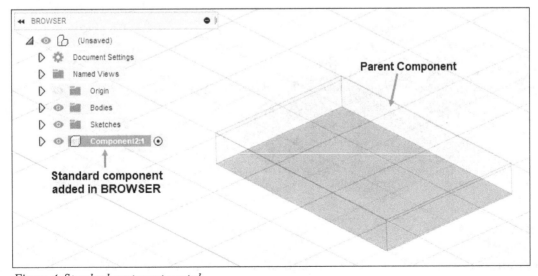

Figure-4. Standard component created

- If you want to create an internal component directly in the current design then select **Internal** radio button from the dialog box. The options related to **Internal** radio button will be displayed in the **NEW COMPONENT** dialog box; refer to Figure-5 .

Figure-5. Options on selecting Internal radio button

- Select **From Bodies** check box from **NEW COMPONENT** dialog box to convert existing bodies into individual components. The updated **NEW COMPONENT** dialog box will be displayed; refer to Figure-6.

Figure-6. Updated NEW COMPONENT dialog box

- The **Select** button of **Bodies** section is active by default. Click on the parts/bodies from model to create a new component; refer to Figure-7. You can select multiple bodies by holding the **CTRL** key.

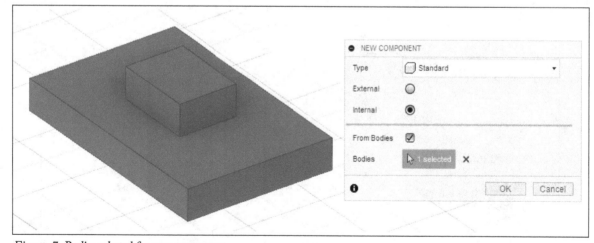

Figure-7. Bodies selected for new component

- After selecting the body, click on the **OK** button from **NEW COMPONENT** dialog box. The new component will be created and action will be displayed in **BROWSER**.

Sheet Metal Component

In this topic, we will discuss the procedure of creating sheet metal component.

- Select the **Sheet Metal** option from **Type** drop-down of **NEW COMPONENT** dialog box. The **NEW COMPONENT** dialog box with sheet metal options will be displayed; refer to Figure-8.

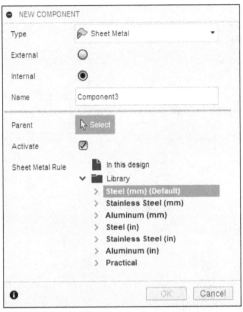

Figure-8. NEW COMPONENT dialog box with sheet metal options

- Select desired sheet metal rule already in use in this design or from the cloud library available in the **Sheet Metal Rule** section of the dialog box.
- The other parameters in the **NEW COMPONENT** dialog box have been discussed earlier.
- After specifying parameters for sheet metal component, click on the **OK** button from **NEW COMPONENT** dialog box. A sheet metal component will be created; refer to Figure-9.

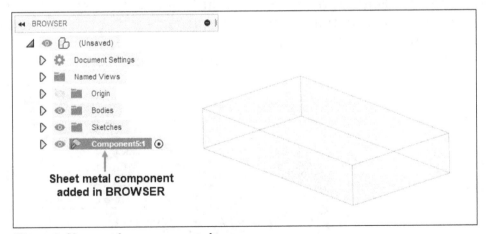

Figure-9. Sheet metal component created

JOINT

The **Joint** tool is used to define joints between components to allow movement of components relative to each other. To apply **Joint** tool, you need to create the component of bodies. The procedure to use this tool is discussed next.

- Click on the **Joint** tool of **ASSEMBLE** drop-down from **Toolbar**; refer to Figure-10. The **JOINT** dialog box will be displayed; refer to Figure-11.

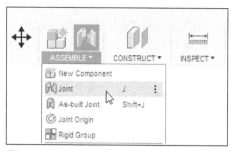

Figure-10. Joint tool

Figure-11. JOINT dialog box

- Select desired option from **Mode** section of **Component 1** area in the **Position** tab of the dialog box. Select **Simple** option from **Mode** section to select a **Snap** point to place the joint origin. Select **Between Two Faces** option from **Mode** section to select **Plane 1** and **Plane 2** to center the joint origin between them then select a **Snap** point. Select **Two Edge Intersection** option from **Mode** section to select **Edge 1** and a non-parallel **Edge 2** to locate the joint origin at the extended intersection.
- **Select** button of **Snap** section from **Component 1** area is active by default. Click on the first component at desired location to assemble; refer to Figure-12. You will be asked to select a location on second component.

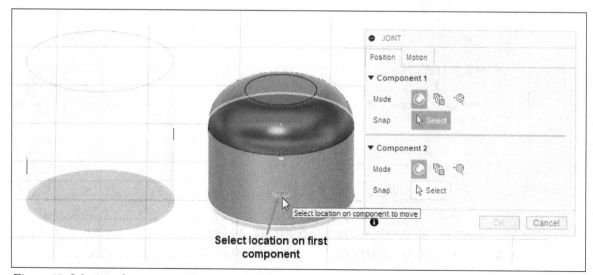

Figure-12. Selection of component 1 location

- Select desired option from **Mode** section of **Component 2** area in the **Position** tab of the dialog box. The options of **Mode** section have been discussed earlier.
- Click on the second component on mating location to select; refer to Figure-13.

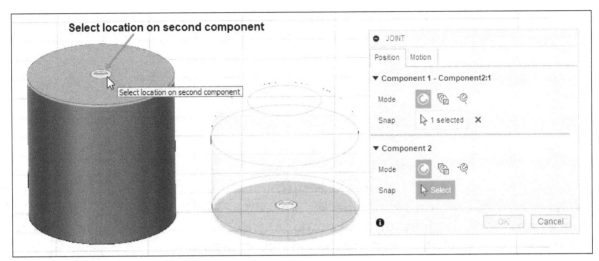

Figure-13. Selection of component 2 location

- On selecting the second component, the selected two component will be joined to each other at the locations selected on them and the updated **JOINT** dialog box will be displayed; refer to Figure-14.

Figure-14. Components joined with updated JOINT dialog box

- Click in the **Angle** edit box from **Joint Alignment** area in **Position** tab of the dialog box and specify the angle of rotation of component.
- If you want to specify offset distance between components then click in the **Offset X** , **Offset Y**, and **Offset Z** edit boxes as required and enter the value. You can also set the offset distance by moving the drag handles.
- Click on the **Flip** button from **Joint Alignment** area of the dialog box, if you want to flip the direction of second component.

Motion

There are seven motion types in **Motion** tab of **JOINT** dialog box which we can apply to the component as required; refer to Figure-15. These motion types are described next.

Figure-15. Types of motion in JOINT dialog box

Rigid Motion

- Select the **Rigid** option from **Type** section in the **Motion** tab of the **JOINT** dialog box, if you want the joint of two component to be rigid to each other and there should be no motion between the two joined component.

Revolute Motion

- Select the **Revolute** option from **Type** section in the **Motion** tab of the **JOINT** dialog box, if you want to revolve the component about other at the joint location. On selecting the **Revolute** option, a revolve drag handle will be displayed. You can also revolve the component by moving the drag handle; refer to Figure-16.

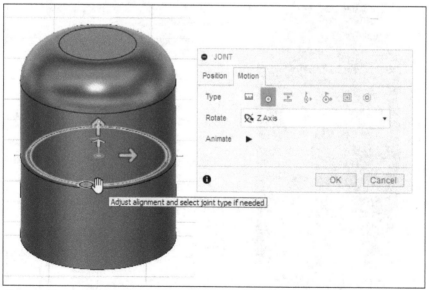

Figure-16. Revolute motion

Slider Motion

- Click on the **Slider** option from **Type** section in the **Motion** tab of the **JOINT** dialog box, if you want to assemble the component in such a way that selected component can slide over the other; refer to Figure-17.

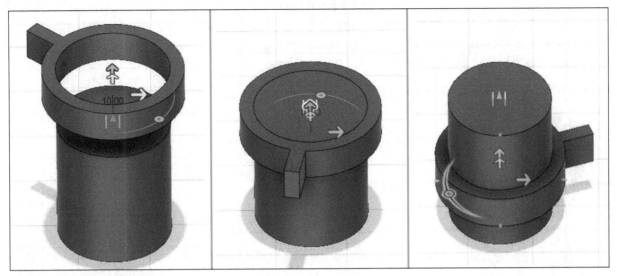

Figure-17. Slider motion

Cylindrical Motion

- Click on the **Cylindrical** option from **Type** section in the **Motion** tab of the **JOINT** dialog box, if you want to assemble the component in such a way that it is free to move along the selected axis and free to rotate about the same selected axis; refer to Figure-18.

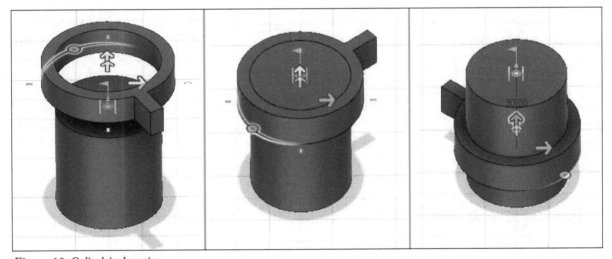

Figure-18. Cylindrical motion

Pin-Slot Motion

- Click on the **Pin-Slot** option from **Type** section in the **Motion** tab of **JOINT** dialog box, if you want to revolve and slide the component together like in nut & bolt locking-unlocking.

Planar Motion

- Click on the **Planar** option from **Type** section in the **Motion** tab of **JOINT** dialog box, if you want to assemble the component in such a way that the component can move over the selected plane only in defined boundary.

Ball Motion

- Click on the **Ball** option from **Type** section in the **Motion** tab of **JOINT** dialog box, if you want to assemble the component in such a way that the component is free to move 360 degree in 3D space pivoted to a point; refer to Figure-19.

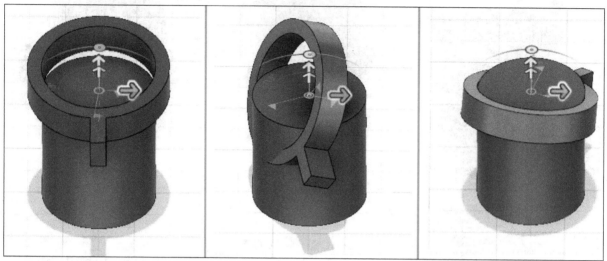

Figure-19. Ball motion

- Click on the **Animate** button from **Motion** tab of **JOINT** dialog box to watch the applied motion by animating **Component 1**. Click the button again to stop the animation.

After specifying the parameters for joint, click on the **OK** button from **JOINT** dialog box to complete the process.

AS-BUILT JOINT

The **As-built Joint** tool allows the joints to be applied to components in their current position as they are built, making it easy to join them in their original orientation. The procedure to use this tool is discussed next.

- Click on the **As-built Joint** tool of **ASSEMBLE** drop-down from **Toolbar**; refer to Figure-20. The **AS-BUILT JOINT** dialog box will be displayed; refer to Figure-21.

Figure-20. As built Joint tool

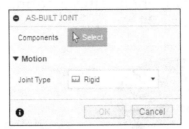

Figure-21. AS BUILT JOINT dialog box

- Select the components from the canvas.
- Select required motion from **Joint Type** drop-down in **Motion** area of the dialog box. The updated **AS-BUILT JOINT** dialog box will be displayed; refer to Figure-22.

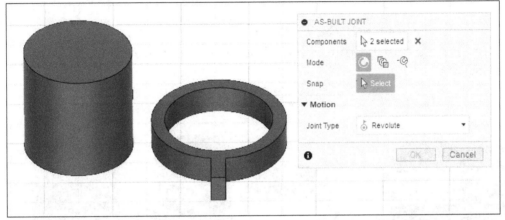

Figure-22. Updated AS BUILT JOINT dialog box

- Select desired option from **Mode** section of the dialog box which have been discussed earlier.
- Select desired references on the components based on the option selected in **Joint Type** drop-down; refer to Figure-23.

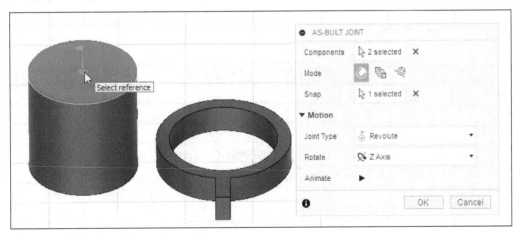

Figure-23. Selecting reference for motion

- Select desired option from **Rotate** drop-down to specify the axis for rotation of the component.
- Click on the **Animate** button from **Motion** area of the dialog box to watch the applied motion by animating the component. Click the button again to stop the animation.
- After specifying the parameters, click on the **OK** button from **AS-BUILT JOINT** dialog box to complete the process.

- If you want to rotate the component after creating joint then double-click on the flag displayed; refer to Figure-24.

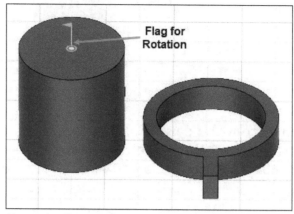

Figure-24. Flag for rotation

- The drag handle will be displayed. Rotate the component as required; refer to Figure-25.

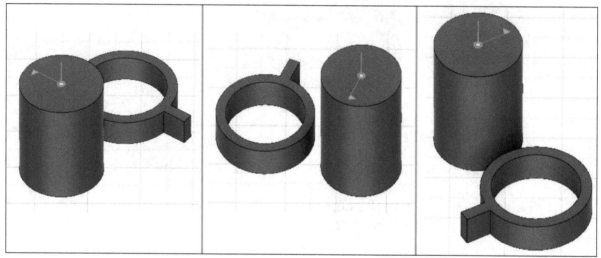

Figure-25. Rotation of component

JOINT ORIGIN

The **Joint Origin** tool is used to define joint origin on a component. The procedure to use this tool is discussed next.

- Click on the **Joint Origin** tool of **ASSEMBLE** drop-down from **Toolbar**; refer to Figure-26. The **JOINT ORIGIN** dialog box will be displayed; refer to Figure-27.

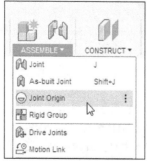

Figure-26. Joint Origin tool

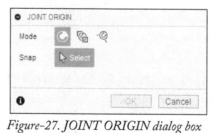

Figure-27. JOINT ORIGIN dialog box

- Select **Simple** option from **Mode** section of **JOINT ORIGIN** dialog box, if you want to place the joint origin on a face.
- Select **Between Two Faces** option from **Mode** section of **JOINT ORIGIN** dialog box if you want to place the joint origin midway between the faces.
- In our case, we are selecting the **Between Two Faces** option from **Mode** section. The updated **JOINT ORIGIN** dialog box will be displayed as shown in Figure-28.

Figure-28. Updated JOINT ORIGIN dialog box

- The **Select** button of **Plane 1** section is active by default. Click on the first face from model to select; refer to Figure-29.

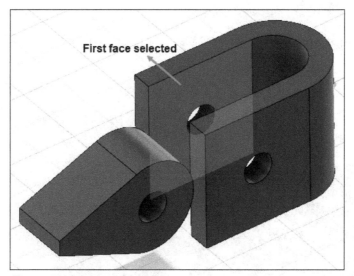

Figure-29. Selection of first face

- On selecting the first face, the **Select** button of **Plane 2** section will be activated. Similarly, select the second face of the model; refer to Figure-30. The updated **JOINT ORIGIN** dialog box will be displayed.

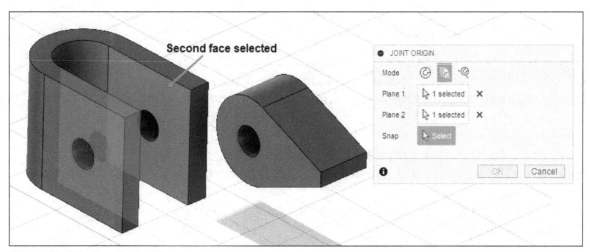

Figure-30. Selection of second face

- The **Select** button of **Snap** option is active by default. Click at desired location to define the snap point; refer to Figure-31.

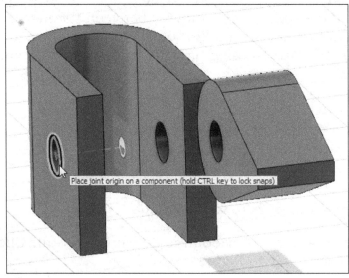

Figure-31. Selection of snap point

- On selecting the snap point, the **JOINT ORIGIN** dialog box will be updated; refer to Figure-32.

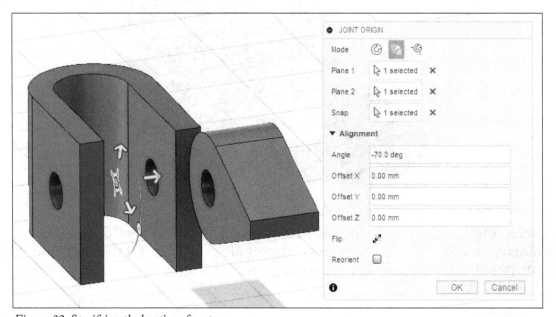

Figure-32. Specifying the location of snap

- If you want to specify the angle of snap then click in the **Angle** edit box of **JOINT ORIGIN** dialog box and specify the angle of rotation of snap.
- If you want to specify the offset distance of snap then click on the **Offset X** edit box and enter the value. You can also set the offset distance by moving the drag handle.
- Similarly, set the offset distance in other edit boxes.
- Select the **Flip** button from **JOINT ORIGIN** dialog box if you want to flip the direction of snap.
- Select the **Reorient** check box from **JOINT ORIGIN** dialog box, if you want to position the axes of origin of geometry.

- After specifying the parameters, click on the **OK** button from **JOINT ORIGIN** dialog box to complete the process.
- Now, you will be able to join the two component with the help of **Joint** tool using newly created joint origin; refer to Figure-33 and Figure-34.

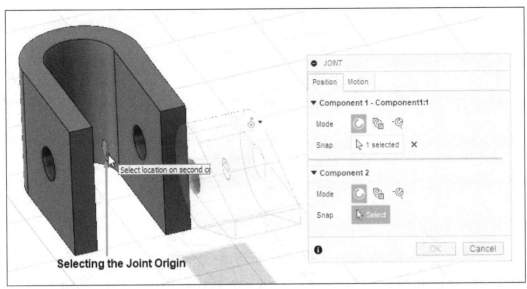

Figure-33. Selection of newly created joint origin

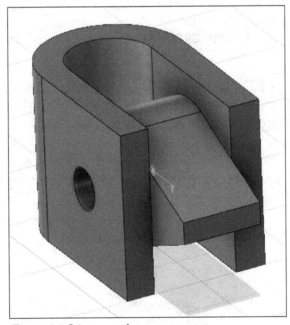

Figure-34. Joint created

RIGID GROUP

The **Rigid Group** tool is used to combine selected components into one group so that the created group act as one object. The procedure to use this tool is discussed next.

- Click on the **Rigid Group** tool of **ASSEMBLE** drop-down from **Toolbar**; refer to Figure-35. The **RIGID GROUP** dialog box will be displayed; refer to Figure-36.

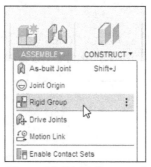

Figure-35. Rigid Group tool

Figure-36. RIGID GROUP dialog box

- The **Select** button of **Components** section is active by default. You need to select the components from the model to create a rigid group.
- Select the **Include Child Components** check box if you want to include the child components into the rigid group.
- Click on **OK** button from **RIGID GROUP** dialog box to complete the process.

DRIVE JOINTS

The **Drive Joints** tool is used to specify the rotation angle or distance value based on joint's degrees of freedom. The procedure to use this tool is discussed next.

- Click on the **Drive Joints** tool of **ASSEMBLE** drop-down from **Toolbar**; refer to Figure-37. The **DRIVE JOINTS** dialog box will be displayed; refer to Figure-38.

Figure-37. Drive Joints tool

Figure-38. DRIVE JOINTS dialog box

- The **Joint Input** selection is active by default. Click on the joint from model. The updated **DRIVE JOINTS** dialog box will be displayed; refer to Figure-39.

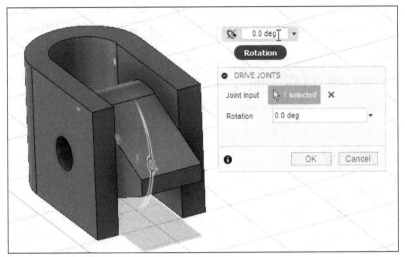

Figure-39. Updated DRIVE JOINTS dialog box

- Click in the **Rotation** edit box and specify the angle of rotation. You can also specify rotation limits by moving the drag handle; refer to Figure-40.

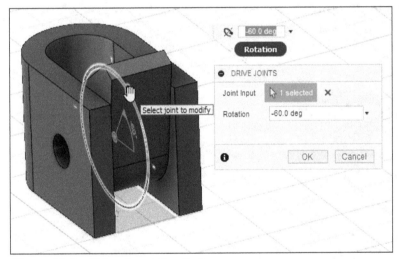

Figure-40. Specifying the rotation limits

- After specifying the parameters, click on the **OK** button from **DRIVE JOINTS** dialog box.

Note that after changing position of joint after driving, the **POSITION** panel is displayed in the **SOLID** tab of **Toolbar**; refer to Figure-41. Select the **Capture Position** tool from the panel if you want to make current position of joint as new start position. Select the **Revert** option from the panel if you want to revert the joint to original position.

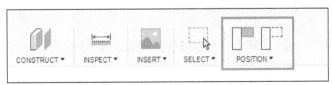

Figure-41. POSITION panel

MOTION LINK

The **Motion Link** tool is used to specify the relationship between motion of two selected joints like relation between two gear rotation or relation between bolt rotation and translation while fastening. The procedure to use this tool is discussed next.

- Click on the **Motion Link** tool of **ASSEMBLE** drop-down from **Toolbar**; refer to Figure-42. The **MOTION LINK** dialog box will be displayed; refer to Figure-43.

Figure-42. Motion Link tool

Figure-43. MOTION LINK dialog box

- **Joints** selection is active by default. Click on desired joints. In our case, we are selecting the slider and revolve joints to create a motion of vice.
- On selecting the two joints, the animation of vice will start and the dialog box will be updated; refer to Figure-44.

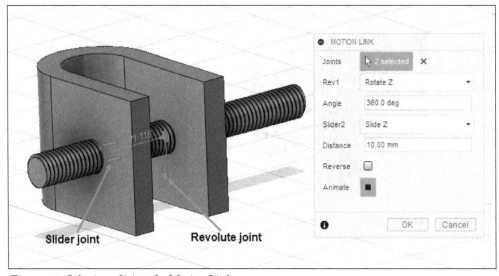

Figure-44. Selection of joints for Motion Link

- Click in the **Distance** edit box of **MOTION LINK** dialog box and specify the value of distance per revolution.
- Click in the **Angle** edit box of **MOTION LINK** dialog box and specify the value of angle of revolution. The combination of distance and angle value define the range of motion like in our case, for one full rotation of 360 degree by handle, the jaw will move 10 mm of distance.
- Select the **Reverse** check box of **MOTION LINK** dialog box, if you want to reverse the direction of revolution.
- Click on the **Animate** button to watch the animation of joints.
- After specifying the parameters, click on **OK** button of **MOTION LINK** dialog box to complete the process; refer to Figure-45.

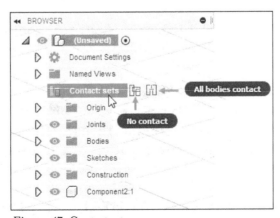

Figure-45. Motion Link created between joints

ENABLE CONTACT SETS

The **Enable Contact Sets** tool is used to enable the physical contact conditions. The procedure to use this tool is discussed next.

- Click on the **Enable Contact Sets** tool of **ASSEMBLE** drop-down from **Toolbar**; refer to Figure-46. The tool will be displayed in **BROWSER**; refer to Figure-47.

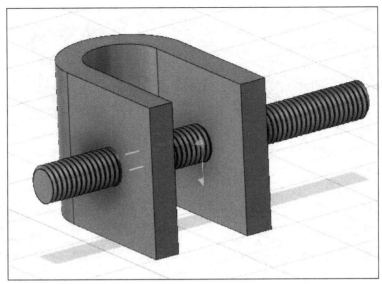

Figure-46. Enable Contact Sets tool

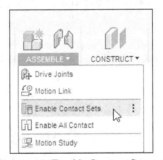

Figure-47. Contact sets

- Click on **All Bodies Contact** button to activate the contact analysis between sets; refer to Figure-48.

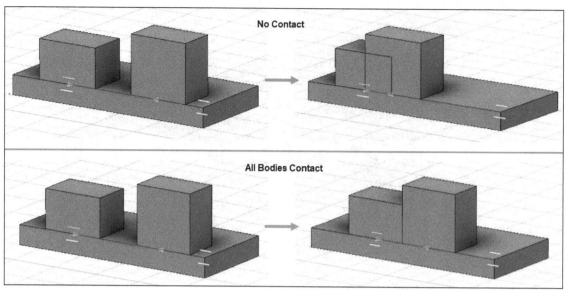

Figure-48. Applying contact sets tool

- Click on **No Contact** button to disable the contact between sets.
- In the above figure, the first case is with contact sets not active and the second case is with contact sets active. So, when contact set is active then physical properties of objects are taken into account like one solid object cannot pass through another solid object.

Enable All Contact

The **Enable All Contact** tool is used to enable all contacts which were applied earlier in the part or model.

- To enable contact, click on the **Enable All Contact** tool of **ASSEMBLE** drop-down from **Toolbar**. All contacts will be enabled and the **ASSEMBLE** drop-down will be updated with two new tools which are discussed next.

Disable Contact

The **Disable Contact** tool is used to disable all contacts which were made earlier in the part or model.

- To disable contact, click on the **Disable Contact** tool of **ASSEMBLE** drop-down from **Toolbar**. All contacts will be disabled.

New Contact Set

The **New Contact Set** tool is used to create new contact set between two selected components. This tool is available when **Enable All Contact** tool is selected. These contact sets prevent parts from passing through one another. The procedure to use this tool is discussed next.

- Click on the **New Contact Set** tool of **ASSEMBLE** drop-down from **Toolbar**; refer to Figure-49. The **NEW CONTACT SET** dialog box will be displayed; refer to Figure-50.

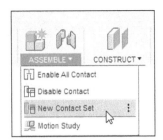

Figure-49. New Contact Set tool

Figure-50. NEW CONTACT SET dialog box

- The **Select** button of **Bodies or Components** section is active by default. Select the two component or bodies to create sets and click on **OK** button from **NEW CONTACT SET** dialog box to complete the process.

MOTION STUDY

The **Motion Study** tool is used to perform kinematic motion analysis based on joints. The procedure to use this tool is discussed next.

- Click on the **Motion Study** tool of **ASSEMBLE** drop-down from **Toolbar**; refer to Figure-51. The **Motion Study** dialog box will be displayed; refer to Figure-52.

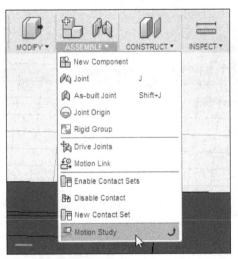

Figure-51. Motion Study tool

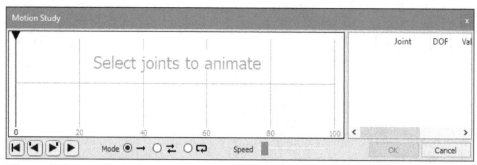

Figure-52. Motion Study dialog box

- You need to select the joint from part or model. Click to select the joint; refer to Figure-53.

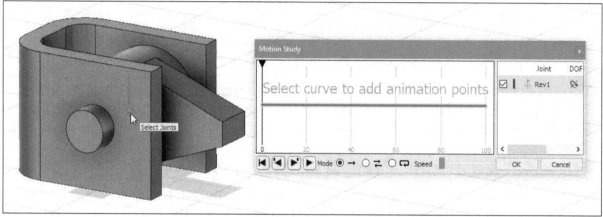

Figure-53. Selecting joint for animation

- After selecting joint, create the points of motion curve as required by clicking in the graph line of motion; refer to Figure-54.

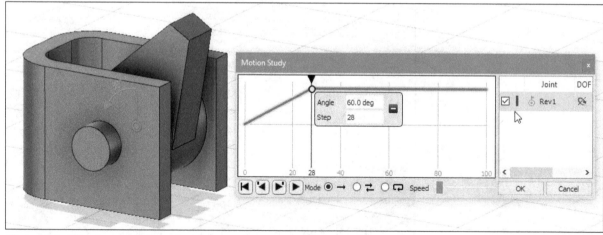

Figure-54. Adjusting points for animation

- Click in the **Angle** edit box and specify required angle for animation point. Note that the options will be displayed based on selected joint.
- After creating the animation points, set the speed of animation from **Speed** bar.
- Select **Play Once** radio button from **Mode** section to play the animation once.
- Select **Play Forward/Backward** radio button from **Mode** section to play the animation forward and then backward.
- Select **Play Loop** radio button from **Mode** section to play the animation in cycle.

- After specifying the parameters for animation, click on **Play** button to check the animation of selected joint.
- Click on **OK** button from **Motion Study** dialog box to exit the motion study of components.

PRACTICAL 1

Assemble the parts of handle as shown in Figure-55. The exploded view of assembly is displayed as shown in Figure-56.

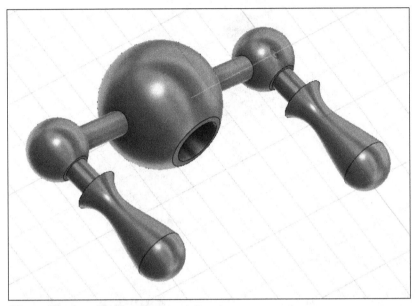

Figure-55. Assembled handle

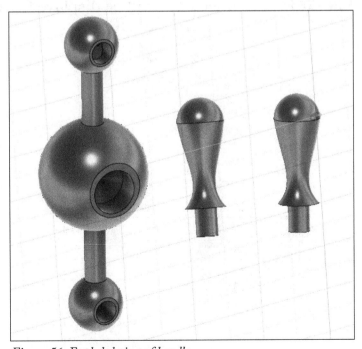

Figure-56. Exploded view of handle

All the part files can be downloaded from resource kit of Fusion 360 software, which is provided at **WWW.CADCAMCAEWORKS.COM**.

The steps to assemble these parts are given next.

Adding file into Fusion 360

- After downloading the resource kit from **CADCAMCAEWORKS.COM**, open the **Chapter 7** folder.

- Double-click on **Handle v0.f3d** file. The file will open in **Autodesk Fusion 360** software; refer to Figure-57.

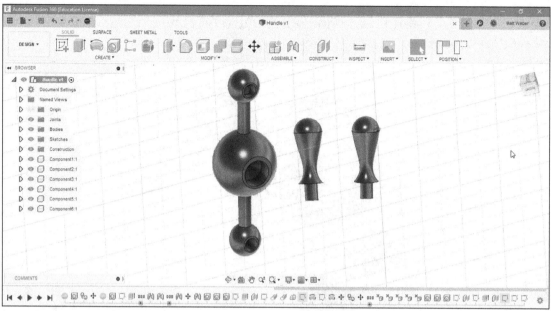

Figure-57. Fusion 360 startup window

Assembling components

- Click on **New Component** tool of **ASSEMBLE** drop-down from **Toolbar**; refer to Figure-58. The **NEW COMPONENT** dialog box will be displayed; refer to Figure-59.

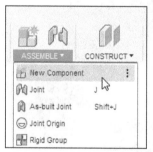

Figure-58. New Component tool

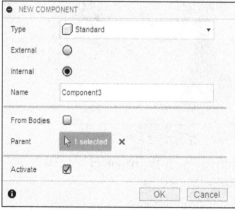

Figure-59. NEW COMPONENT dialog box

- Select the **From Bodies** check box from **NEW COMPONENT** dialog box and select all the bodies to make them component; refer to Figure-60.

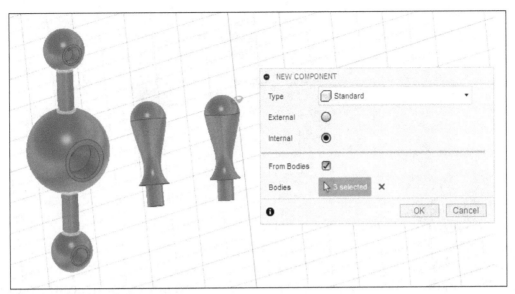

Figure-60. Selecting bodies to create components

- After selecting the bodies, click on **OK** button from **NEW COMPONENT** dialog box. The components will be created and added in **Timeline**.
- Click on **Joint** tool of **ASSEMBLE** drop-down from **Toolbar.** You can also activate the **Joint** tool by pressing **J** key from keyboard. The **JOINT** dialog box will be displayed; refer to Figure-61.

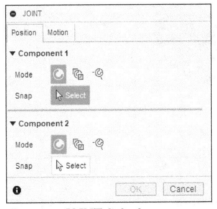

Figure-61. JOINT dialog box

- The **Select** button of **Snap** section from **Component 1** area in **Position** tab is selected by default. Click on the component to join; refer to Figure-62.

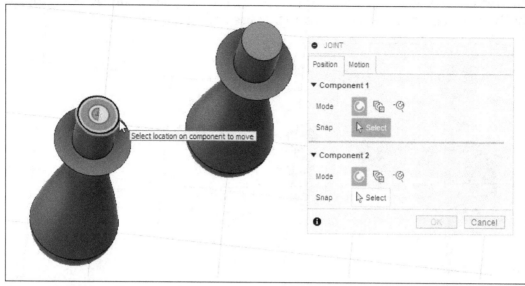

Figure-62. Selection of location on Component1

- After selection of first component, select the second component; refer to Figure-63.

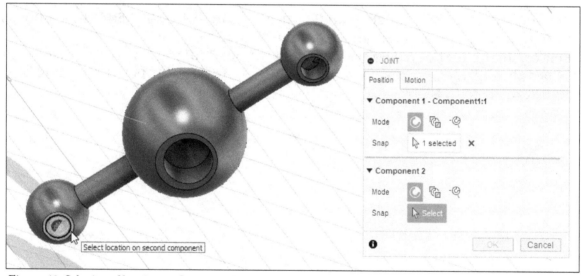

Figure-63. Selection of location on Component2

- Select **Rigid** option from **Type** section in **Motion** tab of the dialog box to create a rigid joint between two component.
- After specifying the parameters, click on **OK** button from **JOINT** dialog box to create a rigid joint.
- Similarly, create another rigid joint between other component.
- After joining these components, the assembly will be displayed as shown in Figure-64.

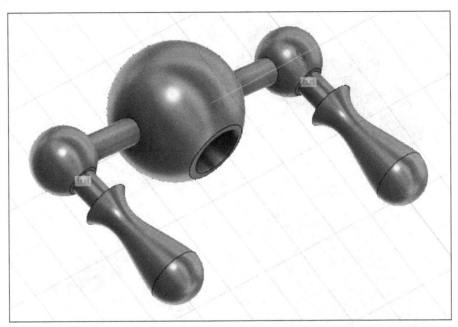

Figure-64. Final assembly of component

PRACTICAL 2

Assemble the parts of **F1 brake balance shifting mechanism** as shown in Figure-65. The exploded view of assembly is displayed as shown in Figure-66.

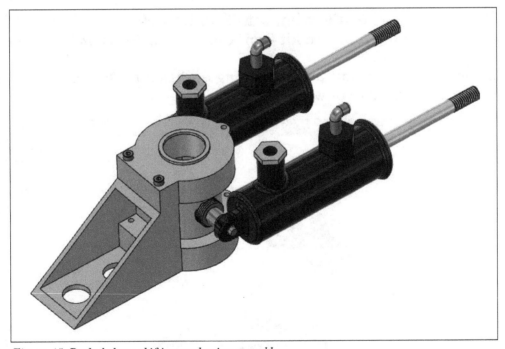

Figure-65. Brake balance shifting mechanism assembly

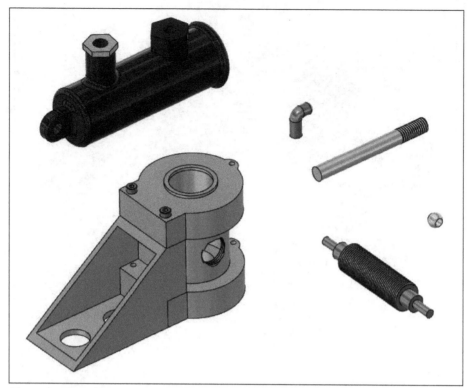

Figure-66. Exploded view of assembly

The steps to assemble these parts are given next.

Adding file into Fusion 360

- After downloading the resource kit from **CADCAMCAEWORKS.COM**, open the **Chapter 7** folder.
- Double-click on **F1 brake balance shifting mechanism.f3d** file. The file will open in **Autodesk Fusion 360** software; refer to Figure-67.

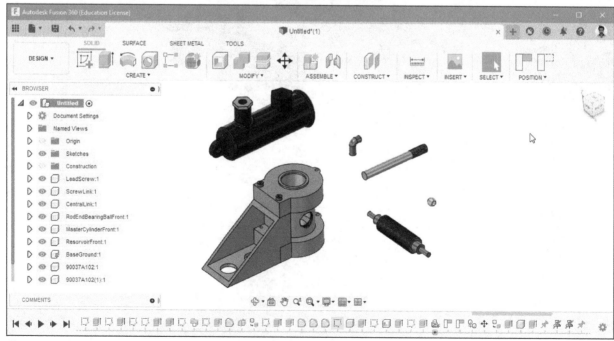

Figure-67. Fusion 360 startup window

Creating New Components

- Select **New Component** tool from **ASSEMBLE** drop-down. The **NEW COMPONENT** dialog box will be displayed.
- Select **From Bodies** check box in the dialog box. You are asked to select the bodies.
- The **Select** button of **Bodies** section is active by default. Select the bodies which you want to convert into components; refer to Figure-68.

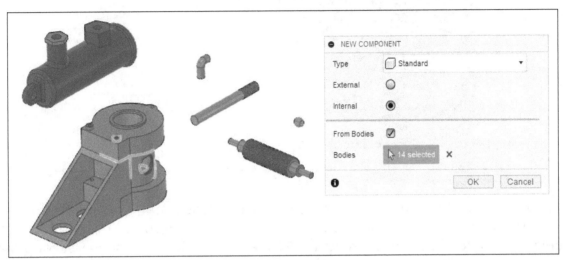

Figure-68. Selecting the bodies to create component

- Click on **OK** button from the dialog box to complete the process.

Assembling the Components

- Select **Joint** tool from **ASSEMBLE** drop-down. The **JOINT** dialog box will be displayed.
- Select **Between Two Faces** option from **Mode** section of **Component 1** area in **Position** tab of the dialog box. You are asked to select the first face.
- The **Select** button of **Plane 1** section is active by default. Select first face of the component as shown in Figure-69. You are asked to select the second face.

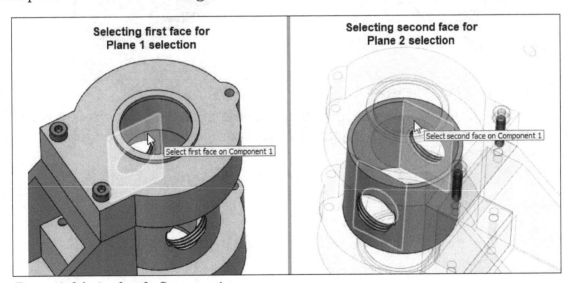

Figure-69. Selecting faces for Component 1

- The **Select** button of **Plane 2** section will become active. Select second face of component as shown in Figure-69. You are asked to select a snap point.
- The **Select** button of **Snap** section will become active. Select the snap point on the component as shown in Figure-70.

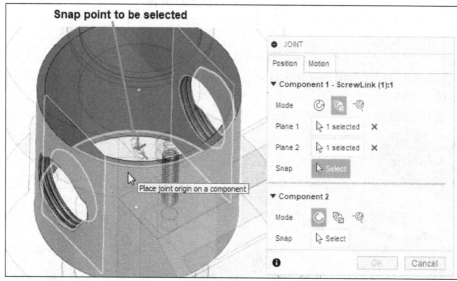

Figure-70. Selecting snap point for Component 1

- Similarly, select **Between Two Faces** option from **Mode** section of **Component 2** area. You are asked to select first and second face of the component.
- Select first face of the component for **Plane 1** selection and select second face of the component for **Plane 2** selection as shown in the Figure-71. You are asked to select the snap point on the component.

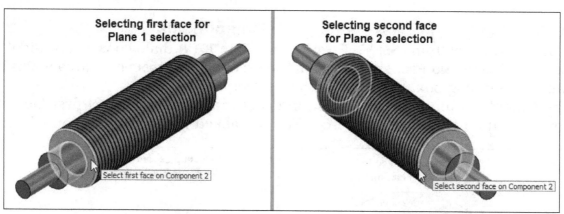

Figure-71. Selecting faces for Component 2

- The **Select** button of **Snap** section will become active. Select the snap point on the component as shown in Figure-72. The components will be joined together; refer to Figure-73.

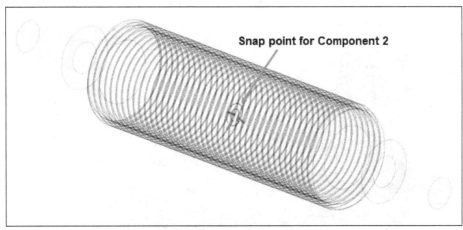

Figure-72. Selecting snap point for Component 2

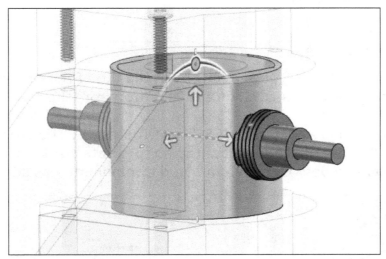

Figure-73. Components joined

- Select **Revolute** option from **Type** section in **Motion** tab of the **JOINT** dialog box.
- After specifying parameters, click on the **OK** button from the dialog box to complete the process.
- Now, select **Joint** tool from **ASSEMBLE** drop-down. The **JOINT** dialog box will be displayed. You are asked to select the snap point on the components.
- Select snap point on both Component 1 and Component 2 as shown in Figure-74.

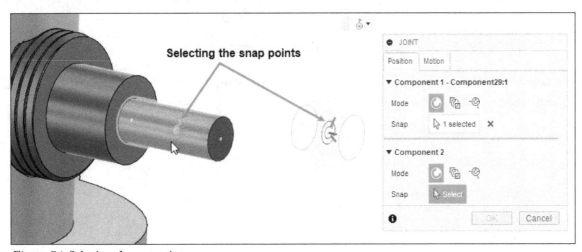

Figure-74. Selecting the snap points

- On selecting the snap point on second component, both the components will be joined; refer to Figure-75.

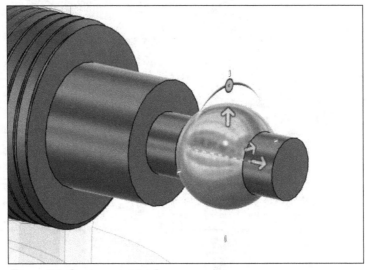

Figure-75. Components joined

- Select **Rigid** option from **Type** section in **Motion** tab of the dialog box.
- After specifying parameters, click on the **OK** button from the dialog box to complete the process.
- Now, select **Joint** tool again from **ASSEMBLE** drop-down. The **Joint** dialog box will be displayed. You are asked to select the snap points on the components.
- Select snap point on both Component 1 and Component 2 as shown in Figure-76.

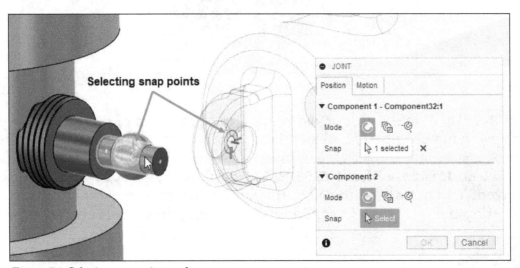

Figure-76. Selecting snap point on the components

- On selecting the snap point on second component, both the components will be joined; refer to Figure-77.

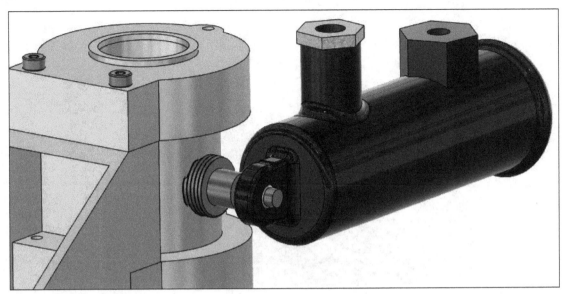

Figure-77. Components joined

- Select **Revolute** option from **Type** section in **Motion** tab of the dialog box.
- After specifying the parameters, click on **OK** button from the dialog box to complete the process.
- Now, select the **Joint** tool from **ASSEMBLE** drop-down. The **JOINT** dialog box will be displayed. You are asked to select the snap point on the components.
- Select snap point on both Component 1 and Component 2 as shown in Figure-78.

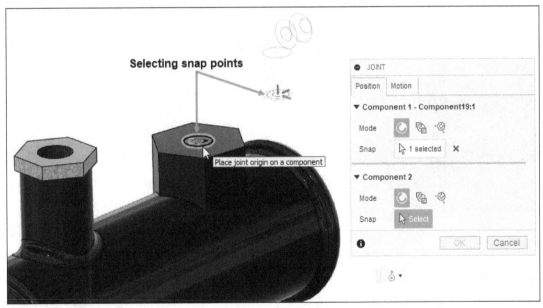

Figure-78. Selecting snap point on the components

- On selecting the snap point on second component, both the components will be joined; refer to Figure-79.

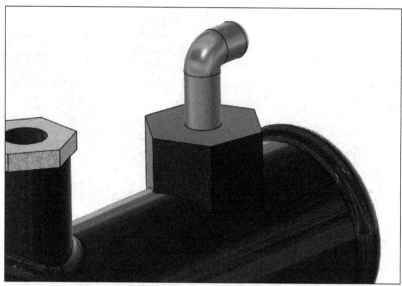

Figure-79. Components joined

- Select **Rigid** option from **Type** section in **Motion** tab of the dialog box.
- After specifying the parameters, click on **OK** button from the dialog box to complete the process.
- Click on **Joint** tool again from **ASSEMBLE** drop-down. The **JOINT** dialog box will be displayed. You are asked to select the snap point on the components.
- Select snap point on both Component 1 and Component 2 as shown in Figure-80.

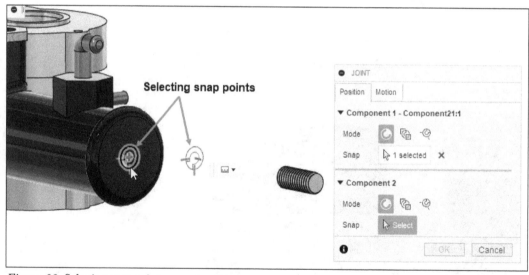

Figure-80. Selecting snap point on components

- On selecting snap point on second component, both the components will be joined; refer to Figure-81.

Figure-81. Components joined

- Select **Slider** option from **Type** section in **Motion** tab of the dialog box.
- After specifying the parameters, click on **OK** button from the dialog box to complete the process.

Copying the Components

- Select the components to be copied while holding the **CTRL** key from **BROWSER** as shown in Figure-82. Right-click on any of the selected components, a shortcut menu will be displayed. Select **Copy** option from the shortcut menu.

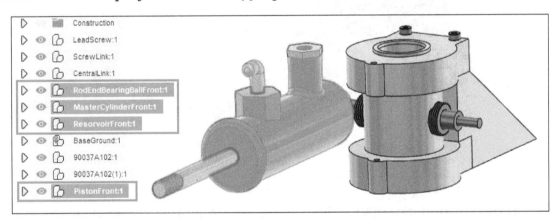

Figure-82. Selecting the components to copy

- Right-click in the empty area of the screen, a **Marking Menu** will be displayed. Select **Paste** option from the **Marking Menu**. The **MOVE/COPY** dialog box will be displayed along with the drag handle; refer to Figure-83.

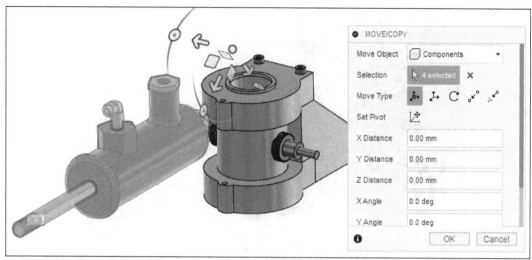

Figure-83. MOVE COPY dialog box along with drag handle

- Drag the copied components with the help of drag handle. The copied components will be displayed; refer to Figure-84.

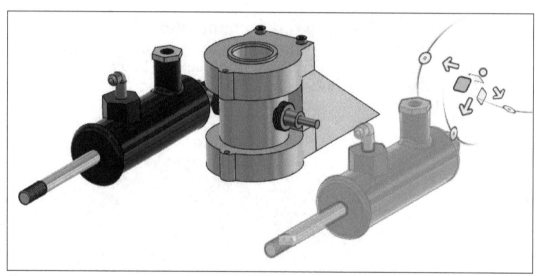

Figure-84. Components copied

- Click on **OK** button from **MOVE/COPY** dialog box to complete the process.

Assembling the Copied Components

- Select **Joint** tool from **ASSEMBLE** drop-down. The **JOINT** dialog box will be displayed. You are asked to select snap point on components.
- Select snap point on both the Component 1 and Component 2 as shown in Figure-85.

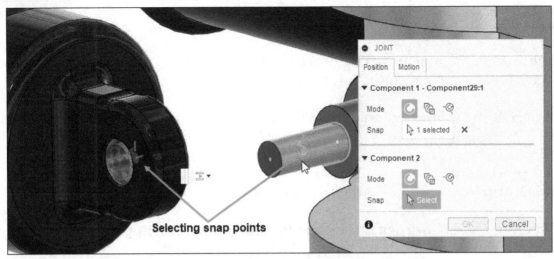

Figure-85. Selecting snap point on components

- On selecting snap point of second component, both the components will be joined.
- Select **Revolute** option from **Type** section in **Motion** tab of the dialog box.
- After specifying the parameters, click on **OK** button from the dialog box to complete the process. The final assembly of components will be displayed; refer to Figure-86.

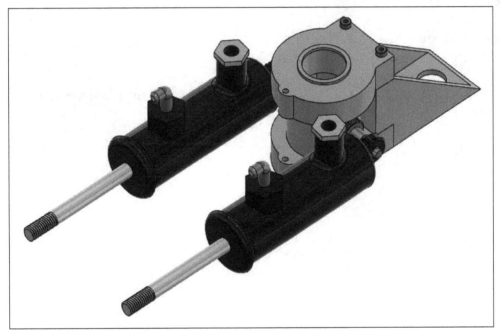

Figure-86. Final assembly of components

SELF ASSESSMENT

Q1. The .... tool is used to create a standard component or sheet metal component from the selected bodies.

Q2. Which of the following tool is used to define motion type of a component in an assembly?

a. New Component   b. Joint
c. Rigid Group   d. Drive Joints

Q3. Which of the following motion type is not available to apply joints in assembly of Autodesk Fusion 360?

a. Slider   b. Cylindrical
c. Pin-slot   d. Symmetry

Q4. Which of the following tools is used to combine multiple components to act as a single component?

a. New Component   b. Joint
c. Rigid Group   d. Drive Joints

Q5. Which of the following tools is used to define relation between motion of two components of assembly?

a. Motion Link   b. Joint
c. Rigid Group   d. Drive Joints

PRACTICE 1

Assemble the parts of motor assembly as shown in Figure-87 and Figure-88. The exploded view of assembly is shown in Figure-89. Note that you can open this practice file from the resource kit downloaded for this book.

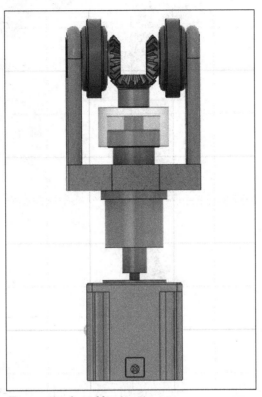

Figure-87. Assembly view 1

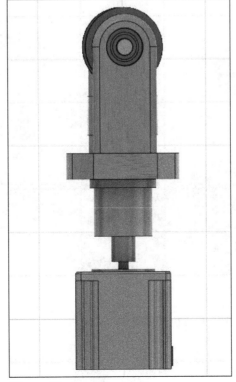

Figure-88. Assembly view 2

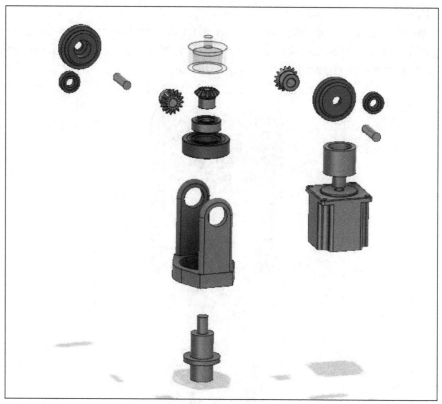

Figure-89. Exploded view of assembly

PRACTICE 2

Assemble the parts of crank assembly as shown in Figure-90. The exploded view in which you will get the model is shown in Figure-91.

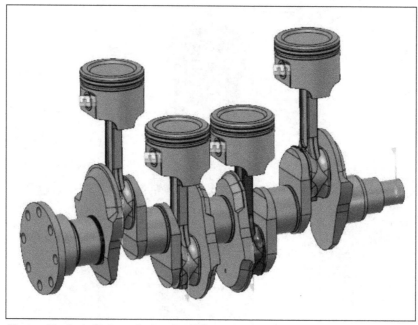

Figure-90. Assembled crank piston model

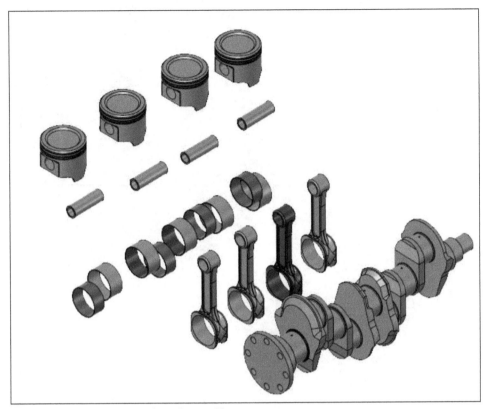

Figure-91. Exploded view of crank assembly

FOR STUDENT NOTES

Chapter 8

Importing Files and Inspection

Topics Covered

The major topics covered in this chapter are:

- *Introduction*
- *Insert Derive*
- *Decal tool*
- *Attached Canvas*
- *Insert Mesh*
- *Insert SVG*
- *Insert DXF*
- *Insert McMaster-Carr Component*
- *Inserting a Manufacturing Part*
- *Insert TraceParts Supplier Components*

- *Measure Tool*
- *Interference Tool*
- *Curvature Comb Analysis*
- *Zebra Analysis*
- *Draft Analysis*
- *Curvature Map Analysis*
- *Accessibility Analysis*
- *Minimum Radius Analysis*
- *Section Analysis*
- *Center of Mass Tool*

INTRODUCTION

In the previous chapter, we have learned about the procedure of creating and assembling component. Till now, we have worked on the components that are/have been created in Autodesk Fusion 360. Now, you will learn to insert external model objects like canvas, decal, dxf, and manufacturer parts. In this chapter, you will learn the procedures of inspecting the part and importing files in **Autodesk Fusion 360** software.

INSERTING FILE

In this section, you will learn the procedure to insert various types of files in **Autodesk Fusion 360**.

Insert Derive

The **Insert Derive** tool is used to insert components, sketches, parameters, or bodies from another design file into the current model file. Note that before using this tool, you need to save the current model file. The procedure to use this tool is discussed next.

- Click on the **Insert Derive** tool of **INSERT** drop-down from **Toolbar**; refer to Figure-1. The **Select Source** dialog box will be displayed; refer to Figure-2.

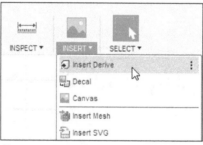

Figure-1. Insert Derive tool

Figure-2. Select Source dialog box

- Select desired source component and click on the **Select** button from **Select Source** dialog box. The selected component will be inserted in the canvas along with the **DERIVE** dialog box; refer to Figure-3.

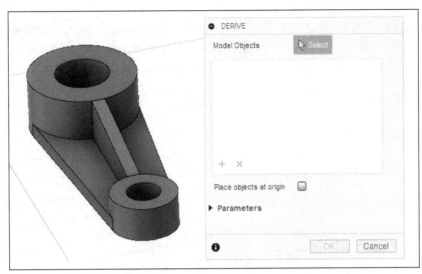

Figure-3. DERIVE dialog box

- The **Select** button of **Model Objects** section is active by default. Click on the inserted component to select the model. The updated **DERIVE** dialog box will be displayed; refer to Figure-4. You can select the sketch and other objects from **BROWSER**.

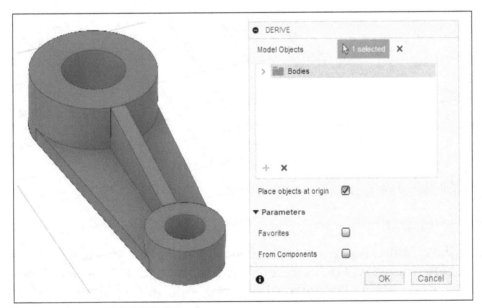

Figure-4. Updated DERIVE dialog box

- Select the **Place objects at origin** check box to place the component origin aligned to model origin.
- Expand the **Parameters** node and select desired check boxes to include parameters in desired part.
- After specifying all the parameters, click on **OK** button from the **DERIVE** dialog box to complete the process.

Decal

The **Decal** tool is used to place an image on the selected face/plane. The procedure to use this tool is discussed next.

- Click on the **Decal** tool of **INSERT** drop-down from **Toolbar**; refer to Figure-5. The **Insert** dialog box will be displayed; refer to Figure-6.

Figure-5. Decal tool

Figure-6. Insert dialog box

- Select required image from **Insert** dialog box and click on **Insert** button from the dialog box. The image will be inserted or click on **Insert from my computer** button from **Insert** dialog box. The **Open** dialog box will be displayed; refer to Figure-7.

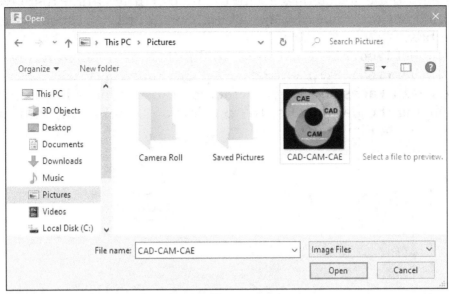

Figure-7. Open dialog box for importing file

- Select desired image which you want to import and click on **Open** button from **Open** dialog box. The **DECAL** dialog box will be displayed with the image inserted; refer to Figure-8.

Figure-8. DECAL dialog box

- The **Select** button of **Face** section is active by default. Click on the face of model to which decal is to be applied. The updated **DECAL** dialog box will be displayed; refer to Figure-9.

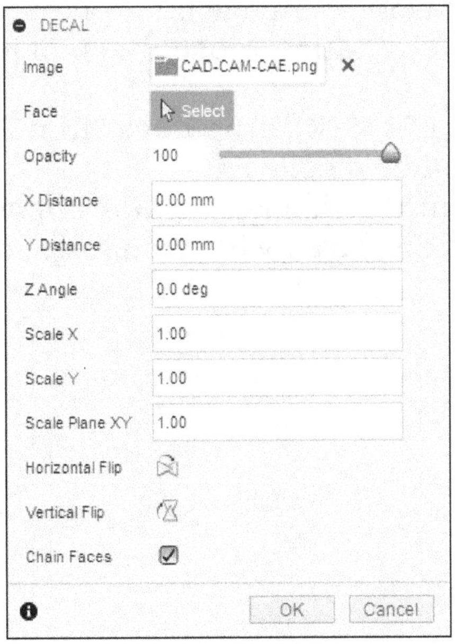

Figure-9. Updated DECAL dialog box

- Set the image on the selected face with the help of drag handle displayed on image. You can also set the image by specifying the values in respective edit boxes in the dialog box.
- Click on the **Horizontal Flip** button to flip the image horizontally.
- Click on the **Vertical Flip** button to flip the image vertically.
- Select the **Chain Faces** check box to include tangent faces in selection.
- After specifying the parameters, the preview of model along with image will be displayed; refer to Figure-10.

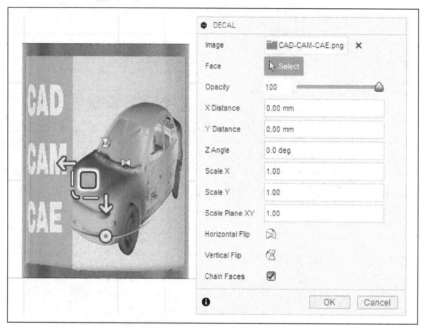

Figure-10. Preview of image on model

If the preview is as desired then click on the **OK** button from the **DECAL** dialog box to complete the process.

Canvas

The **Canvas** tool is used to insert an image on a planer face so that you can create model based on image. The procedure to use this tool is discussed next.

- Click on the **Canvas** tool of **INSERT** drop-down from **Toolbar**; refer to Figure-11. The **Insert** dialog box will be displayed.
- Select desired image from the dialog box and import as discussed earlier. The **CANVAS** dialog box will be displayed with the image inserted; refer to Figure-12.

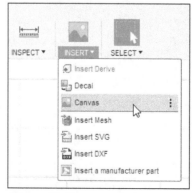

Figure-11. Canvas tool

Figure-12. CANVAS dialog box

- The **Select** button of **Face** section is active by default. Click on the face of model/ plane where you want to place the canvas image. The updated **CANVAS** dialog box will be displayed; refer to Figure-13.

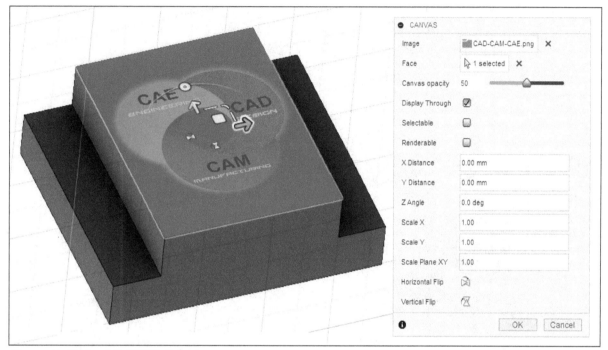

Figure-13. Updated CANVAS dialog box

- Click on **Canvas opacity** edit box and specify the opacity of image. You can also set the opacity by moving the **Canvas opacity** slider.
- Select the **Display Through** check box if you want to make the image visible through the component.
- Select the **Selectable** check box if you want the image to be selectable.
- Select the **Renderable** check box if you want to include the image in renderings of model.
- Click in the **X Distance** edit box and specify a value of distance to move in X direction.
- Click in the **Y Distance** edit box and specify a value of distance to move in Y direction
- Click in the **Z Angle** edit box and specify a value to rotation around z axis.
- Click in the **Scale X** edit box and specify a scale value in X direction.
- Click in the **Scale Y** edit box and specify a scale value in Y direction.
- Click in the **Scale Plane XY** edit box and specify the value of scale in the XY plane.
- Click on the **Horizontal Flip** button to mirror the image horizontally.
- Click on the **Vertical Flip** button to mirror the image vertically.
- After specifying the parameters, click on the **OK** button from **CANVAS** dialog box to complete the process.

Insert Mesh

The **Insert Mesh** tool is used to insert a mesh file in current model. The procedure to use this tool is discussed next.

- Click on the **Insert Mesh** tool of **INSERT** drop-down from **Toolbar**; refer to Figure-14. The **Open** dialog box will be displayed; refer to Figure-15.

Figure-14. Insert Mesh tool

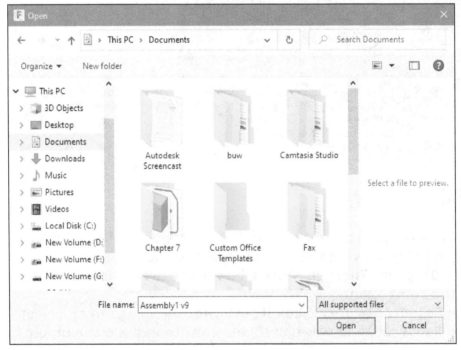

Figure-15. Open dialog box for inserting file

- Select the file and click on **Open** button of **Open** dialog box. The file will be added in application window along with **INSERT MESH** dialog box; refer to Figure-16.

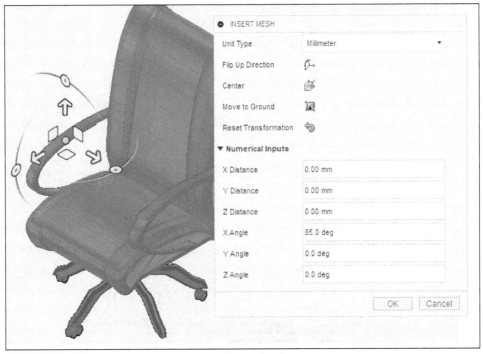

Figure-16. INSERT MESH dialog box

- Click on the **Unit Type** drop-down from **INSERT MESH** dialog box and select required unit to scale the model.
- Click on **Flip Up Direction** button to flip the direction of model.
- Click on the **Center** button to drag the model to center or coordinate system.
- Click on **Move To Ground** button to move the model to ground plane.
- Click on **Reset Transformation** button to reset the model position to the original state.
- If you want to enter the numerical value to move the model then click on the **Numerical Inputs** section and enter desired value. You can also move and rotate the model by using drag handles.
- After specifying the parameters, click on the **OK** button from **INSERT MESH** dialog box to complete the process.

Insert SVG

The **Insert SVG** tool is used to insert an SVG image file into sketch. The procedure to use this tool is discussed next.

- Click on the **Insert SVG** tool of **INSERT** drop-down from **Toolbar**; refer to Figure-17. The **Insert** dialog box will be displayed.
- Select desired file from the dialog box and insert the file as discussed earlier. The **INSERT SVG** dialog box will be displayed with the file selected; refer to Figure-18.

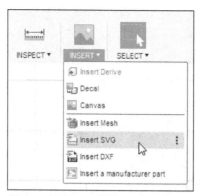

Figure-17. Insert SVG tool

Figure-18. INSERT SVG dialog box

- The **Select** button of **Sketch plane** section is active by default. You need to select the plane for adding SVG file. Click on the plane to select; refer to Figure-19.

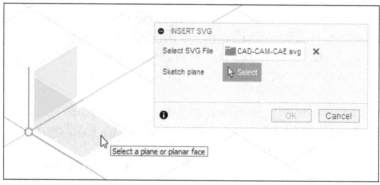

Figure-19. Selecting face or plane

- On selecting the face or plane, the sketching environment will be enabled with updated **INSERT SVG** dialog box; refer to Figure-20.
- Click in the **X Distance** edit box and specify a value of distance to move in X direction.
- Click in the **Y Distance** edit box and specify a value of distance to move in Y direction.
- Click in the **Z Angle** edit box and specify a value to rotation around z axis.
- Click in the **Scale Plane XY** edit box and specify the value of scale in the XY plane.
- Click on the **Horizontal Flip** button to mirror the image horizontally.
- Click on the **Vertical Flip** button to mirror the image vertically.
- You can also use the drag handle to move or flip the model.
- After specifying the parameters, click on the **OK** button from **INSERT SVG** dialog box to complete the process.

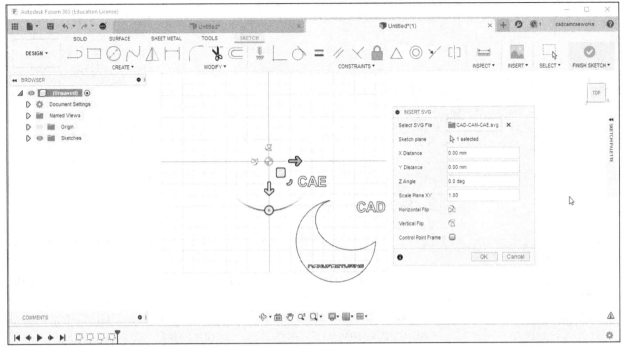

Figure-20. File along with updated INSERT SVG dialog box

Insert DXF

The **Insert DXF** tool is used to insert a DXF file into the active design. The procedure to use this tool is discussed next.

- Click on the **Insert DXF** tool of **INSERT** drop-down from **Toolbar**; refer to Figure-21. The **INSERT DXF** dialog box will be displayed; refer to Figure-22.

Figure-21. Insert DXF tool

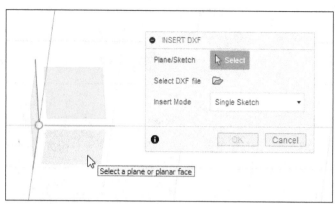

Figure-22. INSERT DXF dialog box

- The **Select** button of **Plane/Sketch** section is active by default. Click on the plane to select.
- Click on **Select DXF File** button from **INSERT DXF** dialog box. The **Open** dialog box will be displayed.
- You need to select the file of **.dxf** format and click on **Open** button. The preview of DXF file will be displayed along with updated **INSERT DXF** dialog box; refer to Figure-23.

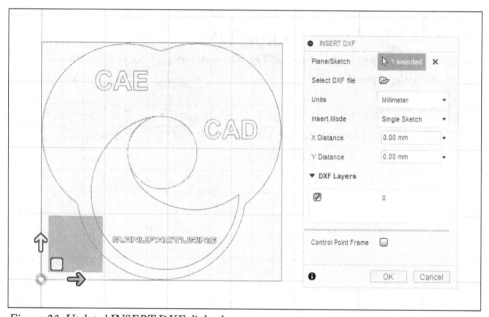

Figure-23. Updated INSERT DXF dialog box

- Click on the **Units** drop-down and select required unit.
- Select **Single Sketch** option from **Insert Mode** drop-down to insert the layers of selected .DXF file into one sketch.
- Select **One Sketch per Layer** option from **Insert Mode** drop-down to insert each sketch of DXF file on different layer.
- Click in the **X Distance** edit box and specify a value of distance to move in X direction.
- Click in the **Y Distance** edit box and specify a value of distance to move in Y direction.
- Select the **DXF Layers** check box to list the layers available in the selected DXF file. Select the layers to import.

- After specifying the parameters, click on the **OK** button from **INSERT DXF** dialog box to complete the process.

Insert McMaster-Carr Component

The **Insert McMaster-Carr Component** tool is used to insert 2D or 3D model of various library components provided by McMaster-Carr. The procedure to use this tool is given next.

- Click on the **Insert McMaster-Carr Component** tool from the **INSERT** drop-down in the **Toolbar**; refer to Figure-24. The **INSERT MCMASTER-CARR COMPONENT** window will be displayed; refer to Figure-25.

Figure-24. Insert McMaster Carr Component tool

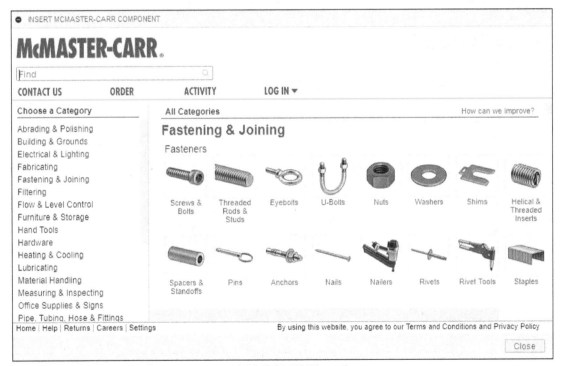

Figure-25. INSERT MCMASTER CARR COMPONENT window

- Select desired component and its model number from the window. The page in window will be displayed as shown in Figure-26.

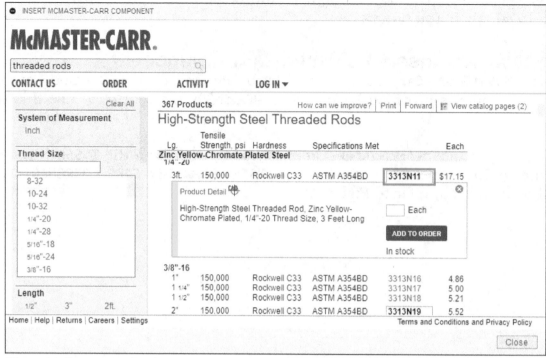

Figure-26. Component selected

- Click on the **Product Detail** link button for selected component. The drawing and model of selected component will be displayed at the bottom in the window; refer to Figure-27.

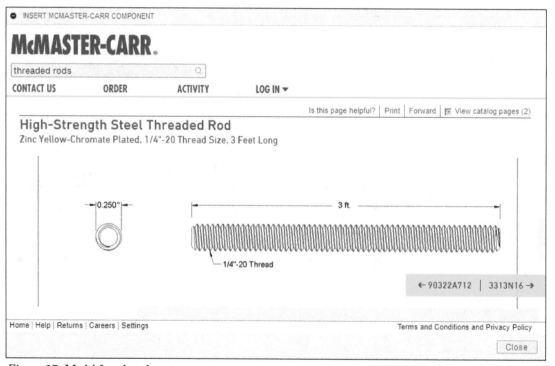

Figure-27. Model for selected component

- Select desired format from drop-down above the drawing and click on the **Save** button to save the model in local drive.
- Close the window by selecting **Close** button.

Insert a manufacturer part

The **Insert a manufacturer part** tool is used for inserting a part of manufacturer or supplier in model through parts4cad. The procedure to use this tool is discussed next.

- Click on the **Insert a manufacturer part** tool of **INSERT** drop-down from **Toolbar**; refer to Figure-28. The Part Community web page will be displayed in default web browser; refer to Figure-29.

Figure-28. Insert a manufacturer part tool

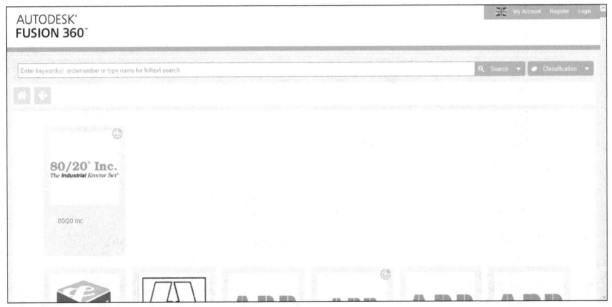

Figure-29. AUTODESK FUSION 360 part community window

- Select the part from manufacturer or supplier and open the file in Autodesk Fusion 360.

Insert TraceParts Supplier Components

The **Insert TraceParts Supplier Components** tool gives you access to millions of supplier components, 3D models, CAD files, and 2D drawings. Simply navigate to what you are looking for, sign-in with your TraceParts account, download, and insert a supplier component directly into Fusion 360 canvas. The procedure to use this tool is discussed next.

- Click on the **Insert TraceParts Supplier Components** tool of **INSERT** drop-down from **Toolbar**; refer to Figure-30. The **AUTODESK FUSION 360 DESIGN LIBRARY** window will be displayed; refer to Figure-31.

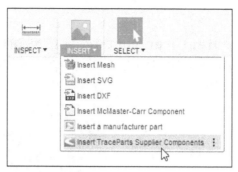

Figure-30. Insert TraceParts Supplier Components tool

Figure-31. AUTODESK FUSION 360 DESIGN LIBRARY window

- Select the part from supplier and open the file in Autodesk Fusion 360.

INSPECT

Till now, we have discussed various tools to insert the objects in Autodesk Fusion 360. Now, we will discuss about measurement and analysis tools available in Autodesk Fusion 360.

Measure

The **Measure** tool is used to measure selected geometry. The procedure to use this tool is discussed next.

- Click on the **Measure** tool of **INSPECT** drop-down from **Toolbar**; refer to Figure-32. The **MEASURE** dialog box will be displayed; refer to Figure-33.

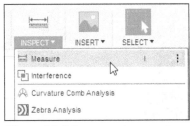

Figure-32. Measure tool

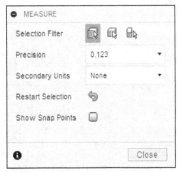

Figure-33. MEASURE dialog box

- Select the **Select Face/Edge/Vertex** button from **Selection Filter** section if you want to select face/edge/vertex for measurement.
- Select the **Select Body** button from **Selection Filter** section if you want to select body for measurement.
- Select the **Select Component** button from **Selection Filter** section if you want to select component for measurement.
- Click on the **Precision** drop-down and select desired precision option for displaying the required number of decimal places in results.
- Select desired secondary unit for the measurement from **Secondary Units** drop-down if you want to display secondary unit as well.
- Click on the **Restart Selection** button to restart the selection process once you have performed measurement.
- In our case, we are selecting **Select Face/Edge/Vertex** button from **Selection Filter** section. Click on the face to measure; refer Figure-34.

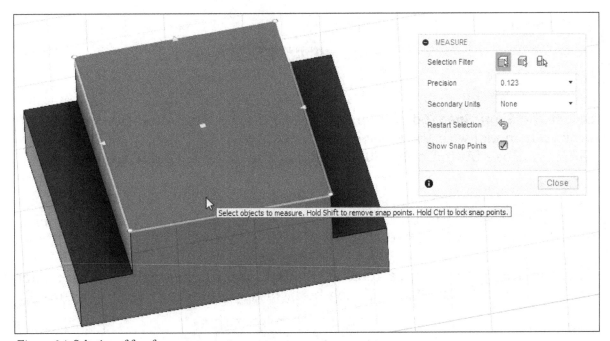

Figure-34. Selection of face for measurement

- On selecting the face, the updated **MEASURE** dialog box will be displayed with the measurement of selected face; refer to Figure-35.

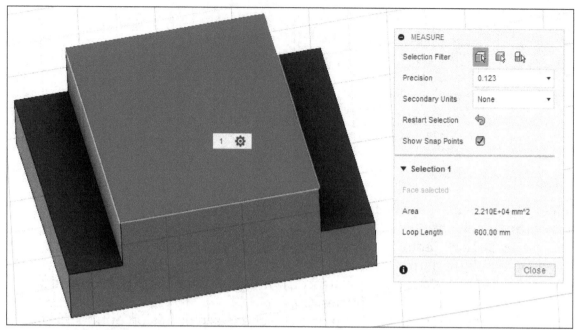

Figure-35. Updated MEASURE dialog box

- If you want to measure the distance between two faces then select the other face.
- Hover the cursor on the respective measurement to copy the value to clipboard; refer to Figure-36.

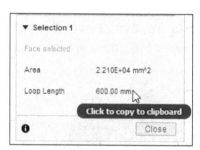

Figure-36. Copying measurement

- Select the **Show Snap points** check box to display snap points when you hover over your model with the **Measure** tool activated.
- Click on **Close** button of **MEASURE** dialog box to exit the **Measure** tool.

Interference

The **Interference** tool is used to display interference of selected components or bodies. The procedure to use this tool is discussed next.

- Click on the **Interference** tool of **INSPECT** drop-down from **Toolbar**; refer to Figure-37. The **INTERFERENCE** dialog box will be displayed; refer to Figure-38.

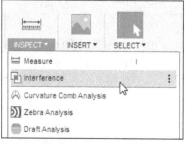

Figure-37. Interference tool

Figure-38. INTERFERENCE dialog box

- The **Select** button of **Select** section is active by default. Click on the model to check interference. You can also use window selection.
- Select the **Include Coincident Faces** check box if you want to include the faces in selection which are touching with each other.
- In our case, we are using window selection to select the component; refer to Figure-39.

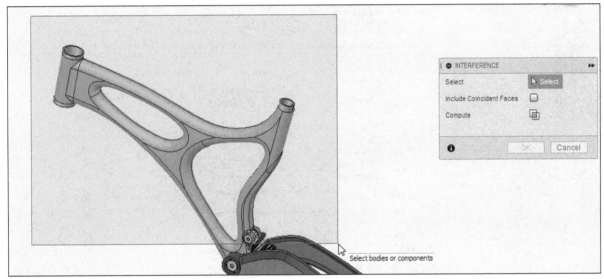

Figure-39. Selection of component for interference

- After selecting the components, click on **Compute** button to calculate the interference. The **Interference Results** dialog box will be displayed; refer to Figure-40.

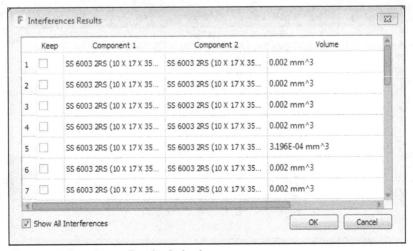

Figure-40. Interference Results dialog box

- Click on the **OK** button from **Interference Results** dialog box to exit the **Interference** tool.

Curvature Comb Analysis

The **Curvature Comb Analysis** tool is used to display a curvature comb on the spline or circular edges to determine the change in curvature. The procedure to use this tool is discussed next.

- Click on the **Curvature Comb Analysis** tool of **INSPECT** drop-down from **Toolbar**; refer to Figure-41. The **CURVATURE COMB ANALYSIS** dialog box will be displayed; refer to Figure-42.

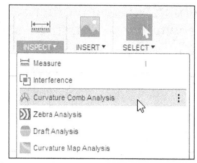

Figure-41. Curvature Comb Analysis tool

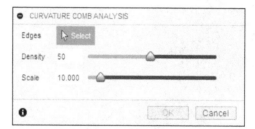

Figure-42. CURVATURE COMB ANALYSIS dialog box

- **Edges** selection of **CURVATURE COMB ANALYSIS** dialog box is active by default. Click on the geometry to select the edge. The preview of comb on edge will be displayed; refer to Figure-43.

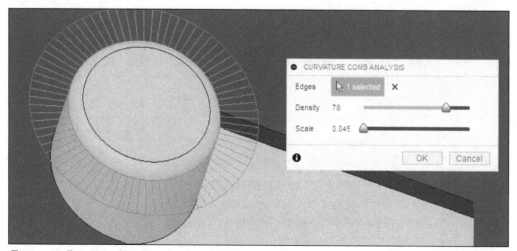

Figure-43. Preview of comb on circular edge

- Click in the **Density** edit box and specify the value. You can also set the comb density by moving the **Density** slider.
- Click in the **Scale** edit box and specify the value. You can also set the comb scale by moving the **Scale** slider.
- After analyzing, click on **OK** button from **CURVATURE COMB ANALYSIS** dialog box to exit the tool.

Zebra Analysis

The **Zebra Analysis** tool is used to analyze continuity between the surfaces of component by displaying stripes on model. The procedure to use this tool is discussed next.

- Click on the **Zebra Analysis** tool of **INSPECT** drop-down from **Toolbar**; refer to Figure-44. The **ZEBRA ANALYSIS** dialog box will be displayed; refer to Figure-45.

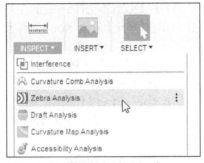

Figure-44. Zebra Analysis tool

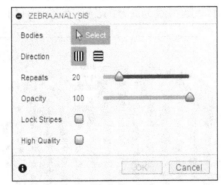

Figure-45. ZEBRA ANALYSIS dialog box

- The **Select** button of **Bodies** section is active by default. Click to select the body for analysis.
- On selecting the body, preview of zebra stripes will be displayed; refer to Figure-46.

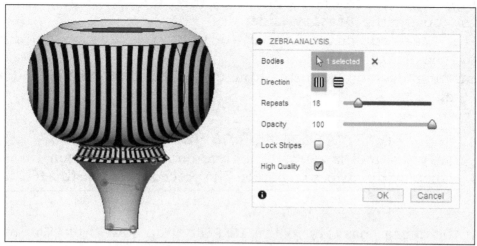

Figure-46. Selection of body for zebra analysis

- Select **Vertical** option from **Direction** section for display of vertical lines on model.
- Select **Horizontal** option from **Direction** section for display of horizontal lines on model.
- Set the number of stripes for model from **Repeats** slider.
- Click on the **Opacity** edit box and enter the value for opacity. You can also set the value of opacity by moving the **Opacity** slider.
- Select the **Lock Stripes** check box of **ZEBRA ANALYSIS** dialog box to freeze the stripes on body.
- Select the **High Quality** check box of **ZEBRA ANALYSIS** dialog box to increase the display quality of stripes.
- After specifying the parameters, click on **OK** button from **ZEBRA ANALYSIS** dialog box. The stripes will be displayed on model; refer to Figure-47.

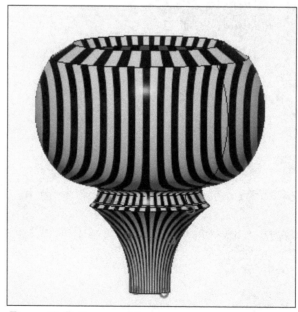

Figure-47. Strips on model

Draft Analysis

The **Draft Analysis** tool is used to display a color gradient on the selected body to evaluate draft angle of various faces of the part. The procedure to use this tool is discussed next.

- Click on the **Draft Analysis** tool of **INSPECT** drop-down from **Toolbar**; refer to Figure-48. The **DRAFT ANALYSIS** dialog box will be displayed; refer to Figure-49.

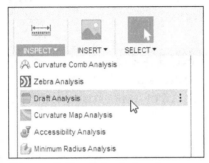

Figure-48. Draft Analysis tool

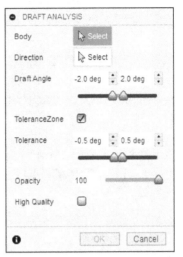

Figure-49. DRAFT ANALYSIS dialog box

- The **Select** button of **Body** section is active by default. Click on the body to perform analysis.
- Click on **Select** button of **Direction** section and select the direction for analysis. On selection, the preview will be displayed; refer to Figure-50.

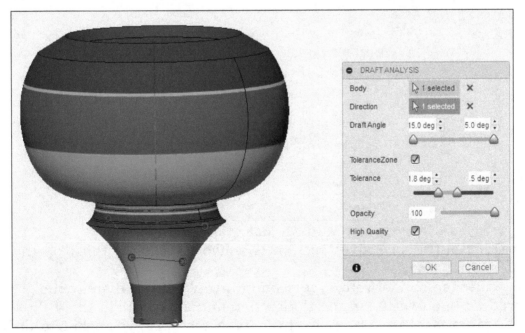

Figure-50. Preview of draft analysis

- Click in **Draft Angle** edit boxes and specify the range of draft angle for which you want to perform analysis. You can also set the draft angle limits by moving the **Draft angle** slider.
- Select the **ToleranceZone** check box if you want to display a tolerance gradient on the results.

- Click on the **Tolerance** edit box and specify the start and stop angles tolerance. You can also set the tolerance angle by moving the slider.
- Click on the **Opacity** edit box and enter the value for opacity for draft colors. You can also set the value of opacity by moving the **Opacity** slider.
- Select the **High Quality** check box of **DRAFT ANALYSIS** dialog box to increase the display quality of gradient.
- After specifying the parameters for analysis, click on **OK** button from **DRAFT ANALYSIS** dialog box to complete the analysis.

Curvature Map Analysis

The **Curvature Map Analysis** tool is used to displays a color gradient on the faces of selected bodies to evaluate the amount of curvature. The procedure to use this tool is discussed next.

- Click on the **Curvature Map Analysis** tool of **INSPECT** drop-down from **Toolbar**; refer to Figure-51. The **CURVATURE MAP ANALYSIS** dialog box will be displayed; refer to Figure-52.

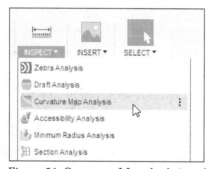

Figure-51. Curvature Map Analysis tool

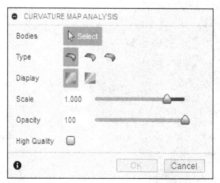

Figure-52. CURVATURE MAP ANALYSIS dialog box

- The **Select** button of **Bodies** section is active by default. Click on the body to select for analysis.
- Select **Gaussian** option from **Type** section if you want to display the gradient as the product of curvature in the U direction and curvature in the V direction.
- Select **Principal Minimum** option from **Type** section if you want to display the gradient highlighting areas of lowest curvature.
- Select **Principal Maximum** option from **Type** section if you want to display the gradient highlighting areas with the highest curvature.
- Select **Smooth** option from **Display** section if you want to display a blend of two colors.

- Select **Bands** option from **Display** section if you want to display no transition between colors.
- Click in the **Scale** edit box of **CURVATURE MAP ANALYSIS** dialog box and enter the value. You can also set the value of scale by moving the **Scale** slider.
- Click on the **Opacity** edit box and enter the value for opacity. You can also set the value of opacity by moving the **Opacity** slider.
- Select the **High Quality** check box of **CURVATURE MAP ANALYSIS** dialog box to increase the display quality of gradient.
- After specifying the parameters for analysis, click on **OK** button from **CURVATURE MAP ANALYSIS** dialog box to exit this analysis; refer to Figure-53.

Figure-53. Curvature map analysis of model

Accessibility Analysis

The **Accessibility Analysis** tool is used to determine whether the area of the model is accessible or inaccessible through selected view of the plane. This analysis is used to perform undercut analysis. The procedure to use this tool is discussed next.

- Click on the **Accessibility Analysis** tool of **INSPECT** drop-down from **Toolbar**; refer to Figure-54. The **ACCESSIBILITY ANALYSIS** dialog box will be displayed; refer to Figure-55.

Figure-54. Accessibility Analysis tool

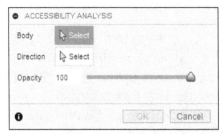

Figure-55. ACCESSIBILITY ANALYSIS dialog box

- The **Select** button of **Body** section is active by default. Click on the face of the body to select; refer to Figure-56. You will be asked to select a direction.

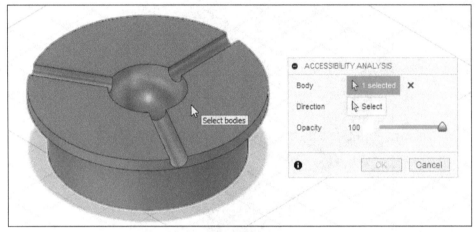

Figure-56. Selecting the body

- Click on the **Select** button of **Direction** section and select the edge, plane, or axis of the model to indicate the direction of analysis. The preview of the accessible areas will be displayed along with updated **ACCESSIBILITY ANALYSIS** dialog box; refer to Figure-57.

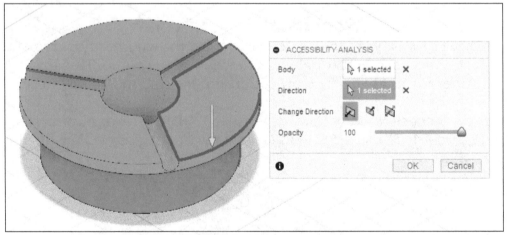

Figure-57. Updated ACCESSIBILITY ANALYSIS dialog box

- On selecting the direction, the areas that are accessible from selected direction are shaded in green and the areas that are inaccessible are shaded in red.
- Select desired option from **Change Direction** section to change the direction of analysis.
- You can also set the opacity by sliding the slider button or edit the opacity in edit box of **Opacity** section.

- The visibility of this analysis can be controlled by expanding the **Analysis** folder in the **BROWSER** and clicking on the eye button to view or hide the analysis; refer to Figure-58.

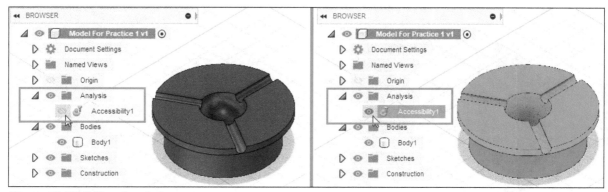

Figure-58. Visibility of analysis

- After specifying the parameters for analysis, click on **OK** button from **ACCESSIBILITY ANALYSIS** dialog box to exit this analysis.

Minimum Radius Analysis

The **Minimum Radius Analysis** tool is used to highlight the areas of the model which are at or below the minimum radius. It also detects the minimum fit for fillets and holes.

- Click on the **Minimum Radius Analysis** tool of **INSPECT** drop-down from **Toolbar**; refer to Figure-59. The **MINIMUM RADIUS ANALYSIS** dialog box will be displayed; refer to Figure-60.

Figure-59. Minimum Radius Analysis tool

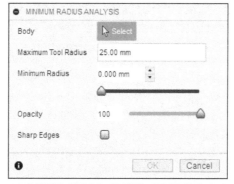

Figure-60. MINIMUM RADIUS ANALYSIS dialog box

- The **Select** button of **Body** section is active by default. Click on the body to select the face. On selecting desired face, the areas of curvature of the model will be highlighted; refer to Figure-61.

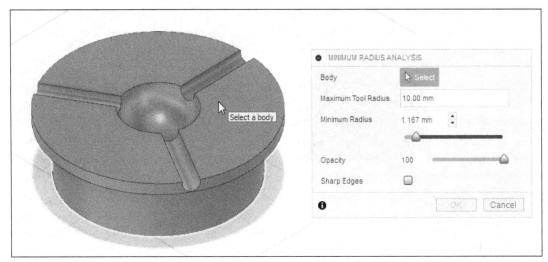

Figure-61. Selection of face of the body

- After selecting the body, set the **Maximum Tool Radius** in the edit box for analysis of the model.
- For tool of **Minimum Radius** 0, the selected parts are all green; refer to Figure-62.

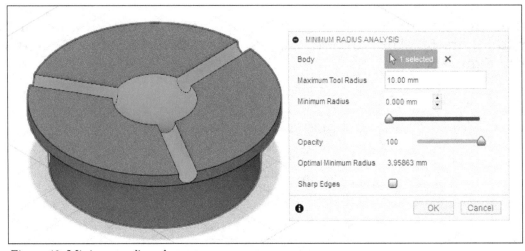

Figure-62. Minimum radius color

- As you increase the minimum radius of the tool, some parts are shaded with red color which indicates that these parts have high curvature to be correctly manufactured by the tool of that radius; refer to Figure-63.

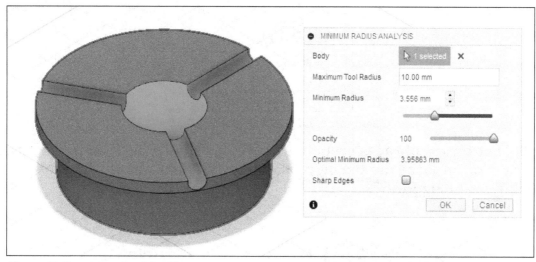

Figure-63. Increasing the tool radius

- Set the opacity of the analysis as you did in the **Accessibility Analysis** tool.
- The visibility of this analysis can be control by expanding the **Analysis** folder in the **BROWSER** and clicking on the eye button to view or hide the analysis; refer to Figure-64.

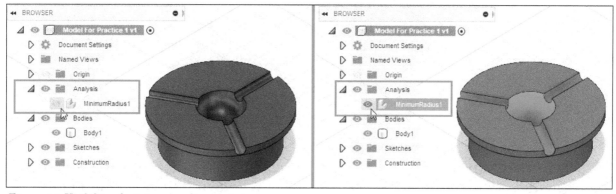

Figure-64. Visibility of minimum radius analysis

- After specifying all the parameters, click on the **OK** button of **MINIMUM RADIUS ANALYSIS** dialog box to exit this analysis.

Section Analysis

The **Section Analysis** tool is used to cut the model at selected plane reference and check the temporary view. The procedure to use this tool is discussed next.

- Click on the **Section Analysis** tool of **INSPECT** drop-down from **Toolbar**; refer to Figure-65. The **SECTION ANALYSIS** dialog box will be displayed; refer to Figure-66.

Figure-65. Section Analysis tool

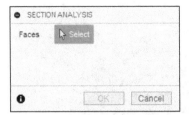

Figure-66. SECTION ANALYSIS dialog box

- The **Select** button of **Faces** section is active by default. Click on the face to select.
- On selection of face, the dialog box will be updated along with the preview of model; refer to Figure-67.

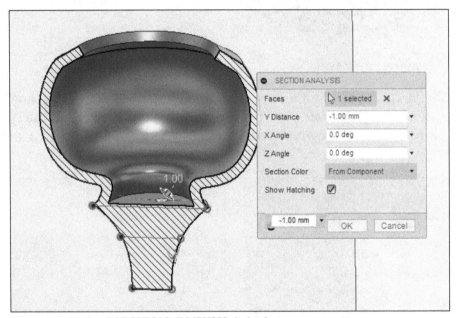

Figure-67. Updated SECTION ANALYSIS dialog box

- Click in the **Y Distance** edit box and specify the location of the section plane along y direction. You can also set the value by moving the drag handle.
- Click in the **X Angle** edit box and specify the angle of plane about x direction. You can also set the value by moving the drag handle.
- Click in the **Z Angle** edit box and specify the angle of plane about y direction. You can also set the value by moving the drag handle.
- Select **From Component** option from **Section Color** drop-down if you want to use the same color as applied to the model.
- Select **Custom** option from **Section Color** drop-down if you want to use the custom color for the section.
- Select the **Show Hatching** check box if you want to show hatching on the section.
- After specifying the parameters for analysis, click on **OK** button from **SECTION ANALYSIS** dialog box to complete the analysis.

Center of Mass

The **Center of Mass** tool is a center point of body used for representing the mean position of matter in a body. The procedure to use this tool is discussed next.

- Click on the **Center of Mass** tool of **INSPECT** drop-down from **Toolbar**; refer to Figure-68. The **CENTER OF MASS** dialog box will be displayed; refer to Figure-69.

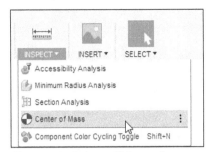

Figure-68. Center of Mass tool

Figure-69. CENTER OF MASS dialog box

• You need to select the body and click on **OK** button from **CENTER OF MASS** dialog box. The center of mass of the body will be displayed; refer to Figure-70.

Figure-70. Center of mass of the body

Component Color Cycling Toggle Tool

The **Component Color Cycling Toggle** tool is used to toggle color of various objects in Autodesk Fusion. You can also press **SHIFT+N** to do the same thing.

SELF ASSESSMENT

Q1. Which of the following tools is used for inserting image on the face of a model?

a) Insert Mesh
b) Insert Derive
c) Decal
d) Insert SVG

Q2. What is the use of **Reset Transformation** button in **INSERT MESH** dialog box?

Q3. What is the use of **Insert Mc-Master Carr Component** tool and what is Mc-Master Carr?

Q4. Write down the procedure of inserting manufacturer or supplier part in the model.

Q5. Write down the procedure to measure the distance between two faces.

Q6. Write down the procedure to find interference between two components.

Q7. The **Curvature Comb Analysis** tool is used to display a curvature comb on the spline or circular edges to determine the ....

Q8. The **Zebra Analysis** tool is used to analyze continuity between the surfaces of component by ....on model.

Q9. Write down the procedure to display zebra stripes on body.

10. The **Draft Analysis** tool is used to display a color gradient on the selected body to .... of various faces of the part.

11. The .... tool is used to displays a color gradient on the faces of selected bodies to evaluate the amount of curvature.

12. What is the use of **Gaussian** option in **Curvature Map Analysis**?

13. The **Section Analysis** tool is used to define the view of component from .... side.

14. Which of the following analyses uses **ToleranceZone** option to display a tolerance gradient on the results?

a) Section Analysis
b) Zebra Analysis
c) Curvature Comb Analysis
d) Draft Analysis

Chapter 9

Surface Modeling

Topics Covered

The major topics covered in this chapter are:

- *Introduction to Surface Designing*
- *Extrude Tool*
- *Patch Tool*
- *Surface Trim and Untrim Tool*
- *Extend Tool*
- *Stitch and Unstitch Tool*
- *Reverse Normal Tool*
- *Practical*
- *Practice*

INTRODUCTION

Surface Designing or surfacing is a technique used to create complex shapes. The most general application of surfacing can be seen in the object that are made with good aerodynamics like car body, aeroplanes, ships, and so on. Sometimes, surface modeling is also used to add complex geometry to the models. The models created by surfacing are called surface models and they do not exist in real world because they have zero thickness. After converting these surfaces to solids, we make real objects.

In Autodesk Fusion 360, there is no separate workspace for creating surface designs. We have all the surfacing tools available in the **SURFACE** tab of **DESIGN** workspace; refer to Figure-1.

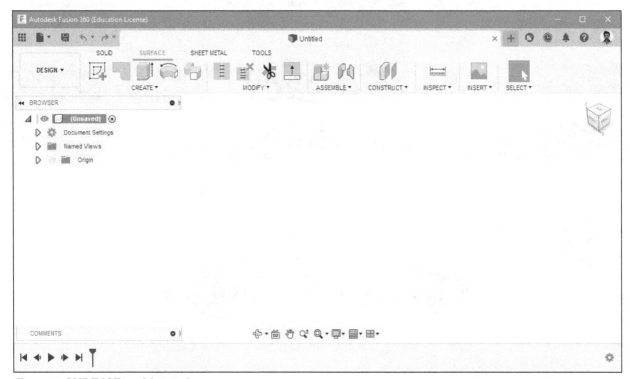

Figure-1. SURFACE modeling tools

Note that we have already used many tools of surface design earlier while creating solid models. So, now we will use the surface modeling tools similar to solid modeling.

Extrude

The **Extrude** tool can also be used to create the extruded surfaces. The procedure to use this tool is discussed next.

*   Click on the **Extrude** tool of **CREATE** drop-down from **Toolbar** in **SURFACE** tab; refer to Figure-2. The **EXTRUDE** dialog box will be displayed; refer to Figure-3.

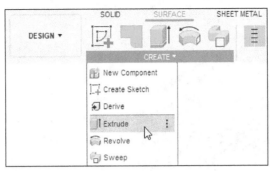

Figure-2. Extrude tool

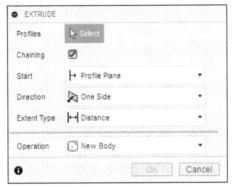

Figure-3. EXTRUDE dialog box

- The **Select** button of **Profiles** section is active by default. Click on the profile to select. The updated **EXTRUDE** dialog box will be displayed; refer to Figure-4.

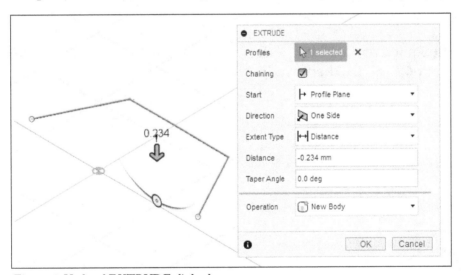

Figure-4. Updated EXTRUDE dialog box

Note that you can select open as well as closed sketch for profile selection to extrude.

- Select the **Chaining** check box if you want to select all the sketch entities in chain.
- The other options of **EXTRUDE** dialog box are same as discussed earlier for solid modeling.
- After specifying the parameters, preview of extrude will be displayed; refer to Figure-5.

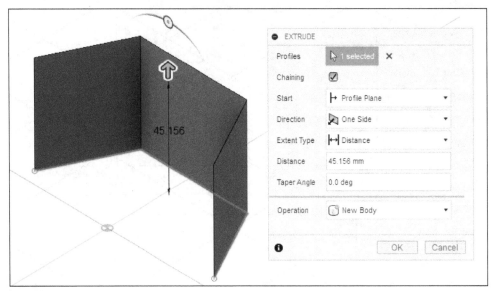

Figure-5. Preview of extrude

- If the preview of extrude is as required then click on **OK** button from **EXTRUDE** dialog box. The extruded surface will be created.

You can use the **Revolve**, **Sweep**, and **Loft** tools in the same way as discussed in previous chapters.

Patch

The **Patch** tool is used to create a surface by connecting edges/curves. The procedure to use this tool is discussed next.

- Click on the **Patch** tool of **CREATE** drop-down from **Toolbar**; refer to Figure-6. The **PATCH** dialog box will be displayed; refer to Figure-7.

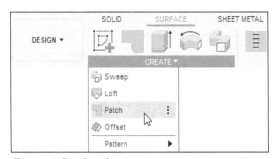

Figure-6. Patch tool

Figure-7. PATCH dialog box

- The **Select** button of **Boundary Edges** section is active by default. Click on the edges/curves forming close loop.
- Select the **Group Edges** check box if you want to select all the edges in a group.
- Select the **Enable Chaining** check box if you want to select all the adjacent edges when clicking on one edge.
- Select **Connected** option from **Continuity** drop-down if you want to create a surface with G0 edges (connected at an angle).
- Select **Tangent** option from **Continuity** drop-down if you want to create a surface with G1 edges (tangential).
- Select **Curvature** option from **Continuity** drop-down, if you want to create a surface with G2 edges (blended with continuous curvature).
- Click on the **Select** button for **Interior Rails/Points** selection and select the curves/points to be used as guide rails/points.
- After specifying the parameters, the preview of patch will be displayed; refer to Figure-8.

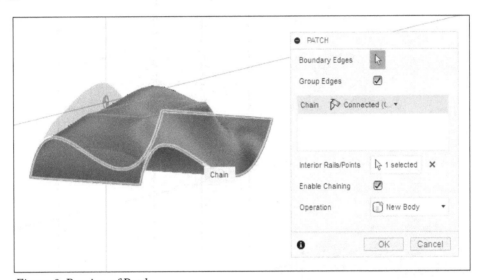

Figure-8. Preview of Patch

- Click on **OK** button from **PATCH** dialog box to complete the process.

Other Solid Modeling Tools For Surfacing

The other tools in the **CREATE** drop-down can be used in the same way as discussed in **Chapter 4: Advanced 3D Modeling** of the book. You can now create surfaces by using **Offset**, **Pattern**, **Mirror**, **Thicken**, and **Boundary Fill** tools by following the procedure discussed earlier.

EDITING TOOLS

In the previous section, we have learned about different surface creation tools used in surface modeling. In this section, we will discuss the surface editing tools.

Trim

The **Trim** tool is used to trim the surface using selected trimming object as cutting tool. The procedure to use this tool is discussed next.

- Click on the **Trim** tool of **MODIFY** drop-down from **Toolbar**; refer to Figure-9. The **TRIM** dialog box will be displayed; refer to Figure-10.

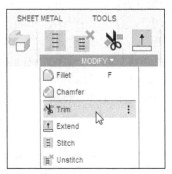

Figure-9. Trim tool

Figure-10. TRIM dialog box

- The **Select** button of **Trim Tool** section is active by default. You need to select an existing sketch to trim the surface. You can also select a plane/face intersecting with the surface as trim tool; refer to Figure-11.

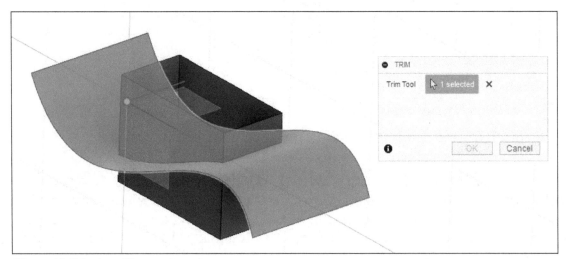

Figure-11. Selection of face as trim tool

- Now, you need to select the area to be removed. Click on desired area of sketch to trim; refer to Figure-12.

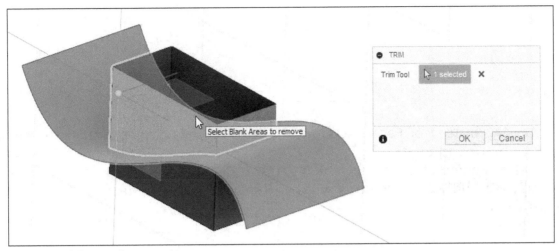

Figure-12. Selection of area to trim

- Click on the **OK** button from **TRIM** dialog box to trim the selected area; refer to Figure-13.

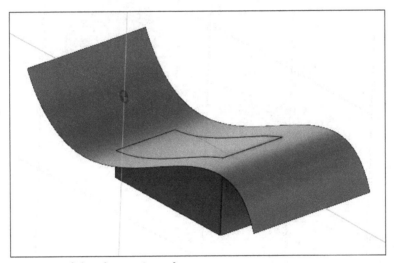

Figure-13. Selected area trimmed

Untrim

The **Untrim** tool is used to extend trimmed surfaces and fill gaps, holes, or empty regions of a surface. The procedure to use this tool is discussed next.

- Click on the **Untrim** tool of **MODIFY** drop-down from **Toolbar**; refer to Figure-14. The **UNTRIM** dialog box will be displayed; refer to Figure-15.

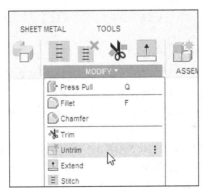

Figure-14. Untrim tool

Figure-15. UNTRIM dialog box

- The **Select** button of **Faces** section is active by default. You need to select the faces to be untrimmed.
- Select desired face of the model which you want to be untrimmed; refer to Figure-16. The updated **UNTRIM** dialog box will be displayed with the face untrimmed; refer to Figure-17.

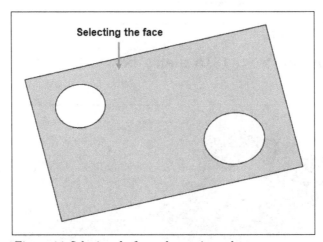

Figure-16. Selecting the face to be untrimmed

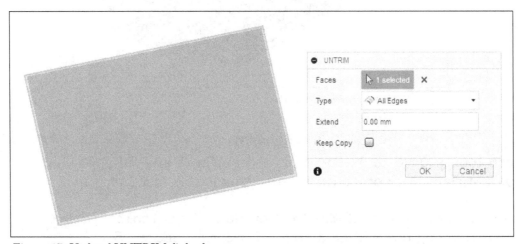

Figure-17. Updated UNTRIM dialog box

- Select **All Edges** option from **Type** drop-down to untrim all internal and external edges of a surface.
- Select **Internal Edges** option from **Type** drop-down to untrim internal edges only.
- Select **External Edges** option from **Type** drop-down to untrim external edges only.
- Select **Manual** option from **Type** drop-down to untrim manually selected edges of a face. Hold **CTRL** button to suppress command preview after selecting first edge and then select multiple edges.
- Enter desired value in **Extend** edit box to extend untrimmed surface.
- Select **Keep Copy** check box to create a copy of original trimmed surface.
- After specifying parameters, click on **OK** button from **UNTRIM** dialog box.

Extend

The **Extend** tool is used to extend the surface edge. The procedure to use this tool is discussed next.

- Click on the **Extend** tool of **MODIFY** drop-down from **Toolbar**; refer to Figure-18. The **EXTEND** dialog box will be displayed; refer to Figure-19.

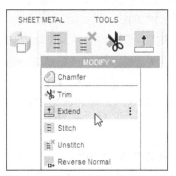

Figure-18. Extend tool

Figure-19. EXTEND dialog box

- The **Select** button of **Edges** section is active by default. You need to select the edge to extend. The updated **EXTEND** dialog box will be displayed; refer to Figure-20.

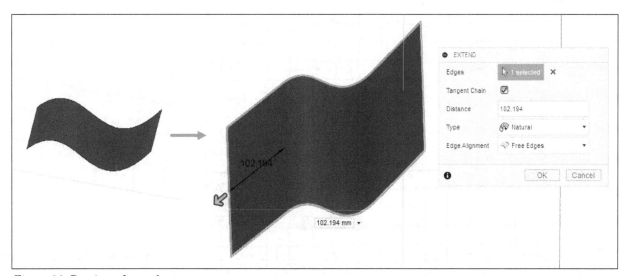

Figure-20. Preview of extend

- Select the **Target Chain** check box if you want to select all tangent or adjacent faces on selection of single face.
- Click in the **Distance** edit box and specify the distance for extend. You can also set the distance by moving the manipulator.
- Select **Natural** option from **Type** drop-down if you want to extend the surface naturally.

- Select **Perpendicular** option from **Type** drop-down if you want to extend the selected edge perpendicularly.
- Select **Tangent** option from **Type** drop-down if you want to extend the selected edge tangentially.
- Select **Free Edges** option from **Edge Alignment** drop-down to apply G0 point continuity to align edges of extended surface.
- Select **Align Edges** option from **Edge Alignment** drop-down to apply G1 tangent continuity to align edges of extended surface.
- After specifying the parameters, click on **OK** button from **EXTEND** dialog box.

Stitch

The **Stitch** tool is used to combine the surface into one body. The procedure to use this tool is discussed next.

- Click on the **Stitch** tool of **MODIFY** drop-down from **Toolbar**; refer to Figure-21. The **STITCH** dialog box will be displayed; refer to Figure-22.

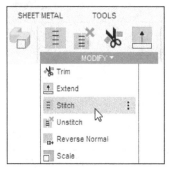

Figure-21. Stitch tool

Figure-22. STITCH dialog box

- The **Select** button of **Stitch Surfaces** section is active by default. You need to select the surfaces of model to be stitched; refer to Figure-23.

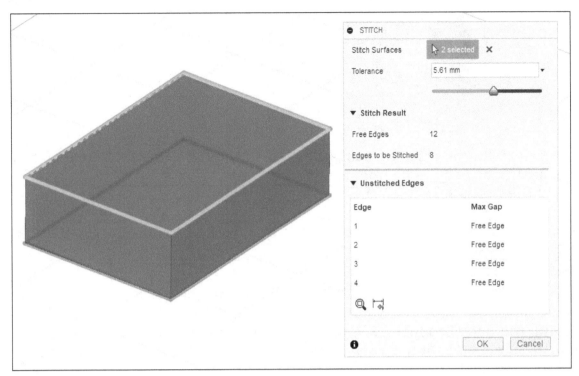

Figure-23. Selection of face to stitch

- The green line on the model shows the stitched surface and red line shows the loose surface connection.
- You need to increase the tolerance from **Tolerance** edit box or **Tolerance** slider to stitch the remaining surfaces which have higher gap between consecutive edges.
- You can also check the result of stitched and unstitched edges from **Stitch Result** section.
- After specifying the parameters, click on the **OK** button from **STITCH** dialog box to complete the process.

Unstitch

The **Unstitch** tool is used to unstitch the earlier stitched surface from model. The procedure to use this tool is discussed next.

- Click on the **Unstitch** tool of **MODIFY** drop-down from **Toolbar**; refer to Figure-24. The **UNSTITCH** dialog box will be displayed; refer to Figure-25.

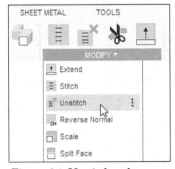

Figure-24. Unstitch tool

Figure-25. UNSTITCH dialog box

- The **Select** button of **Faces/Bodies** section is active by default. You need to select the surface or body to be unstitched.
- Select the **Chain Selection** check box to select the adjacent surface of the selected surface.
- After selecting the surfaces, click on **OK** button from **UNSTITCH** dialog box to unstitch the selected surface.

Reverse Normal

The **Reverse Normal** tool is used to flip the normal direction of selected surfaces. When modifying the body or model, some of the surfaces of model may have unexpected normal directions. To reverse their normal direction, this tool is used. The procedure to use this tool is discussed next.

- Click on the **Reverse Normal** tool of **Modify** drop-down from **Toolbar**; refer to Figure-26. The **REVERSE NORMAL** dialog box will be displayed; refer to Figure-27.

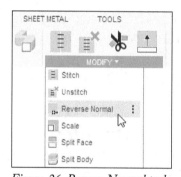
Figure-26. Reverse Normal tool

Figure-27. REVERSE NORMAL dialog box

- The **Select** button of **Faces** section is active by default. You need to select the face to flip; refer to Figure-28.

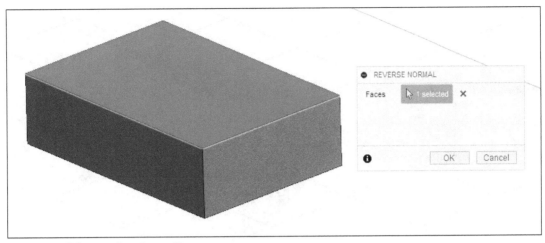

Figure-28. Selection of surface to flip

- After selection of surfaces, click on the **OK** button from **REVERSE NORMAL** dialog box. The selected surfaces will be flipped; refer to Figure-29.

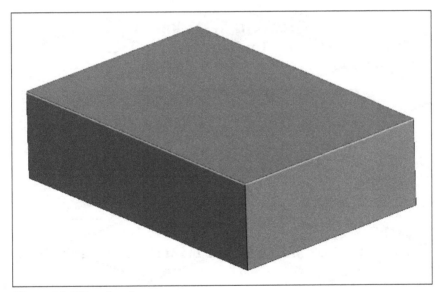

Figure-29. Flipped surface

The other tools of **SURFACE** tab have been discussed earlier in this book.

Practical 1

Create the model of aircraft toy as shown in Figure-30. (This is an open dimension problem, you need to get the shape of an aircraft using surfaces.)

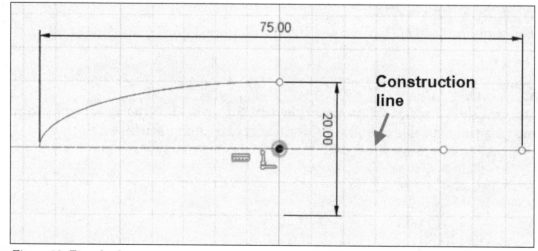

Figure-30. Practical 1 model

Creating first sketch

*   Click on the **SURFACE** tab from **Toolbar**; refer to Figure-31. The tools to create surface will be activated.

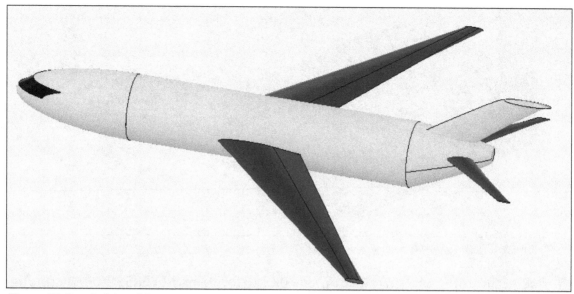

Figure-31. Tools to create surface

*   Create the sketch on **Top** plane using **Ellipse**, **Line**, and **Trim** tools; refer to Figure-32.

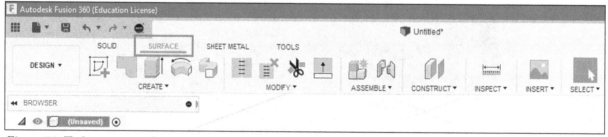

Figure-32. First sketch

Creating Surface

- Click on **Revolve** tool of **CREATE** drop-down from **Toolbar**. The **REVOLVE** dialog box will be displayed; refer to Figure-33.

Figure-33. REVOLVE dialog box

- You need to select the sketch and axis of sketch to apply the revolve feature; refer to Figure-34.

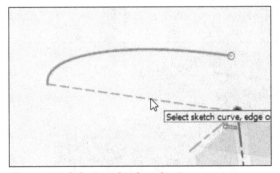

Figure-34. Selecting sketch and axis

- Revolve the sketch to 360 degree to create the nose of aircraft.
- Start a new sketch and project the round edge of nose on **Right** plane as shown in Figure-35. Exit the sketching environment.

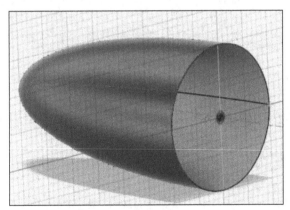

Figure-35. Edge projected on plane

- Click on **Extrude** tool of **CREATE** drop-down from **Toolbar**. The **EXTRUDE** dialog box will be displayed.
- Select the sketch as displayed to extrude; refer to Figure-36.

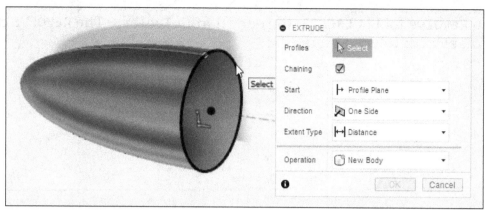

Figure-36. Selecting sketch to extrude

- Click in the **Distance** edit box and enter the value as **90**.
- After specifying the parameters, click on **OK** button from **EXTRUDE** dialog box to complete the extrusion process.
- Click on **Offset Plane** tool of **CONSTRUCT** drop-down from **Toolbar**. The **OFFSET PLANE** dialog box will be displayed.
- Select the **YZ** plane as shown in Figure-37.

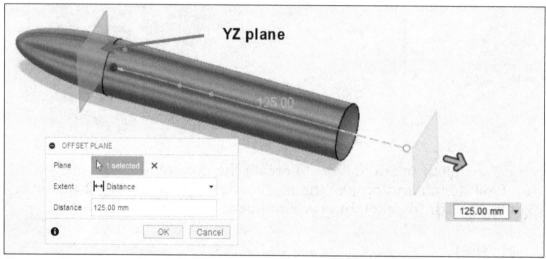

Figure-37. Creating offset plane

- Click in the **Distance** edit box of **OFFSET PLANE** dialog box and enter the value as **125**.
- After specifying the parameters, click on **OK** button from **OFFSET PLANE** dialog box to create the plane.
- Start a new sketch using newly created plane and click on the **Center Diameter Circle** tool from **CREATE** drop-down.
- Create the circle as shown in Figure-38.

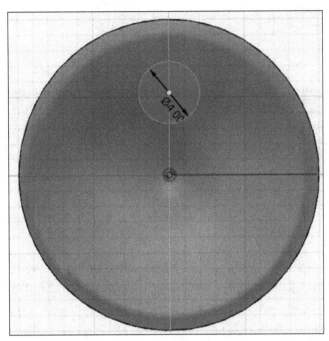

Figure-38. Creating circle

- After creating the sketch, click on **Finish Sketch** button from **Toolbar**.
- Click on **Loft** tool of **CREATE** drop-down from **Toolbar**. The **LOFT** dialog box will be displayed.
- You need to select the two sketch to create loft feature; refer to Figure-39.

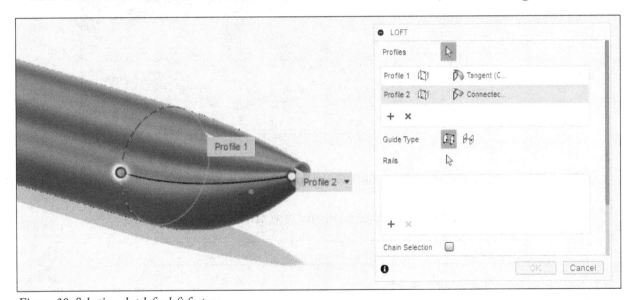

Figure-39. Selecting sketch for loft feature

- Specify the parameters as desired and click on **OK** button from **LOFT** dialog box to complete the process.
- Click on the **Create Sketch** tool and select the **XZ** plane as sketching plane. Create the sketch with the help of **Ellipse** and **Line** tool as shown in Figure-40.

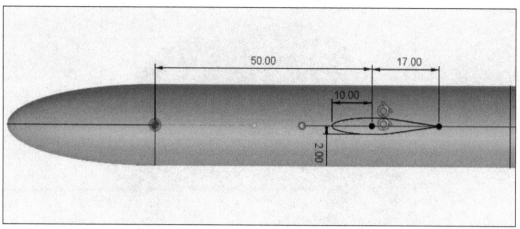

Figure-40. Sketch for wing

- After creating the sketch, click on **Finish Sketch** button from **Toolbar**.
- Now, create the offset plane at a distance of **75** from the **XZ** plane; refer to Figure-41.

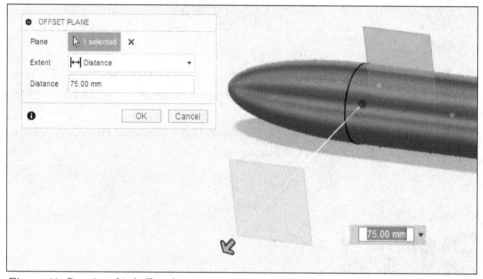

Figure-41. Creating third offset plane

- Click on the **Ellipse** tool from **CREATE** drop-down and select the recently created plane.
- Create the ellipse on current plane as shown in Figure-42.

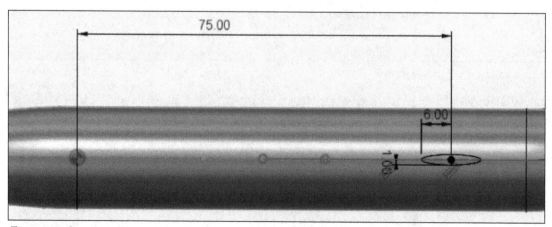

Figure-42. Creating ellipse

- Click on the **Finish Sketch** tool from **Toolbar**.
- Click on the **Loft** tool from **CREATE** drop-down. The **LOFT** dialog box will be displayed.
- Select the recently created sketches (ellipse forms) and create the loft; refer to Figure-43. Make sure the **Chain Selection** check box is selected.

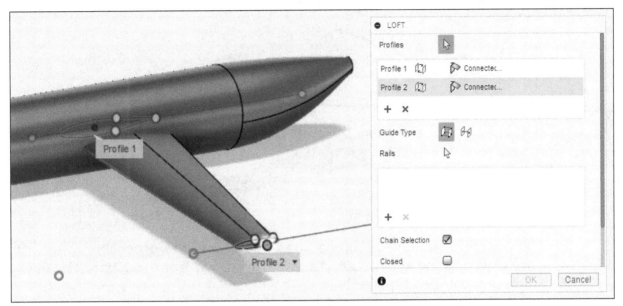

Figure-43. Creating wing

- After selection of profiles for loft, click on the **OK** button from **LOFT** dialog box to complete the loft process.
- Click on **Mirror** tool of **CREATE** drop-down from **Toolbar**. The **MIRROR** dialog box will be displayed; refer to Figure-44.

Figure-44. MIRROR dialog box

- Click on **Type** drop-down from **MIRROR** dialog box and select **Bodies** option.
- Click on **Select** button of **Objects** section and select the recently created wing.
- Click on **Select** button of **Mirror Plane** section and select the **XZ** plane; refer to Figure-45.

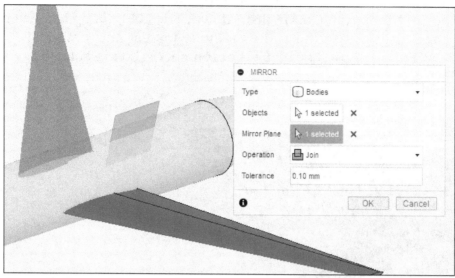

Figure-45. Creating Mirror copy of wing

- Click on **OK** button from **MIRROR** dialog box to complete the process.
- Click on the **Offset Plane** tool of **CONSTRUCT** drop-down from **Toolbar**. The **OFFSET PLANE** dialog box will be displayed.
- Create an offset plane at distance of **23** mm from **XZ** plane as shown in Figure-46.

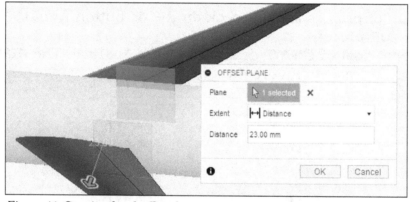

Figure-46. Creating fourth offset plane

- Create an ellipse on the **XZ** plane using **Ellipse** tool as shown in Figure-47.

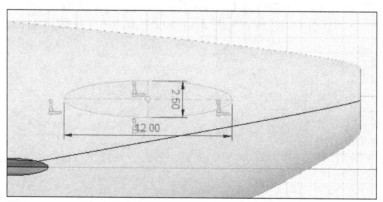

Figure-47. Creating sketch for rear wing

- After creating sketch, click on **Finish Sketch** button from **Toolbar**.
- Create another ellipse on the newly created offset plane as shown in Figure-48.

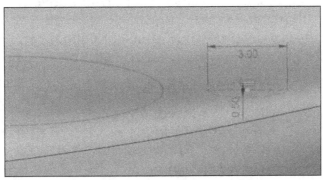

Figure-48. Sketch for rear wing

- Click on **Finish Sketch** button from **Toolbar** to exit the sketch.
- Click on **Loft** tool of **CREATE** drop-down from **Toolbar**. The **LOFT** dialog box will be displayed.
- You need to select two recently created sketch in **Profiles** section of **LOFT** dialog box; refer to Figure-49.

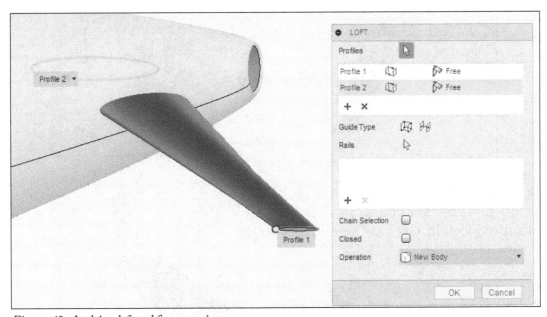

Figure-49. Applying loft tool for rear wing

- After specifying the parameters, click on **OK** button from **LOFT** dialog box to complete the process.
- To fill the open ends of wings, we need to apply the **Patch** tool.
- Click on the **Patch** tool of **CREATE** drop-down from **Toolbar**. The **PATCH** dialog box will be displayed; refer to Figure-50.

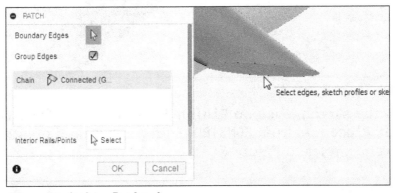

Figure-50. Applying Patch tool

- The **Select** button of **Boundary Edges** section is active by default. Click on the area displayed in the above figure.
- Click on the **Continuity** drop-down and select **Connected** option.
- After specifying the parameters, click on the **OK** button from **PATCH** dialog box to complete the process.
- To create other rear wing, click on **Mirror** tool of **CREATE** drop-down from **Toolbar**. The **MIRROR** dialog box will be displayed.
- Select the recently created wing in **Objects** section of **MIRROR** dialog box.
- Click on **Select** button of **Mirror Plane** section of **MIRROR** dialog box and select the **XY** plane as reference; refer to Figure-51.

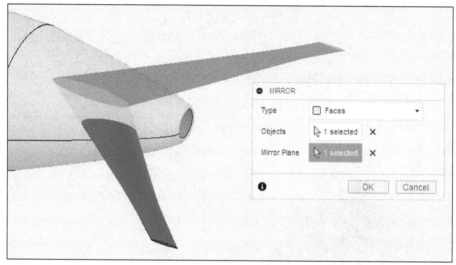

Figure-51. Creating other rear wing

- Click on **OK** button from **MIRROR** dialog box to create the copy.
- Click on **Create Sketch** tool of **CREATE** drop-down and select the **XY** plane for creation of sketch. Create the ellipse as shown in Figure-52.

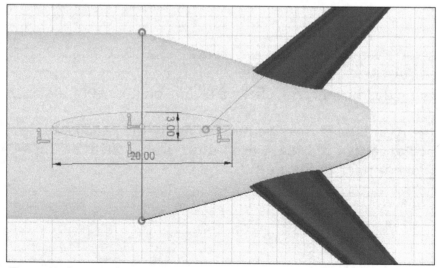

Figure-52. Creating sketch for third wing

- After creating the sketch, click on **Finish Sketch** button from **Toolbar**.
- Click on **Offset Plane** tool from **CONSTRUCT** drop-down and create a plane with **XY** plane as reference; refer to Figure-53.

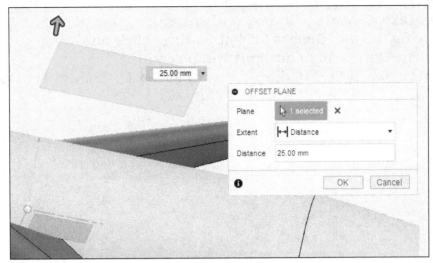

Figure-53. Creating offset plane for third rear wing

- Click on the **Create Sketch** tool of **CREATE** drop-down and select the recently created plane for sketch.
- Create the sketch as shown in Figure-54.

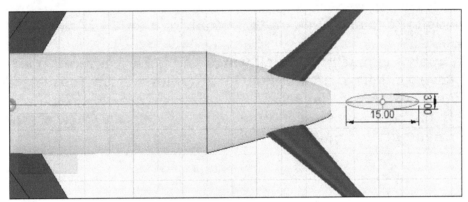

Figure-54. Creating another sketch for rear wing

- Click on **Finish Sketch** button from **Toolbar** to exit the sketch.
- Click on **Loft** tool from **CREATE** drop-down and select the recently created sketch under **Profiles** section; refer to Figure-55.

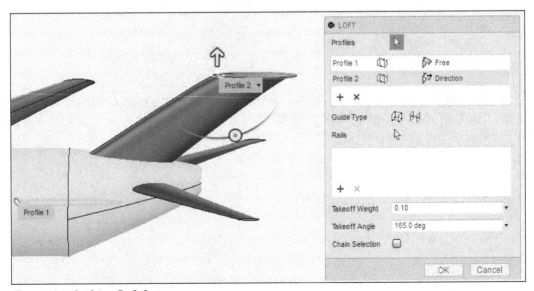

Figure-55. Applying Loft feature

- Click in the **Takeoff Weight** edit box and enter the value as **0.10**.
- Click in the **Takeoff Angle** edit box and enter the value as **165**.
- After specifying the parameters for loft feature, click on the **OK** button from **LOFT** dialog box. The third arm of air craft will be created.
- Click on **Patch** tool of **CREATE** drop-down in **SURFACE** tab from **Toolbar**. The **PATCH** dialog box will be displayed; refer to Figure-56.

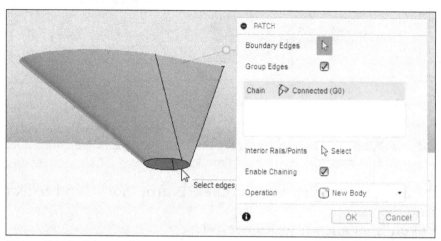

Figure-56. Applying patch on wing

- Select the open edge and click on **OK** button from **PATCH** dialog box to complete the process.
- Similarly, apply the patch tool to all wings and back of aircraft body.
- Click on **Reverse Normal** tool of **MODIFY** drop-down from **Toolbar**. The **REVERSE NORMAL** dialog box will be displayed.
- The **Select** button of **Faces** section is active by default. Click on the faces with wrong normal; refer to Figure-57.

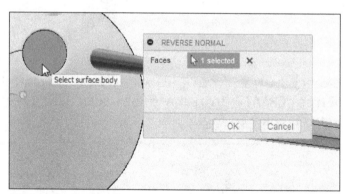

Figure-57. Applying Reverse Normal tool

- Click on the **OK** button from **Reverse Normal** dialog box to complete the process.
- Click on **Center Rectangle** tool of **CREATE** drop-down and select the **YZ** plane for sketch; refer to Figure-58.

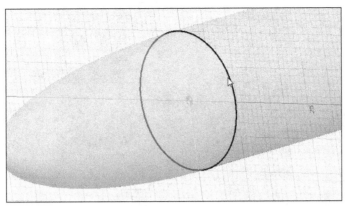

Figure-58. Selecting plane for sketch

- Create a rectangle on the selected plane as shown in Figure-59.

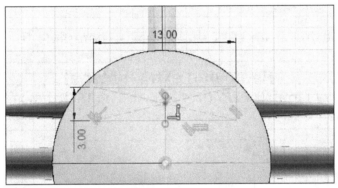

Figure-59. Sketch for windshield

- Click on **Finish Sketch** button from **Toolbar** to exit the sketch.
- Click on **Trim** tool of **MODIFY** drop-down from **Toolbar**. The **TRIM** dialog box will be displayed.
- The **Select** button of **Trim Tool** section is active by default. Select the recently created rectangular sketch and click on the inner area of trim preview; refer to Figure-60.

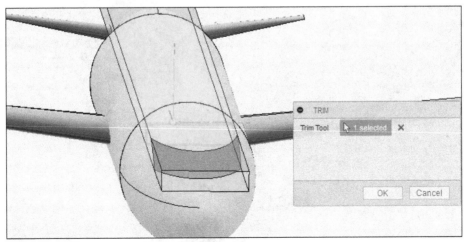

Figure-60. Selection of area for trim

- After selecting, click on the **OK** button of **TRIM** dialog box to exit the **Trim** tool.
- Click on **Loft** tool from **CREATE** drop-down. The **LOFT** dialog box will be displayed.
- Select the area in **Profiles** section as displayed in Figure-61.

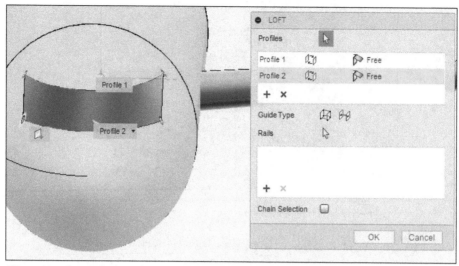

Figure-61. Creating windshield

- Click on the **OK** button from **LOFT** dialog box to complete the process.

Removing extra material

- Click on the **Trim** tool from **MODIFY** drop-down. The **TRIM** dialog box will be displayed; refer to Figure-62.

Figure-62. TRIM dialog box

- The **Select** button of **Trim Tool** section is active by default. Click on the extra material to remove; refer to Figure-63.

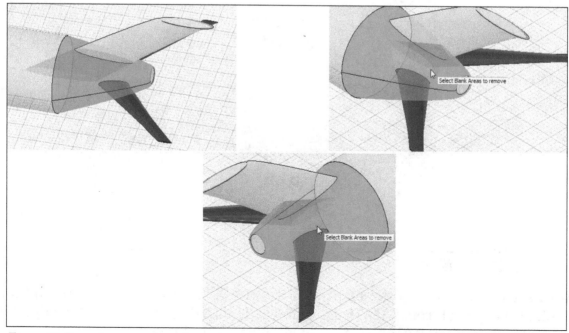

Figure-63. Removing extra material

- After selection of material to remove, click on the **OK** button from **TRIM** dialog box.
- After applying all the tools from above section. The aircraft will be created completely; refer to Figure-64.

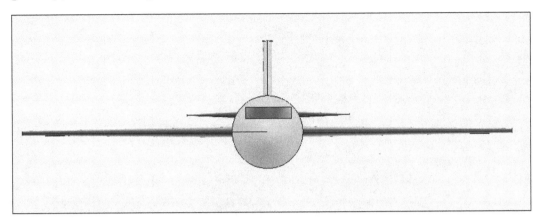

Figure-64. Front view of aircraft

PRACTICAL 2

Create the model of Jug as shown in Figure-65 and Figure-66.

Figure-65. Model of Jug

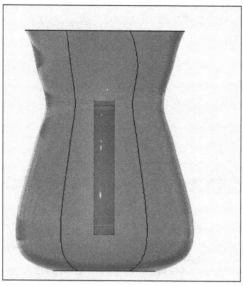

Figure-66. Another view of Jug

Starting Autodesk Fusion 360

- Double-click on the **Autodesk Fusion 360** software icon from desktop if the software has not started yet.
- Click on the **SURFACE** tab from **Toolbar** in **DESIGN** workspace. The tools to create surface will be activated.

Creating Sketch

The first step to create jug is to create sketches. The procedure is discussed next.

- Click on the **Create Sketch** tool of **CREATE** drop-down from **Toolbar** and select the **Top** plane as reference for sketch.
- Create the sketch as shown in Figure-67. Click on the **Finish Sketch** tool to exit.

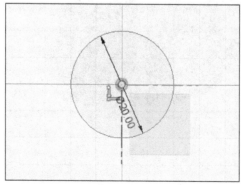

Figure-67. Creating base sketch

- Click on **Offset Plane** tool of **CONSTRUCT** drop-down from **Toolbar**. The **OFFSET PLANE** dialog box will be displayed.
- The **Select** button of **Plane** section is active by default. Select the **Top** plane as reference.
- Click in the **Distance** edit box and enter the value as **2**; refer to Figure-68.

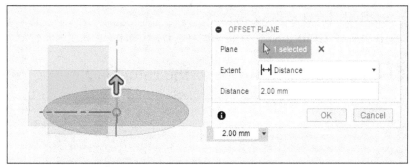

Figure-68. Creating first offset plane

- After specifying the parameters, click on the **OK** button from **OFFSET PLANE** dialog box to complete the process of creating plane.
- Create a center diameter circle on recently created offset plane; refer to Figure-69.

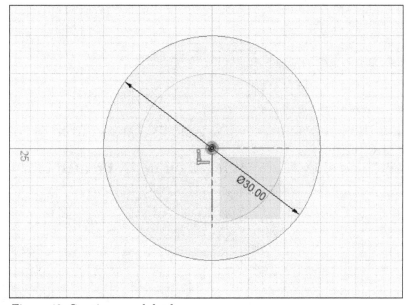

Figure-69. Creating second sketch

- Click on the **Patch** tool from **CREATE** drop-down. The **PATCH** dialog box will be displayed.
- The **Select** button of **Boundary Edges** section from **PATCH** dialog box is active by default. You need to select the first sketch to patch.
- After specifying the parameters, click on the **OK** button from **PATCH** dialog box to complete the process; refer to Figure-70.

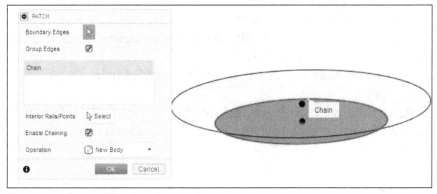

Figure-70. Creating Patch of first sketch

- Click on the **Offset Plane** tool of **CONSTRUCT** drop-down from **Toolbar**. The **OFFSET PLANE** dialog box will be displayed.
- The **Select** button of **Plane** section is active by default. Select the first offset plane as reference.
- Click in the **Distance** edit box and enter the value as **22**; refer to Figure-71.

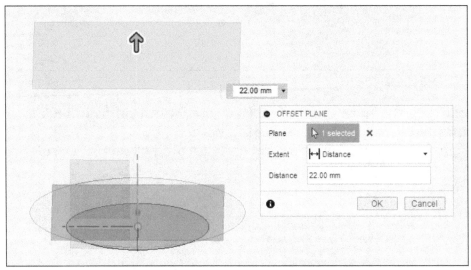

Figure-71. Creating third offset plane

- After specifying the parameter, click on the **OK** button from **OFFSET PLANE** dialog box to create the plane.
- Create the sketch as shown in Figure-72 on newly created offset plane.

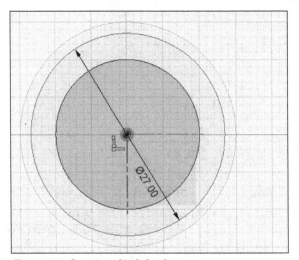

Figure-72. Creating third sketch

- Click on **Finish Sketch** button from **Toolbar** to exit the sketch.
- Click on the **Offset Plane** tool of **CONSTRUCT** drop-down from **Toolbar**. The **OFFSET PLANE** dialog box will be displayed.
- The **Select** button of **Plane** section is active by default. Select the second offset plane as reference.
- Click in the **Distance** edit box and enter the value as **7**; refer to Figure-73. You can also set the distance by moving the drag handle.

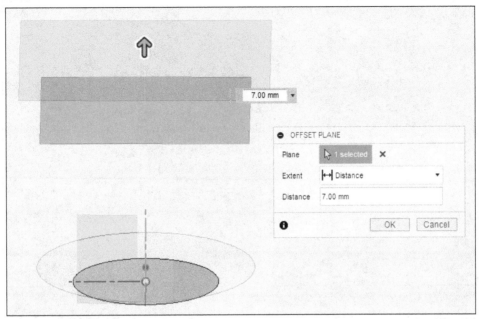

Figure-73. Creation of third offset plane

- After specifying the parameter, click on the **OK** button from **OFFSET PLANE** dialog box to create the plane.
- Create the sketch as shown in Figure-74 on newly created offset plane.

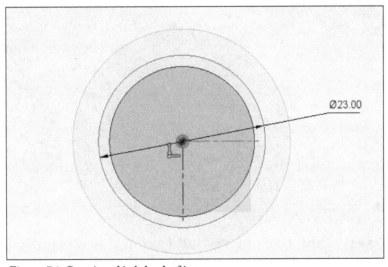

Figure-74. Creating third sketch of jug

- Click on the **Offset Plane** tool of **CONSTRUCT** drop-down from **Toolbar**. The **OFFSET PLANE** dialog box will be displayed.
- The **Select** button of **Plane** section is active by default. Select the third offset plane as reference.
- Click in the **Distance** edit box and enter the value as **8**; refer to Figure-75. You can also set the distance by moving the drag handle.

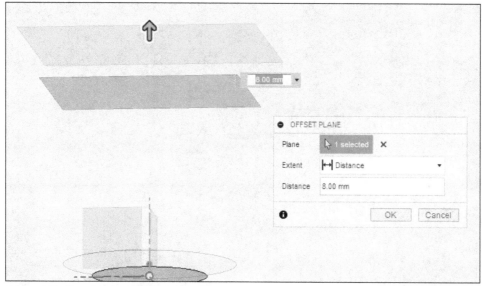

Figure-75. Creating fourth offset plane for jug

- After specifying the parameter, click on the **OK** button from **OFFSET PLANE** dialog box to create the plane.
- Create the sketch as shown in Figure-76 on newly created offset plane.

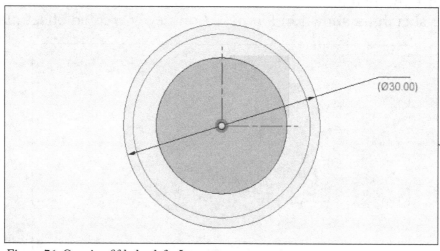

Figure-76. Creating fifth sketch for Jug

- Click on the **Offset Plane** tool of **CONSTRUCT** drop-down from **Toolbar**. The **OFFSET PLANE** dialog box will be displayed.
- The **Select** button of **Plane** option is active by default. Select the third offset plane as reference.
- Click in the **Distance** edit box and enter the value as **8**; refer to Figure-77. You can also set the distance by moving the drag handle.

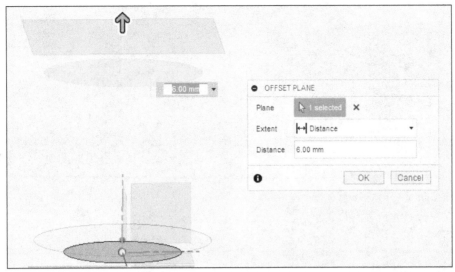

Figure-77. Creating fifth offset plane

- After specifying the parameter, click on the **OK** button from **OFFSET PLANE** dialog box to create the plane.
- Click on the **Circle** tool of **CREATE** drop-down from **Toolbar** and select the recently created plane as reference.
- Create the sketch as shown in Figure-78 with the help of **Circle**, **Trim**, and **Arc** tool.

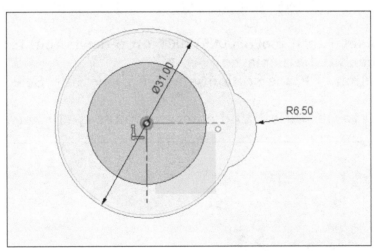

Figure-78. Creating sixth sketch

Creating Loft feature

Till now, we have created the sketch for jug. Now, we will apply the loft feature to join all the sketches. The procedure is discussed next.

- Click on the **Loft** tool of **CREATE** drop-down from **Toolbar**. The **LOFT** dialog box will be displayed.
- You need to select all the sketch from bottom to top in **Profiles** section of **LOFT** dialog box; refer to Figure-79.

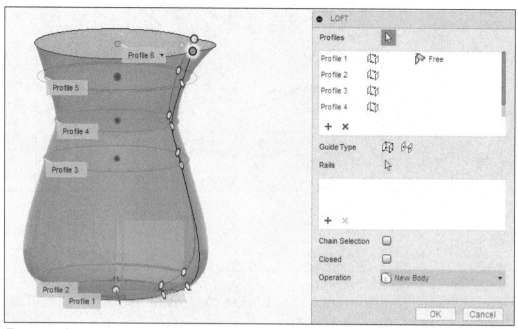

Figure-79. Creating loft feature of jug

- After selecting the profiles, click on the **OK** button from **LOFT** dialog box to complete the loft feature.

Here loft feature is completed. Now, we will create the handle of jug.

- Click on the **Offset Plane** tool of **CONSTRUCT** drop-down from **Toolbar**. The **OFFSET PLANE** dialog box will be displayed.
- The **Select** button of **Plane** section is active by default. Select the **YZ** plane as reference.
- Click in the **Distance** edit box and enter the value as **-10**; refer to Figure-80.

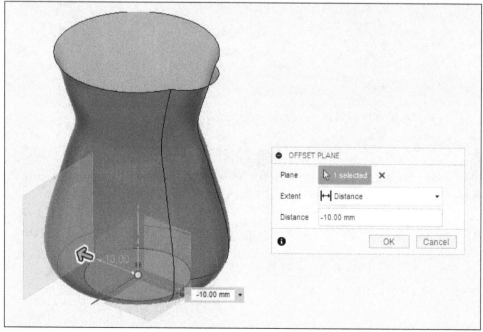

Figure-80. Creating sixth offset plane

- After specifying the parameter, click on the **OK** button from **OFFSET PLANE** dialog box to create the plane.
- Create the sketch as shown in Figure-81 on newly offset plane.

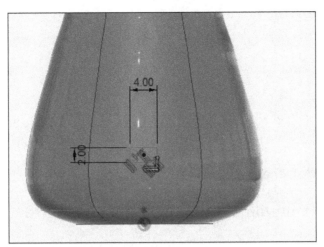

Figure-81. Creating seventh sketch

- Click on the **Create Sketch** tool from **CREATE** drop-down and select the **XZ** plane.
- Create the sketch as shown in Figure-82.

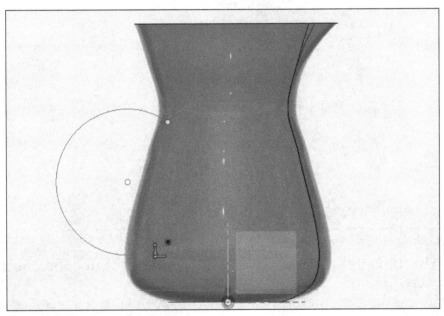

Figure-82. Creating eighth sketch

Creating Sweep feature

- Click on the **Sweep** tool from **CREATE** drop-down. The **SWEEP** dialog box will be displayed; refer to Figure-83.

Figure-83. SWEEP dialog box

- The **Select** button of **Profile** section is active by default. You need to select the recently created rectangular sketch.
- Click on the **Select** button of **Path** section and select the recently created arc; refer to Figure-84.

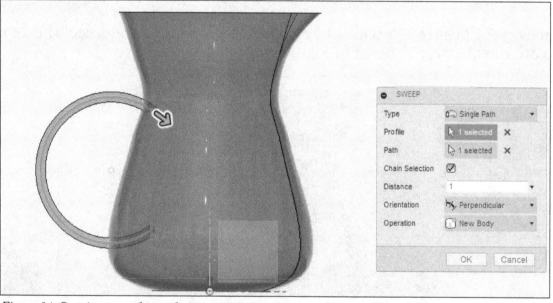

Figure-84. Creating sweep feature for jug

- Click on the **Orientation** drop-down of **SWEEP** dialog box and select the **Perpendicular** option.
- After specifying the parameters, click on the **OK** button from **SWEEP** dialog box to complete the feature.

Till here, we have created the model of jug but we need to apply the trim tool on the jug to remove the unwanted parts.

Trimming Extra Entities

- Click on the **Trim** tool of **MODIFY** drop-down from **Toolbar**. The **TRIM** dialog box will be displayed.
- The **Select** button of **Trim Tool** section is active by default. Click on the interior of jug and select the unwanted parts; refer to Figure-85.

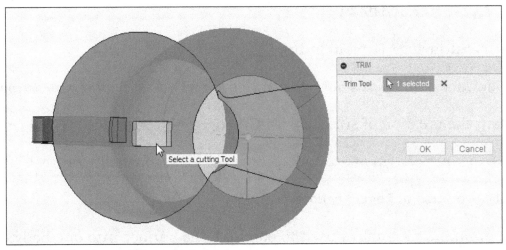

Figure-85. Applying the Trim tool

- After selecting, click on the **OK** button from **Trim** dialog box to trim the material.
- The model of jug is created successfully; refer to Figure-86.

Figure-86. Model of Jug created

- You can apply the physical material or appearance as required to the jug.
- You can also create the model of jug with the help of **Revolve** tool in place of **Loft** tool.

SELF ASSESSMENT

Q1. What is Surfacing? Explain in brief.

Q2. Why surface models can not exist in real world ? And how it can be made real?

Q3. Explain the creation of surface with G0 edges, G1 edges and G2 edges.

Q4. How can you perform an extrude protrusion of an open sketch?

Q5. What is the use of **Reverse Normal** tool?

6. The .... tool is used to combine the surface into one body.

PRACTICE 1

Create the model of mouse as shown in Figure-87 and Figure-88. Assume appropriate dimensions. If you are working on your system using a computer mouse then you can use that as reference for creating model.

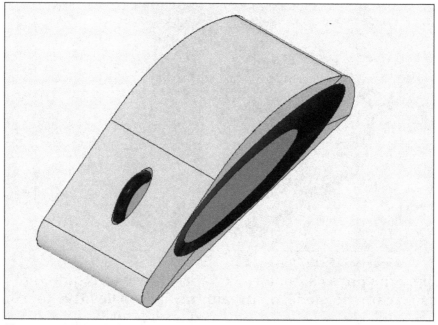

Figure-87. Model of mouse

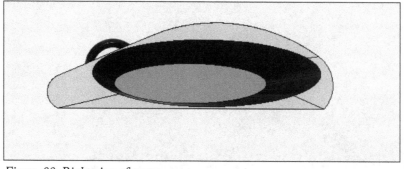

Figure-88. Right view of mouse

PRACTICE 2

Create the model of water bottle as shown in Figure-89 and Figure-90. The drawing view of water bottle is shown in Figure-91. (You are free to apply your artistic mind for improving design of bottle and when at it, why not start a competition in your class for designing bottle.)

Figure-89. Model of Water Bottle

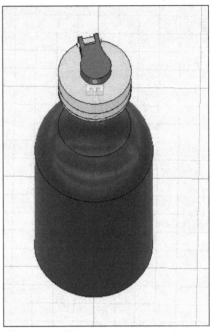

Figure-90. Top view of water bottle

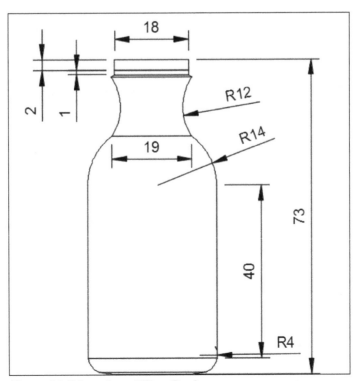

Figure-91. Dimensions of Water Bottle

FOR STUDENTS NOTES

Chapter 10

Rendering and Animation

Topics Covered

The major topics covered in this chapter are:

- **Introduction to Rendering**
- **Appearance Tool**
- **Scene Settings Tool**
- **Environment Library Tab**
- **Texture Map Control**
- **In-Canvas Render Setting**
- **Capture Image Tool**
- **Render**
- **Introduction to Animation Workspace**
- **New Storyboard**
- **Transform Components**

- **Restore Home Tool**
- **Auto Explode : One Level**
- **Auto Explode : All Levels**
- **Manual Explode**
- **Show/Hide Tool**
- **Create Callout Tool**
- **View Tool**
- **Publish Video Tool**

INTRODUCTION

When a designer is working on a 3D Model, the model he/she creates is generally a mathematical representation of points and surfaces in three-dimensions. Sometimes, a realistic 2D image of 3D model is required for presentation to the client. The conversion of 3D model with specified environmental scenario through mathematical approximation to a finalized 2D image is known as Rendering. During this process, the entire scene's textural and lighting information are combined to determine the color value of each pixel in the finalized image. To perform rendering, select the **RENDER** option from the **Change Workspace** drop-down in the **Ribbon**.

INTRODUCTION TO RENDER WORKSPACE

The tools in **Render Workspace** are used to create the photo-realistic image from a scene of model. There are various tools used for rendering process which are discussed next.

Appearance Tool

The **Appearance** tool is used to apply different materials to selected bodies/components/faces. The procedure to use this tool has been discussed earlier in the book. So, we are not repeating the tool here.

Scene Settings

The **Scene Settings** tool is used to control the lighting, background color, ground effects, and camera rendering. The procedure to use this tool is discussed next.

*   Click on the **Scene Settings** tool of **SETUP** panel from **Toolbar**; refer to Figure-1. The **SCENE SETTINGS** dialog box will be displayed; refer to Figure-2.

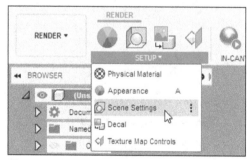

Figure-1. Scene Settings tool

Figure-2. SCENE SETTINGS dialog box

Environment section

- Click on the **Brightness** edit box and specify the value of brightness for model. You can also move the **Brightness** slider to adjust the brightness for the model.
- If you want to control the position and rotation of lights then click on the **Position** button of **Environment** section from **Settings** tab. The drag handle along with some tools will be displayed on the canvas screen; refer to Figure-3.

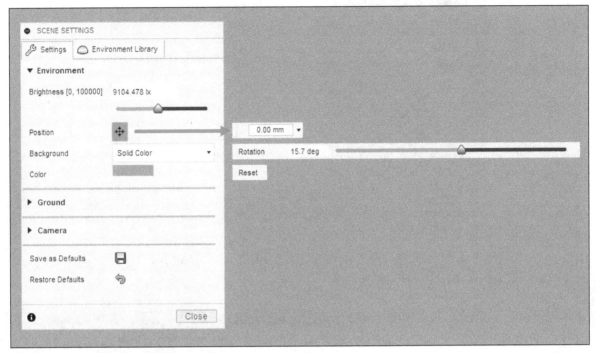

Figure-3. Tools of Position button

- Click in the floating window and enter the position of lights. You can also set the position by moving the drag handle from model.
- Click in the **Rotation** edit box and enter the value of rotation of lights. You can also adjust the value by moving the **Rotation** slider.
- If you want to reset all the changed settings, then click on the **Reset** button from rotation tools.
- Select **Environment** option from **Background** drop-down if you want to use environment image. The environments are available in **Environment Library** tab of this dialog box.
- Select **Solid Color** option from **Background** drop-down if you want to set a particular color to the background.
- To select a solid color, click on the **Color** button. A color box will be displayed; refer to Figure-4.

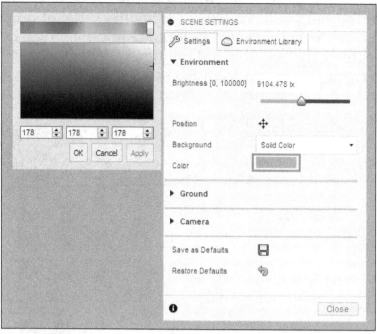

Figure-4. Color box

- Select required color from color box and click on the **Apply** button to set the selected color as background color.

Ground section

- Select the **Ground Plane** check box of **Ground** section from **Settings** tab if you want to display the shadow and reflection of model on the ground canvas.
- Select the **Flatten Ground** check box of **Ground** section from **Settings** tab if you want to enable a "textured" ground plane where the environment image is mapped as a texture.
- Select the **Reflections** check box of **Ground** section from **Settings** tab if you want to display the reflection of model on the ground.
- Click in the **Roughness** edit box and specify the value of sharpness of reflection. You can also set the sharpness of reflection by moving the **Roughness** slider.

Camera section

- Select the **Orthographic** option of **Camera** drop-down from **Camera** section if you want to set the orthographic view of model.
- Select the **Perspective** option of **Camera** drop-down from **Camera** section if you want to set the perspective view of model.
- Select the **Perspective with Ortho Faces** option from **Camera** drop-down if you want to set the perspective view of model with ortho faces.
- Click in the **Exposure** edit box of **Camera** section from **Settings** tab and enter the value of camera exposure. You can also set the camera exposure by moving the **Exposure** slider.
- Select the **Depth of Field** check box if you want to apply depth of field effect. Depth of field is used to focus on specific area of the model. This effect is only enabled when **Ray Tracing** is enabled. On selecting the **Depth of Field** check box, the dialog box will be updated; refer to Figure-5.

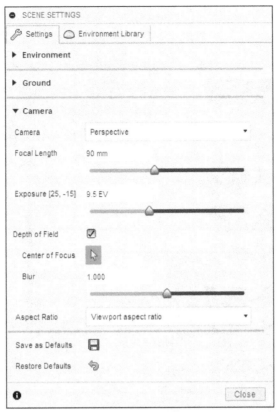

Figure-5. Updated dialog box on selecting Depth of Field check box

- The **Center of Focus** selection button is active by default. You need to select a location on model for focusing.
- Click in the **Blur** edit box and enter the value of blur the model except the selected center of focus. You can also set the value of blur by moving the **Blur** slider.
- Click on the **Aspect Ratio** drop-down and select required aspect ratio of canvas screen.
- Click on **Save as Defaults** button to save the setting as default setting.
- Click on the **Restore Defaults** button to restore the changed settings to default.

Environment Library tab

- Click on the **Environment Library** tab from the **SCENE SETTINGS** dialog box. The tab will be displayed; refer to Figure-6.

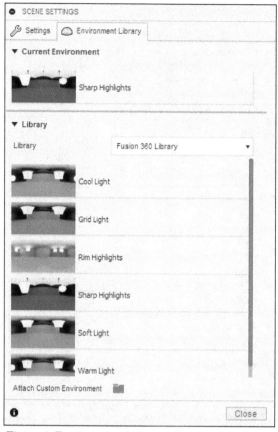

Figure-6. Environment Library tab

- The **Current Environment** section shows the applied environment on the current model.
- If you want to change the environment of the model then drag a specific environment from **Library** section and drop it to the current model or in the **Current Environment** section. You can download the environments in library by using the **Download** button next to them in the list.
- If you want to apply the custom environment for model then click on the **Attach Custom Environment** button of **Environment Library** tab from **SCENE SETTINGS** dialog box. The **Open** dialog box will be displayed; refer to Figure-7.

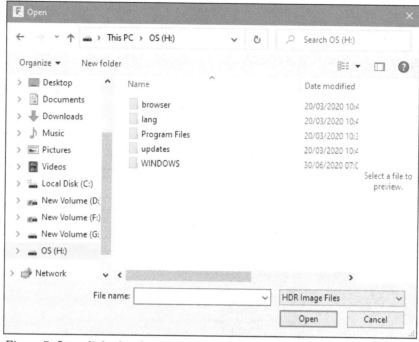

Figure-7. Open dialog box for attaching custom environment

- Select required file and click on **Open** button. The environment will be applied to model.
- After selecting required environment from **Environment Library** tab, click on **Close** button to exit the **Scene Settings** tool.

Note- There is no **Timeline** in the **Render** workspace. The tools applied in the **RENDER** workspace will not be recorded.

Decal tool has been discussed earlier in the book.

Texture Map Controls

The **Texture Map Controls** tool is used to set the orientation of the texture applied to the face or model. The procedure to use this tool is discussed next.

- Click on the **Texture Map Controls** tool of **SETUP** panel from **Toolbar**; refer to Figure-8. The **TEXTURE MAP CONTROLS** dialog box will be displayed; refer to Figure-9.

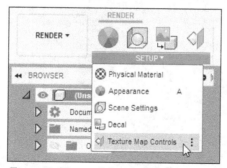

Figure-8. Texture Map Controls tool

Figure-9. TEXTURE MAP CONTROLS dialog box

- The **Select** button of **Selection** option is selected by default. You need to select the texture of model. The updated **TEXTURE MAP CONTROLS** dialog box will be displayed; refer to Figure-10.

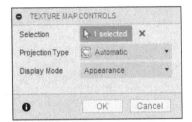

Figure-10. Updated TEXTURE MAP CONTROLS dialog box

Automatic

- Select the **Automatic** option from **Projection Type** drop-down if you want to adjust the texture on the model using model topology.

Planar

- Select the **Planar** option from **Projection Type** drop-down if you want to map the image or texture on the model as projection.
- On selecting the **Planar** option, you need to select the axis for adjusting the texture. On selecting the texture, the dialog box will be updated; refer to Figure-11.

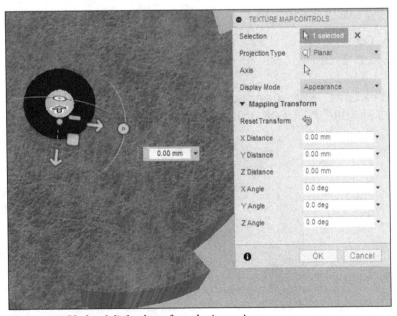

Figure-11. Updated dialog box after selecting axis

- Adjust the image or texture by using the manipulator or by specify the values in related edit boxed in **TEXTURE MAP CONTROLS** dialog box.

- After adjusting the texture, click on the **OK** button from **TEXTURE MAP CONTROLS** dialog box.

Box

- Click on the **Box** option from **Projection Type** drop-down if you want to map the image on model into box like objects.
- On selecting the **Box** option, the manipulator and updated dialog box will be displayed; refer to Figure-12.

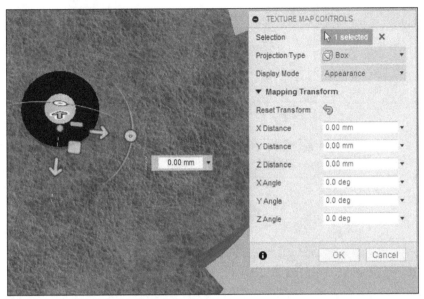

Figure-12. Dialog box on selecting Box-option

- Click in the specific edit box and enter desired value. You can also set the value by moving the arrow of manipulator.
- If you want to reset the values of **Mapping Transform** or manipulator then click on the **Reset Transform** button.

Spherical

- Select the **Spherical** option from **Projection Type** drop-down if you want to map texture into a spherical object.
- On selecting the **Spherical** option, the manipulator along with updated dialog box will be displayed.
- The options were discussed in last topic.

Cylindrical

- Select the **Cylindrical** option from **Projection Type** drop-down if you want to map texture into a cylindrical object.
- On selecting the **Cylindrical** option, the manipulator along with updated dialog box will be displayed.
- The options were discussed in last topic.
- After specifying the parameters, click on the **OK** button from **TEXTURE MAP CONTROLS** dialog box to exit the tool.

In-canvas Render

The **In-canvas Render** button is used to start and stop the In-canvas rendering process. The procedure to use this tool is discussed next.

- Click on the **In-canvas Render** button from **IN-CANVAS RENDER** drop-down; refer to Figure-13. The render quality slider will be displayed to display at the bottom right area of canvas screen; refer to Figure-14.

Figure-13. In-canvas Render tool

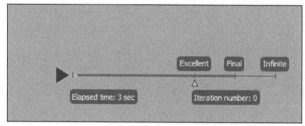

Figure-14. Render quality slider

- The current scene of the model will be changed into the photo-realistic 2D image.
- Clicking again on the **In-canvas Render Stop** button will stops the rendering process.

In-Canvas Render Setting

The **In-canvas Render Setting** tool is used to change the setting of rendering process. The procedure to use this tool is discussed next.

- Click on the **In-Canvas Render Settings** tool from **IN-CANVAS RENDER** drop-down; refer to Figure-15. The **IN-CANVAS RENDER SETTINGS** dialog box will be displayed; refer to Figure-16.

Figure-15. In-canvas Render Settings tool

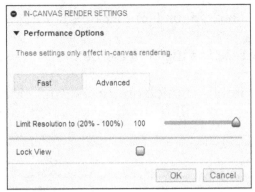

Figure-16. IN-CANVAS RENDER SETTINGS dialog box

- Set the quality of rendering by adjusting the **Limit Resolution to** slider. The higher quality rendering takes more time to calculate rendering process and lower quality rendering takes less time to calculate the rendering process.
- Select the **Lock View** check box to lock the current view of model otherwise clear the check box.
- After specifying the parameters, click on the **OK** button from **IN-CANVAS RENDER SETTINGS** dialog box.

Capture Image

The **Capture Image** tool is used to capture the current canvas as an image. The procedure to use this tool is discussed next.

- Click on the **Capture Image** tool of **IN-CANVAS RENDER** drop-down in **Toolbar**; refer to Figure-17. The **Image Options** dialog box will be displayed; refer to Figure-18.

Figure-17. Capture Image tool

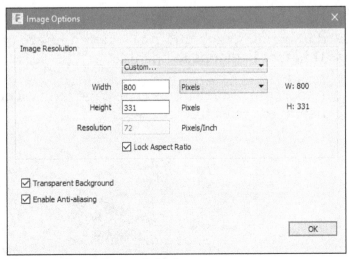

Figure-18. Image Options dialog box

- Click on the **Image Resolution** drop-down and select required resolution of image.
- Select the **Transparent Background** check box if you want to set the transparent background of model.
- Select the **Enable Anti-aliasing** check box if you want to activate the anti-aliasing property.
- After specifying the parameters, click on the **OK** button from **Image Options** dialog box. The **Save As** dialog box will be displayed; refer to Figure-19.

Figure-19. Save As dialog box

- Click in the **Name** edit box of **Save As** dialog box and enter the specific name for image.
- Click on the **Type** drop-down and select required format of image file.
- Click on the **Location** drop-down button and select the location of saving image.
- Select the **Save to my computer** check box if you want to save the file to computer.
- Select the **Save to a project in the cloud** check box to save the image file on cloud.
- After specifying the parameters for image, click on the **Save** button from **Save As** dialog box. The image will be saved to the specified location.

Render

The **Render** tool is used to create a high quality rendered image of the current scene. The procedure to use this tool is discussed next.

- Click on the **Render** tool from **Toolbar**; refer to Figure-20. The **RENDER SETTINGS** dialog box will be displayed; refer to Figure-21.

Figure-20. Render tool

Figure-21. RENDER SETTINGS dialog box

WEB

- Click on the **WEB** tab of **RENDER SETTINGS** dialog box. Various options under **WEB** tab will be displayed; refer to Figure-22.

Figure-22. WEB tab

- Three highlighted options are the size of image in which your rendered file will be saved. Click on required size.
- Click on the **Cloud Renderer** option from **RENDER WITH** section, if you want to render the file from Autodesk cloud. On selecting the **Cloud Renderer** button, the dialog box will be updated; refer to Figure-23.

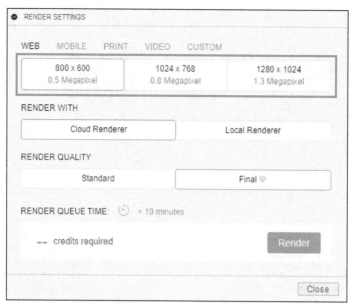

Figure-23. Updated WEB tab

- With **Local Renderer** option selected, click on the **Advanced settings** toggle button from **WEB** tab to manually set the render quality value. The **RENDER QUALITY** slider will be displayed; refer to Figure-24.

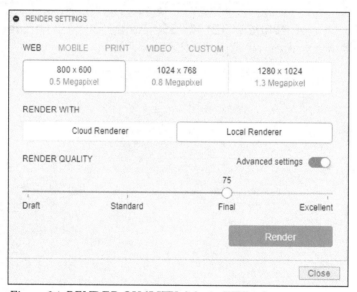

Figure-24. RENDER QUALITY slider in WEB tab

- Set the value of quality from slider as required. If you want to automatically set the quality of render then again click on the **Advanced Settings** toggle button. The **Standard(50)** and **Final(75)** option will be displayed.
- Click on required option from **RENDER QUALITY** section and click on the **Render** button from **WEB** tab. The rendering process will start.
- The completed rendered file will be saved in **RENDERING GALLERY**; refer to Figure-25.

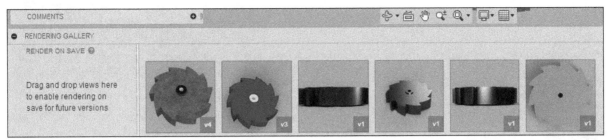

Figure-25. RENDERING GALLERY

Render Gallery

- To view the enlarged size of rendered image, click on that particular image. The image will be displayed in **RENDERING GALLERY** dialog box; refer to Figure-26.

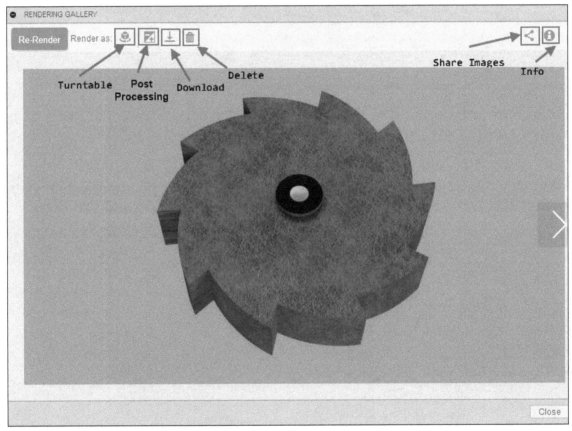

Figure-26. RENDERING GALLERY dialog box

- If you want to create a video of the rotating model then click on the **Turntable** button from **RENDERING GALLERY** dialog box. The **Render Settings - Turntable** dialog box will be displayed; refer to Figure-27.

Figure-27. Render Settings-Turntable dialog box

- Specify desired parameters and click on the **Render** button. The file will be re-created and displayed in **RENDERING GALLERY**.

Post Processing

- If you want to adjust the exposure of model then click on the **Post-processing** button. The **Post-processing** dialog box will be displayed; refer to Figure-28.

Figure-28. Post processing dialog box

- The value of exposure can be adjusted by using the slider for **Exposure Value**.
- Select desired option from the **Preset** drop-down. The options in the dialog box will be automatically adjusted presentation settings for rendering.
- Similarly, you can set the other parameters in the dialog box. After setting the parameters, click on the **Apply** button.

Downloading Render Image File

- If you want to download the current file then click on the **Download** button from **RENDERING GALLERY** dialog box. The drop-down will be displayed as shown in Figure-29.

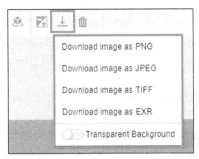

Figure-29. Download drop-down

- Select the **Transparent Background** toggle button if you want to make background of model transparent. On selecting this toggle button, you will be able to save the image file in PNG or TIFF format only.
- Select desired **Download image as...** button. The **Save File** dialog box will be displayed; refer to Figure-30.

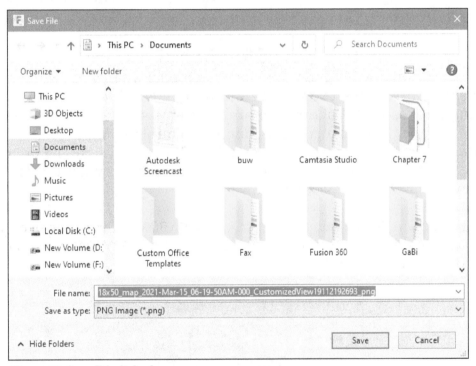

Figure-30. Save File dialog box

- Select the location for saving file and click on the **Save** button from the dialog box to save the file.
- If you want to delete the current file from **RENDERING GALLERY** then click on the **Delete** button from **RENDERING GALLERY** dialog box. The **Delete Image** dialog box will be displayed; refer to Figure-31.

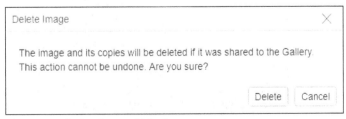

Figure-31. Delete Image dialog box

- If you are sure about deleting the particular file then click on the **Delete** button from **Delete Image** dialog box otherwise click on **Cancel** button.
- If you want to share the image to Fusion 360 Gallery or A360 then click on the **Share Images** button at the top right in **RENDERING GALLERY** dialog box and select the respective option.
- If you want to see the information of the current file then click on the **Info** button from **RENDERING GALLERY** dialog box.
- After rendering the file, click on the **Close** button from **RENDERING GALLERY** dialog box to exit the rendering process.

ANIMATION

The **Animation** workspace is used to animate the joints of assembly. This workspace is also used to create exploded view of assembly. You can create the exploded view automatically as well as manually in this workspace. To activate the workspace, click on the **ANIMATION** option from the **Workspace** drop-down. The tools to perform animation will be displayed.

New Storyboard

A storyboard is automatically generated for each model when you switch to **ANIMATION** workspace. In the process of creating animation, additional storyboard can be added at any time as well. Each new storyboard has a default name "Storyboardx" where x is a number. You can edit this name by double-clicking on it at the bottom in application window. Storyboard acts as base for animation. To create different animations of same assembly, you can create different storyboards. The procedure to create new storyboard is discussed next.

- Click on the **New Storyboard** tool from **STORYBOARD** drop-down of **Toolbar** in **ANIMATION** workspace; refer to Figure-32. The **NEW STORYBOARD** dialog box will be displayed; refer to Figure-33.

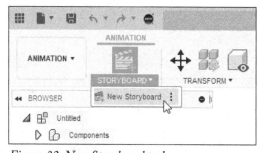

Figure-32. New Storyboard tool

Figure-33. NEW STORYBOARD dialog box

- Select the **Clean** option from **Storyboard Type** drop-down if you want to delete all previous actions of animation and want to create a new storyboard.

- Select the **Start from end of previous** option from **Storyboard Type** drop-down if you want to use position of components at the end of previous animation as the starting position for the new animation.

- After setting desired options, click on the **OK** button from the dialog box. The created storyboard will be added at the bottom of **ANIMATION TIMELINE**; refer to Figure-34.

Figure-34. ANIMATION TIMELINE

Transform Components

The **Transform Components** tool is used to move and rotate the component. Note that the movement and rotation performed in this workspace will be recorded in animation. The procedure to use this tool is discussed next.

- Click on the **Transform Components** tool of **TRANSFORM** drop-down; refer to Figure-35. The **TRANSFORM COMPONENTS** dialog box will be displayed; refer to Figure-36.

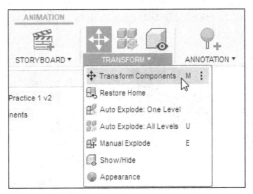

Figure-35. Transform Components tool

Figure-36. TRANSFORM COMPONENTS dialog box

- The **Select** button of **Components** section is active by default. Click on the component that you want to move or rotate. On selection of component, a manipulator will be displayed on the selected component along with updated **TRANSFORM COMPONENTS** dialog box; refer to Figure-37.

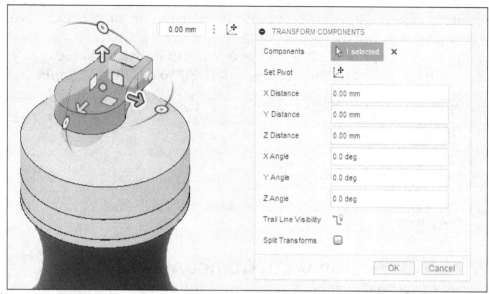

Figure-37. Manipulator on selected component

- The actions like moving or rotating the components are registered as animation in storyboard. To register the movement, drag the playhead in **ANIMATION TIMELINE** area to time value upto which the selected transformation operation will be completed. For example, if you want to move the component shown in figure by 60 mm in 2 seconds then first move the playhead of timeline to 2 seconds and then move the component by 60 mm; refer to Figure-38.

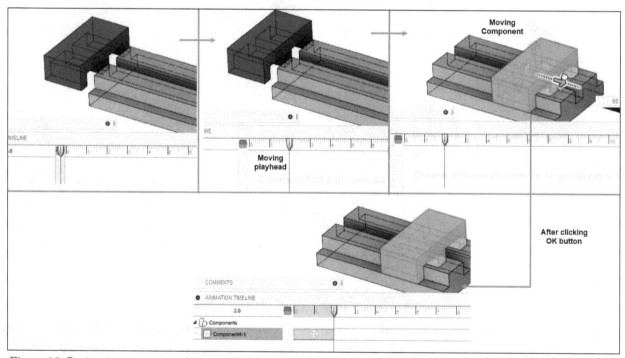

Figure-38. Registering movement of component in timeline

- You can specify desired values for movement in the **TRANSFORM COMPONENTS** dialog box or you can set the parameters by moving the manipulator.
- After specifying the parameters, click on the **OK** button from **TRANSFORM COMPONENTS** dialog box to add the action in animation timeline.

- To play the animation, you need to click on the **Play** button from **ANIMATION TIMELINE.**

Restore Home

The **Restore Home** tool is used to restore the model position as it is at the end of animation.

- Click on the **Restore Home** tool from **TRANSFORM** drop-down; refer to Figure-39. The position of model will be restored.

Figure-39. Restore Home tool

Auto Explode: One Level

The **Auto Explode: One Level** tool is used to automatically separate all selected components down to one level in hierarchy. The procedure to use this tool is discussed next.

- Select the parts of assembly to be exploded (separated from each other) while holding the **CTRL** key. You can also select all components using window selection; refer to Figure-40.

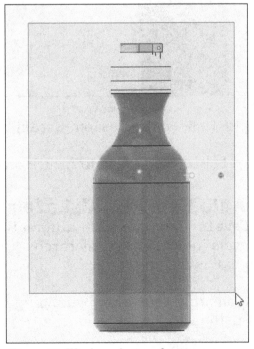

Figure-40. Window selection of component

- Click on the **Auto Explode: One Level** tool from **TRANSFORM** drop-down; refer to Figure-41. The components will be separated on the canvas screen; refer to Figure-42.

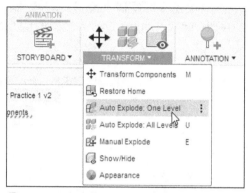

Figure-41. Auto Explode One Level tool

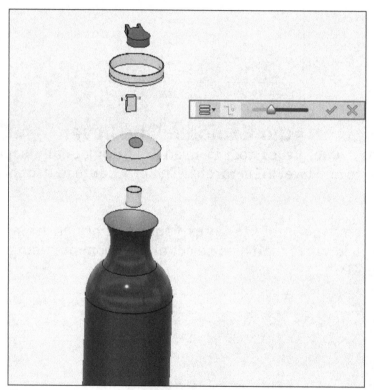

Figure-42. Preview of Auto Explode One Level

- By moving the highlighted slider (**Explosion Scale**), you can set the distance between component.
- After adjusting the space between components, click on the **OK** button from slider.

Auto Explode: All Levels

The **Auto Explode: All Levels** tool is used to automatically separate all selected components down to all levels in component hierarchy (including sub-assemblies). The procedure to use this tool is discussed next.

- Select the parts of components to separate by holding the **CTRL** key. You can also select the components using window selection.

- Click on the **Auto Explode: All Levels** tool from **TRANSFORM** drop-down; refer to Figure-43. The components of selected model will be separated.

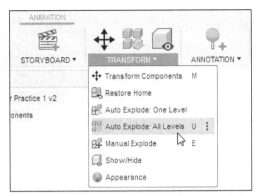

Figure-43. Auto Explode All Levels tool

- Click on the screen to view the separated component properly; refer to Figure-44.

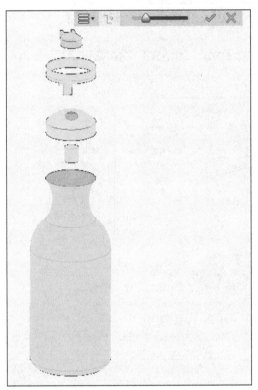

Figure-44. Preview of Explode

- Move the slider to adjust the distance between separated components of model.
- After adjusting the components, click on the ✅ **OK** button. The exploded view of assembly will be displayed; refer to Figure-45.

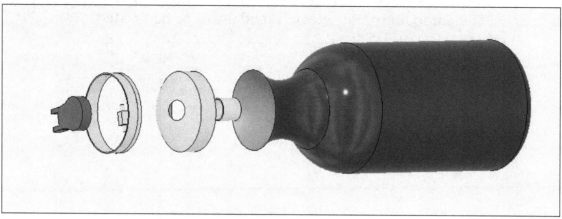

Figure-45. Exploded view of water bottle assembly

Manual Explode

The **Manual Explode** tool is used to manually separate the selected component. The procedure to use this tool is discussed next.

- Click on the **Manual Explode** tool of **TRANSFORM** drop-down from **Toolbar**; refer to Figure-46.

Figure-46. Manual Explode tool

- You are asked to select the components to be moved during exploding. Click on the component to separate; refer to Figure-47.

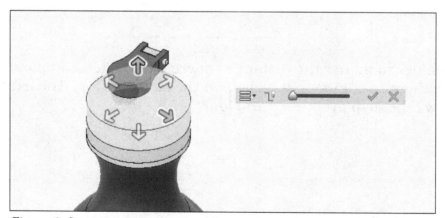

Figure-47. Separating component manually

- On selecting a component, number of arrows will be displayed on the selected component. You need to click on the arrow of direction in which you want to move the selected component.

- Move the **Explosion Scale** slider to separate the selected component from remaining model; refer to Figure-48.

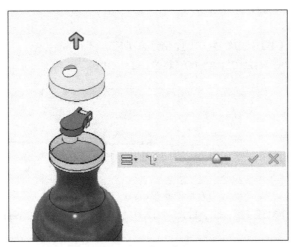

Figure-48. Separating the selected component

Show/Hide

The **Show/Hide** tool is used to show or hide the selected component. The procedure to use this tool is discussed next.

- Click on desired component from model to show or hide.
- Now, click on the **Show/Hide** tool from **TRANSFORM** drop-down; refer to Figure-49. The selected component will be shown or hidden based on its condition.

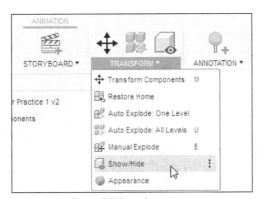

Figure-49. Show/Hide tool

- To display the hidden component, click on the crossed eye button from **Browser**; refer to Figure-50.

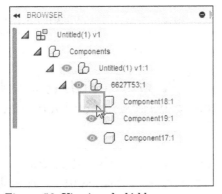

Figure-50. Viewing the hidden component

Create Callout

The **Create Callout** tool is used to add annotation to the animation. The procedure to use this tool is discussed next.

- Click on the **Create Callout** tool from **ANNOTATION** panel; refer to Figure-51. The callout symbol will be attached to the cursor.

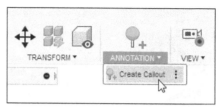

Figure-51. Create Callout tool

- Click on the component to add the callout. A text box will be displayed; refer to Figure-52.

Figure-52. Callout text box

- Type desired text in the text box and click on **OK** button. The annotation callout will be added to the selected component.

View

The **View** tool is used to start and stop the recording of the actions or commands occurring on the model or component. When **View** tool is activated then it captures the actions like move for animation and when the **View** tool is deactivated it does not record the process; refer to Figure-53.

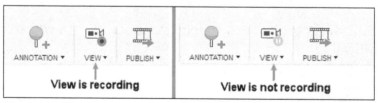

Figure-53. View tool

Publish Video

The **Publish Video** tool is used to publish the video. The procedure to use this tool is discussed next.

- Click on the **Publish Video** tool from the **PUBLISH** panel in the **Toolbar**; refer to Figure-54. The **Video Options** dialog box will be displayed; refer to Figure-55.

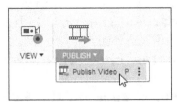

Figure-54. Publish Video tool

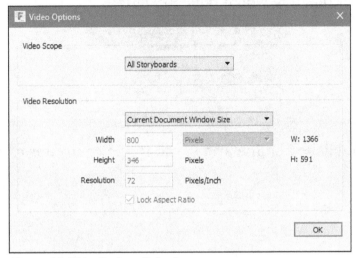

Figure-55. Video Options dialog box

- Select **Current Storyboard** option from **Video Scope** section if you want to create a video of current storyboard animation only.
- Select **All Storyboards** option from **Video Scope** section if you want to create a video of all storyboard animations combined.
- Click on the **Current Document Window Size** drop-down from **Video Resolution** section or select required video resolution.
- After specifying the parameters, click on the **OK** button from **Video Options** dialog box. The **Save As** dialog box will be displayed; refer to Figure-56.

Figure-56. Save As dialog box for Publish Video tool

- Click in the **Name** edit box and specify the name of file.
- Select the **Save to a project in the cloud** check box if you want to save the file online or in your Autodesk account.
- Select the **Save to my computer** check box if you want to save the file in your computer.
- After specifying the parameters, click on the **Save** button from **Save As** dialog box. The **Publish Video** progress dialog box will be displayed which tells about the progress of publishing video; refer to Figure-57.

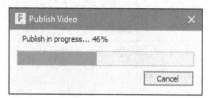

Figure-57. Publish Video progress dialog box

The video will be published online and saved to the computer at preferred location.

SELF ASSESSMENT

Q1. What is Rendering?

Q2. The .... is used to create the photo-realistic image from a scene of model.

Q3. The .... tool is used to control the lighting, background color, ground effects, and camera rendering.

Q4. From where can you get the environments for background as well as for model?

Q5. How can you display the shadow and reflections of model on ground canvas?

Q6. A textured ground plane where the environment image is mapped as a texture can be enabled by selecting **Ground Plane** check box. (T/F)

Q7. The reflection of model on the ground can be displayed by selecting the **Reflections** check box. (T/F)

Q8. What is Depth of Field? And how it can be enabled?

Q9. If you want to change the environment of the model then drag a specific environment from .... section and drop it to the current model or in the **Current Environment** section.

Q10. The **Appearance** tool is used to set the orientation of the texture applied to the face or model. (T/F)

Q11. What is the use of **Reset Transform** button in **Box** dialog box?

Q12. What happens when **In-canvas Render** button is selected?

Q13. What is use of Animation Workspace?

Q14. Which tool is used to add annotation to the animation?

FOR STUDENT NOTES

Chapter 11

Drawing

Topics Covered

The major topics covered in this chapter are:

- **Introduction to Drawing Workspace**
- **Specifying Drawing Preferences**
- **Drawing View**
- **Projected view**
- **Section View**
- **Detail View**
- **Break View**
- **Moving and Rotating Views**
- **Center line**
- **Center Mark and Center Mark Pattern**
- **Edge Extension**
- **Creating Sketch**

- **Dimensioning**
- **Annotation Settings**
- **Editing Dimensions**
- **Creating Text and Note**
- **Applying Surface Texture**
- **Feature Control Frame**
- **Datum Identifier**
- **Generating Tables and Balloon**
- **Bend Identifier and Table**
- **Renumbering and Aligning Balloons**
- **Output DWG, CSV, DXF, and PDF**

INTRODUCTION

Drawing is the engineering representation of a model on the paper. For manufacturing a model in the real world, we need some means by which we can tell the manufacturer what to manufacture. For this purpose, we create drawings from the models. These drawings have information like dimensions, material, tolerances, objective, precautions, and so on.

STARTING DRAWING WORKSPACE

In Autodesk Fusion 360, we create drawings in the **Drawing** workspace. The procedure to open a file in **DRAWING** workspace is discussed next.

- Open desired model from **Data Panel** or saved directory.
- Click on **From Design** option from **DRAWING** cascading menu in **Change Workspace** drop-down. The **CREATE DRAWING** dialog box will be displayed along with the model; refer to Figure-1.

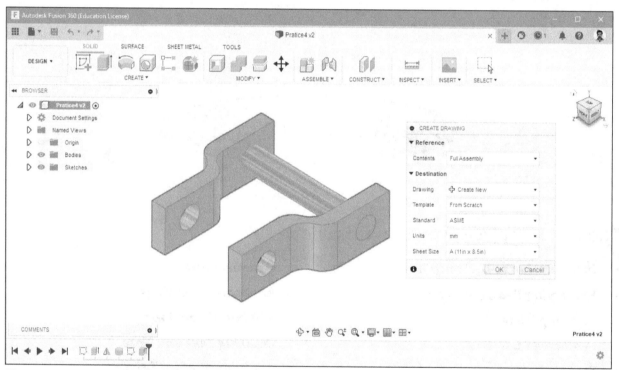

Figure-1. CREATE DRAWING dialog box

- Select the **Full Assembly** option from **Contents** drop-down of **Reference** area in the dialog box if you want to create drawing using whole assembly. Select the **Visible Only** option from **Contents** drop-down to automatically select only visible components of assembly. Select the **Select** option from **Contents** drop-down to manually select components in the canvas or browser.
- Click on the **Standard** drop-down from **CREATE DRAWING** dialog box and select the standard for the drawing. **ASME** option can use the third angle projection as well as first angle projection but **ISO** option uses first angle projection for placing views. There are also many other differences between two standards which you can find better in relevant books. Note that these projections are default settings but you can override them using **Preferences** dialog box.
- Click on the **Units** drop-down from **CREATE DRAWING** dialog box and select the unit for drawing.

- Click on the **Sheet Size** drop-down from **CREATE DRAWING** dialog box and select the size of sheet for drawing.
- After specifying the parameters, click on the **OK** button from **CREATE DRAWING** dialog box. The **DRAWING** workspace will be activated along with **DRAWING VIEW** dialog box in a new drawing file; refer to Figure-2.
- The model will be attached to the cursor. Click on the drawing sheet to place the base view.

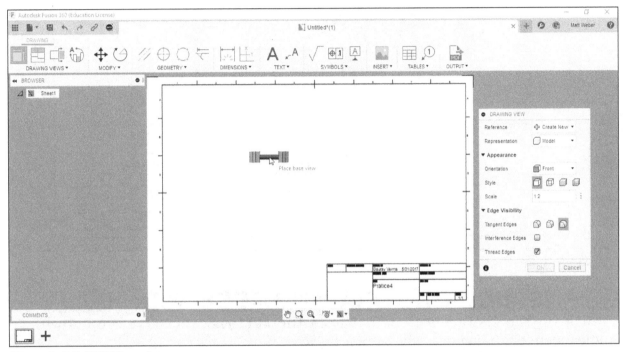

Figure-2. DRAWING workspace

Parameters in DRAWING VIEW dialog box

- Click on the **Orientation** drop-down and select the base view of model. Here **Front** view is default option. When you change the view from **Front** to **Top**, the view of model will automatically changed.
- Select the style of model from the **Style** section.
- Click in the **Scale** drop-down and select required scaling factor for model.
- Select required option from **Tangent Edges** section to apply the edges with the selected view.
- Select the **Interference Edges** check box if you want to keep interfering edges within the selected view.
- Select the **Thread Edges** check box if you want to keep the threads within the selected view.
- After specifying the parameters in **DRAWING VIEW** dialog box, click on **OK** button. The selected view of model will be placed; refer to Figure-3.

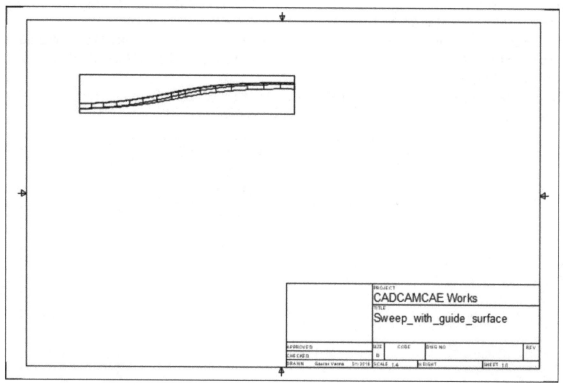

Figure-3. Top view of model

SPECIFYING DRAWING PREFERENCES

• Click on the **Preferences** option from the **Autodesk Account** drop-down. The **Preferences** dialog box will be displayed.
• Select the **Drawing** option from the **General** node at the left in the dialog box. The options will be displayed as shown in Figure-4.

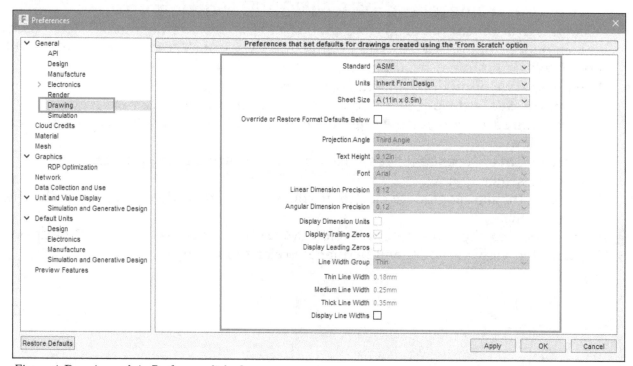

Figure-4. Drawing node in Preferences dialog box

- Select desired option from the **Standard** drop-down to define which standard to use for drafting. You can select **ASME** or **ISO** option from the drop-down to set it default for drafting or you can select the **Inherit From Design** to define the standard by 3D Model.
- Select desired unit of length from the **Units** drop-down. You can select inch or mm as unit.
- Select desired default size of sheet from the **Sheet Size** drop-down.
- Select the **Override or Restore Format Defaults Below** check box if you want to use parameters different from selected standard. The options below this check box will be activated; refer to Figure-5.

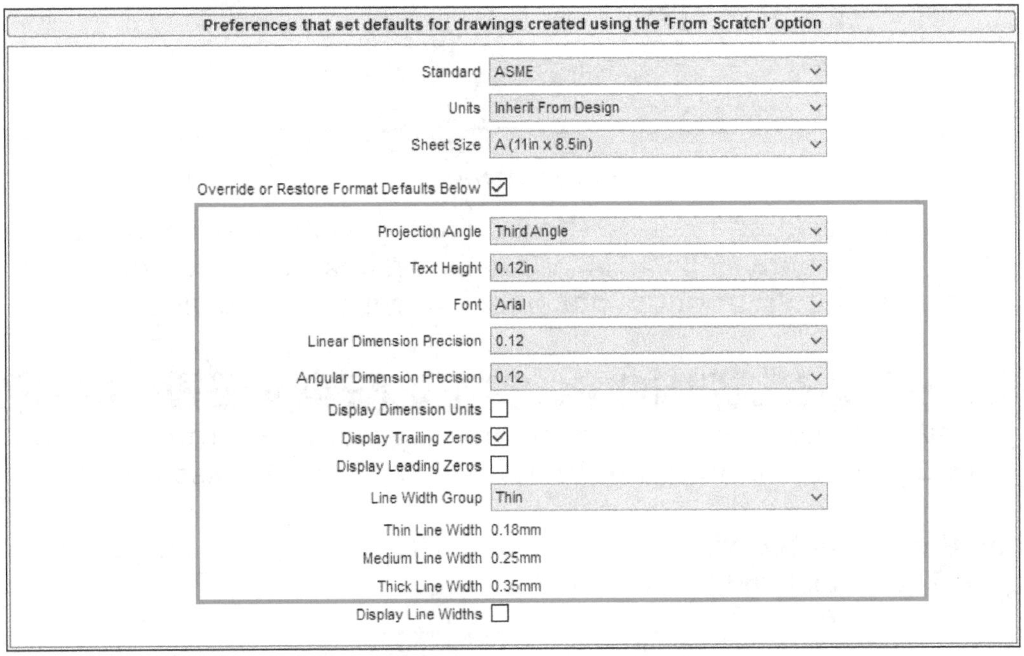

Figure-5. Override options

- Select desired option from the **Projection Angle** drop-down to define the view projection method for drawing.
- Select desired font from **Font** drop-down to define font of all the annotations in drawing area.
- Select desired value for height of annotations in **Text Height** drop-down.
- Select desired precision level for linear and angular dimensions in the **Linear Dimension Precision** and **Angular Dimension Precision** drop-downs, respectively.
- Select the **Display Dimension Units** check box to display unit of annotation in drawing.
- Select the **Display Trailing Zeros** check box to display zeros after decimal places if the dimension is not up to specified precision level.
- Select the **Display Leading Zeros** check box if you want to display zeros before the dimension values in the annotations.
- Select the **Display Line Widths** check box if you want to display lines in drawing by their specified widths rather then using same width for all lines.
- Click on the **OK** button from the **Preferences** dialog box.

CREATING BASE VIEW

The **Base View** tool is used to create base view derived directly from a 3D model. You can create various views of your model using this view. Note that you should have only one base view and other views should be projection of that view. The procedure to use this tool is discussed next.

- Click on the **Base View** tool from **DRAWING VIEWS** panel; refer to Figure-6. The **DRAWING VIEW** dialog box will be displayed.

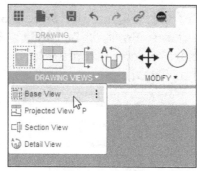

Figure-6. Base View tool

- Click at desired location to place the view and specify desired parameters in the dialog box. The options of **DRAWING VIEW** dialog box have been discussed earlier.
- Click on the **OK** button from the dialog box to generate the view.

CREATING PROJECTED VIEW

The **Projected View** tool is used to create drawing views generated from an existing base or drawing view. The procedure to use this tool is discussed next.

- Click on the **Projected View** tool from **DRAWING VIEWS** panel in the **Toolbar**; refer to Figure-7. The tool will be activated.

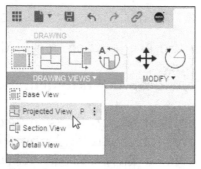

Figure-7. Projected View tool

- Click on the pre-existing base view from drawing sheet to select the view as parent view; refer to Figure-8. The projected view will be attached to the cursor; refer to Figure-9.

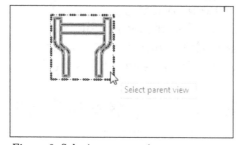

Figure-8. Selecting parent view

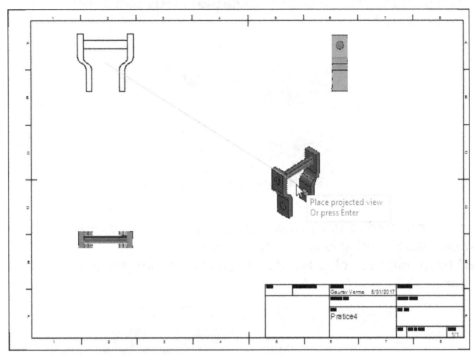

Figure-9. Placing projected view

- Place the projected views of model at desired locations and press **ENTER**. The projected view(s) are created on drawing sheet; refer to Figure-10.

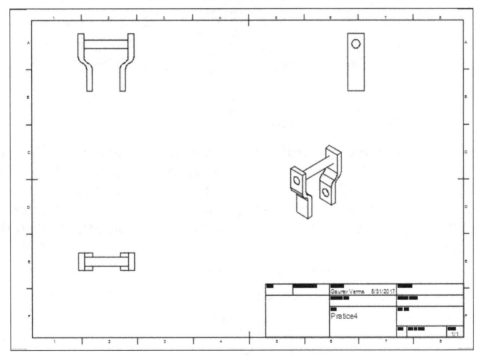

Figure-10. Projected views created

CREATING SECTION VIEW

The **Section View** tool is used to create full, half, offset, or aligned section views. The procedure to use this tool is discussed next.

- Click on the **Section View** tool from **DRAWING VIEWS** panel; refer to Figure-11. The tool will be activated.

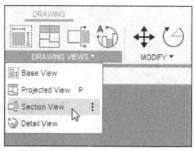

Figure-11. Section View tool

- Click on the projected or base view from drawing sheet to be used for section view. The **DRAWING VIEW** dialog box will be displayed.
- You need to divide the selected parent view into two parts to create half section; refer to Figure-12.

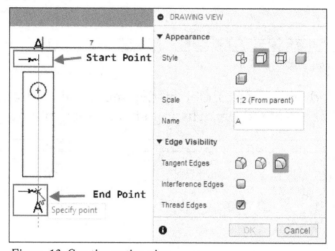

Figure-12. Creating section view

- Click on the parent view to specify the start point and end point. You can create different shape of section line to create desired type of section. Note that you can use snap point in right-click shortcut menu to select points for section line; refer to Figure-13. After specifying the start point, intermediate points, and end point, click on **ENTER** key. The section view will be created and attached to the cursor; refer to Figure-14.

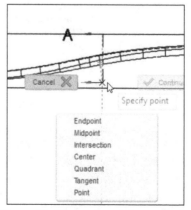

Figure-13. Shortcut menu for snap points

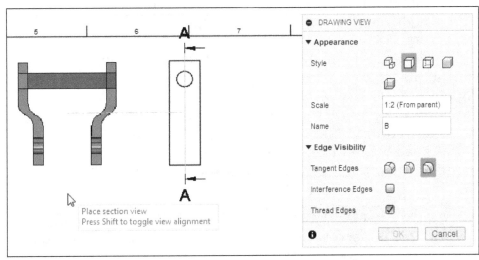

Figure-14. Section view attached to the cursor

- Place the section view at desired location and click on **OK** button from **DRAWING VIEW** dialog box. The section view will be created on drawing sheet; refer to Figure-15.

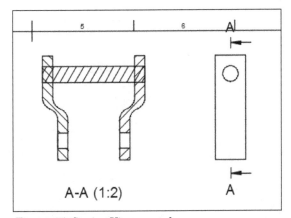

Figure-15. Section View created

- Note that the hatching is created automatically in the section view. To edit hatching, double-click on it. The **HATCH** dialog box will be displayed. Specify desired values and click on the **Close** button.
- If you want to edit the section annotation then double-click on it. The **TEXT** dialog box will be displayed; refer to Figure-16.

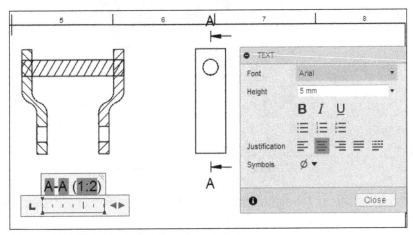

Figure-16. TEXT dialog box for annotation

- Click on the **Font** drop-down and select required font from the list.
- Click in the **Height** edit box and specify the height of text and select the writing style of text from **TEXT** dialog box.
- Click on desired text alignment from **Justification** section.
- If you want to insert any special symbol then click on it from **Symbols** section of the dialog box.
- After specifying the parameters for the text, click in the text box from drawing sheet and enter required text.
- Click on the **Close** button from **TEXT** dialog box to complete the process. The text will be modified according to parameters specified.

CREATING DETAIL VIEW

The **Detail View** tool is used to create detail view from any projected or base view. The procedure to use this tool is discussed next.

- Click on the **Detail View** tool of **DRAWING VIEWS** panel; refer to Figure-17. The tool will be activated and you will be asked to select a view whose selected section is to be included in detail view.

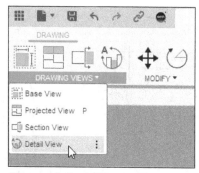

Figure-17. Detail View tool

- Select desired view from drawing sheet; refer to Figure-18. The **DRAWING VIEW** dialog box will be displayed.

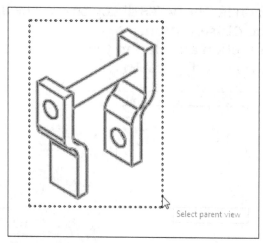

Figure-18. Selecting parent view for detail view

- Click at desired location to specify the center point for boundaries of detail view; refer to Figure-19.
- Create the boundary on parent view as required; refer to Figure-20.

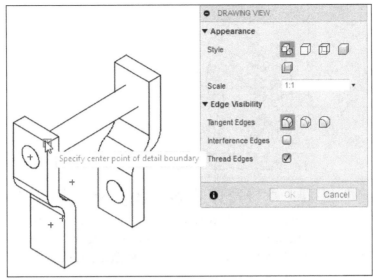

Figure-19. Specifying center point for boundary

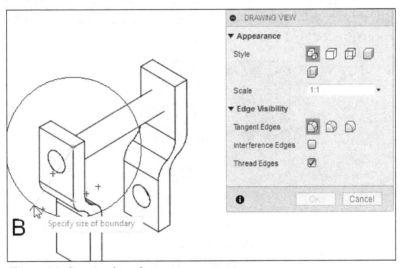

Figure-20. Creating boundary

- On creating the circular boundary, the detail view of model will be attached to the cursor; refer to Figure-21. Place it as desired.

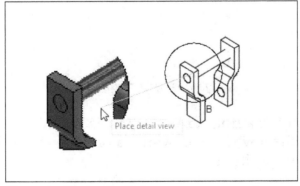

Figure-21. Placing detail view

- After placing the view, set the parameters of detail view in **DRAWING VIEW** dialog box and click on **OK** button. The detailed view will be created; refer to Figure-22.

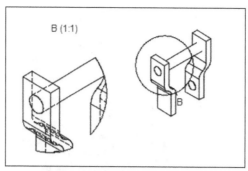

Figure-22. Detailed view created

CREATING BREAK VIEW

The **Break View** tool is used to shorten an existing drawing view by removing a portion of the design and indicating the missing section in the drawing. Broken view is generally created to place long views. The procedure to use this tool is discussed next.

- Click on the **Break View** tool from **DRAWING VIEWS** panel; refer to Figure-23. The **BREAK VIEW** dialog box will be displayed and you will be asked to select the drawing view to be broken; refer to Figure-24.

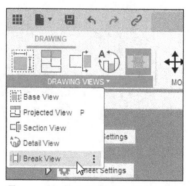

Figure-23. Break View tool

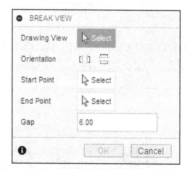

Figure-24. BREAK VIEW dialog box

- The **Select** button of **Drawing View** section is active by default. Select desired view from drawing sheet which you want to break; refer to Figure-25. You will be asked to select break position.

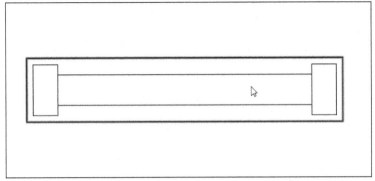

Figure-25. Selecting drawing view for broken view

- Select the start point of break position using the **Start Point** selection button; refer to Figure-26. You will be asked to select end point of break position.

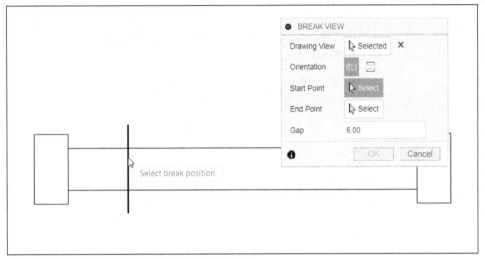

Figure-26. Selecting start point of break position

- Move the cursor away and click at desired location to specify end position of break position in the **End Point** selection button; refer to Figure-27.

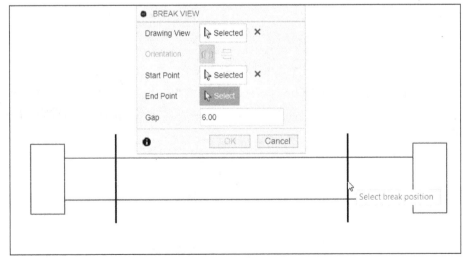

Figure-27. Selecting end point of break position

- Specify desired value in **Gap** edit box of the dialog box to specify distance for the gap between the two sections.
- After specifying parameters for break view, click on the **OK** button from the dialog box. The broken view will be created; refer to Figure-28.

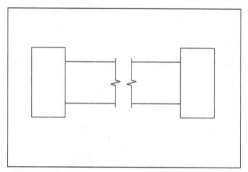

Figure-28. Broken view created

MOVING VIEWS

The **Move** tool is used to move selected view from one place to another. The procedure to use this tool is discussed next.

- Click on the **Move** tool from **MODIFY** drop-down; refer to Figure-29. The **MOVE** dialog box will be displayed; refer to Figure-30.

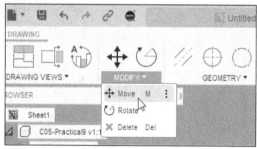

Figure-29. Move tool

Figure-30. MOVE dialog box

- The **Select** button of **Selection** section in **MOVE** dialog box is active by default. Click on the view to select the view to be moved. The updated **MOVE** dialog box will be displayed; refer to Figure-31.

Figure-31. Updated MOVE dialog box

- Click on **Transform** button from **MOVE** dialog box and select a base point from the selected view. The selected view will get attached to the cursor to move the view from one place to another; refer to Figure-32.

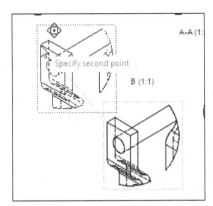
Figure-32. Specifying second point

• Click on the sheet to place the view. The view will be moved from previous position to new position.
• Click on the **OK** button from **MOVE** dialog box to complete the process.

ROTATING VIEWS

The **Rotate** tool is used to rotate the selected view as desired. The procedure to use this tool is discussed next.

• Click on the **Rotate** tool from **MODIFY** panel; refer to Figure-33. The **ROTATE** dialog box will be displayed; refer to Figure-34.

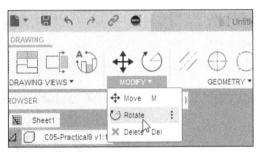
Figure-33. Rotate tool

Figure-34. ROTATE dialog box

• The **Select** button of **Selection** section in **ROTATE** dialog box is active by default. Click on the view to be rotated.
• Click on **Transform** button from **ROTATE** dialog box and select the base point from selected view. The view will be attached to the cursor; refer to Figure-35.

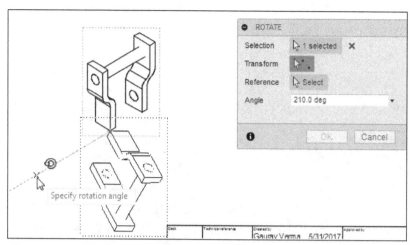

Figure-35. Rotating view

- You can adjust the angle of rotation of view by moving the cursor around parent view and place as required. If you want to specify the value of rotation then click in the **Angle** edit box of **ROTATE** dialog box and specify the value.
- You can also set the reference for rotation by clicking on **Select** button of **Reference** section and selecting the reference point from parent view.
- After specifying the parameters, click on **OK** button from **ROTATE** dialog box to complete the process.

CREATING CENTERLINE

The **Centerline** tool is used to create centerline. The procedure to use this tool is discussed next.

- Click on the **Centerline** tool from **GEOMETRY** panel; refer to Figure-36. The tool will be activated.

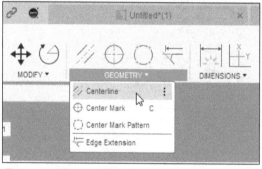

Figure-36. Centerline tool

- You need to select parallel edges from a view to create a center line. Click on the desired edge to select; refer to Figure-37. You will be asked to select the second edge.

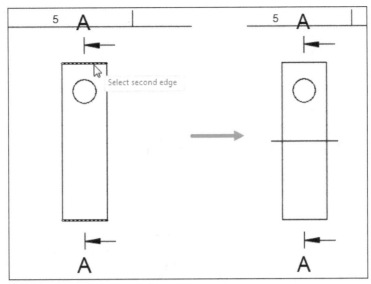

Figure-37. Creating centerline

- On selecting second edge, the center line will be created.

CREATING CENTER MARK

The **Center Mark** tool is used to create center mark for rounds and circles. The procedure to use this tool is discussed next.

- Click on the **Center Mark** tool from **GEOMETRY** panel; refer to Figure-38. The tool will be activated. You need to select a round edge from any view.

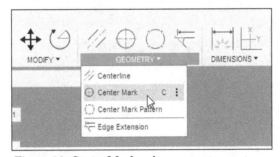

Figure-38. Center Mark tool

- On selecting the edge, the center mark will be created for selected round edge; refer to Figure-39.

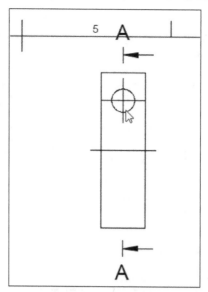

Figure-39. Creating center mark

CREATING CENTER MARK PATTERN

The **Center Mark Pattern** tool is used to generate pattern of center marks in the drawing view. The procedure to use this tool is given next.

• Click on the **Center Mark Pattern** tool from the **GEOMETRY** panel of **Toolbar**; refer to Figure-40. The **CENTER MARK PATTERN** dialog box will be displayed; refer to Figure-41. You will be asked to select a hole or round edge whose center mark is to be created.

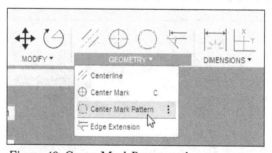

Figure-40. Center Mark Pattern tool

Figure-41. CENTER MARK PATTERN dialog box

• Click on desired holes one by one. If the holes are in circular pattern then the dialog box will be displayed as shown in Figure-42. Select the **Center Mark** check box to generate center mark for pitch circle. Click on the **OK** button to generate center marks.

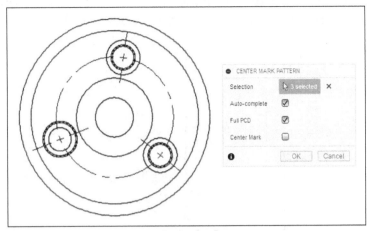

Figure-42. Holes selected in circular pattern

- In cases where you want to specify the PCD reference manually, click on the **Reference Edge** selection button after selecting holes and then select the reference edge; refer to Figure-43.

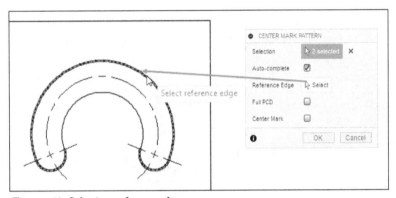

Figure-43. Selecting reference edge

- In case of selecting holes in linear direction, you can select maximum 3 holes for center mark.

EDGE EXTENSION

The **Edge Extension** tool is used to create an extension of associative edge for two intersecting (unequal) edges within an existing drawing view. The procedure to use this tool is discussed next.

- Click on the **Edge Extension** tool from **GEOMETRY** panel; refer to Figure-44. The tool will be activated.

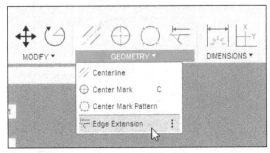

Figure-44. Edge Extension tool

- You need to select the two non-parallel (going to intersect) edges to create the extension. Click on the edge to select; refer to Figure-45.

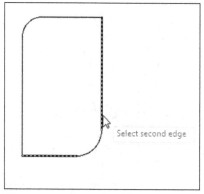

Figure-45. Selecting second edge

- On selecting the second edge, the extended edge will be displayed; refer to Figure-46.

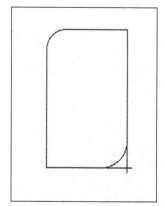

Figure-46. Edge extension created

CREATING SKETCH

The **Create Sketch** tool enters the sketch mode and creates a new sketch on the current sheet. The procedure to use this tool is discussed next.

- Click on the **Create Sketch** tool from **GEOMETRY** panel in the **Toolbar**; refer to Figure-47. The **SKETCH** contextual tab will be displayed along with **SKETCH** dialog box; refer to Figure-48.

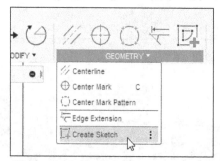

Figure-47. Create Sketch tool

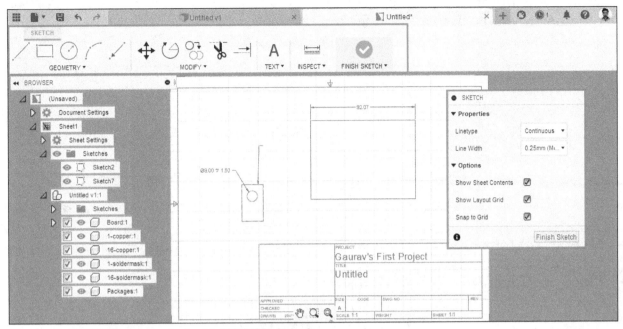

Figure-48. SKETCH contextual tab with SKETCH dialog box

- Select desired option from **Linetype** drop-down of **Properties** area in the dialog box to change the linetype of new or selected sketch geometry.
- Select desired option from **Line Width** drop-down of **Properties** area in the dialog box to change the line width of new or selected sketch geometry.
- Select **Show Sheet Contents** check box from **Options** area to show the contents of the sheet as you edit the sketch.
- Select **Show Layout Grid** check box from **Options** area to show the layout grid in the canvas as you edit the sketch.
- Select **Snap to Grid** check box from **Options** area to snap to the layout grid in the canvas as you edit the sketch.
- After specifying parameters in the **SKETCH** dialog box, create desired sketch from **SKETCH** tab in the **Toolbar**.
- After creating the sketch, click on **Finish Sketch** button from **Toolbar** or from the dialog box to exit the tab.

DIMENSIONING

The **Dimension** tool is used to apply dimension to model. The procedure to use this tool is discussed next.

- Click on the **Dimension** tool from **DIMENSIONS** panel; refer to Figure-49. The tool will be activated.

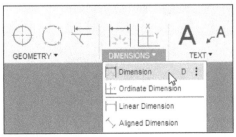

Figure-49. Dimension tool

*   When you hover the cursor over on any geometry, the measurement of that dimension will be displayed; refer to Figure-50.

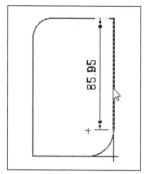

Figure-50. Checking dimension

*   If you want to mention that dimension then click on the line. The dimension will be attached to the cursor.
*   Place the dimension at desired distance by clicking.

Ordinate Dimensioning

The **Ordinate Dimension** tool is used to apply ordinate dimensioning. The procedure to use this tool is discussed next.

*   Click on the **Ordinate Dimension** tool from **DIMENSIONS** panel; refer to Figure-51. The tool will be activated and you will be asked to specify origin for ordinate dimensioning.

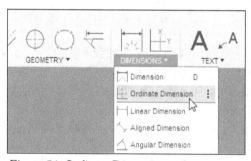

Figure-51. Ordinate Dimension tool

*   Click at 0desired location to place origin of ordinate dimensioning. The ordinate dimension will be attached to the cursor; refer to Figure-52.

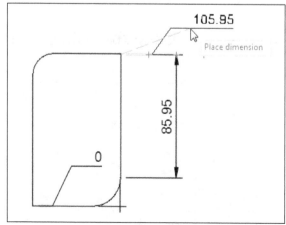

Figure-52. Calculating ordinate dimension

- Click at desired location to place dimension text. Similarly, you can dimension horizontally/vertically.

Creating Linear Dimension

The **Linear Dimension** tool is used to create linear dimensions in horizontal/vertical directions. The procedure to use this tool is given next.

- Click on the **Linear Dimension** tool from the **DIMENSIONS** panel in the **Toolbar**; refer to Figure-53. You will be asked to select points.

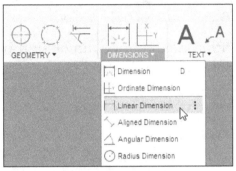

Figure-53. Linear Dimension tool

- Select the two points one by one to create horizontal/vertical distance dimension. The dimension will get attached to cursor.
- Move the cursor in desired direction to create vertical or horizontal dimension if the two points are not in aligned vertically/horizontally; refer to Figure-54.

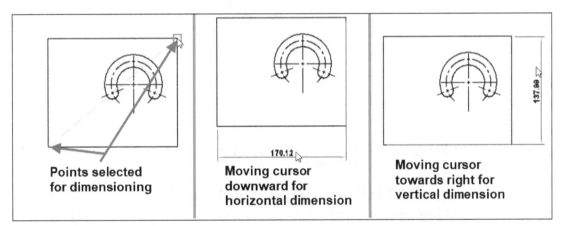

Figure-54. Creating linear dimension

- Click at desired location to place the dimension.

Creating Aligned Dimension

The **Aligned Dimension** tool is used to create aligned dimension between two selected points. The procedure is given next.

- Click on the **Aligned Dimension** tool from **DIMENSIONS** panel in the **Toolbar**; refer to Figure-55. You will be asked to select points.

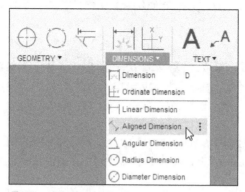

Figure-55. Aligned Dimension tool

- Click at desired two points. The distance dimension between two selected points will get attached to cursor; refer to Figure-56.

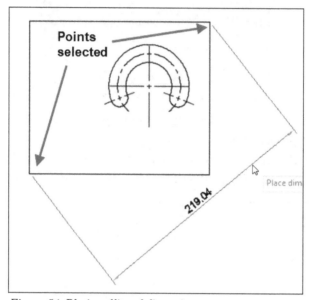

Figure-56. Placing Aligned dimension

- Click at desired location to place the dimension.

Creating Angular Dimension

The **Angular Dimension** tool is used to create angle dimension between two selected lines/edges. If you select an arc then its angular span is created as dimension. The procedure to use this tool is given next.

- Click on the **Angular Dimension** tool from the **DIMENSIONS** panel in the **Toolbar**; refer to Figure-57. You will be asked to select entities.

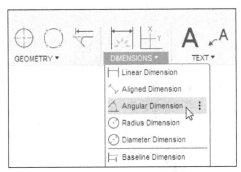

Figure-57. Angular Dimension tool

- Select two lines between which you want to create angular dimension or select an arc whose angular dimension is to be created. The dimension will get attached to cursor; refer to Figure-58.

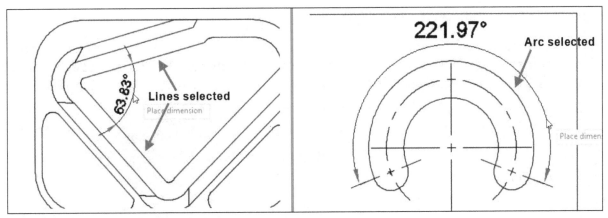

Figure-58. Creating angular dimension

- Click at desired location to place the dimension.

Creating Radius Dimension

The **Radius Dimension** tool is used to apply radius dimension to selected arc, circle, and circular edges. The procedure to use this tool is given next.

- Click on the **Radius Dimension** tool in **DIMENSIONS** panel of **Toolbar**; refer to Figure-59. You will be asked to select circular edge or circle in the drawing.

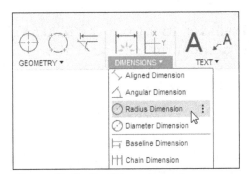

Figure-59. Radius Dimension tool

- Click at desired entity. The radius dimension will get attached to cursor; refer to Figure-60.

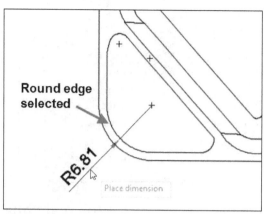

Figure-60. Creating radius dimension

- Click at desired location to place the dimension.

Creating Diameter Dimension

The **Diameter Dimension** tool is used to create diameter dimension for selected round edges, circles, and fillets in the drawing; refer to Figure-61. The procedure to use this tool is similar to **Radius Dimension** tool.

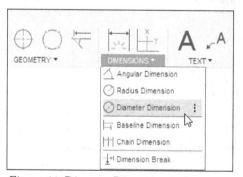

Figure-61. Diameter Dimension tool

Baseline Dimensioning

The baseline dimensioning is performed when you want to create linear dimensions with respect to one common base reference. The procedure to create baseline dimensions is given next.

- Click on the **Linear Dimension** tool from the **DIMENSIONS** panel of **Toolbar**; refer to Figure-62, and create a linear dimension which will be used as base line; refer to Figure-63.

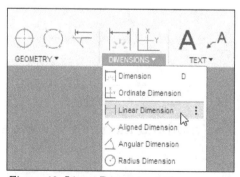

Figure-62. Linear Dimension tool

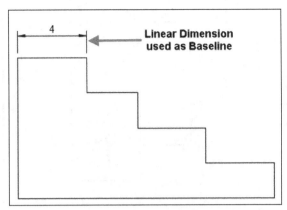

Figure-63. Linear dimension used as baseline

- Click on the **Baseline Dimension** tool from the **DIMENSIONS** panel of the **Toolbar**; refer to Figure-64, and select one of the leg of base dimension.

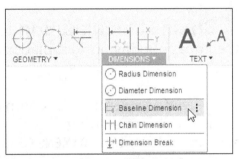

Figure-64. Baseline Dimension tool

- A new dimension will be generated with one end point fixed to selected leg and other end point of dimension attached to cursor; refer to Figure-65.

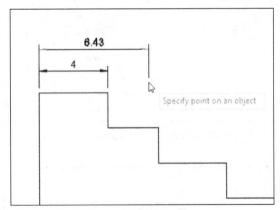

Figure-65. Placing baseline dimension

- Select desired point of sketch to create the dimension. You will be asked to specify end point for another dimension. Select desired points to create required dimensions. Once you have created required dimensions, press **ENTER** from keyboard to complete the process. The dimensions will be generated; refer to Figure-66.

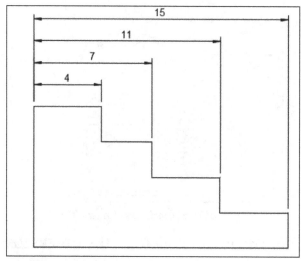

Figure-66. Baseline dimension created

Chain Dimensioning

In chain dimensioning, all the linear dimensions are placed side by side. The procedure to use this tool is given next.

- Create a linear dimension to used as reference for chain dimensioning as discussed in previous section for baseline dimensioning.
- Click on the **Chain Dimension** tool from the **DIMENSIONS** panel of the **Toolbar**; refer to Figure-67, and select the base dimension. You will be asked to specify end point of new dimension; refer to Figure-68.

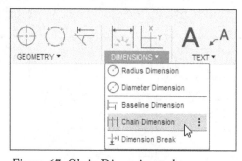

Figure-67. Chain Dimension tool

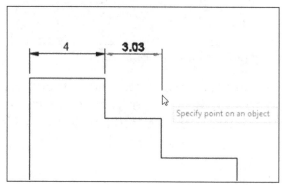

Figure-68. Placing chain dimensions

- Click at desired location to specify end point. You will be asked to specify end point of another dimension.

- Click at desired locations to create required dimensions and then press **ENTER** to exit. The dimensions will be created; refer to Figure-69.

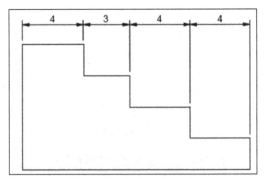

Figure-69. Chain dimensions created

Dimension Break

The **Dimension Break** tool is used to resolve intersecting dimensions in the drawing. This tool adds/removes break in selected dimension. The procedure to use this tool is given next.

- Click on the **Dimension Break** tool from the **DIMENSIONS** panel of the **Toolbar**; refer to Figure-70. The **DIMENSION BREAK** dialog box will be displayed; refer to Figure-71, and you will be asked to select the dimension to add or remove break.

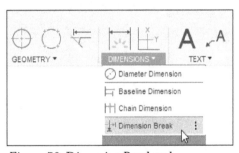

Figure-70. Dimension Break tool

Figure-71. DIMENSION BREAK dialog box

- Select desired dimension and then select the **Add Break** or **Remove Break** option from the **Operation** drop-down to perform respective action. In our case, we are selecting **Add Break** option from the drop-down to add break in intersecting dimension.
- Click on the **OK** button from the dialog box. The selected operation will be performed on the dimension; refer to Figure-72.

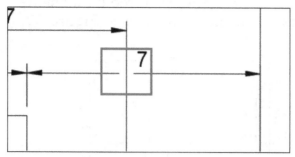

Figure-72. Dimension after applying Add Break

EDITING DIMENSIONS

Once you have generated dimensions, there are some cases where you need to provide special symbols with dimensions or you want to modify the value of dimension. The procedure to edit dimension is given next.

• Double-click on the dimension in drawing that you want to edit. The **DIMENSION** dialog box will be displayed; refer to Figure-73.

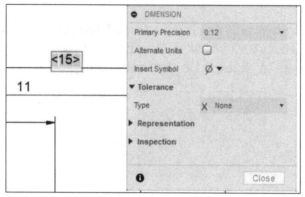

Figure-73. DIMENSION dialog box

• Select desired precision level from **Primary Precision** drop-down.
• Select the **Alternate Units** check box to also display dimensions in inches if they are in mm and vice-versa; refer to Figure-74.

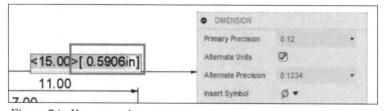

Figure-74. Alternate units

• Click at desired place in dimension box and select desired symbol to be inserted. The symbol will be added to dimension; refer to Figure-75.

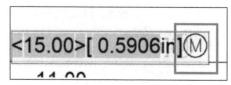

Figure-75. Symbol added to dimension

- Select desired type of tolerance from the **Type** drop-down in **Tolerance** section to apply tolerance to dimension. Select the **Symmetrical** option from the **Type** drop-down if tolerance is equal in both negative and positive directions. Select the **Deviation** option from the **Type** drop-down if tolerance is different in positive and negative directions. Select the **Limit** option from the **Type** drop-down if you want to display upper and lower limit of dimension; refer to Figure-76.

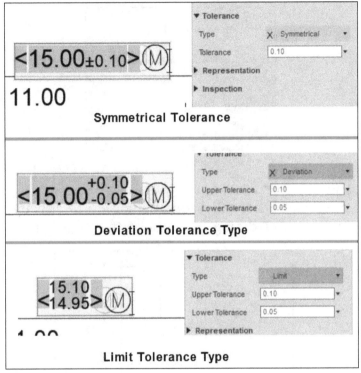

Figure-76. Tolerance types

- After selecting desired option from drop-down, specify desired value of tolerance in edit boxes below it in **Tolerance** section.
- Expand the **Representation** node and select desired button to make the dimension as reference dimension, not to scale, and theoretically exact; refer to Figure-77.

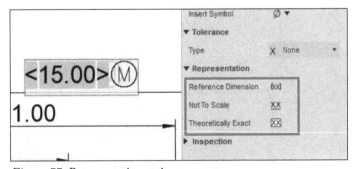

Figure-77. Representation options

- Similarly, expand the **Inspection** node and set desired parameters if you want to define inspection rate of current dimension.
- After specifying parameters, click on the **Close** button or click anywhere in blank space of drawing to exit the tool.

CREATING TEXT

The **Text** tool is used to insert text into the active drawing. The procedure to use this tool is discussed next.

- Click on the **Text** tool from **TEXT** panel; refer to Figure-78. The **Text** tool is activated and you will be asked to draw a text box.

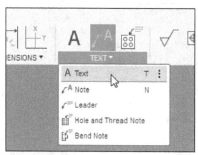

Figure-78. Text tool

- Click to specify the first corner.
- After specifying the first corner, click on the screen to specify the second corner; refer to Figure-79. The **TEXT** dialog box will be displayed; refer to Figure-80.

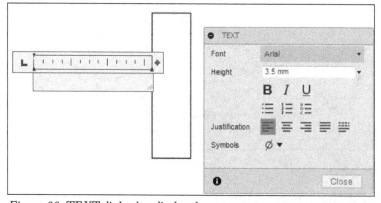

Figure-79. Specifying second corner

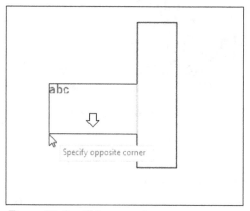

Figure-80. TEXT dialog box displayed

- Click on the **Font** drop-down and select required font for text.
- Click in the **Height** edit box and specify the height of text.
- Set the other parameters as required and type desired text in the **Text** box.
- Click on **Close** button. The text will be added in drawing sheet.

CREATING NOTE

The **Note** tool automatically creates different type of note based on the type of object you select on the current sheet. The procedure to use this tool is discussed next.

*   Click on the **Note** tool from **TEXT** panel in the **Toolbar**; refer to Figure-81. The **NOTE** dialog box will be displayed; refer to Figure-82.

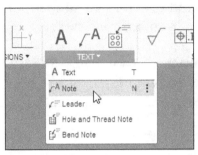

Figure-81. Note tool

Figure-82. NOTE dialog box

*   The **Automatic** option in **Type** drop-down of **NOTE** dialog box is selected by default.
*   Select **Leader** option from **Type** drop-down to create a multiline text object with a leader that associates the note to a component or feature.
*   Select **Hole and Thread Note** option from **Type** drop-down to create a hole or thread note associated with a hole or thread feature.
*   Select **Bend Note** option from **Type** drop-down to create a bend note associated with a sheet metal flat pattern bend.
*   Here, we selected **Leader** option in **Type** drop-down of dialog box or click on **Leader** tool from **TEXT** panel of **Toolbar**; refer to Figure-81. The tool will be activated.
*   The **Select** button of **Select** section is active by default. Click on the drawing to select the start point. An arrow will be attached to the cursor.
*   Click on the screen to place text box. The text box will be displayed along with **TEXT** dialog box; refer to Figure-83.

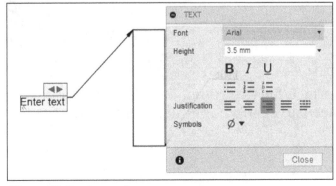

Figure-83. Leader text box

- Set desired parameter in **TEXT** dialog box and type the text in text box.
- After specifying the parameters, click on **Close** button from **TEXT** dialog box.

Similarly, you can use the **Hole and Thread Note** tool and **Bend Note** tool.

APPLYING SURFACE TEXTURE

The **Surface Texture** tool is used to apply surface finish symbol to objects in drawing. The procedure to use this tool is discussed next.

- Click on the **Surface Texture** tool from **SYMBOLS** panel of **Toolbar;** refer to Figure-84. The tool will be activated.

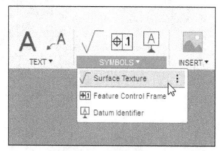

Figure-84. Surface Texture tool

- Click on the model in a drawing view to select object. An arrow will be displayed on the selected part.
- Click to select the second point for leader or press **ENTER** key to place the symbol directly on selected face. The **SURFACE TEXTURE** dialog box will be displayed; refer to Figure-85.

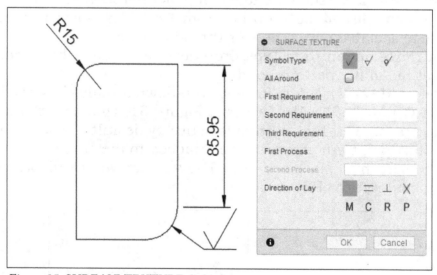

Figure-85. SURFACE TEXTURE dialog box

- Select desired surface finish symbol button from **Symbol Type** section.
- Select the **All Around** check box from **SURFACE TEXTURE** dialog box if you want

to apply surface finish to all the surfaces in current view.

- Click in the **First Requirement** edit box of **SURFACE TEXTURE** dialog box and enter the first parameter of surface finish symbol. Similarly, specify the other parameters in the edit boxes of the dialog box.
- Select the lay direction button as desired from **Direction of Lay** section.
- After specifying the parameters, click on the **OK** button from **SURFACE TEXTURE** dialog box; refer to Figure-86.

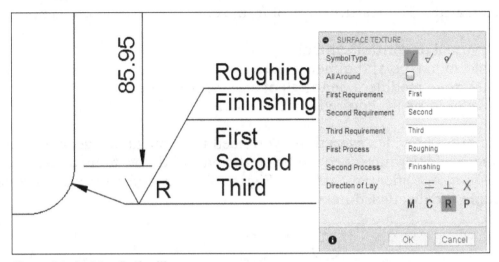

Figure-86. Applying Surface Texture

Surface Texture Symbols or Surface Roughness Symbols

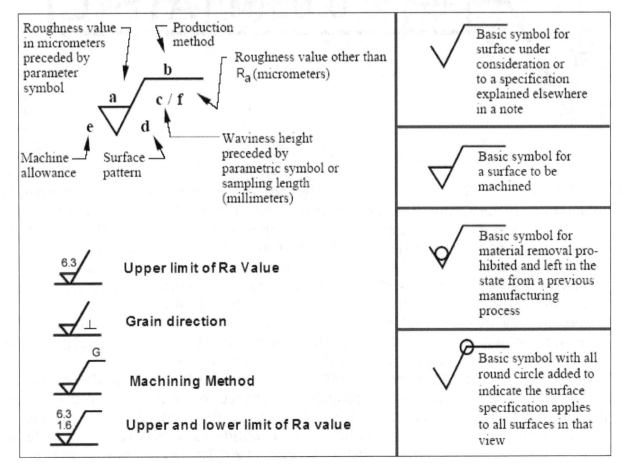

CREATING FEATURE CONTROL FRAME

The **Feature Control Frame** contains data for controlled specifications. It shows the characteristics symbol, datum reference, and tolerance value. It is a rectangular frame which is divided into two or more sections. The feature control frame is also known as GD&T box in laymen's language. The method to insert Feature Control Frame in drawing is same as discussed for **Surface Texture** symbol. In GD&T, a feature control frame is required to describe the conditions and tolerances of a geometric control on a part's feature. The feature control frame consists of four pieces of information:

1. GD&T symbol or control symbol
2. Tolerance zone type and dimensions
3. Tolerance zone modifiers: features of size, projections...
4. Datum references (if required by the GD&T symbol)

This information provides everything you need to know about geometry of part like what geometrical tolerance needs to be on the part and how to measure or determine if the part is in specification; refer to Figure-87. The common elements of feature control frame are discussed next.

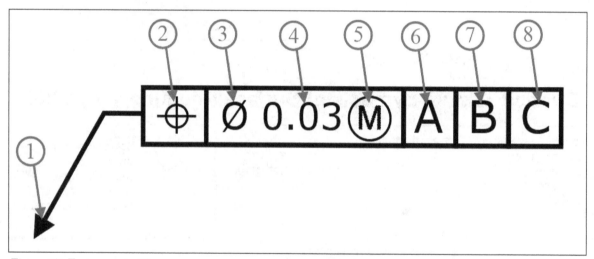

Figure-87. Feature control frame

1. **Leader Arrow** – This arrow points to the feature then the geometric control is placed on. If the arrow points to a surface then the surface is controlled by the GD&T. If it points to a diametric dimension then the axis is controlled by GD&T. The arrow is optional but helps clarify the feature being controlled.
2. **Geometric Symbol** – This is where your geometric control is specified.
3. **Diameter Symbol (if required)** – If the geometric control is a diametrical tolerance then the diameter symbol (Ø) will be in front of the tolerance value.
4. **Tolerance Value** – If the tolerance is a diameter you will see the Ø symbol next to the dimension signifying a diametric tolerance zone. The tolerance of the GD&T is in same unit of measure that the drawing is written in.
5. **Feature of Size or Tolerance Modifiers (if required)** – This is where you call out max material condition or a projected tolerance in the feature control frame.
6. **Primary Datum (if required)** – If a datum is required, this is the main datum used for the GD&T control. The letter corresponds to a feature somewhere on the part which will be marked with the same letter. This is the datum that must

be constrained first when measuring the part. Note: The order of the datum is important for measurement of the part. The primary datum is usually held in three places to fix 3 degrees of freedom

7. **Secondary Datum (if required)** – If a secondary datum is required, it will be to the right of the primary datum. This letter corresponds to a feature somewhere on the part which will be marked with the same letter. During measurement, this is the datum fixated after the primary datum.

8. **Tertiary Datum (if required)** – If a third datum is required, it will be to the right of the secondary datum. This letter corresponds to a feature somewhere on the part which will be marked with the same letter. During measurement, this is the datum fixated last.

Reading Feature Control Frame

The feature control frame forms a kind of sentence when you read it. Below is how you would read the frame in order to describe the feature.

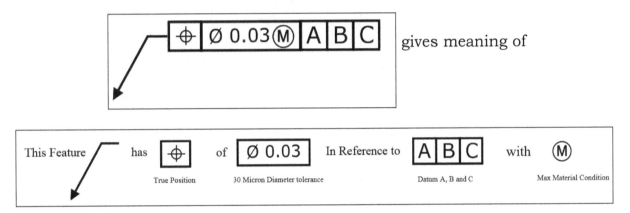

Meaning of various geometric symbols are given in Figure-88.

SYMBOL	CHARACTERISTICS	CATEGORY
—	Straightness	Form
▱	Flatness	
○	Circulatity	
⌗	Cylindricity	
⌒	Profile of a Line	Profile
⌓	Profile of Surface	
∠	Angularity	Orientation
⊥	Perpendicularity	
//	Parallelism	
⊕	Position	Location
◎	Concentricity	
≡	Symmetry	
↗	Circular Runout	Runout
↗↗	Total Runout	

Figure-88. Geometric Symbols

Figure-89 and Figure-90 shows the use of geometric tolerances in real-world.

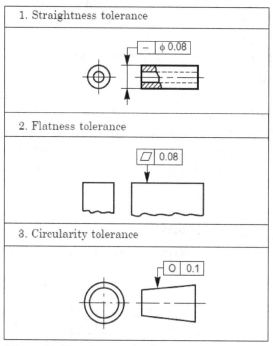

Figure-89. Use of geometric tolerance 1

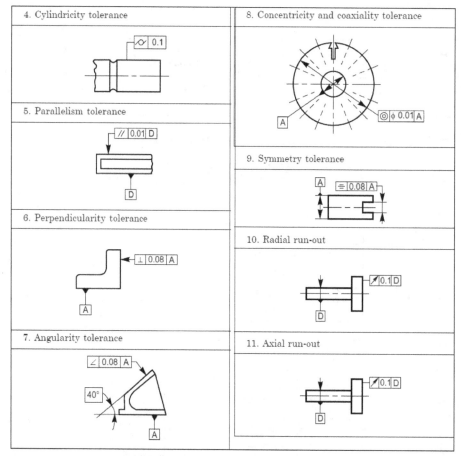

Figure-90. Use of geometric tolerance 2

Note that in applying most of the geometrical tolerances, you need to define a datum plane like in Perpendicularity, Parallelism, and so on. There are a few dimensioning symbols also used in geometric dimensioning and tolerances, which are given in Figure-91.

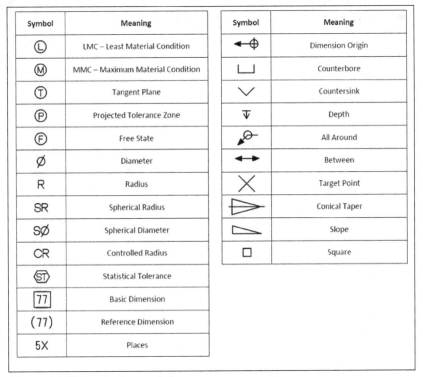

Symbol	Meaning	Symbol	Meaning
Ⓛ	LMC – Least Material Condition	←⊕	Dimension Origin
Ⓜ	MMC – Maximum Material Condition	⊔	Counterbore
Ⓣ	Tangent Plane	∨	Countersink
Ⓟ	Projected Tolerance Zone	⊽	Depth
Ⓕ	Free State	⌀	All Around
⌀	Diameter	←→	Between
R	Radius	✕	Target Point
SR	Spherical Radius	▷	Conical Taper
SØ	Spherical Diameter	◺	Slope
CR	Controlled Radius	☐	Square
⟨ST⟩	Statistical Tolerance		
77	Basic Dimension		
(77)	Reference Dimension		
5X	Places		

Figure-91. Dimensioning symbols

The procedure to use this tool is discussed next.

- Click on the **Feature Control Frame** tool from **SYMBOLS** panel of **Toolbar**; refer to Figure-92. The tool will be activated.

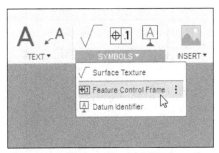

Figure-92. Feature Control Frame tool

- Click on the drawing to select the object. An arrow will be attached to the cursor; refer to Figure-93.

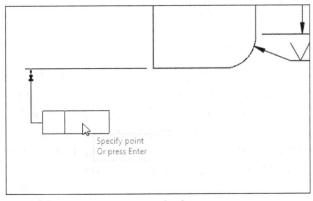

Figure-93. Placing the rectangular box

- Click on the screen to place the rectangular box and press **ENTER** key. The **FEATURE CONTROL FRAME** dialog box will be displayed; refer to Figure-94.

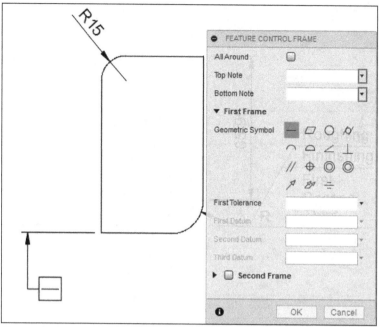

Figure-94. FEATURE CONTROL FRAME dialog box

- Click in the **Top Note** edit box and enter the note. If you want to add any symbol then click on the drop-down button. A list of symbols will be displayed; refer to Figure-95.

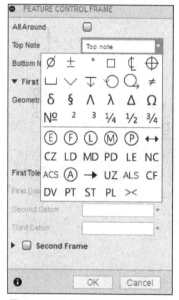

Figure-95. Adding symbols

- Click on required symbol. The selected symbol will be added over the feature control frame box. Similarly, you can set the symbol for **Bottom Note** edit box.
- Similarly, select required symbol for **Geometric Symbol** section. The selected symbol will be displayed in the feature frame rectangular box.
- Click in the **First Tolerance** edit box and enter the value of tolerance. You can also add the required symbol; refer to Figure-96.

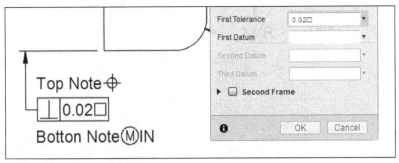

Figure-96. Adding tolerance

- Click on the **First Datum** edit box and enter the value of datum as required; refer to Figure-97.

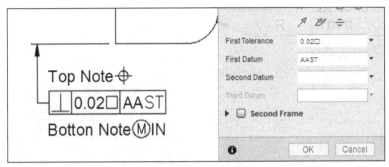

Figure-97. Specifying datum value

- Similarly enter the remaining parameters in respective edit boxes.
- Select the **Second Frame** check box of **FEATURE CONTROL FRAME** dialog box if you want to add second frame for the drawing. Feature Control Frames with two frames is also called Composite Feature Control Frame.
- After specifying the parameters, click on the **OK** button from **FEATURE CONTROL FRAME** dialog box to complete the process.

CREATING DATUM IDENTIFIER

The **Datum Identifier** tool identifies a datum feature for a feature control frame symbol. Datum identifier are circular frames divided in two parts by a horizontal line. The lower half represents the datum feature and the upper half is for additional information, such as dimensions of the datum target area; refer to Figure-98.

Figure-98. Datum target area

The procedure to use this tool is discussed next.

- Click on the **Datum Identifier** tool from **SYMBOLS** panel of **Toolbar**; refer to Figure-99. The tool will be activated.

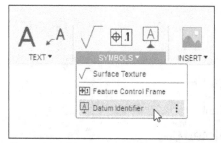

Figure-99. Datum Identifier tool

- Click on the drawing to select object. An arrow will be attached to the cursor.
- Specify the point and press **ENTER** key. The datum ID symbol will be displayed along with **DATUM IDENTIFIER** dialog box; refer to Figure-100.

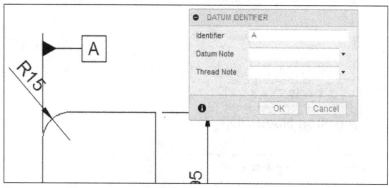

Figure-100. DATUM IDENTIFIER dialog box

- Click in the **Datum Note** edit box and enter desired note.
- Click in the **Thread Note** edit box and enter desired text. You can also add the required symbol from **Symbol** drop-down.
- After specifying the parameters, click on the **OK** button from **DATUM IDENTIFIER** dialog box; refer to Figure-101.

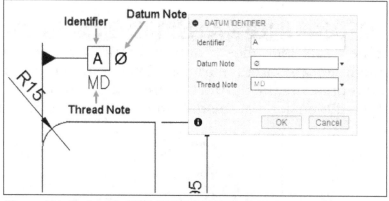

Figure-101. Updated DATUM IDENTIFIER dialog box

CREATING WELDING SYMBOL
AND OTHER SYMBOLS

The **Welding** tool in **SYMBOLS** drop-down of **Ribbon** is used to insert welding symbols applied to selected edge/face. Using this tool, you can apply annotation to the edges where weld is to be performed. The procedure to place a weld symbol is similar to placing Feature Control Frame discussed earlier. Similarly, you can use the **Taper and Slope** tool from the **SYMBOLS** drop-down in the **DRAWING** tab of **Ribbon** to apply symbol defining taper/slope on selected edge.

GENERATING TABLES

The **Table** tool is used to generate empty table, parts list, and bend table of the assembly model linked with drawing. The procedure to use this tool is discussed next.

- Click on the **Table** tool from **TABLES** drop-down in the **Toolbar**; refer to Figure-102. The **TABLE** dialog box will be displayed along with attached empty table; refer to Figure-103.

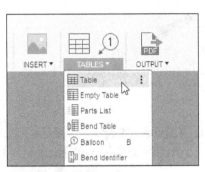

Figure-102. Table tool

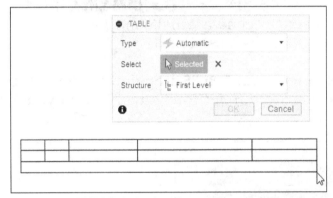

Figure-103. TABLE dialog box along with empty table

- The **Automatic** option in **Type** drop-down of **TABLE** dialog box is selected by default. The **Automatic** option automatically creates a parts list or bend table, based on the contents of the current sheet.
- Select **Empty Table** from **Type** drop-down to create an empty table.
- Select **Parts List** option from **Type** drop-down to create a parts list based on an assembly, storyboard, component, or sheet metal component reference. Note that Parts list is also called Bill of Materials or BOM.
- Select **Bend Table** option from **Type** drop-down to create a bend table based on a sheet metal flat pattern reference.
- Here, we are selecting the **Parts List** option from **Type** drop-down of dialog box or click **Parts List** tool from **TABLES** drop-down in the **Toolbar**; refer to Figure-102.
- The **Select** button of **Select** section is active by default. Select desired view on the current sheet.
- Select **First Level** option from **Structure** drop-down to include first level components only.
- Select **All Levels** option from **Structure** drop-down to include all components in the assembly.
- Click at desired location in the drawing to place the parts list. Parts list will be generated with balloons; refer to Figure-104.

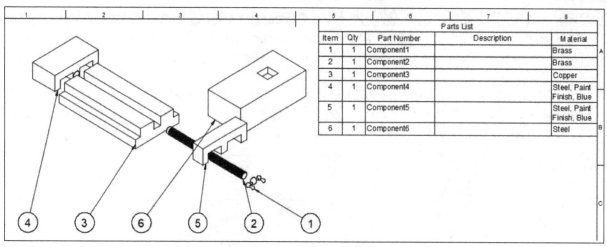

Figure-104. Parts List of assembly

Similarly, you can use **Custom Table (Empty Table)** tool and **Bend Table** tool. Note that you can edit the fields of empty table by double-clicking in them.

GENERATING BALLOON

The **Balloon** tool is used to generate balloons for components in drawing. The procedure to use this tool is discussed next.

- Click on the **Balloon** tool from **TABLES** drop-down; refer to Figure-105. The **BALLOON** dialog box will be displayed; refer to Figure-106 and you will be asked to select the component.

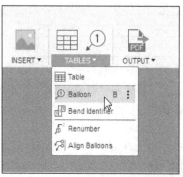

Figure-105. Balloon tool

Figure-106. BALLOON dialog box

- By default, the **Standard** option is selected in the dialog box so straight lines are created with balloon. If you want to create spline attached to balloon in place of line then select the **Patent** option from the **Type** drop-down.
- Click on desired edge of component. On selection of edge, an arrow will be attached to the cursor.
- Click at desired location to place the balloon; refer to Figure-107 and Figure-108.

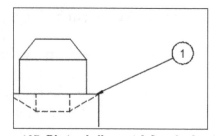

Figure-107. Placing balloon with Standard option

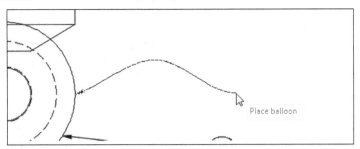

Figure-108. Placing balloon with Patent option

BEND IDENTIFIER AND TABLE

The **Table** tool discussed earlier works in a different way for sheet metal component. If you are generating drawing for a sheetmetal part then **Table** tool will generate bend table for the part and like balloons in assembly, bend identifiers are used to identify bends as per the table. The process is given next.

- After starting drawing environment, using flat pattern of a sheet metal part, click on the **Bend Identifier** tool from the **TABLES** drop-down in the **Toolbar**; refer to Figure-109. You will be asked to select the bends in flat pattern.

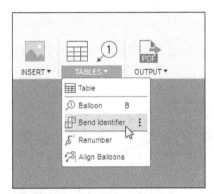

Figure-109. Bend Identifier tool

- Click on the sheet metal bends in drawing one by one. Bend identifiers will be assigned to the bends; refer to Figure-110.

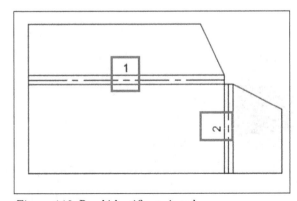

Figure-110. Bend identifier assigned

- Press **ESC** to exit the tool.
- Click on the **Table** tool from **TABLES** drop-down in the **Toolbar** and place the bend table at desired location; refer to Figure-111.

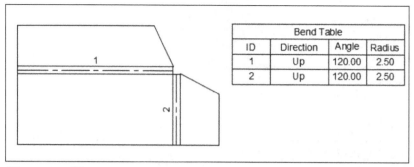

Figure-111. Flat pattern with bend table

RENUMBERING BALLOONS

The **Renumber** tool is used to renumber the existing balloons. The procedure to use this tool is discussed next.

- Click on the **Renumber** tool from the **TABLES** drop-down in the **Toolbar**; refer to Figure-112. The **RENUMBER** dialog box will be displayed; refer to Figure-113.

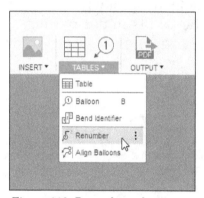

Figure-113. RENUMBER dialog box

Figure-112. Renumber tool

- Click in the **Starting Number** edit box and enter desired number.
- Click on **Select** button from **Selection** section and click on the balloons as per their sequences to renumber them. The balloon numbers will be changed; refer to Figure-114.

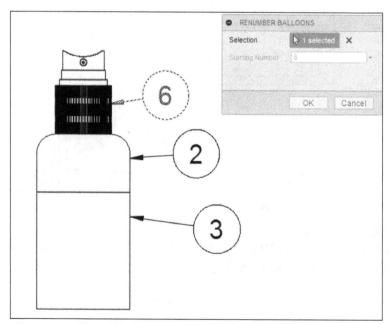

Figure-114. Changed balloon number

- After specifying the parameters, click on **OK** button from **RENUMBER** dialog box to complete the process.

ALIGNING BALLOONS

The **Align Balloons** tool is used to align all the pre-existing balloons. The procedure to use this tool is discussed next.

- Click on the **Align Balloons** tool from **TABLES** drop-down in **Toolbar**; refer to Figure-115. The tool will be activated.

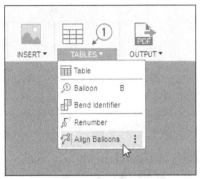

Figure-115. Align Balloons tool

- Click to select all the balloons to aligned; refer to Figure-116 and press **ENTER**. You will be asked to specify starting point of line to be used for alignment of balloons.

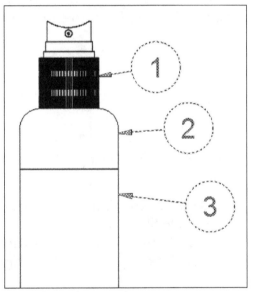

Figure-116. Selection of balloons for allignment

- Click on the screen to specify the start point. End point of alignment line will be attached to cursor.
- Click on the screen at desired location to define alignment of balloons; refer to Figure-117.

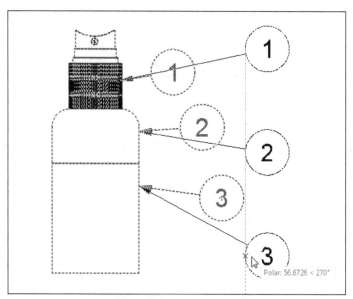

Figure-117. Performing alignment process

EXPORT PDF

The **Export PDF** tool is used to output the drawing in PDF format. The procedure to use this tool is discussed next.

- Click on the **Export PDF** tool of **Export** drop-down in **Toolbar**; refer to Figure-118. The **OUTPUT PDF** dialog box will be displayed; refer to Figure-119.

Figure-118. Export PDF tool

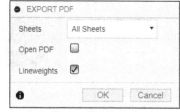

Figure-119. Export PDF dialog box

- Click on the **Sheets** drop-down from the **Export PDF** dialog box and select required sheet for output. You can also use **SHIFT** key or **CTRL** key to select multiple sheets.
- Select the **Open PDF** check box if you want to open the output PDF file.
- Select the **Lineweights** check box if you want to make bold lines of drawing otherwise clear it.
- After specifying the parameters, click on the **OK** button from **Export PDF** dialog box. The **Export PDF** dialog box will be displayed to save the file; refer to Figure-120.

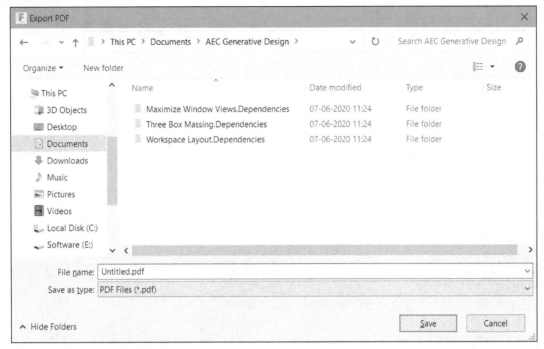

Figure-120. Output PDF dialog box for saving file

- Click in the **File name** edit box and specify desired name for pdf file.
- Browse to desired location where you want to save the file and click on the **Save** button. The PDF file will be created and opened if you had selected the **Open PDF** check box from **EXPORT PDF** dialog box.

EXPORTING DWG

The **Export DWG** tool is used to create an output file of drawing in DWG format. The procedure to use this tool is discussed next.

- Click on the **Export DWG** tool from **OUTPUT** drop-down in the **Toolbar**; refer to Figure-121. The **Export DWG** dialog box will be displayed to save the file; refer to Figure-122.

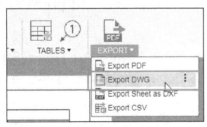

Figure-121. Export DWG tool

Figure-122. Export DWG dialog box

- Select **Simplified DWG** option from **Format** drop-down of dialog box to export the contents of the current sheet as a DWG file, including views, title block, border, and all annotations as 2D geometry in Model Space.
- Select **AutoCAD DWG** option from **Format** drop-down to export the contents of all sheets to a DWG file. Each sheet is exported as a separate Layout tab.
- Click on the **Sheets** drop-down from **OUTPUT DWG** dialog box and select required sheet for output. You can also use **SHIFT** key or **CTRL** key to select multiple sheets.
- After specifying the parameters, click on the **OK** button from **OUTPUT DWG** dialog box. The **Export DWG** dialog box will be displayed to save the file; refer to Figure-123.

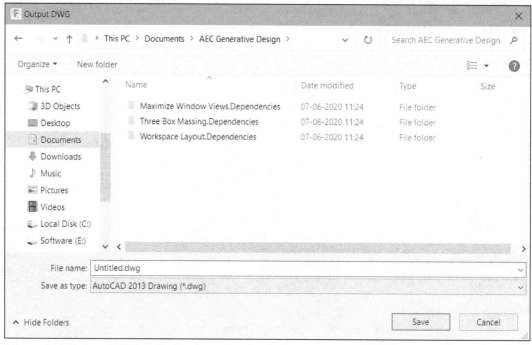

Figure-123. Output DWG dialog box for saving the file

- Select desired location to save the file.
- After specifying the parameter, click on the **Save** button from **Output DWG** dialog box to save the file.

EXPORTING DXF FILE

The **Export Sheet as DXF** tool from **OUTPUT** drop-down in the **Toolbar** exports the sheets in a drawing as a DXF file. The procedure to use this tool is similar to the **Simplified DWG** option in the **Format** drop-down of **OUTPUT DWG** dialog box.

EXPORTING CSV FILE

The **Export CSV** tool is used to create the output of the current file in CSV format. CSV is a format used by database management software and point cloud software. The procedure to use this tool is discussed next.

- Click on the **Export CSV** tool from **EXPORT** drop-down in the **Toolbar**; refer to Figure-124. The **Output Table** dialog box will be displayed; refer to Figure-125.

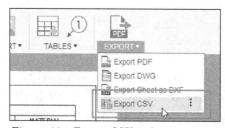

Figure-124. Export CSV tool

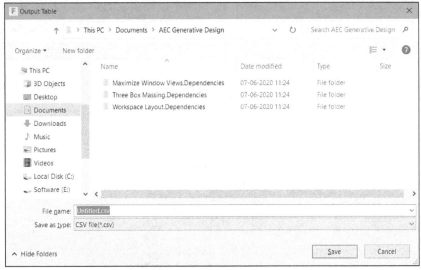

Figure-125. Output Table dialog box

- Select desired name and location of file.
- After specifying the parameters, click on the **Save** button to save the file.

PRACTICAL 1

Generating part drawing for the model as shown in Figure-126.

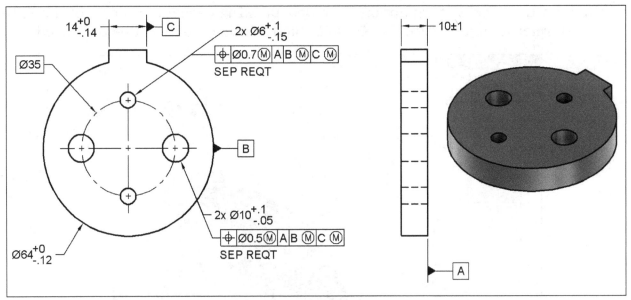

Figure-126. Practical 1 Drawing

Starting a New Drawing using Part

- Set the parameters as shown in Figure-127 for preference.

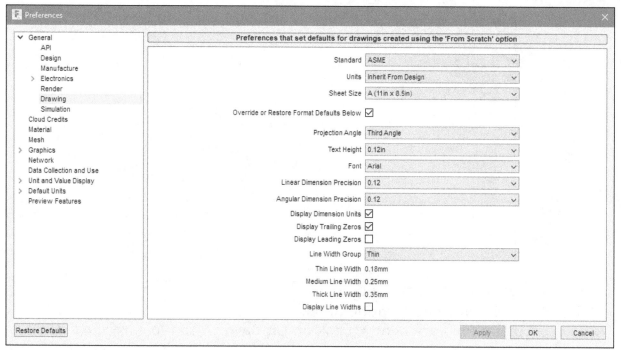

Figure-127. Specifying drawing preferences

- Open the model file for Practical 1 of Chapter 11 in the resource kit. The model will be displayed in the canvas.
- Click on the **From Design** option from **New Drawing** cascading menu of the **File** menu; refer to Figure-128. The **CREATE DRAWING** dialog box will be displayed.

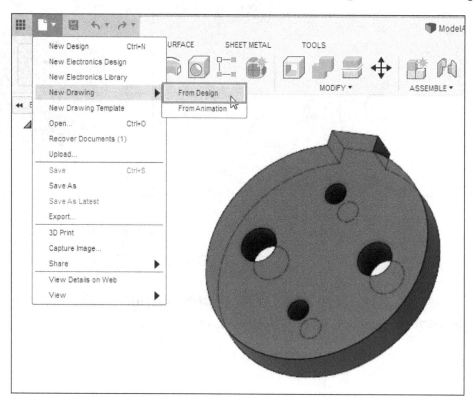

Figure-128. Starting new drawing

• Make sure **ASME** is selected in the **Standard** drop-down and click on the **OK** button. The **DRAWING VIEW** dialog box will be displayed with preview of drawing view; refer to Figure-129.

Figure-129. Preview of drawing view

Placing Views

• Select the **Top** option from the **Orientation** drop-down, **Visible and Hidden Edges** button from **Style** section, and **1:1** in **Scale** drop-down from the dialog box.
• Click at desired location to place the view; refer to Figure-130.

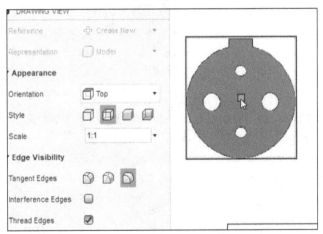

Figure-130. Placing base view

• Click on the **OK** button from the dialog box to create the view.
• Click on the **Projected View** tool from the **DRAWING VIEWS** panel in the **Toolbar**. You will be asked to select the base view.
• Select the view earlier placed and move the cursor towards right. The view will get attached to the cursor; refer to Figure-131.

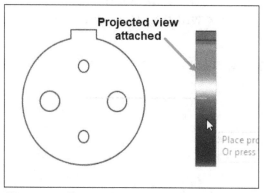

Figure-131. Projected view attached to cursor

- Click at desired location to place the view and press **ESC** to exit.

Applying Annotations

- Click on the **Center Mark Pattern** tool from the **GEOMETRY** drop-down in the **Toolbar**. The **CENTER MARK PATTERN** dialog box will be displayed.
- Click on one of the holes in the base view and select the **Center Mark** check box. The preview of center mark pattern will be displayed; refer to Figure-132.

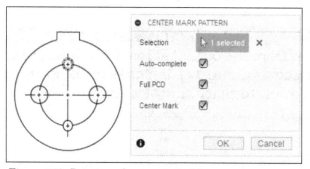

Figure-132. Preview of center mark pattern

- Click on the **OK** button from the dialog box.
- Click on the **Dimension** tool from the **DIMENSIONS** panel in the **Toolbar** and apply dimensions as shown in Figure-133.

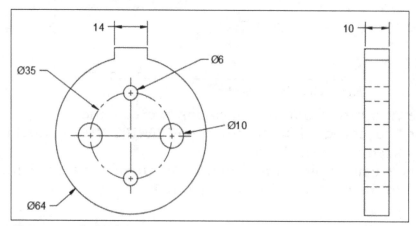

Figure-133. Applying dimensions

- Click on the **Datum Identifier** tool from the **SYMBOLS** panel in the **Toolbar** and place the datum identifiers as shown in Figure-134.

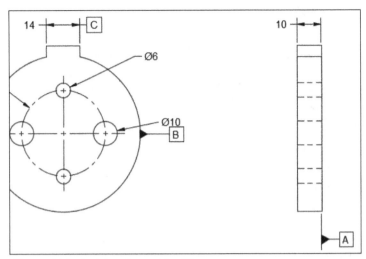

Figure-134. Applying datum identifier

Adding Tolerances to Dimension

- Double-click on the dimension with **10** value. The **DIMENSION** dialog box will be displayed; refer to Figure-135.

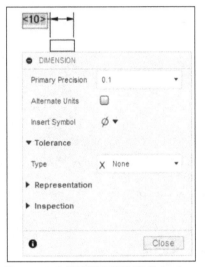

Figure-135. DIMENSION dialog box

- Select the **Symmetrical** option from the **Type** drop-down in **Tolerance** node of dialog box and specify the value as **1** in **Tolerance** edit box. Click on the **Close** button to exit the dialog box.
- Similarly, set the tolerances for other dimensions; refer to Figure-136.

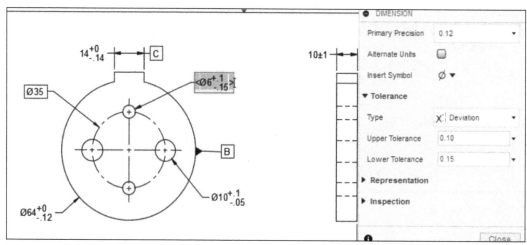

Figure-136. Applying tolerances

- Click on the **Feature Control Frame** tool from the **SYMBOLS** panel in the **Toolbar** and select the hole of diameter **6**. The control frame will get attached to cursor.
- Click at desired location to place the control frame and press **ENTER**. The **FEATURE CONTROL FRAME** dialog box will be displayed; refer to Figure-137.

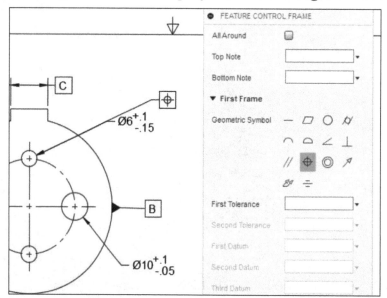

Figure-137. Placing feature control frame

- Specify the parameters as shown in Figure-138 and click on the **Close** button.

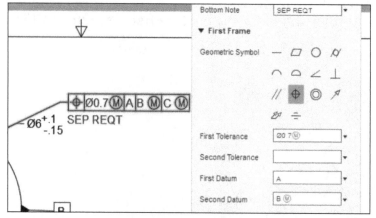

Figure-138. Creating feature control frame

Similarly, define position feature control frame for other hole and place another base view as 3D NE Isometric view; refer to Figure-139.

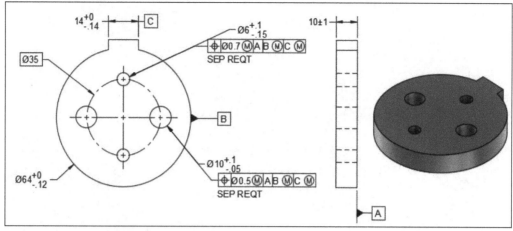

Figure-139. Completed drawing

PRACTICE

Create drawings for the models discussed in practical and practice sections of Chapter 5 of this book.

SELF ASSESSMENT

Q1. The .... tool is used to create drawing views generated from an existing base or drawing view.

Q2. What is the difference between **ASME** and **ISO** option in **CREATE DRAWING** dialog box?

Q3. In which view of drawing, hatching is automatically created?

Q4. Text can be aligned by **Justification** section. (T/F)

Q5. The drawing view can be move from one place to another with the help of **Move** button. (T/F)

Q6. What is the use of **Edge extension** tool?

Q7. Two parallel edges should be selected to create the extension. (T/F)

Q8. When is the baseline dimensioning required?

Q9. What is Feature Control Frame?

Q10. What is the use of **Datum Identifier** tool?

Q11. The .... tool is used to create an output file of drawing in DWG format.

FOR STUDENT NOTES

Chapter 12

Sculpting

Topics Covered

The major topics covered in this chapter are:

- **Introduction to Sculpting**
- **Box**
- **Plane**
- **Cylinder**
- **Sphere**
- **Torus**
- **Quadball**
- **Pipe**
- **Face**
- **Extrude**
- **Revolve**
- **Edit Form and Edit By Curve Tool**
- **Insert Edge and Insert Point**

- **Subdivide and Bridge**
- **Merge Edge**
- **Fill Hole**
- **Erase and Fill**
- **Weld Vertices and Unweld Edges**
- **Crease and Uncrease**
- **Bevel Edges and Slide Edges**
- **Smooth and Cylindrify**
- **Pull and Flatten**
- **Straighten and Interpolate**
- **Match and Thicken**
- **Freeze and UnFreeze**

INTRODUCTION

Sculpting is a workspace that offers tools to push, pull, grab, or pinch objects to modify their shapes. In this chapter, we will learn various commands and tool of **FORM** mode which are used to create or modify the object.

OPENING THE FORM MODE

The components or parts created in Form mode are easily converted in solid bodies. Generally, the **FORM** mode is not a workspace so it will not be available in **Change Workspace** drop-down. It is displayed in **DESIGN** workspace. The procedure to activate this mode is discussed next.

- Click on the **DESIGN** option from **Change Workspace** drop-down. The **DESIGN** workspace will be displayed in **Autodesk Fusion 360** window.
- Click on **Create Form** tool of **CREATE** drop-down from **Toolbar**; refer to Figure-1. The **FORM** mode will be displayed with updated **Toolbar**; refer to Figure-2.

Figure-1. Create Form tool

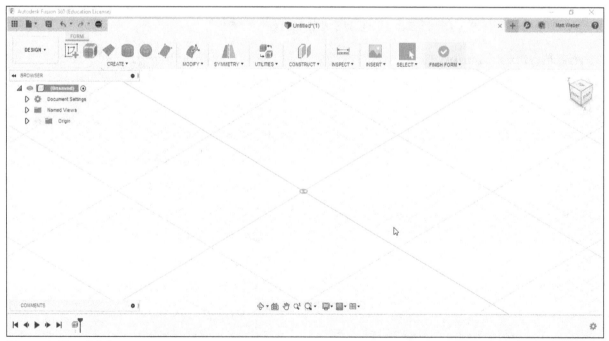

Figure-2. Form Mode

CREATION TOOLS

In this section, we will discuss the creation tools used to create form.

Box

The **Box** tool is used to create a rectangular body on the selected plane or face. The procedure to use this tool is discussed next.

- Click on the **Box** tool of **CREATE** drop-down from **Toolbar**; refer to Figure-3. The **BOX** dialog box will be displayed; refer to Figure-4.

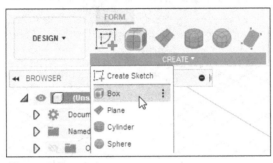

Figure-3. Box tool

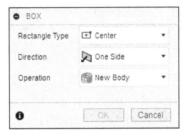

Figure-4. BOX dialog box

- You will be asked to select the plane or face. Click to select plane or face.
- Select **Center** option from **Rectangle Type** drop-down if you want to create a center rectangle as reference for box.
- Select **2-Point** option from **Rectangle Type** drop-down if you want to create a 2-Point rectangle as a reference for creating box. In our case, we are selecting the **Center** option.
- Specify desired parameters in **Direction** and **Operation** drop-down as discussed earlier in **DESIGN Workspace**.
- Click on the screen to specify the center point of rectangle.
- Enter desired dimension of length and width in the respective floating window and click on the screen. The preview of box will be displayed along with updated **BOX** dialog box; refer to Figure-5.

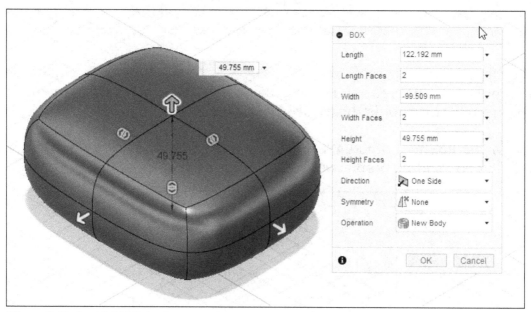

Figure-5. Updated BOX dialog box

- If you want to change the length of box then click in the **Length** edit box and enter desired value.
- Click in the **Length Faces** edit box of **BOX** dialog box and enter the number of faces in which surface of box will be divided along the length.
- If you want to change the width of plane then click in the **Width** edit box and enter desired value.
- Click in the **Width Faces** edit box of **BOX** dialog box and enter the number of faces in which surface of box will be divided along the width.
- Click in the **Height** edit box and enter desired value of height of box.
- Click in the **Height Faces** edit box and enter the number of faces in which height of plane will be divided along the height.
- Select the **Mirror** option from **Symmetric** drop-down if you want to create a symmetric box. On selecting the **Mirror** option, the updated dialog box will be displayed; refer to Figure-6.

Figure-6. Updated BOX dialog box on selecting Mirror Symmetry

- Select the **Length Symmetry** check box if you want to apply mirror symmetry along length of sculpt object; refer to Figure-7.

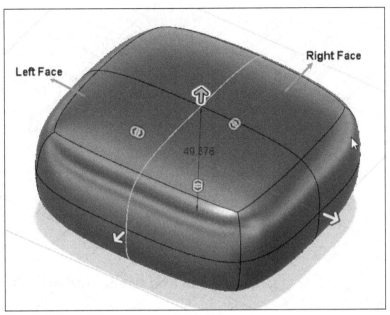

Figure-7. Length Symmetry

- Select the **Width Symmetry** check box if you want to apply mirror symmetry along width of sculpt object; refer to Figure-8. Note that later while editing, if you will move one face then the other symmetric face will also move accordingly.

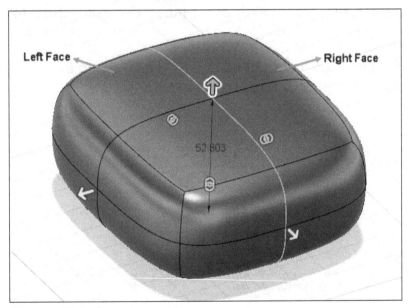

Figure-8. Width Symmetry

- Select the **Height Symmetry** check box if you want to apply mirror symmetry between upper and lower faces of sculpt object; refer to Figure-9.

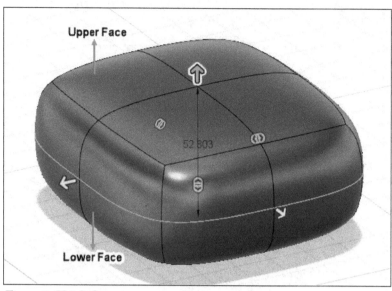

Figure-9. Height Symmetry

- Specify desired parameters in **Direction** and **Operation** drop-down as discussed earlier in **DESIGN Workspace**.
- Note that you can also change shape of box by using the drag handles. After specifying the parameters, click on **OK** button from **BOX** dialog box to complete the process.

Plane

The **Plane** tool is used to create T-Spline plane. The procedure to use this tool is discussed next.

- Click on the **Plane** tool from **CREATE** drop-down; refer to Figure-10. The **PLANE** dialog box will be displayed; refer to Figure-11.

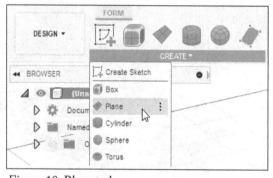

Figure-10. Plane tool

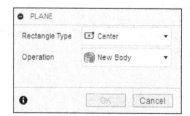

Figure-11. PLANE dialog box

- The options of **Rectangle Type** drop-down from **PLANE** dialog box are same as discussed earlier in last section.

- Click on the screen to specify the center point.
- Enter desired dimension of length and width in respective input boxes and click on the screen. The preview of plane will be displayed along with updated **PLANE** dialog box; refer to Figure-12.

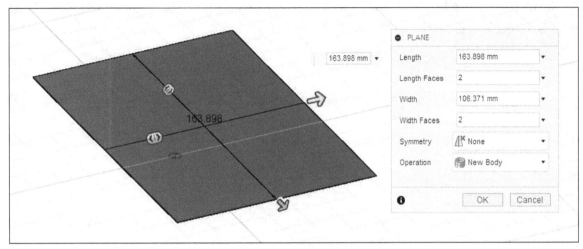

Figure-12. Updated PLANE dialog box

- If you want to change the length of plane then click in the **Length** edit box and enter desired value or use drag handles.
- Click in the **Length Faces** edit box of **PLANE** dialog box and enter the number of faces in which surface of plane will be divided along the length.
- If you want to change the width of plane then click on the **Width** edit box and enter desired value.
- Click in the **Width Faces** edit box of **PLANE** dialog box and enter the number of faces in which surface of plane will be divided along the width.
- After specifying the parameters, click on the **OK** button from **PLANE** dialog box to complete the process of creating plane. The plane will be displayed; refer to Figure-13.

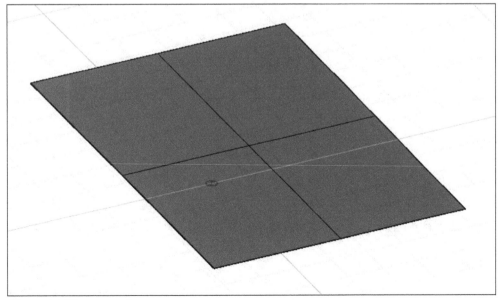

Figure-13. Plane created

Cylinder

The **Cylinder** tool is used to create a cylindrical body by defining diameter and depth. The procedure to use this tool is discussed next.

- Click on the **Cylinder** tool from **CREATE** drop-down; refer to Figure-14. The **CYLINDER** dialog box will be displayed; refer to Figure-15.

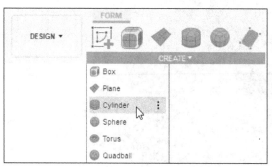

Figure-14. Cylinder tool

Figure-15. CYLINDER dialog box

- You need to select the base plane or face to create the cylinder. Click on the plane to select.
- Click on the screen to specify the center point and enter desired radius of cylinder; refer to Figure-16.

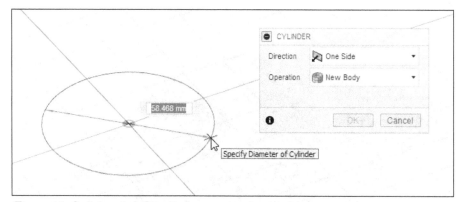

Figure-16. Creating circle for cylinder

- After specifying the diameter, click on the screen. The preview of cylinder will be displayed along with updated **CYLINDER** dialog box; refer to Figure-17.

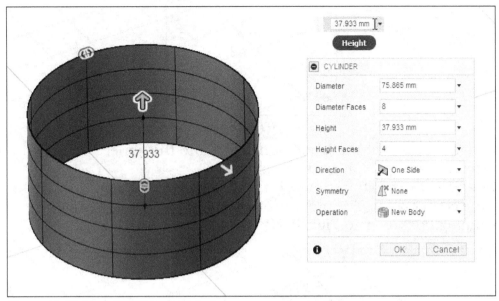

Figure-17. Updated CYLINDER dialog box

- If you want to change the diameter of cylinder then click on the **Diameter** edit box of **CYLINDER** dialog box and enter desired diameter.
- Click in the **Diameter Faces** edit box and enter the number of faces in which the round surface of cylinder will be divided.
- Click in the **Height** edit box and enter desired value of height of cylinder.
- Click in the **Height Faces** edit box and enter the number of faces in which height of cylinder will be divided along the height.
- The options in the **Direction** drop-down have been discussed earlier.
- Select **Circular** option from **Symmetry** drop-down if you want to apply circular symmetry between the faces of sculpt object.
- Click on the **Symmetric Faces** edit box and enter desired value of circular symmetry on round surface.
- After specifying the parameters, click on the **OK** button from **CYLINDER** dialog box to complete the process.

Sphere

The **Sphere** tool is used to create a T-Spline sphere. The procedure to use this tool is discussed next.

- Click on the **Sphere** tool from **CREATE** drop-down; refer to Figure-18. The **SPHERE** dialog box will be displayed; refer to Figure-19.

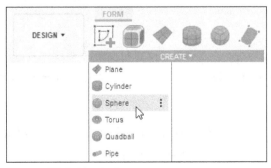

Figure-18. Sphere tool

Figure-19. SPHERE dialog box

- You are asked to select the base plane to create the sphere. Click on the plane/ face where you want to place the sphere.
- You are asked to specify centerpoint of the sphere. Click on desired location to specify the center point. The preview of sphere will be displayed along with updated **SPHERE** dialog box; refer to Figure-20.

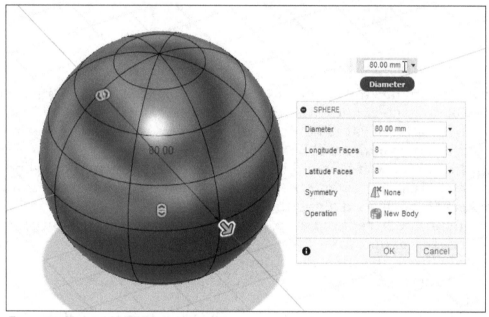

Figure-20. Updated SPHERE dialog box

- Click in the **Diameter** edit box and enter desired value of sphere diameter.
- Click in the **Longitude Faces** edit box and enter desired number of faces displayed on longitude of sphere.
- Click in the **Latitude Faces** edit box and enter desired number of faces displayed on latitude of sphere.
- The options of **Symmetry** drop-down have been discussed earlier in **Box** and **Cylinder** tools.
- After specifying the parameters, click on the **OK** button from **SPHERE** dialog box to complete the process.

Torus

The **Torus** tool is used to create a T-Spline torus. The procedure to use this tool is discussed next.

- Click on the **Torus** tool from **CREATE** drop-down; refer to Figure-21.The **TORUS** dialog box will be displayed; refer to Figure-22.

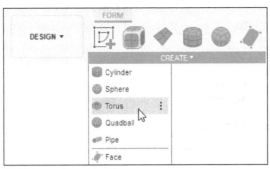

Figure-21. Torus tool

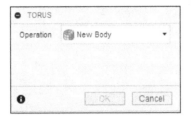

Figure-22. TORUS dialog box

- You need to select the base plane to create the torus. Click on the screen to select.
- Click on the screen to specify the center point and enter desired diameter of torus in the input box. The preview of torus will be displayed along with updated **TORUS** dialog box; refer to Figure-23.

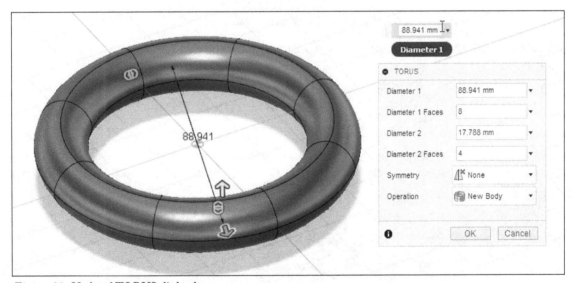

Figure-23. Updated TORUS dialog box

- Click in the **Diameter 1** edit box of **TORUS** dialog box and enter desired value of diameter for center circle of torus.
- Click in the **Diameter 1 Faces** edit box and enter the number of faces in which round torus will be divided horizontally.
- Click in the **Diameter 2** edit box and enter the diameter of torus tube.
- Click in the **Diameter 2 Faces** edit box and enter the number of faces in which round torus will be divided vertically.
- After specifying the parameters, click on the **OK** button from **TORUS** dialog box.

Quadball

The **Quadball** tool is used to create a T-Spline quadball. The procedure to use this tool is discussed next.

- Click on the **Quadball** tool from **CREATE** drop-down; refer to Figure-24. The **QUADBALL** dialog box will be displayed; refer to Figure-25.

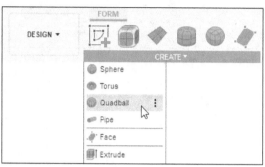

Figure-24. Quadball tool

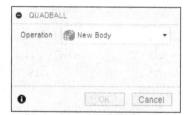

Figure-25. QUADBALL dialog box

- You need to select the base plane to create the quadball. Click on desired face/plane to select.
- Click on the screen to specify the center point. The preview of quadball will be displayed along with updated **QUADBALL** dialog box; refer to Figure-26.

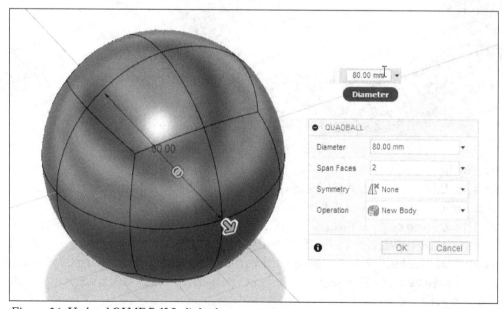

Figure-26. Updated QUADBALL dialog box

- Click in the **Diameter** edit box and enter desired value of quadball diameter.
- Click in the **Span Faces** dialog box and enter the number of faces in which quadball will be divided.

- The options of **Symmetry** and **Operation** drop-down were discussed earlier.
- After specifying the parameters, click on **OK** button from **QUADBALL** dialog box. The quadball will be displayed; refer to Figure-27.

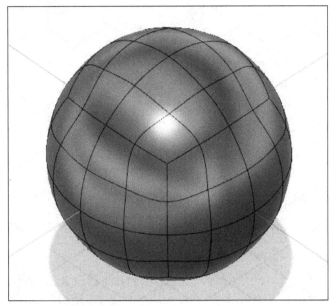

Figure-27. Quadball created with 4 span faces

Pipe

The **Pipe** tool is used to create complex pipe based on selected sketch. The procedure to use this tool is discussed next.

- Click on the **Pipe** tool from **CREATE** drop-down; refer to Figure-28. The **PIPE** dialog box will be displayed; refer to Figure-29.

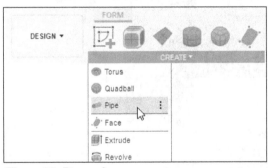

Figure-28. Pipe tool

Figure-29. PIPE dialog box

- The **Path** selection button is active by default. Click on the path to select. You can select sketch lines/curves or edges of the model. You can also use window selection to create pipe.
- Select the **Chain selection** check box if you want to select the nearby geometries in chain while selecting the one.
- On selection of sketch, the preview of pipe will be displayed; refer to Figure-30.

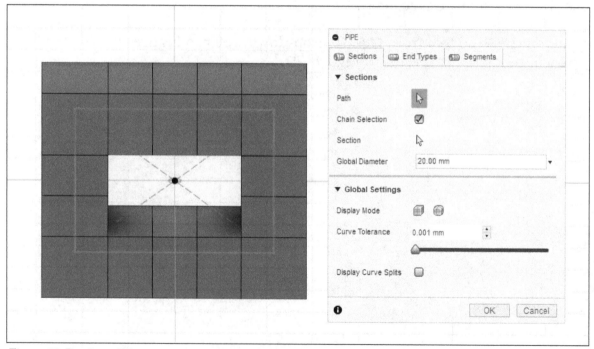

Figure-30. Preview of Pipe

- Click on the **Section** button of **Sections** tab if you want to modify the sections of pipe. On selecting the button, the sections of pipe will be displayed; refer to Figure-31.

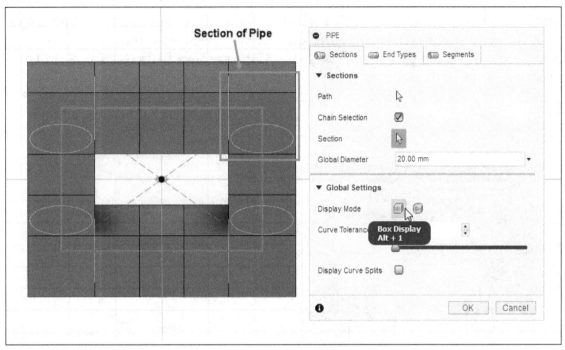

Figure-31. Sections of pipe

- Click on desired section to change the diameter, angle, and position.
- Click on the **Diameter** edit box and enter desired value of particular section diameter; refer to Figure-32.

Figure-32. Section options in PIPE dialog box

- Similarly, specify the value of **Angle** and **Position** in their respective edit boxes as desired. You can also move the arrow displayed on the selected section to specify the value.
- Click on the **Reset Section** button to reset all the changed value of sections.

- Click on the **Remove Section** button to remove the selected section.
- Click on **Box Display** button of **Display Mode** section from **Global Settings** area to display the rectangular T-Spline pipe of the selected sketch.
- Select **Smooth Display** button of **Display Mode** section from **Global Settings** area to display a smooth circular pipe; refer to Figure-33.

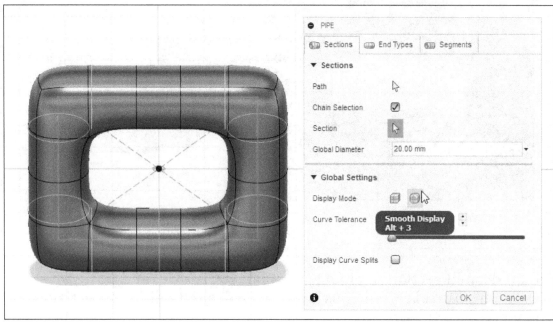

Figure-33. Smooth pipe

- Click in the **Curve Tolerance** edit box and enter the value of pipe tolerance. You can also set the tolerance by adjusting the **Curve Tolerance** slider.
- Select the **Display Curve Splits** check box if you want to see the curve splits of pipe.
- Click on the **End Types** tab in the dialog box. The **Handle** selection button of **End Types** tab is active by default and you are asked to select the end handle of pipe to modify its shape.
- Select desired end handle from the model. Select **Open** option from **End Type** drop-down of **End Types** tab if you want to keep all the ends of pipe open; refer to Figure-34.

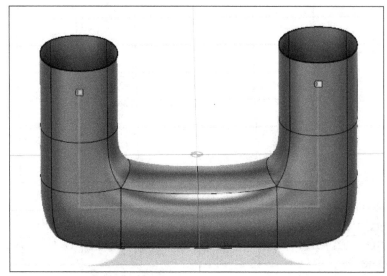

Figure-34. Open End Type

- Select **Square** option from **End Type** drop-down of **End Types** tab if you want to close all the ends of pipe in square like structure; refer to Figure-35.

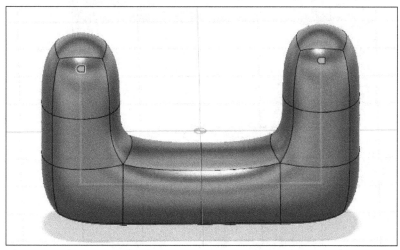

Figure-35. Square End Type

- Select **Spike** option from **End Type** drop-down of **End Types** tab if you want to close all the ends of pipe in spike like structure; refer to Figure-36.

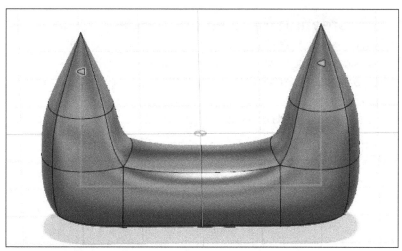

Figure-36. Spike End Type

- Click on the **Segments** tab in the dialog box. The **Segment** selection button of **Segments** tab is active by default.
- Move the **Density** slider to increase or decrease the density of pipe segment.
- After specifying the parameters, click on the **OK** button from **PIPE** dialog box to complete the process.

Face

The **Face** tool is used to create individual faces. The procedure to use this tool is discussed next.

- Click on the **Face** tool from **CREATE** drop-down; refer to Figure-37. The **FACE** dialog box will be displayed; refer to Figure-38.

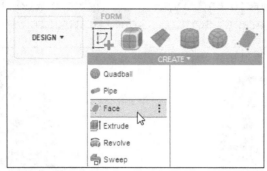

Figure-37. Face tool

Figure-38. FACE dialog box

Creating Face using Simple button

- Select **Simple** button from **Mode** section of **FACE** dialog box if you want to create a simple face by selecting four vertices on a specific plane.
- On selecting the **Simple** button, you need to select plane for creating the face. Click to select the plane.
- Now, specify the four corner on the plane as desired; refer to Figure-39.

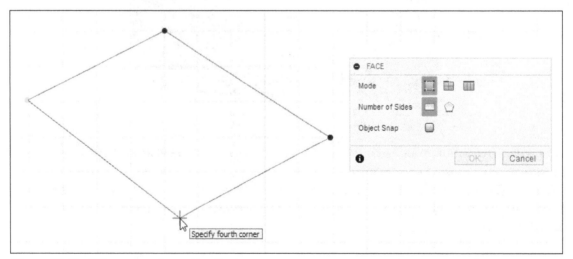

Figure-39. Specifying corners for creating face

- On selecting the fourth corner, the face will be created; refer to Figure-40.

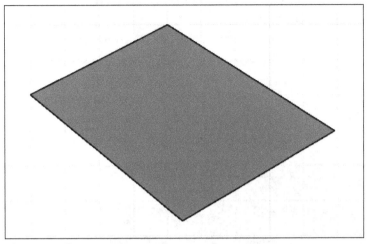

Figure-40. Face created

Creating Face using Edge button

The **Edge** button is used to create a face by selecting an edge. The procedure is discussed next.

- Click on the **Edge** button from **Mode** section of **FACE** dialog box. You need to select the edge of an object.
- Now, you need to select the plane on which you want to create a face. Click to select the plane.
- Click on the screen to specify third and fourth corner.
- On creating fourth corner, the face will be created; refer to Figure-41.

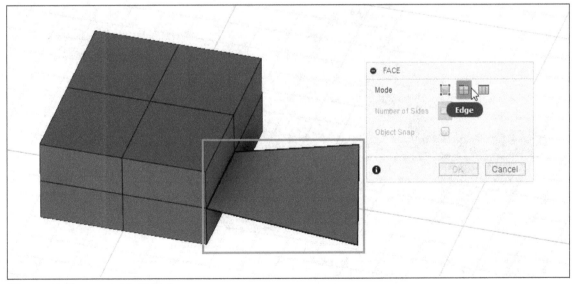

Figure-41. Face created by Edge button

Creating Chain of Faces using Edges

The **Chain** button is used to create multiple faces continuously. The procedure is discussed next.

- Click on the **Chain** button from **Mode** section of **FACE** dialog box. You will be asked to select an edge for creating face.

- Select desired edge. Now, you need to select the plane on which you want to create a face. Click to select the plane.
- Specify third and fourth corner to create first face; refer to Figure-42.

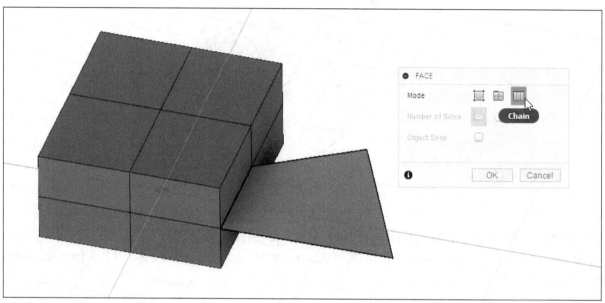

Figure-42. First face created

- After creating first face, specify third and fourth corner to create second face.
- On creating second face, click on the screen to create next faces as desired; refer to Figure-43.

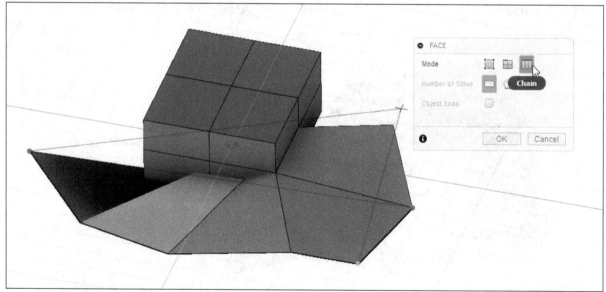

Figure-43. Multiple faces created using Chain Mode

- After creating desired number of faces, click on the **OK** button from **FACE** dialog box to complete the process.

Extrude

The **Extrude** tool is used to extrude the selected sketch up to the desired depth or height. The procedure to use this tool is discussed next.

* Click on the **Extrude** tool from **CREATE** drop-down; refer to Figure-44. The **EXTRUDE** dialog box will be displayed; refer to Figure-45.

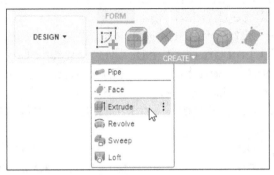

Figure-44. Extrude tool

Figure-45. EXTRUDE dialog box

* The **Select** button of **Profile** section in **EXTRUDE** dialog box is active by default. You are asked to select the sketch for extrude. On selecting the sketch; the updated **EXTRUDE** dialog box will be displayed; refer to Figure-46.

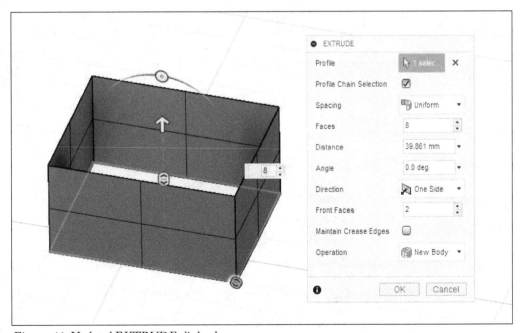

Figure-46. Updated EXTRUDE dialog box

- Select the **Profile Chain Selection** check box if you want to select the nearby geometries.
- Select **Uniform** button from **Spacing** drop-down if you want to position the faces evenly around the profile.
- Select **Curvature** button from **Spacing** drop-down if you want to position the faces based on the curvature of the profile. The more the curvature area more the faces.
- Click in the **Faces** edit box and enter desired number of faces in which curvature of extruded sketch will be divided.
- Click in the **Distance** edit box and enter the distance for extrusion.
- Click in the **Angle** edit box and enter desired angle of extrusion.
- Click in the **Front Faces** edit box and enter desired number of faces in which surface of extrude will be divided along the height.
- Select the **Maintain Crease Edges** check box from **EXTRUDE** dialog box if you want to keep the crease at merged edges.
- After specifying the parameters, click on the **OK** button from **EXTRUDE** dialog box to complete the extrusion process. Note that you can also select face of any other sculpt body to extrude but some of the options in this dialog box will not be available in that case.

Revolve

The **Revolve** tool is used to create a revolve feature by sweeping the selected sketch around the selected axis. The procedure to use this tool is discussed next.

- Click on the **Revolve** tool from **CREATE** drop-down; refer to Figure-47. The **REVOLVE** dialog box will be displayed; refer to Figure-48.

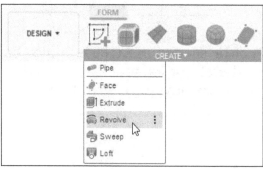

Figure-47. Revolve tool

Figure-48. REVOLVE dialog box

- The **Select** button of **Profile** section in **REVOLVE** dialog box is active by default. Click on the sketch to apply revolve feature.
- Click on **Select** button of **Axis** section and select the axis of rotation of sketch; refer to Figure-49.

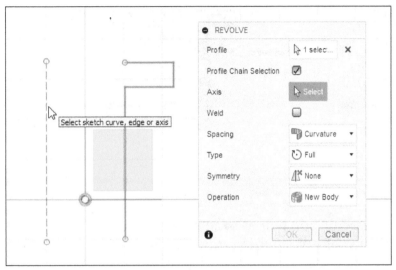

Figure-49. Selection of axis for sketch

- On selecting the axis, the preview of revolve feature will be displayed along with updated **REVOLVE** dialog box; refer to Figure-50.

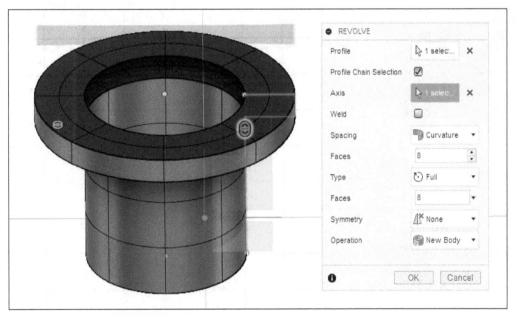

Figure-50. Updated REVOLVE dialog box

- Click in the **Faces** edit box below **Spacing** section and enter desired number of faces in which curvature of revolved sketch will be divided along flat faces; refer to Figure-51.

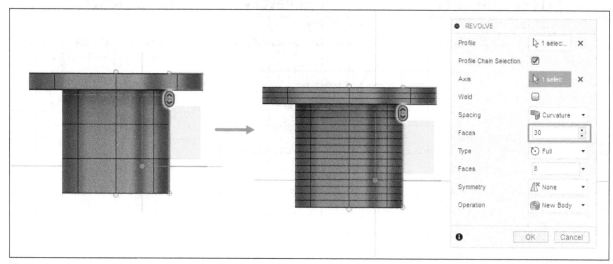

Figure-51. Entering the value of first faces option

- Click in the **Faces** edit box below **Type** section and enter desired number of faces in which curvature of revolved sketch will be divided along round faces; refer to Figure-52.

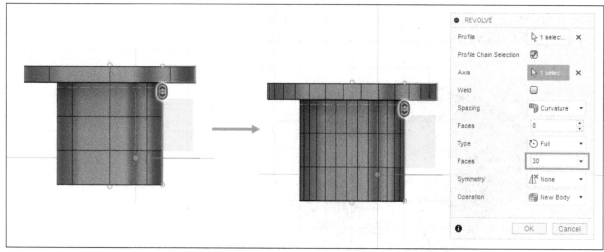

Figure-52. Entering the value of second faces option

- After specifying the parameters, click on the **OK** button from **REVOLVE** dialog box to complete the process.

Similarly, you can use other tools of **CREATE** drop-down as discussed in **DESIGN Workspace**.

MODIFYING TOOLS

Till now, we have learned about various tools to create the sculpt object. In this section, we will learn various tools for modifying the sculpt object.

Edit Form

The **Edit Form** tool is used to move, scale, or rotate the selected geometry. The procedure to use this tool is discussed next.

- Click on the **Edit Form** tool of **MODIFY** drop-down; refer to Figure-53. The **EDIT FORM** dialog box will be displayed; refer to Figure-54.

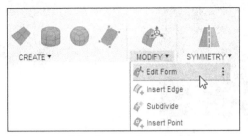

Figure-53. Edit Form tool

Figure-54. EDIT FORM dialog box

- The **Select** button of **T-Spline Entity** section is active by default. Click on any face of the sculpt object to edit.
- Click on **Multi** button from **Transform Mode** section to edit the face with the help of 3D Manipulator; refer to Figure-55.

Figure-55. Multi manipulator

- Click on **Translation** button from **Transform Mode** section to edit the face with the help of translation manipulator; refer to Figure-56.

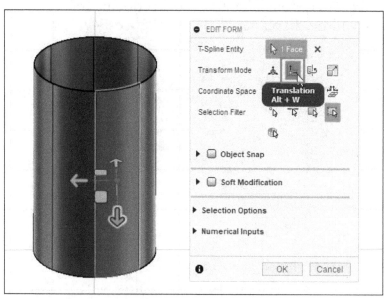

Figure-56. Translation manipulator

- Click on **Rotation** button from **Transform Mode** section to edit the face with the help of rotation manipulator; refer to Figure-57.

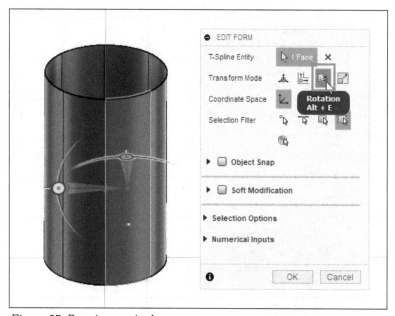

Figure-57. Rotation manipulator

- Click on **Scale** button from **Transform Mode** section to edit the face with the help of scale manipulator; refer to Figure-58.

Figure-58. Scale manipulator

- Select the **World Space** button from **Coordinate Space** section to orient the manipulator relative to the origin of model.
- Select the **View Space** button from **Coordinate Space** section to orient the manipulator relative to the current view of model.
- Select the **Selection Space** button from **Coordinate Space** section to orient the manipulator relative to the selected face.
- Select the **Local Per Entity** button from **Coordinate Space** section to orient the manipulator relative to the selected object.
- Select **Vertex** button from **Selection Filter** section to select the vertex of object for editing purpose. On selecting this button, only vertices will available for selection.
- Select **Edge** button from **Selection Filter** section to select the edge of object. On selecting this button, only edges of object will be available for selection.
- Select **Face** button from **Selection Filter** section to select the face of object. On selecting this button, only faces of object will be available for selection.
- Select **All** button from **Selection Filter** section to select the edge, vertex, or face from an object. On selecting this button, edge, vertex, and face will available for selection.
- Select **Body** button from **Selection Filter** section to select the body for editing.
- Select the **Object Snap** check box and specify desired value in the **Offset** edit box to specify the limit within which the selected vertex can move.

Soft Modification

Select the **Soft Modification** check box if you want to control the influence of changes on the surrounding area of object. On selecting the check box, the updated **EDIT FORM** dialog box will be displayed; refer to Figure-59.

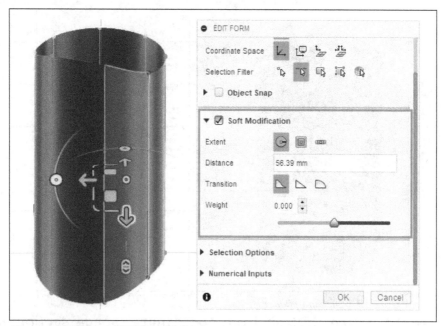

Figure-59. Updated EDIT FORM dialog box

- When selected, the vertices of body is visually represented with red and white vertices highlighting that can be adjusted with the gradient slider.
- Select **Distance** button from **Extent** section to specify the distance in the **Distance** edit box for controlling the influence of round region. With the help of vertices displayed on object, you can control the amount of change they undergo and the shape of the affected region.
- Select **Face Count** button from **Extent** section to specify the number of faces far from selected object.
- Select **Rectangular Face Count** button from **Extent** section to specify the number of width and length faces.
- Select desired shape of the affected region from **Transition** option.
- Click in the **Weight** edit box and enter the value of amount of influence applied in the affected region. The value of weight will be in between -1 to 1. You can also adjust the value of weight by moving the **Weight** slider.

Selection Options

- Click on the **Selection Options** node from **EDIT FORM** dialog box. The options of **Selection Options** node will be displayed; refer to Figure-60.

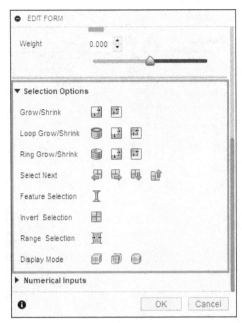

Figure-60. Selection Options

- The **Grow/Shrink** option is used to contract and expand the selected region.
- The **Loop Grow/Shrink** option is used to expand and contract the selected loop for the current location.
- The **Ring Grow/Shrink** option is used to expand and contract the ring after selecting the adjacent edge of the selected edge.
- The **Select Next** button is used to move the selected vertex, edge, or face to the next adjacent object. The movement to adjacent faces is based upon the current camera position.
- The **Feature Selection** button is used to select all the faces of a strut or a hole.
- The **Invert Selection** button is used for reverse the selection. It means all the selected object will be deselected and deselected will be selected.
- The **Range Selection** button is used to select all the faces between two selected faces.
- The **Display Mode** section is used to select **Box display**, **Control Frame display**, and **Smooth Display** mode for the object.

Numerical Inputs

- Click on the **Numerical Inputs** node from **EDIT FORM** dialog box to enter the numerical value of various parameters, refer to Figure-61.

Figure-61. Numerical Inputs

- After specifying the various parameters, click on **OK** button from **EDIT FORM** dialog box to complete the process.

Edit By Curve

The **Edit By Curve** tool is used to manipulate T-Spline edges using a driving curve. The procedure to use this tool is discussed next.

- Click on the **Edit By Curve** tool from **MODIFY** drop-down; refer to Figure-62. The **EDIT BY CURVE** dialog box will be displayed; refer to Figure-63.

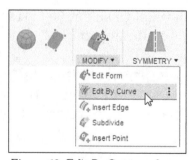

Figure-62. Edit By Curve tool

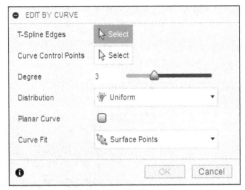

Figure-63. EDIT BY CURVE dialog box

- The **Select** button of **T-Spline Edges** section is active by default. Click on the edges from T-Spline body to be used for modification; refer to Figure-64.

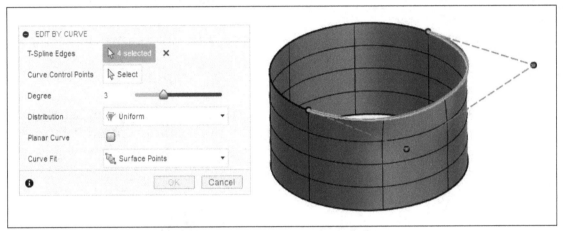

Figure-64. Selecting the T Spline edges

- Click on **Select** button of **Curve Control Points** section of the dialog box and select the points to be used for editing on earlier selected curve; refer to Figure-65. The handles to modify select point will be displayed.

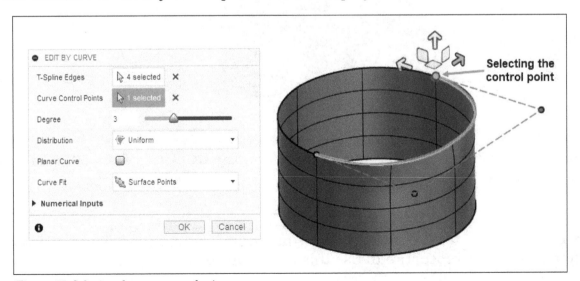

Figure-65. Selecting the curve control points

- Specify desired value in **Degree** edit box or specify the degree value by sliding the **Degree** slider to use 1-7 degrees of control on the driving curve. This will generate specified number of control points on the curve.

- Select **Uniform** option from **Distribution** drop-down to distribute control points uniformly along curve.
- Select **Edge Length** option from **Distribution** drop-down to distribute control points in proportion to lengths of T-Spline edges.
- Select **Planar Curve** check box to restrict modification of driving curves in best fit plane; refer to Figure-66.

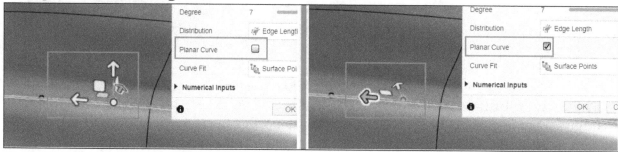

Figure-66. Effect of Planar Curve check box

- Select the **Surface Points** option from **Curve Fit** drop-down to fit vertices on smooth surface to driving curve.
- Select the **Control Points** option from **Curve Fit** drop-down to fit control points of selected edges to driving curve.
- Expand the **Numerical Inputs** node and specify numerical values of various parameters or use the distance, planar, scale, and rotation manipulator handles in the canvas to adjust the control points; refer to Figure-67.

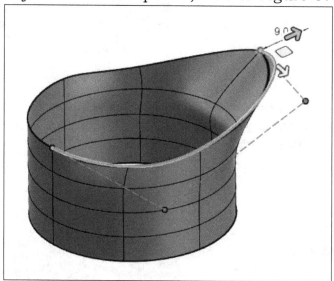

Figure-67. Adjusting the control points using manipulator

- After specifying desired parameters, click on the **OK** button from **EDIT BY CURVE** dialog box to complete the process.

Insert Edge

The **Insert Edge** tool is used to insert an edge at a specified distance from the selected edge. The procedure to use this tool is discussed next.

- Click on the **Insert Edge** tool from **MODIFY** drop-down; refer to Figure-68. The **INSERT EDGE** dialog box will be displayed; refer to Figure-69.

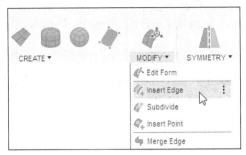

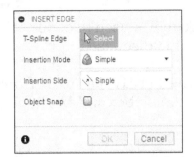

Figure-68. Insert Edge tool   *Figure-69. INSERT EDGE dialog box*

- The **Select** button of **T-Spline Edge** section is active by default. Click on the edge from model to select. The preview of inserted edge will be displayed in green color; refer to Figure-70.

Figure-70. Preview of inserted edge

- Select **Simple** option from **Insertion Mode** drop-down if you do not want to move any point but may be it change the surface shape.
- Select **Exact** option from **Insertion Mode** drop-down to maintain the exact surface shape by adding points.
- Select **Single** option from **Insertion Side** drop-down to insert the edge in only one side to the selected edge.
- Select **Both** option from **Insertion Side** drop-down to insert the edge on both side to the selected edge.
- Click in the **Insert Location** edit box and enter the value of location for inserting the edge. The value of **Insert Location** lie between 0 to 1.
- Select the **Object Snap** check box to move the new vertices to the closest point. These vertices can snap to solid, surface, and mesh bodies.
- Click in the **Offset** edit box and enter desired value.
- After specifying the parameters, click on the **OK** button from **INSERT EDGE** dialog box to complete the process.

Subdivide

The **Subdivide** tool is used to divide selected face in four or more faces. The procedure to use this tool is discussed next.

- Click on the **Subdivide** tool from **MODIFY** drop-down; refer to Figure-71. The **SUBDIVIDE** dialog box will be displayed; refer to Figure-72.

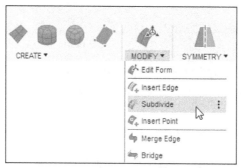

Figure-71. Subdivide tool

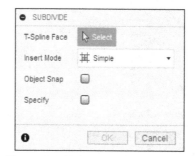

Figure-72. SUBDIVIDE dialog box

- The **Select** button of **T-Spline Face** section is active by default. Click on the face from model to select. Selected face will be divided into four face; refer to Figure-73.

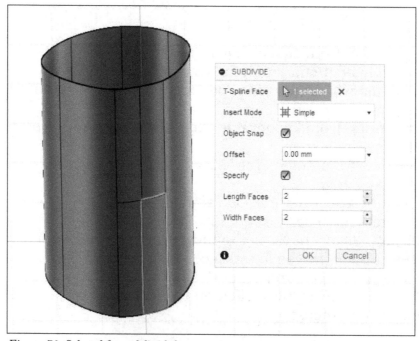

Figure-73. Selected face subdivided

- Select **Simple** option from **Insert Mode** drop-down to add the minimum number of control points to the subdivided face.
- Select **Exact** option from **Insert Mode** drop-down if you do not want to change the surface body and maintain the extra shape control points are added.
- Select the **Object Snap** check box to move the new vertices to the closest point. These vertices can snap to solid, surface, and mesh bodies.
- Click in the **Offset** edit box and enter desired value.
- Select the **Specify** check box to enter the number of faces along length and width of the selected face.
- After specifying the parameters, click on the **OK** button from **SUBDIVIDE** dialog box to complete the process.

Insert Point

The **Insert Point** tool is used to insert control points at the selected locations. The procedure to use this tool is discussed next.

- Click on the **Insert Point** tool from **MODIFY** drop-down; refer to Figure-74. The **INSERT POINT** dialog box will be displayed; refer to Figure-75.

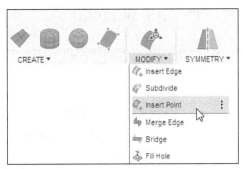

Figure-74. Insert Point tool

Figure-75. INSERT POINT dialog box

- The **Insertion Point** button is active by default. You need to click on the edge of the model to insert point; refer to Figure-76.

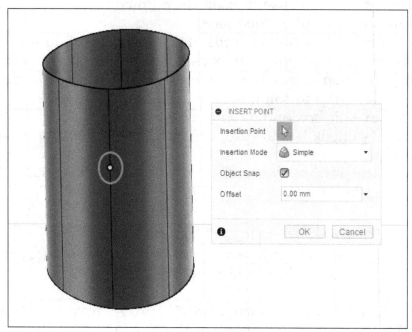

Figure-76. Inserting point

- The options of **Insertion Mode** and **Object Snap** were discussed earlier in this book.
- After specifying the parameters, click on the **OK** button from **INSERT POINT** dialog box to complete the process.

Merge Edge

The **Merge Edge** tool is used to connect two bodies by joining their edges. The procedure to use this tool is discussed next.

- Click on the **Merge Edge** tool from **MODIFY** drop-down; refer to Figure-77. The **MERGE EDGE** dialog box will be displayed; refer to Figure-78.

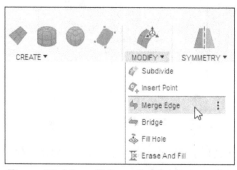

Figure-77. Merge Edge tool

Figure-78. MERGE EDGE dialog box

- The **Select** button of **Edge Group One** section is active by default. You need to select the edges or boundaries of a body.
- Click on the **Select** button of **Edge Group Two** section and select the boundary of other group; refer to Figure-79.

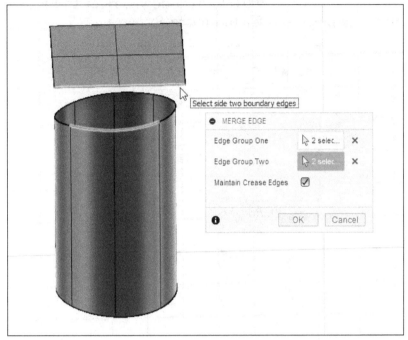

Figure-79. Selecting edges to merge

- Double-click on the edge of a face to select the adjacent edges also.
- Select the **Maintain Crease Edges** check box from **MERGE EDGE** dialog box to keep the crease at the merged edges.
- After specifying the parameters, click on the **OK** button from **MERGE EDGE** dialog box to complete the process; refer to Figure-80.

Figure-80. Edges merged

Bridge

The **Bridge** tool is used to connect two bodies by adding their intermediate faces. The procedure to use this tool is discussed next.

- Click on the **Bridge** tool from **MODIFY** drop-down; refer to Figure-81. The **BRIDGE** dialog box will be displayed; refer to Figure-82.

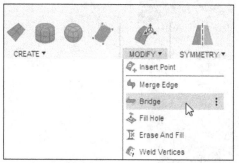

Figure-81. Bridge tool

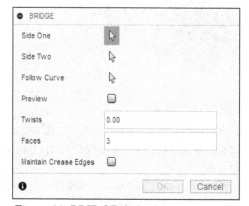

Figure-82. BRIDGE dialog box

- Click on the **Side One** button from **BRIDGE** dialog box and select desired faces.
- Click on the **Side Two** button from **BRIDGE** dialog box and select the faces from the model to join.
- Click on the **Follow Curve** button from **BRIDGE** dialog box and select a curve for the bridge to follow.
- Select the **Preview** check box to display a mesh preview of the bridge from the selected faces; refer to Figure-83.

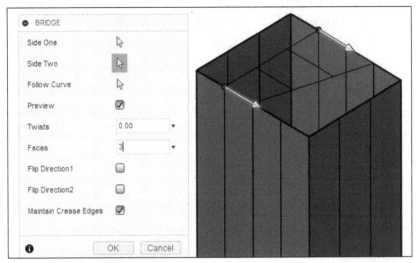

Figure-83. Preview of bridge

- Click in the **Twists** edit box and specify the number of rotation of bridge between two selected different faces (side one and side two).
- Click in the **Faces** edit box and specify the number of faces created between two selected sides.
- Select the **Flip Direction1** check box to flip the bridge direction of side one.
- Select the **Flip Direction2** check box to flip the bridge direction of side two.
- Select the **Maintain Crease Edges** check box from **BRIDGE** dialog box to keep the crease at the merged edges.
- After specifying the parameters, click on the **OK** button from **BRIDGE** dialog box to complete the process; refer to Figure-84.

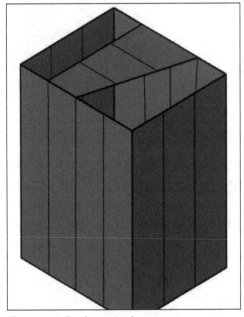

Figure-84. Bridge created

Fill Hole

The **Fill Hole** tool is used to close the opening of a T-Spline model. The procedure to use this tool is discussed next.

- Click on the **Fill Hole** tool from **MODIFY** drop-down; refer to Figure-85. The **FILL HOLE** dialog box will be displayed; refer to Figure-86.

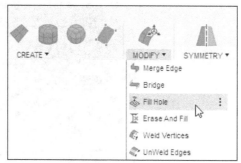

Figure-85. Fill Hole tool

Figure-86. FILL HOLE dialog box

- The **Select** button of **T-Spline Edge** section is active by default. Click on the edge of hole to select. A preview will be displayed; refer to Figure-87.

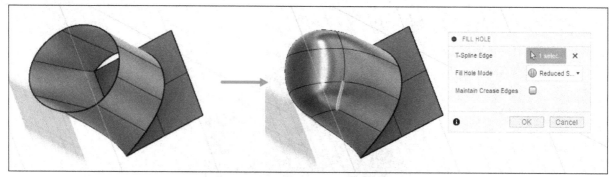

Figure-87. Preview of fill hole

- Select **Reduced Star** option from **Fill Hole Mode** drop-down to fill the hole by adding minimum number of star points of face.
- Select **Fill Star** option from **Fill Hole Mode** drop-down to fill the hole using single face. Due to this option, the star points will created at each vertex; refer to Figure-88.

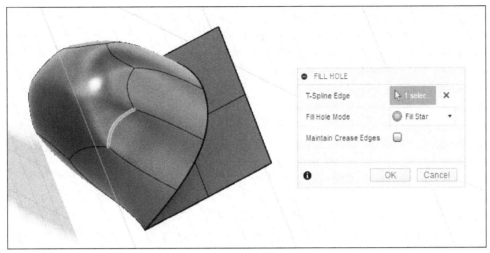

Figure-88. Fill Star option

- Select **Collapse** option from **Fill Hole Mode** drop-down to fill the hole by collapsing all the vertices of selected edge to the center point of hole face; refer to Figure-89.

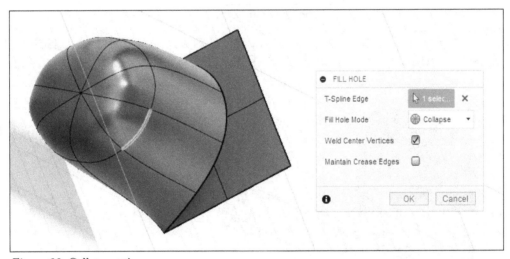

Figure-89. Collapse option

- Select the **Weld Center Vertices** check box from **FILL HOLE** dialog box to weld the vertices at the center of hole face. The **Weld Center Vertices** option is available when **Collapse** option is selected in the **Fill Hole Mode** drop-down.
- After specifying the parameters, click on the **OK** button from **FILL HOLE** dialog box to complete the process.

Erase And Fill

The **Erase And Fill** tool is used to delete a part of T-Spline geometry and fill new gaps with faces. The procedure to use this tool is discussed next.

- Click on the **Erase And Fill** tool from **MODIFY** drop-down; refer to Figure-90. The **ERASE AND FILL** dialog box will be displayed; refer to Figure-91.

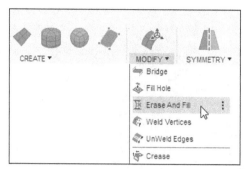

Figure-90. Erase And Fill tool

Figure-91. ERASE AND FILL dialog box

• The **Select** button of **T-Spline Faces** section is active by default. Select the faces to be erased and filled; refer to Figure-92.

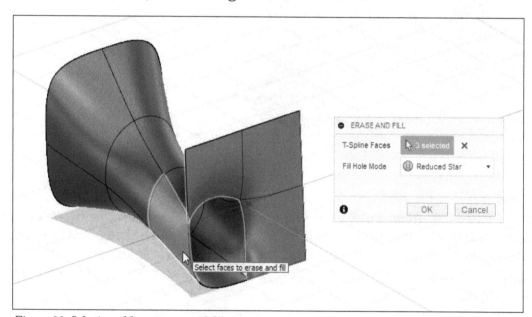

Figure-92. Selection of faces to erase and fill

• Select **Reduced Star** option from **Fill Hole Mode** drop-down to create the faces using minimum number of star points.
• Select **Fill Star** option from **Fill Hole Mode** drop-down to fill the hole with a single face. This tool creates star points at each vertex.
• After specifying the parameters, click on the **OK** button from **ERASE AND FILL** dialog box to complete the process; refer to Figure-93.

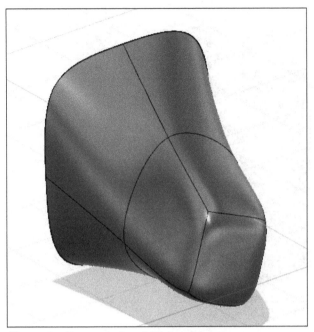

Figure-93. Faces erased and filled

Weld Vertices

The **Weld Vertices** tool is used to join two vertices into single vertex. The procedure to use this tool is discussed next.

- Click on the **Weld Vertices** tool from **MODIFY** drop-down; refer to Figure-94. The **WELD VERTICES** dialog box will be displayed; refer to Figure-95.

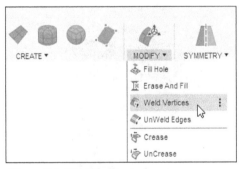

Figure-94. Weld Vertices tool

Figure-95. WELD VERTICES dialog box

- The **Select** button of **T-Spline Vertices** section is active by default. Click on the vertices to join together; refer to Figure-96.

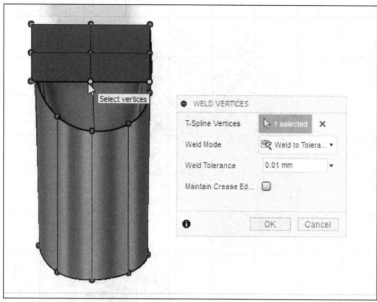

Figure-96. Selecting vertices

- Select **Vertex to Vertex** option from **Weld Mode** drop-down to move the first vertex to the second vertex.
- Select **Vertex to Midpoint** option from **Weld Mode** drop-down to move the two vertices to the midpoint of the selections.
- Select **Weld to Tolerance** option from **Weld Mode** drop-down to weld all the visible and invisible vertices together with in a specified tolerance.
- Click in the **Weld Tolerance** edit box and enter the value of tolerance as required.
- After specifying the parameters, click on the **OK** button from **WELD VERTICES** dialog box to complete the process; refer to Figure-97.

Figure-97. Vertices welded

UnWeld Edges

The **UnWeld Edges** tool is used to detach an edge or loop. The procedure to use this tool is discussed next.

- Click on the **UnWeld Edges** tool from **MODIFY** drop-down; refer to Figure-98. The **UNWELD EDGES** dialog box will be displayed; refer to Figure-99.

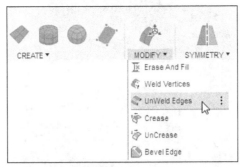

Figure-98. UnWeld Edges tool

Figure-99. UNWELD EDGES dialog box

- The **Select** button of **T-Spline Edges** section is active by default. Click on the edge or loop to select; refer to Figure-100. Double-click on the edge to select the loop.

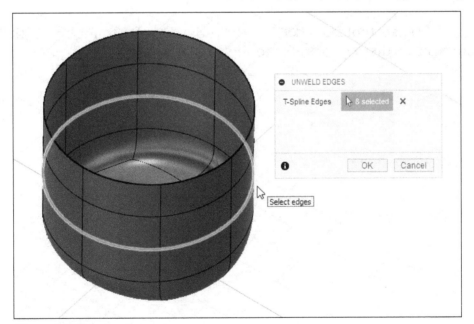

Figure-100. Selecting loop

- After selecting the edge or loop, click on the **OK** button from **UNWELDED EDGES** dialog box to complete the process; refer to Figure-101.

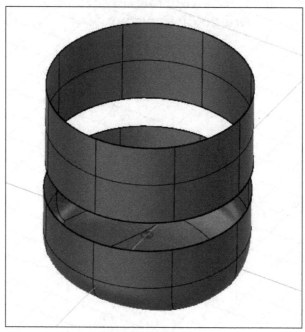

Figure-101. Edges unwelded

Crease

The **Crease** tool is used to add a sharp crease on the selected T-Spline body. The procedure to use this tool is discussed next.

- Click on the **Crease** tool from **MODIFY** drop-down; refer to Figure-102. The **CREASE** dialog box will be displayed; refer to Figure-103.

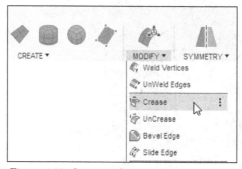

Figure-102. Crease tool

Figure-103. CREASE dialog box

- The **Select** button of **T-Spline Vertices or Edges** section is active by default. Click to select the edge. You can also use window selection to select multiple edges; refer to Figure-104.

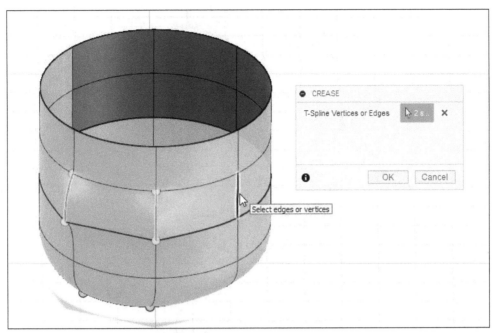

Figure-104. Selection of edges for crease

- After selection of edges for crease, click on **OK** button from **CREASE** dialog box. The selected edges will be converted into creased edges; refer to Figure-105.

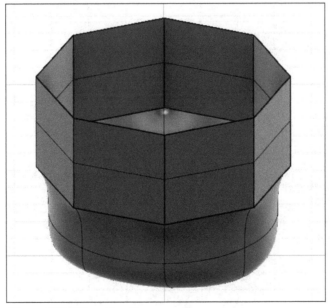

Figure-105. Edges creased

UnCrease

The **UnCrease** tool is used to remove crease from the selected body, edge, or vertex. The procedure to use this tool is discussed next.

- Click on the **UnCrease** tool from **MODIFY** drop-down; refer to Figure-106. The **UNCREASE** dialog box will be displayed; refer to Figure-107.

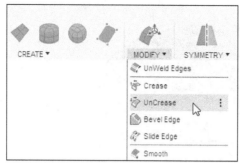

Figure-106. UnCrease tool

Figure-107. UNCREASE dialog box

- The **Select** button of **T-Spline Vertices or Edges** section is active by default. Click to select the edge. You can also use window selection to select multiple edges; refer to Figure-108.

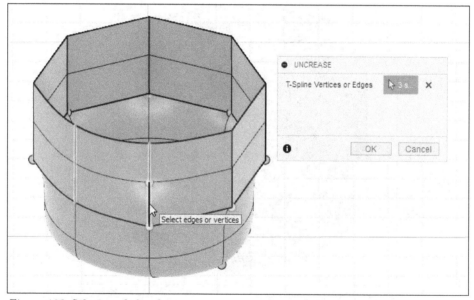

Figure-108. Selection of edges for uncrease

- After selection of edges for uncrease, click on **OK** button from **UNCREASE** dialog box to complete the process; refer to Figure-109.

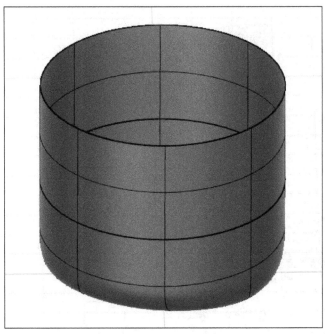

Figure-109. Edges uncreased

Bevel Edge

The **Bevel Edge** tool is used to flatten the area of body by selecting a specific edge. The procedure to use this tool is discussed next.

- Click on the **Bevel Edge** tool from **MODIFY** drop-down; refer to Figure-110. The **BEVEL EDGE** dialog box will be displayed; refer to Figure-111.

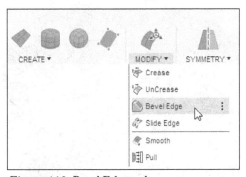

Figure-110. Bevel Edge tool

Figure-111. BEVEL EDGE dialog box

- The **Select** button of **T-Spline Edge** section is active by default. Click on the edge of a body to select; refer to Figure-112.

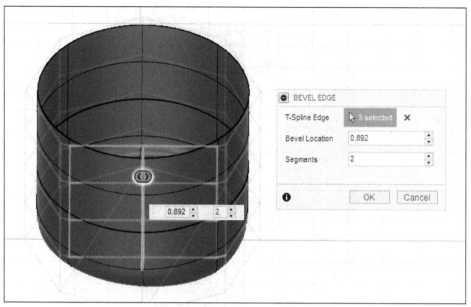

Figure-112. Selection of edges for bevel edge

- Click in the **Bevel Location** edit box and enter the value to position the new edge in decimal percentage. You can also set the value of **Bevel Location** option by moving the manipulator displayed on the selected edge.
- Click in the **Segments** edit box and enter desired number of faces to be inserted in between new edges.
- After specifying the parameters, click on the **OK** button from **BEVEL EDGE** dialog box. The selected edge will be flattened; refer to Figure-113.

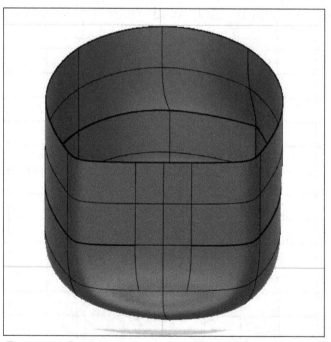

Figure-113. Bevel edge created

Slide Edge

The **Slide Edge** tool is used to move two edges closer together or farther apart. The procedure to use this tool is discussed next.

- Click on the **Slide Edge** tool from **MODIFY** drop-down; refer to Figure-114. The **SLIDE EDGE** dialog box will be displayed; refer to Figure-115.

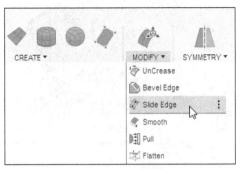

Figure-114. Slide Edge tool

Figure-115. SLIDE EDGE dialog box

- The **Select** button of **T-Spline Edge** section is active by default. Click on the edge of a body to select; refer to Figure-116.

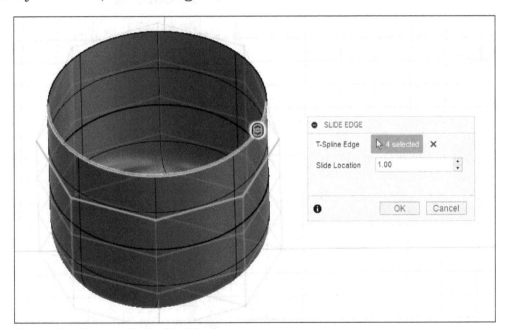

Figure-116. Selection of edges for slide edge

- Click in the **Slide Location** edit box and enter the value to position the new edge in decimal percentage. You can also set the value of **Slide Location** option by moving the manipulator displayed on the selected edge.
- After specifying the parameters, click on the **OK** button from **SLIDE EDGE** dialog box. The slide edge will be created; refer to Figure-117.

Figure-117. Slide edge created

Smooth

The **Smooth** tool is used to smooth an area of the T-Spline geometry by selecting the desired faces. The procedure to use this tool is discussed next.

- Click on the **Smooth** tool from **MODIFY** drop-down; refer to Figure-118. The **SMOOTH** dialog box will be displayed; refer to Figure-119.

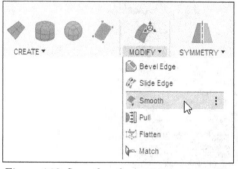

Figure-118. Smooth tool

Figure-119. SMOOTH dialog box

- The **Select** button of **Faces** section is active by default. Select the faces around desired area; refer to Figure-120.

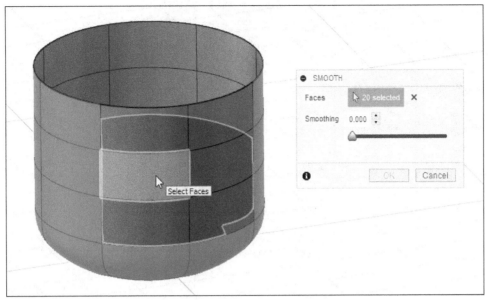

Figure-120. Selection of faces for smoothing

- Click in the **Smoothing** edit box and enter the value of smoothing rate from 0 to 1. You can also set the value of **Smoothing** option by moving the slider.
- After specifying the parameters, click on the **OK** button from **SMOOTH** dialog box to complete the process; refer to Figure-121.

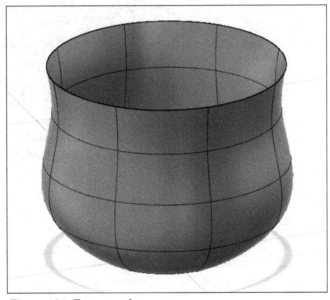

Figure-121. Faces smoothen

Cylindrify

The **Cylindrify** tool is used to form uneven T-Spline geometry into a smooth cylindrical shape. The procedure to use this tool is discussed next.

- Click on the **Cylindrify** tool from **MODIFY** drop-down; refer to Figure-122. The **CYLINDRIFY** dialog box will be displayed; refer to Figure-123.

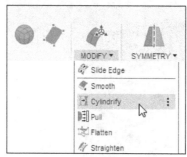

Figure-122. Cylindrify tool

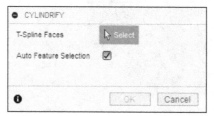

Figure-123. CYLINDRIFY dialog box

- The **Select** button of **T-Spline Faces** section is active by default. Select the faces around desired area; refer to Figure-124.

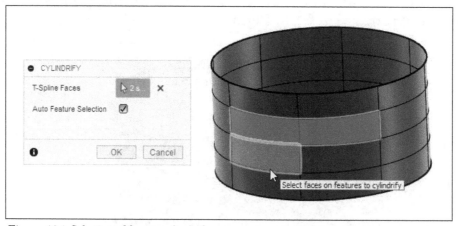

Figure-124. Selection of faces to cylindrify

- Select **Auto Feature Selection** check box to automatically select and preview entire features.
- After specifying the parameters, click on the **OK** button from the **CYLINDRIFY** dialog box to complete the process; refer to Figure-125.

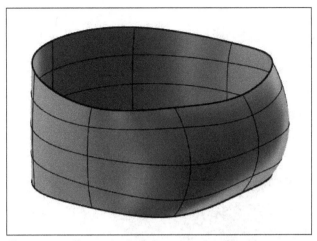

Figure-125. Faces cylindrified

Pull

The **Pull** tool is used for moving the selected vertices to the target body. The procedure to use this tool is discussed next.

- Click on the **Pull** tool from **MODIFY** drop-down; refer to Figure-126. The **PULL** dialog box will be displayed; refer to Figure-127.

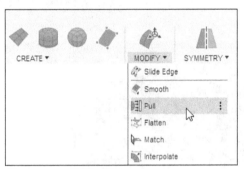

Figure-126. Pull tool

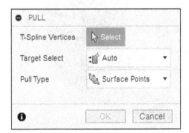

Figure-127. PULL dialog box

- The **Select** button of the **T-Spline Vertices** section is active by default. You need to click on the specific vertex to pull the vertex up to the targeted body; refer to Figure-128.

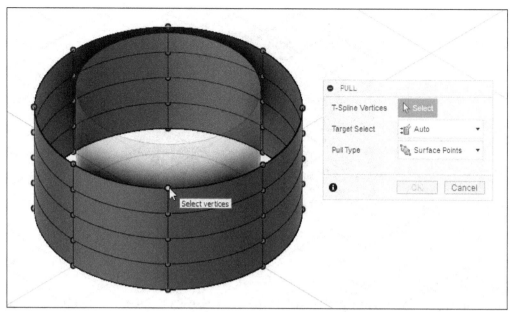

Figure-128. Pulling the vertex

- Select **Auto** option from **Target Select** drop-down to automatically pull the selected vertex.
- Select the **Select Targets** option from **Target Select** drop-down to manually select the target body. The updated **PULL** dialog box will be displayed; refer to Figure-129.

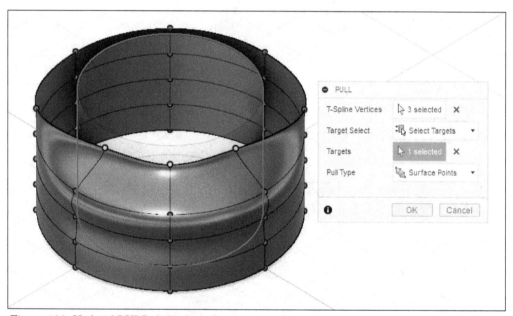

Figure-129. Updated PULL dialog box

- The **Select** button of **Targets** section is active by default. You need to click on the target body to pull the selected vertex.
- Select **Surface Points** option from **Pull Type** drop-down to move the surface points to the target body.
- Select **Control Points** option from **Pull Type** drop-down to move the control points to the target body.
- After specifying the parameters, click on the **OK** button from **PULL** dialog box to complete the process.

Flatten

The **Flatten** tool is used for moving the selected control point to the plane for flatten the selected surface. The procedure to use this tool is discussed next.

- Click on the **Flatten** tool from **MODIFY** drop-down; refer to Figure-130. The **FLATTEN** dialog box will be displayed; refer to Figure-131.

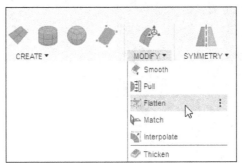

Figure-130. Flatten tool

Figure-131. FLATTEN dialog box

- The **Select** button of **T-Spline Vertices** section is active by default. Select the vertices of the face which you want to flatten; refer to Figure-132.

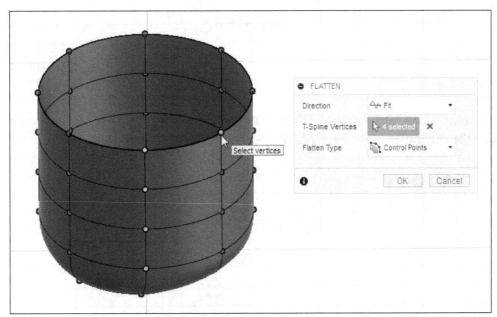

Figure-132. Selecting vertices for flatten surface

- On selecting all the vertices, the preview of flatten surface will be displayed; refer to Figure-133.

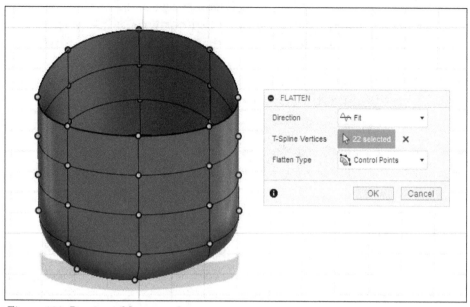

Figure-133. Preview of flatten surface

- Select **Fit** option from **Direction** drop-down to select the best fit plane for flatten the selected surface.
- Select the **Select Plane** option from **Direction** drop-down to manually select the plane to flatten all control points. You are required to select the plane.
- Select the **Select Parallel Plane** option from **Direction** drop-down to move the control points to the selected parallel plane. You are required to select the plane.
- Select desired option from the **Flatten** Type drop-down as discussed earlier in this chapter.
- After specifying the parameters, click on the **OK** button from **FLATTEN** dialog box to complete the process.

Straighten

The **Straighten** tool is used to align vertices of a T-Spline body into best fit line, existing line, or a line through two selected points. The procedure to use this tool is discussed next.

- Click on the **Straighten** tool from **MODIFY** drop-down; refer to Figure-134. The **STRAIGHTEN** dialog box will be displayed; refer to Figure-135.

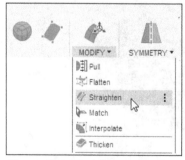

Figure-134. Straighten tool

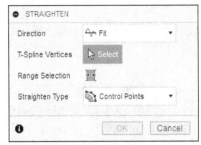

Figure-135. STRAIGHTEN dialog box

- The **Select** button of **T-Spline Vertices** section is active by default. Select the vertices of the face which you want to straighten; refer to Figure-136.

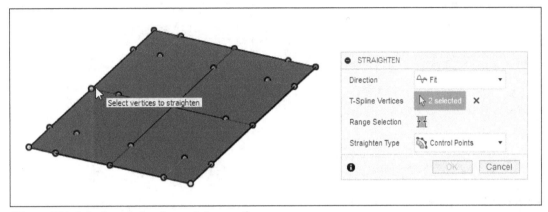

Figure-136. Selecting vertices for straighten surface

- On selecting all the vertices, the preview of straighten surface will be displayed; refer to Figure-137.

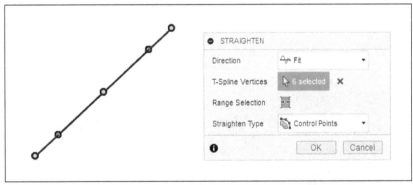

Figure-137. Preview of straighten surface

- Select **Fit** option from **Direction** drop-down to select the best fit plane for flatten the selected surface.
- Select the **Select Line** option from **Direction** drop-down to manually select the lines to straighten all control points. You are required to select the line.
- Select the **Select Parallel Line** option from **Direction** drop-down to select the parallel lines to straighten the control points. You are required to select the line.
- Select the **Select Two Points** option from **Direction** drop-down to select two points to define a line to straighten control points. You are required to select two points.
- Click on the **Range Selection** button to select all vertices or tangency handles between two selected vertices or tangency handles in the same row.
- Select desired option from **Straighten Type** drop-down which have been discussed

earlier.

*   After specifying the parameters, click on the **OK** button from **STRAIGHTEN** dialog box to complete the process.

Match

The **Match** tool is used to align the selected T-Spline edge with a sketch, face, or edge. The procedure to use this tool is discussed next.

*   Click on the **Match** tool from **MODIFY** drop-down; refer to Figure-138. The **MATCH** dialog box will be displayed; refer to Figure-139.

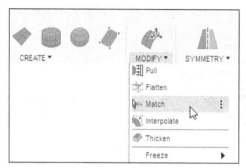

Figure-138. Match tool

Figure-139. MATCH dialog box

*   The **Select** button of **T-Spline Edges** section is active by default. You need to select the edge. You can also select the loop by double clicking on the edge; refer to Figure-140.

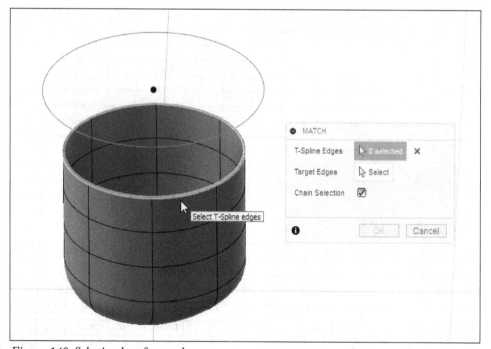

Figure-140. Selecting loop for match

*   Click on the **Select** button of **Target Edges** section and select the target edge or sketch.
*   On selecting, the preview of alignment will be displayed along with updated **MATCH**

dialog box; refer to Figure-141.

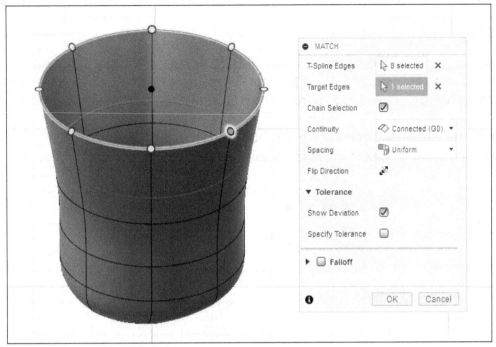

Figure-141. Updated MATCH dialog box

- Click on the **Flip Direction** button from **MATCH** dialog box to flip the direction of alignment; refer to Figure-142.

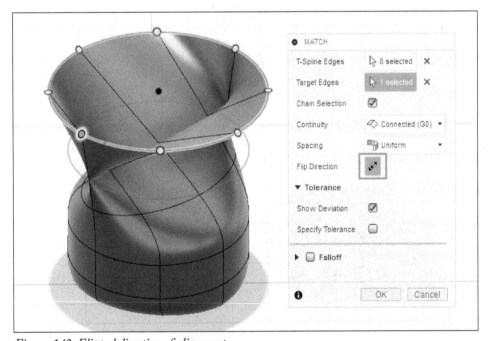

Figure-142. Flipped direction of alignment

- Select the **Show Deviation** check box from **Tolerance** node to display the amount and location of maximum deviation.
- Select the **Specify Tolerance** check box from **Tolerance** node to specifies the value of how close the T-Spline edges need to be to the target edge.
- Select the **Falloff** check box from **MATCH** dialog box to determine the value of surface affected by the match.
- Click in the **Falloff Range** edit box and enter the value of surface affected by

the match.
• After specifying the parameters, click on the **OK** button from **MATCH** dialog box to complete the process.

Interpolate

The **Interpolate** tool is used to move the T-Spline control points or surface points for improving fitting. The procedure to use this tool is discussed next.

• Click on the **Interpolate** tool from **MODIFY** drop-down; refer to Figure-143. The **INTERPOLATE** dialog box will be displayed; refer to Figure-144.

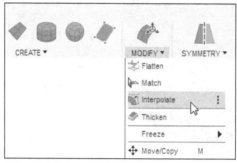

Figure-143. Interpolate tool

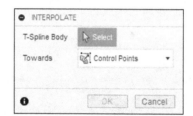

Figure-144. INTERPOLATE dialog box

• The **Select** button of **T-Spline Body** section is active by default. Click on the body to select; refer to Figure-145.

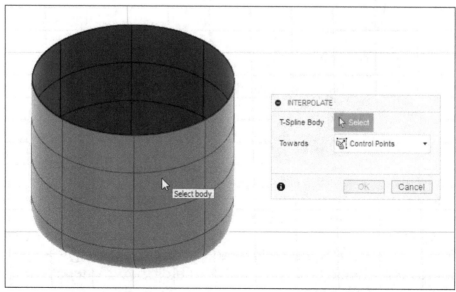

Figure-145. Selecting the body

- On selecting the body, the preview will be displayed; refer to Figure-146.

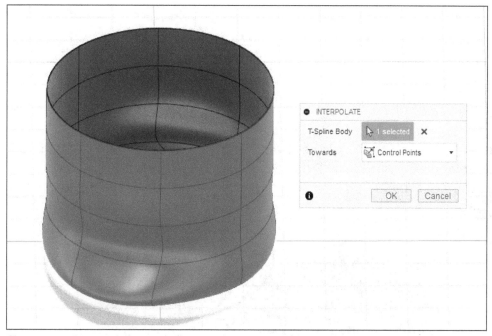

Figure-146. Preview of Interpolate

- Select **Surface Points** option from **Towards** drop-down to move the control points towards the surface.
- Select **Control Points** option from **Towards** drop-down if you want to fit the surface through existing control points.
- After specifying the parameters, click on the **OK** button from **INTERPOLATE** dialog box to complete the process.

Thicken

The **Thicken** tool is used to apply thickness to the sculpt faces. The procedure to use this tool is discussed next.

- Click on the **Thicken** tool from **MODIFY** drop-down; refer to Figure-147. The **THICKEN** dialog box will be displayed; refer to Figure-148.

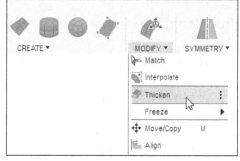

Figure-147. Thicken tool

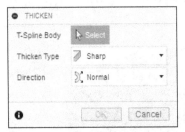

Figure-148. THICKEN dialog box

- The **Select** button of **T-Spline Body** section is active by default. Click on the body to select. The updated **THICKEN** dialog box will be displayed.
- Click in the **Thickness** edit box and enter desired thickness. You can also move the arrow displaying on the selected model to adjust the thickness; refer to Figure-149.

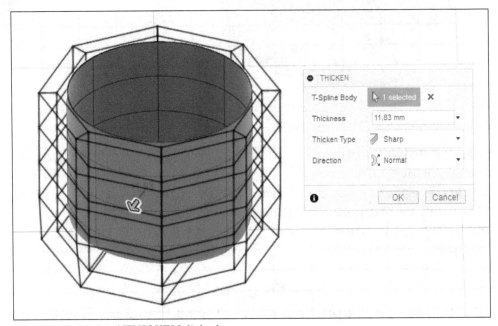

Figure-149. Updated THICKEN dialog box

- Select the **Sharp** button from **Thicken Type** drop-down to connect the surface with the straight face.
- Select the **Soft** button from **Thicken Type** drop-down to connect the surface with the round face.
- Select the **No Edge** button from **Thicken Type** drop-down to not connect the surfaces.
- Select **Normal** button from **Direction** drop-down to create a new surface perpendicular to the selected surface.
- Select **Axis** button from **Direction** drop-down to create a new surface perpendicular to a selected axis.
- After specifying the parameters, click on the **OK** button from **THICKEN** dialog box. The selected body will be thicken; refer to Figure-150.

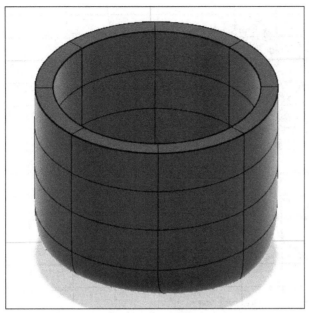

Figure-150. Surface thickened

Freeze

The **Freeze** tool is used to freeze the selected edges or faces to prevent changes. The procedure to use this tool is discussed next.

- Click on the **Freeze** tool of **Freeze** cascading menu from **MODIFY** drop-down; refer to Figure-151. The **FREEZE** dialog box will be displayed; refer to Figure-152.

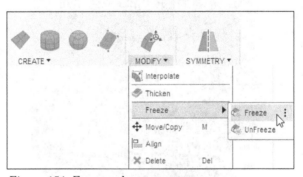

Figure-151. Freeze tool

Figure-152. FREEZE dialog box

- The **Select** button of **T-Spline Faces/Edges** section is active by default. Click on the edges/ faces to select; refer to Figure-153.

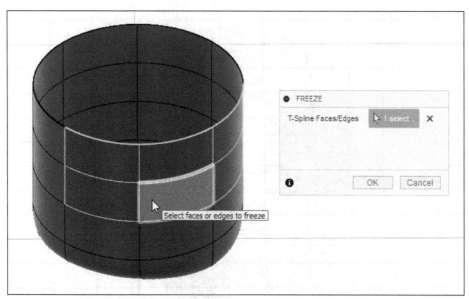

Figure-153. Selection of faces and edges to freeze

- After selection of edges, click on the **OK** button from **FREEZE** dialog box. The selected edges or faces will be frozen; refer to Figure-154.

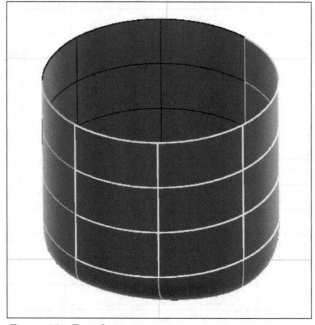

Figure-154. Faces frozen

UnFreeze

The **UnFreeze** tool is used to unfreeze the frozen edges of faces. The procedure to use this tool is discussed next.

- Click on the **UnFreeze** tool of **Freeze** cascading menu from **MODIFY** drop-down; refer to Figure-155. The **UNFREEZE** dialog box will be displayed; refer to Figure-156.

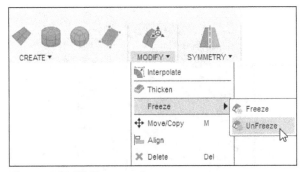

Figure-155. UnFreeze tool

Figure-156. UNFREEZE dialog box

- The **Select** button of **T-Spline Faces/Edges** section is active by default. Click on the edges/faces to unfreeze; refer to Figure-157.

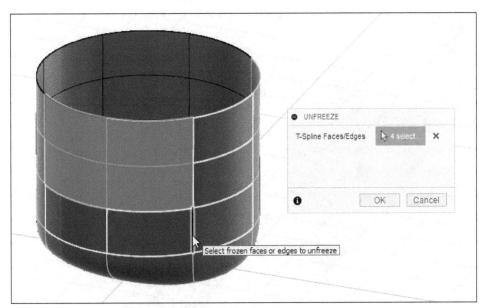

Figure-157. Selection of faces and edges to unfreeze

- After selecting, click on the **OK** button from **UNFREEZE** dialog box. The frozen edges will be unfreeze; refer to Figure-158.

Figure-158. Faces and edges unfroze

PRACTICAL

In this practical, we will create the model shown in Figure-159.

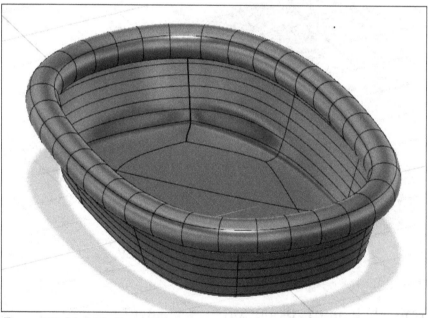

Figure-159. Model for Practical 1

Creating Sketch

- Open the **FORM** mode as discussed earlier.
- Click on the **Create Sketch** tool from **CREATE** drop-down in the **Toolbar** and select a plane on which you want to create sketch.
- Click on the **Center Rectangle** tool of **Rectangle** cascading menu from **CREATE** drop-down and create a rectangle as shown in Figure-160.

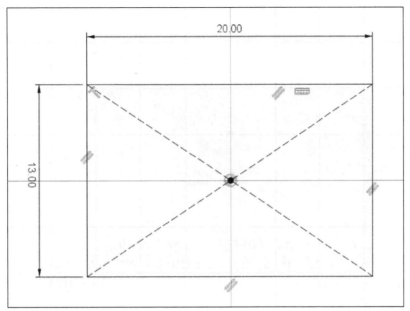

Figure-160. Creating first sketch

- Click on the **Fillet** tool of **MODIFY** drop-down from **Toolbar** and apply the fillet of **3** as shown in Figure-161.

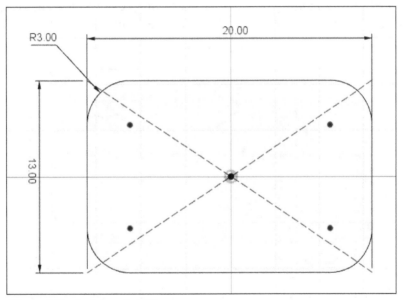

Figure-161. Applying the Fillet

- Click on the **Finish Sketch** button from **Toolbar** to exit the sketch.

Creating plane

- Click on the **Offset Plane** tool of **CONSTRUCT** drop-down from **Toolbar**. The **OFFSET PLANE** dialog box will be displayed.
- The **Select** button of **Plane** section is active by default. You need to select the recently created sketch plane as reference; refer to Figure-162.

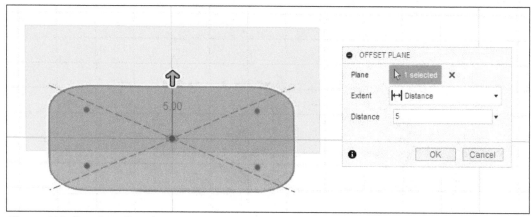

Figure-162. Creating offset plane1

- Click in the **Distance** edit box of **OFFSET PLANE** dialog box and enter the value as **5**.
- After specifying the parameters, click on the **OK** button from **OFFSET PLANE** dialog box. The plane will be created and displayed above the first sketch.

Creating second sketch

- Click on the **Create Sketch** tool from **CREATE** drop-down in the **Toolbar** and select the recently created plane as reference to create sketch.
- Click on the **Center Rectangle** tool of **Rectangle** cascading menu from **CREATE** drop-down and create a sketch as shown in Figure-163.

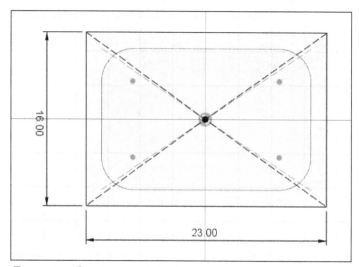

Figure-163. Creating second sketch

- Click on the **Fillet** tool of **MODIFY** drop-down from **Toolbar** and apply the fillet of **4** as shown in Figure-164.

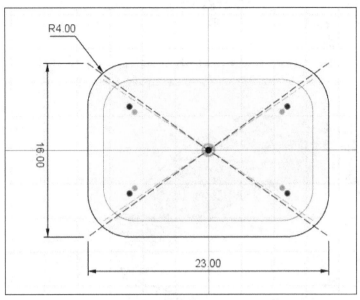

Figure-164. Applying fillet in second sketch

- Click on the **Finish Sketch** button from **Toolbar** to exit the sketch.

Creating Loft Feature

- Click on the **Loft** tool of **CREATE** drop-down from **Toolbar**. The **LOFT** dialog box will be displayed.
- Click on the **Chain Selection** check box of **LOFT** dialog box to select the edges in chain.
- Select the first and second created sketch to select in **Profiles** section; refer to Figure-165.

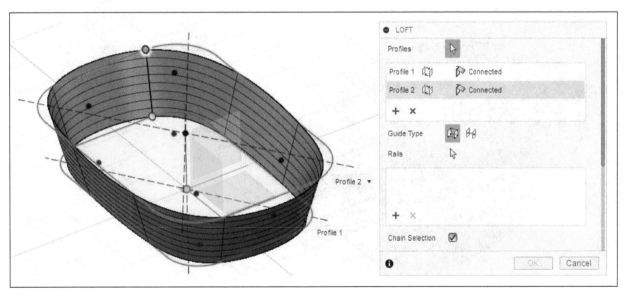

Figure-165. Selection of profiles for loft

- After specifying the parameters displayed on above figure, click on the **OK** button from **LOFT** dialog box.

Applying Fill Hole

- Click on the **Fill Hole** tool of **MODIFY** drop-down from **Toolbar**. The **FILL HOLE** dialog box will be displayed.
- Select the edge from the model as displayed; refer to Figure-166.

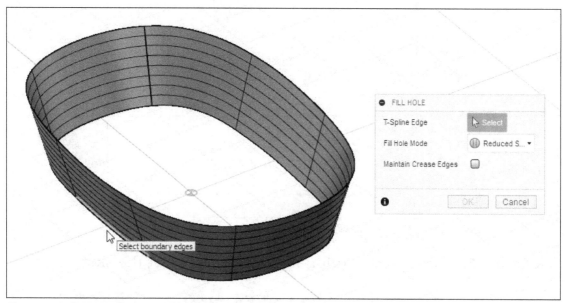

Figure-166. Selecting the edge for fill hole

- The selected hole will be filled.
- Click on the **PIPE** tool of **CREATE** drop-down from **Toolbar**. The **PIPE** dialog box will be displayed.
- Click on the edges of model as shown in figure to select as a path for pipe; refer to Figure-167.

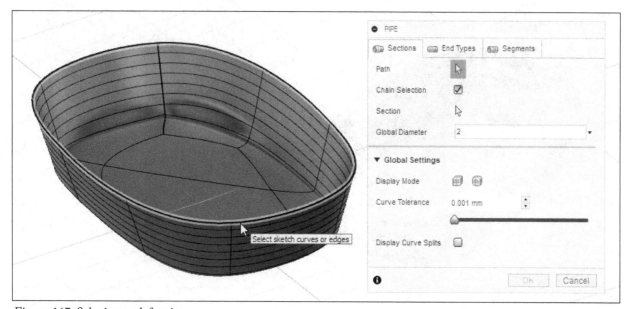

Figure-167. Selecting path for pipe

- Click on the **Smooth Display** button of **Display Mode** section in **Global Settings** area from **Sections** tab of **PIPE** dialog box and enter the parameters as displayed in above figure.
- After specifying the parameters, the model will be displayed as shown in Figure-168.

Figure-168. Practical1 created

SELF ASSESSMENT

Q1. What is Sculpting?

Q2. Select the .... check box if you want to see the curve splits of pipe.

Q3. The .... button is used to create multiple faces continuously.

Q4. What is the use of **Edit Form** tool?

Q5. The **Feature** selection button is used to select all the faces of a selected hole. (T/F)

Q6. The **Merge Edge** tool is used to connect two bodies by joining their edges. (T/F)

PRACTICE 1

Create a wooden tool as displayed in Figure-169 and Figure-170. As a primitive structure, use the Sculpt cylinder.

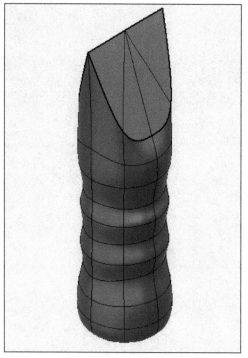

Figure-169. First view of Practice 1

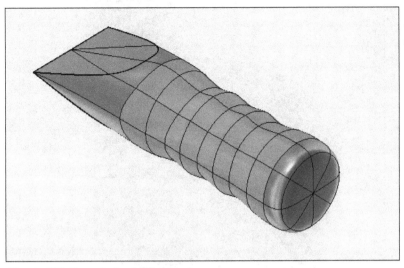

Figure-170. Second view of Practice 1

PRACTICE 2

Create the model as shown in Figure-171 and Figure-172 by using the tools of **FORM** mode.

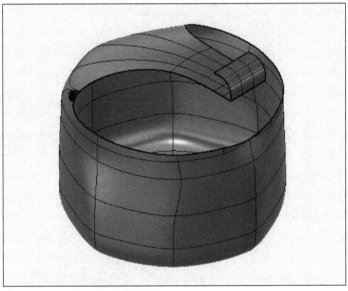

Figure-171. First view of Practice 2

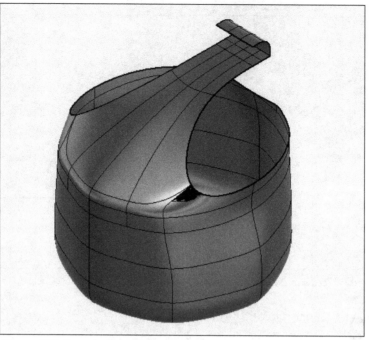

Figure–172. Second view of Practice 2

FOR STUDENT NOTES

Chapter 13

Sculpting-2

Topics Covered

The major topics covered in this chapter are:

- *Mirror Internal*
- *Mirror Duplicate*
- *Circular Internal*
- *Circular Duplicate*
- *Clear Symmetry*
- *Isolate Symmetry*
- *Display Mode*
- *Repair Body*
- *Make Uniform*
- *Convert Tools*

INTRODUCTION

In the last chapter, we have learned to create and modify the sculpt object by using various tools. In this chapter, we will discuss the symmetry and utilities tool used in **FORM** mode.

SYMMETRY TOOLS

Symmetry tools are used to create symmetric copies of the selected sculpt features in the **FORM Mode**. These tools are discussed next.

Mirror - Internal

The **Mirror - Internal** tool is used to create an internal mirror symmetry in the T-Spline body on selecting an edge, face, and vertex. The procedure to use this tool is discussed next.

- Click on the **Mirror - Internal** tool from **SYMMETRY** drop-down; refer to Figure-1. The **MIRROR - INTERNAL** dialog box will be displayed; refer to Figure-2.

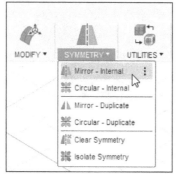

Figure-1. Mirror - Internal tool

Figure-2. MIRROR- INTERNAL dialog box

- The **Select** button of **Select Face** section is active by default. Click on the face from master side; refer to Figure-3.

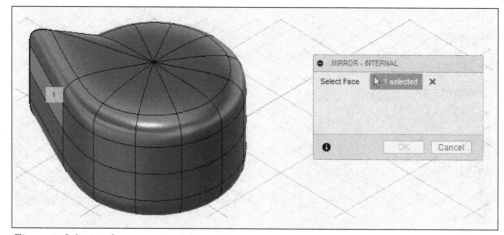

Figure-3. Selecting face on master side

- Click on the face to be made mirror symmetric. The preview of mirror will be displayed; refer to Figure-4.

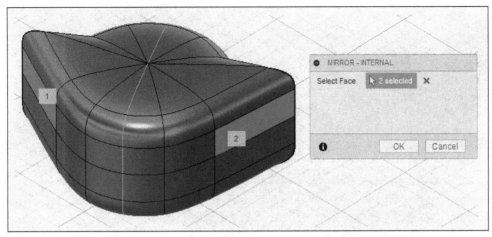

Figure-4. Preview of mirror

- If the preview of mirror symmetry is as required then click on the **OK** button from **MIRROR - INTERNAL** dialog box to complete the process.

Circular - Internal

The **Circular - Internal** tool is used to create an internal circular symmetry based on selected face, edge, or vertex. The procedure to use this tool is discussed next.

- Click on the **Circular - Internal** tool from **SYMMETRY** drop-down; refer to Figure-5. The **CIRCULAR - INTERNAL** dialog box will be displayed; refer to Figure-6.

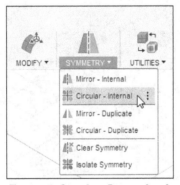

Figure-5. Circular- Internal tool

Figure-6. CIRCULAR- INTERNAL dialog box

- The **Select** button of **Select Face** section is active by default. Click on the face to select for symmetry; refer to Figure-7.

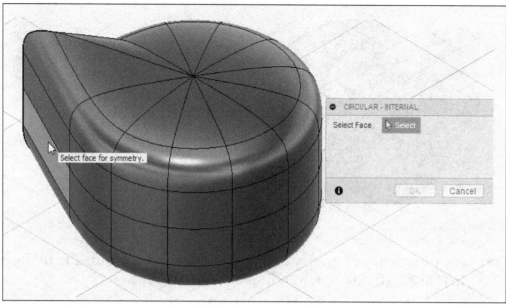

Figure-7. Selecting face for symmetry

- After selecting, the updated **CIRCULAR - INTERNAL** dialog box will be displayed along with the preview; refer to Figure-8.

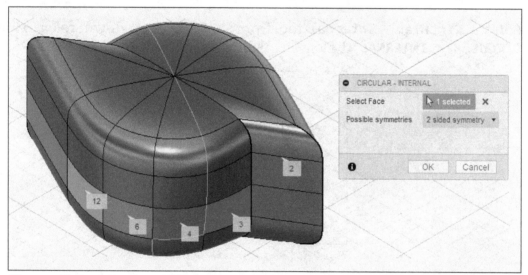

Figure-8. Updated CIRCULAR INTERNAL dialog box

- Select desired option from **Possible symmetries** drop-down to define number of symmetric sections and click on the **OK** button from **CIRCULAR - INTERNAL** dialog box to complete the process; refer to Figure-9.

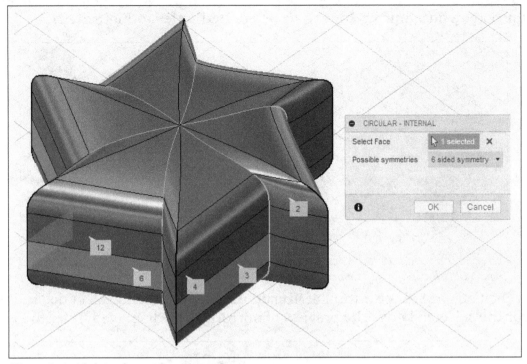

Figure-9. 6-sided circular symmetry created

Mirror - Duplicate

The **Mirror - Duplicate** tool is used to create a new T-Spline body or surface based on the selected plane or face. The procedure to use this tool is discussed next.

* Click on the **Mirror - Duplicate** tool from **SYMMETRY** drop-down; refer to Figure-10. The **MIRROR - DUPLICATE** dialog box will be displayed; refer to Figure-11.

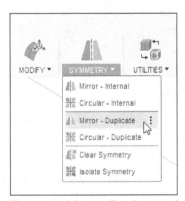

Figure-10. Mirror- Duplicate tool

Figure-11. MIRROR- DUPLICATE dialog box

- The **Select** button of **T-Spline Body** section is active by default. Click on the sculpt body whose mirror copy is to be created; refer to Figure-12.

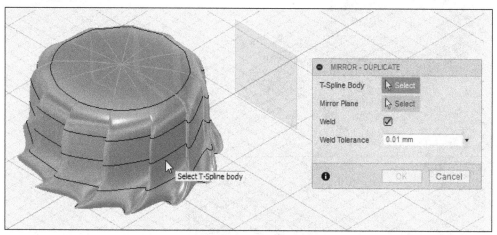

Figure-12. Selection of body for mirror

- Now, click on the **Select** button of **Mirror Plane** section and select desired plane to mirror the selected body. The preview of mirror will be displayed; refer to Figure-13.

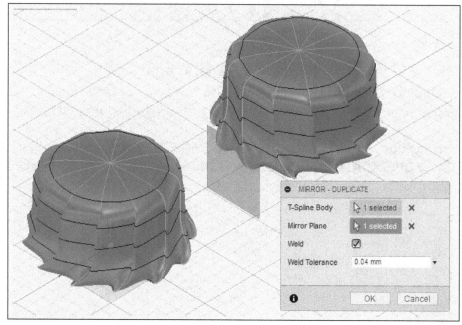

Figure-13. Preview of mirror-duplicate

- Select the **Weld** check box from **MIRROR - DUPLICATE** dialog box to weld the symmetric edges together.
- Click in the **Weld Tolerance** edit box and enter desired value of tolerance.
- After specifying the parameters, click on the **OK** button from **MIRROR - DUPLICATE** dialog box to complete the process.

Circular - Duplicate

The **Circular - Duplicate** tool is used to create circular symmetric copies of the selected body around an axis. The procedure to use this tool is discussed next.

- Click on the **Circular - Duplicate** tool from **SYMMETRY** drop-down; refer to Figure-14. The **CIRCULAR - DUPLICATE** dialog box will be displayed; refer to Figure-15.

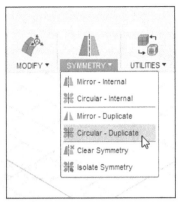

Figure-14. Circular- Duplicate tool

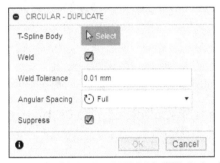

Figure-15. CIRCULAR- DUPLICATE dialog box

- The **Select** button of **T-Spline Body** section is active by default. Click on the body from canvas to select.
- Click to select the axis; refer to Figure-16.

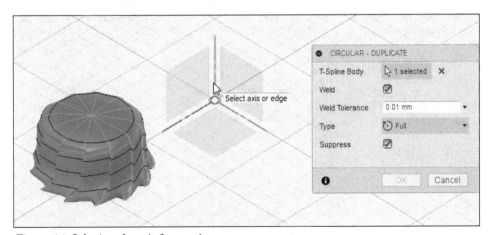

Figure-16. Selecting the axis for rotation

- Select the **Weld** check box if required.
- Click in the **Quantity** edit box and enter desired number of duplicate copies you want to create; the preview of duplicate copies will be displayed; refer to Figure-17.

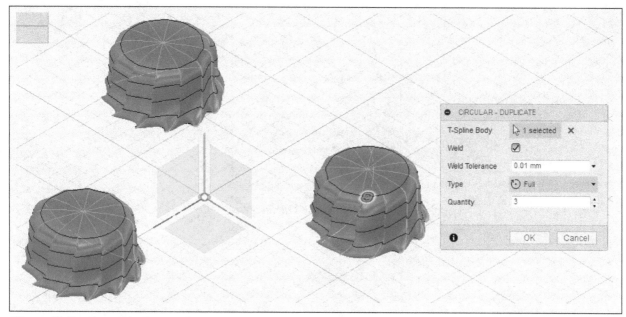

Figure-17. Preview of duplicate copies

- After specifying the parameters, click on the **OK** button from **CIRCULAR -DUPLICATE** dialog box to complete the process.

Clear Symmetry

The **Clear Symmetry** tool is used to delete the symmetry created earlier on the body. The procedure to use this tool is discussed next.

- Click on the **Clear Symmetry** tool from **SYMMETRY** drop-down; refer to Figure-18. The **CLEAR SYMMETRY** dialog box will be displayed; refer to Figure-19.

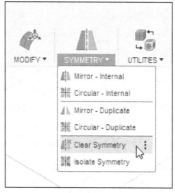

Figure-18. Clear Symmetry tool

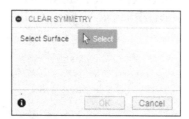

Figure-19. CLEAR SYMMETRY dialog box

- The **Select** button of **Select Surface** section is active by default. Click on the symmetry bodies for removing the symmetry constraint; refer to Figure-20.

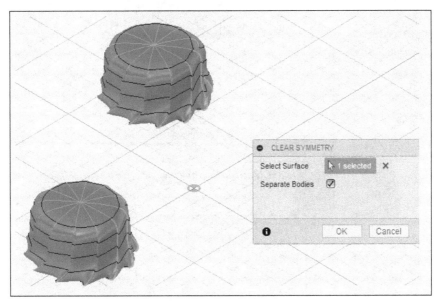

Figure-20. Selecting bodies for clearing symmetry

- Select the **Separate Bodies** check box to create a new bodies for disjointed surface.
- After specifying the parameters, click on the **OK** button from **CLEAR SYMMETRY** check box to complete the process; refer to Figure-21.

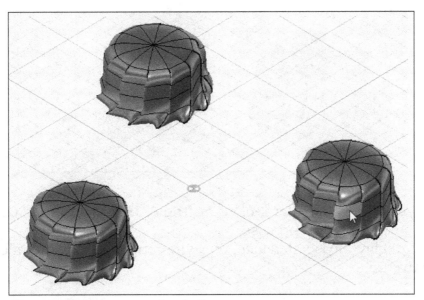

Figure-21. Deleted symmetry from bodies

Isolate Symmetry

The **Isolate Symmetry** tool is used to remove the symmetry condition from the selected face, edge, or vertex but the selected geometry will still symmetric to the other duplicate bodies. The procedure to use this tool is discussed next.

- Click on the **Isolate Symmetry** tool from **SYMMETRY** drop-down; refer to Figure-22. The **ISOLATE SYMMETRY** dialog box will be displayed; refer to Figure-23.

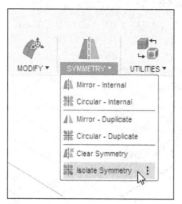

Figure-22. Isolate Symmetry tool

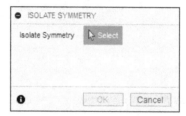

Figure-23. ISOLATE SYMMETRY dialog box

- The **Select** button of **Isolate Symmetry** section is active by default. Click on the symmetry to select; refer to Figure-24.

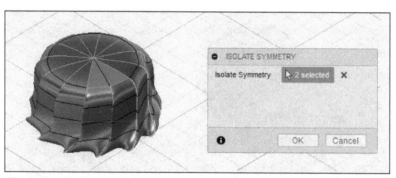

Figure-24. Selecting faces

- After specifying the parameters, click on the **OK** button from **ISOLATE SYMMETRY** dialog box to complete the process.
- If you edit or modify the isolated face, the symmetry faces will modified accordingly; refer to Figure-25.

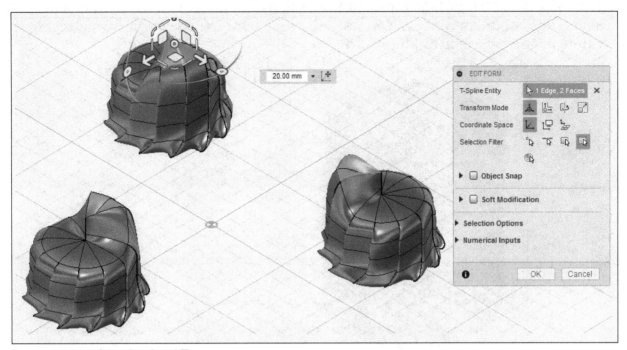

Figure-25. Modifying the selected Face

UTILITIES

Till now, we have discussed various symmetric tool. In this section, we will discuss various utility tools used in **FORM Mode**.

Display Mode

The **Display Mode** tool is used to switches the view of selected body to box or smooth display. The procedure to use this tool is discussed next.

* Click on the **Display Mode** tool from **UTILITIES** drop-down; refer to Figure-26. The **DISPLAY MODE** dialog box will be displayed; refer to Figure-27.

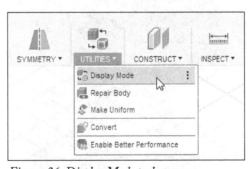

Figure-26. Display Mode tool

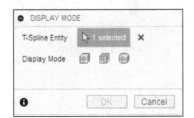

Figure-27. DISPLAY MODE dialog box

- The **Select** button of **T-Spline Entity** is active by default. You need to select the face, edge, body, or vertex. You can also use window selection for selecting the whole geometry; refer to Figure-28.

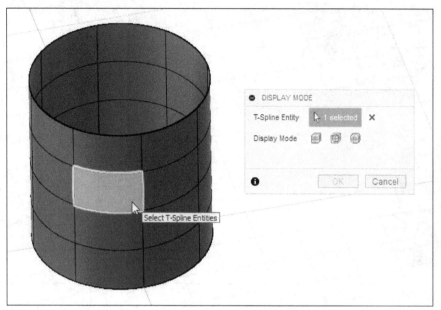

Figure-28. Selecting the face

- Select **Box Display** button from **Display Mode** section to display the control points of the T-Spline body.
- Select **Control Frame Display** button from **Display Mode** section to display the rounded frame body with the control frame around it.
- Select **Smooth Display** button from **Display Mode** section to display the rounded shape of the T-Spline body.
- After selecting the required display, click on the **OK** button from **DISPLAY MODE** dialog box to complete the process; refer to Figure-29.

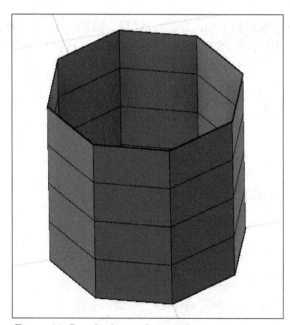

Figure-29. Box display mode created

Repair Body

The **Repair Body** tool is used for displaying the information about the mesh of sculpt body. This tool also repairs error star points and error T points. The procedure to use this tool is discussed next.

- Click on the **Repair Body** tool from **UTILITIES** drop-down; refer to Figure-30. The **REPAIR BODY** dialog box will be displayed; refer to Figure-31.

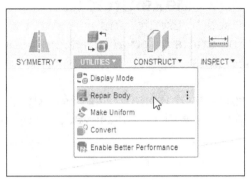

Figure-30. Repair Body tool

Figure-31. REPAIR BODY dialog box

- The **Select** button of **T-Spline Body** section is active by default. Click on the body to select.
- Click on the **AutoRepair** button from **REPAIR BODY** dialog box to repair error star points, error T points, and free edges.
- Select the **Error Star** check box of **Error Labels** section to display a red star on star points with an error.
- Select **Error T Points** check box of **Error Labels** section to display a red T on T point with an error.
- Select **Free Edges** check box to highlight the open edges on the body.
- Click in the **Weld Tolerance** edit box and specify the distance between edges to weld when using **AutoRepair** button.
- Select the **NGons** check box from **Geometry Labels** section to display the NGons with the number of edges on the selected model. NGons are faces which less than or more than 4 edges.

- Select the **T Points** check box to display a yellow T on T points of model.
- Select the **L Points** check box to display a yellow L on L Points of the selected model.
- Select the **Star Points** check box to display a yellow star on star points of model.
- After specifying the parameters, click on the **OK** button from **REPAIR BODY** dialog box to complete the process.

Make Uniform

The **Make Uniform** tool is used to create uniform surface of the selected body. This tool is used for making all the knots interval of selected body uniform. The procedure to use this tool is discussed next.

- Click on the **Make Uniform** tool from **UTILITIES** drop-down; refer to Figure-32. The **MAKE UNIFORM** dialog box will be displayed; refer to Figure-33.

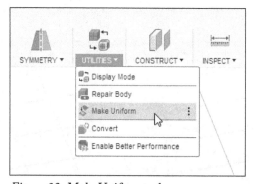

Figure-32. Make Uniform tool

Figure-33. MAKE UNIFORM dialog box

- The **Select** button of **T-Spline Body** section is active by default. Click on the body to select and click on the **OK** button. The tool will be applied.

Convert

The **Convert** tool is used to convert a sculpt object into other forms. The type of body created depends on selected body. The procedure to use this tool is discussed next.

- Click on the **Convert** tool from **UTILITIES** drop-down; refer to Figure-34. The **CONVERT** dialog box will be displayed; refer to Figure-35.

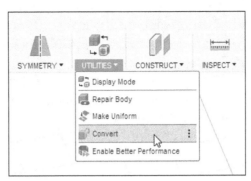

Figure-34. Convert tool

Figure-35. CONVERT dialog box

- Select **T-Splines to BRep** option from **Convert Type** drop-down to convert a T-Spline body into solid body.
- Select **BRep Face to T-Splines** option from **Convert Type** drop-down to convert a surface/face into sculpt.
- Select **Quad Mesh to T-Splines** option from **Convert Type** drop-down to convert a mesh body to a T-Spline body (sculpt).
- In our case, we are converting a T-Spline body to a solid body. The **Select** button of **Selection** section is active by default. Click on the sculpt body to convert; refer to Figure-36.

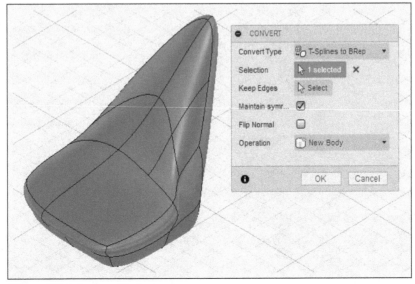

Figure-36. Selection of body to convert

- Click on the **Select** button from **Keep Edges** section and select required edges to maintain the selected edges in converted body.
- Select the **Maintain symmetry** check box from **CONVERT** dialog box to maintain the symmetry of T-Spline body after conversion.
- Select the **Flip Normal** check box to change the normal direction of selected bodies.
- After specifying the parameters, click on the **OK** button from **CONVERT** dialog box; refer to Figure-37. The converted body will be displayed in **DESIGN Workspace**.

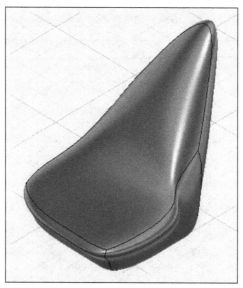

Figure-37. Converted body

Enable Better Performance

The **Enable Better Performance** tool is used to toggle between better performance or better display. The better display shows the bodies at highest quality and better performance calculates modification by applying G0 conditions at star points.

SELF ASSESSMENT

Q1. Which of the following tools is used to replicate changes made in one side of sculpt body to another side?

a. Mirror - Duplicate   b. Mirror - Internal
c. Isolate Symmetry   d. Circular - Internal

Q2. Which of the following tool is used to create replica of sculpt body with respect to a plane?

a. Mirror - Duplicate   b. Mirror - Internal
c. Isolate Symmetry   d. Circular - Internal

Q3. Which of the following tool is used to create symmetrical copies of selected sculpt body about an axis?

a. Mirror - Duplicate   b. Mirror - Internal
c. Circular Duplicate   d. Circular - Internal

Q4. Which of the following tool is used to remove symmetric conditions applied selected faces of sculpt body?

a. Clear Symmetry   b. Make Uniform
c. Isolate Symmetry   d. Erase and Fill

Q5. Which of the following tool is used to check geometry labels of sculpt mesh like L points, T points, and Star points?

a. Display Mode   b. Repair Body
c. Make Uniform   d. Enable Better Performance

FOR STUDENT NOTES

Chapter 14

Mesh Design

Topics Covered

The major topics covered in this chapter are:

- *Introduction to Mesh Workspace*
- *Insert Mesh*
- *BRep to Mesh*
- *Remesh*
- *Reduce*
- *Make Closed Mesh*
- *Erase and Fill*
- *Smooth*
- *Plane Cut*
- *Reverse Normal*
- *Delete Faces*
- *Separate and Merge Bodies*
- *Face Groups*

INTRODUCTION

A model of mesh consists of vertices, edges, and faces that use polygonal representation, including triangles and quadrilaterals, to define a 3D shape. Mesh models has no mass properties but they can be used as frame reference to create solid model with mass properties. Using mesh model, allows to use manipulation techniques that are not available in solid modeling like applying crease, split, smoothness level, and so on. Note that 3D printers use mesh model to create the 3D print.

OPENING THE MESH WORKSPACE

The tools in **MESH** workspace are used to create mesh models. At the time of writing this book, the Mesh workspace is available in Preview mode so, you need to select the **Mesh Workspace** check box from **Preview Features** node in the **Preferences** dialog box; refer to Figure-1. The procedure to open **Mesh** workspace is discussed next.

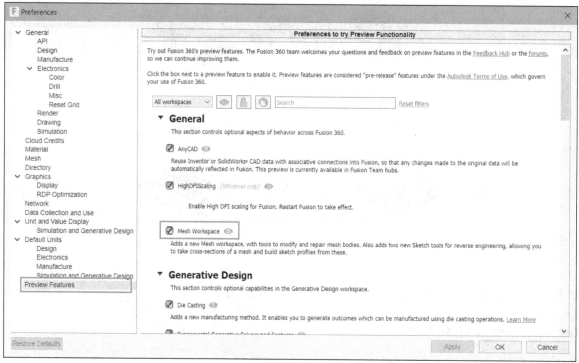

Figure-1. Mesh Workspace check box

- Click on the **DESIGN** option from **Change Workspace** drop-down. The **Design Workspace** will be displayed in **Autodesk Fusion 360** window.
- Click on **Create Mesh** tool from **CREATE** drop-down in the **Toolbar**; refer to Figure-2. The **MESH Workspace** will be displayed with **MESH PALETTE**; refer to Figure-3.

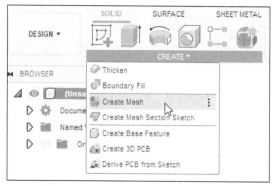

Figure-2. Create Mesh tool

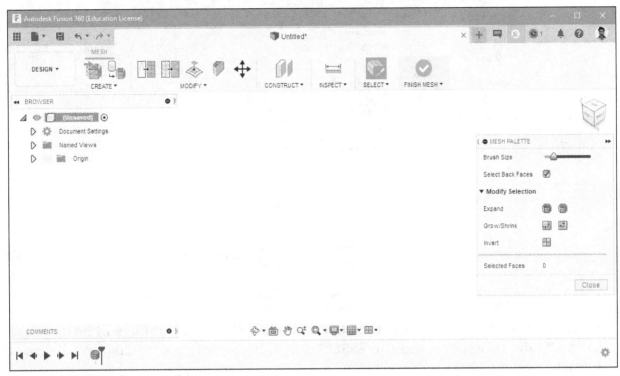

Figure-3. MESH Workspace window

- If **MESH PALETTE** is not displayed by default then select the **Mesh Palette** check box from **SELECT** drop-down in the **Toolbar** of Mesh environment. The tools in **MESH PALETTE** are used to manage selection range for mesh objects. Like, increasing the brush size using **Brush Size** slider, will expand the range of selection and allow selecting more mesh elements per click. Buttons in **Modify Selection** section are used to increase or decrease the number of elements in current selection.

INSERTING FILE

After starting Mesh workspace, the next step is to import a mesh model or convert an existing solid to mesh model. In this section, we will learn to insert selected file into Mesh workspace and convert a solid model to mesh model.

Insert Mesh

The **Insert Mesh** tool is used for inserting a .OBJ or .STL mesh file into current design. The procedure to use this tool is discussed next.

- Click on the **Insert Mesh** tool from **CREATE** drop-down; refer to Figure-4. The **Open** dialog box will be displayed; refer to Figure-5.

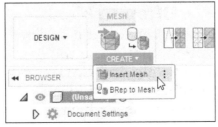

Figure-4. Insert Mesh tool

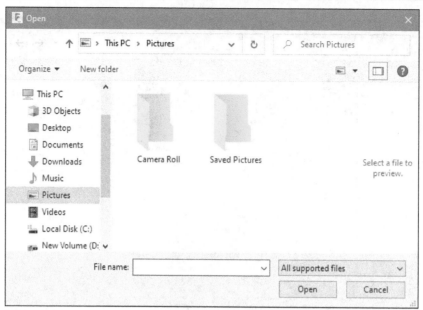

Figure-5. Open dialog box for inserting mesh file

- Select desired file and click on the **Open** button. The selected file will be displayed in the **MESH** workspace and the **INSERT MESH** dialog box will be displayed; refer to Figure-6.

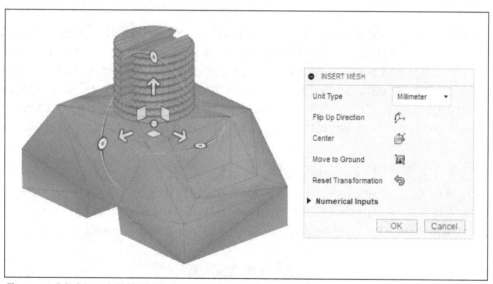

Figure-6. Model with INSERT MESH dialog box

- Set desired unit & orientation of model and then click on the **OK** button from the dialog box to insert the mesh. You can also use the handles displayed on model to modify orientation of mesh model.

BRep to Mesh

The **BRep to Mesh** tool is used to convert the selected solid body into mesh body. A BRep (Boundary representation) method is the one in which object is defined by its boundary limits. In Solid modeling and CAD, Solids and surfaces are generally created by BRep method. Sometimes, solids and surfaces are collectively called BRep. The procedure to use this tool is discussed next.

- Make sure you have a solid/surface model created in **Design** workspace or open a model and then switch to **MESH Workspace** to convert the body to mesh body.
- Click on the **BRep to Mesh** tool from **CREATE** drop-down; refer to Figure-7. The **BREP TO MESH** dialog box will be displayed; refer to Figure-8.

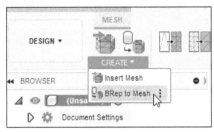

Figure-7. BRep to Mesh tool

Figure-8. BREP TO MESH dialog box

- The **Select** button of **Body** section is active by default. Click on the body to be converted to mesh; refer to Figure-9.

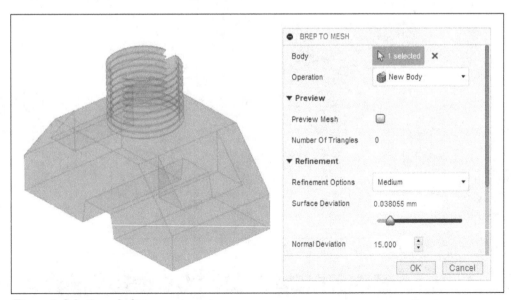

Figure-9. Selecting a body

- Select the **Preview Mesh** check box under **Preview** node from **BREP TO MESH** dialog box to display the preview of mesh body of the selected solid body. The preview of mesh body will be displayed along with **Number of Triangles**; refer to Figure-10.

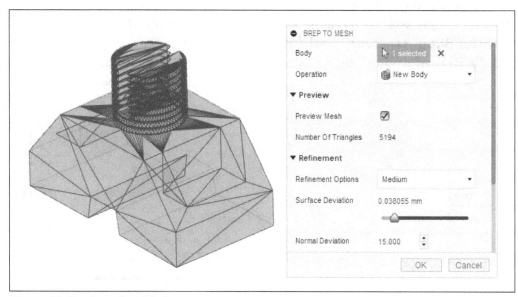

Figure-10. Preview of mesh body

- Select the **High**, **Medium**, or **Low** option from **Refinement Options** drop-down to set the refinement of mesh body automatically.
- If you want to set the refinement manually then click on the **Custom** option from **Refinement Options** drop-down.
- Move the **Surface Deviation**, **Normal Deviation**, **Maximum Edge Length**, and **Aspect Ratio** sliders to set the respective values.
- After specifying the parameters, click on the **OK** button from **BREP TO MESH** dialog box to complete the process; refer to Figure-11.

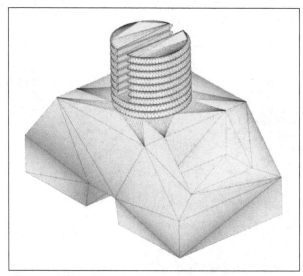

Figure-11. Created mesh body

MODIFICATION TOOLS

In this section, we will discuss various tools which are used to modify the mesh body.

Remesh

The **Remesh** tool is used to refine selected mesh faces or body to form regular-shaped triangular faces. Note that in FEM, there can be different shaped elements in a mesh like tetrahedra, hexahedra, and so on. The procedure to use this tool is discussed next.

- Click on the **Remesh** tool from **MODIFY** drop-down; refer to Figure-12. The **REMESH** dialog box will be displayed; refer to Figure-13.

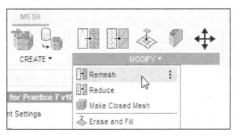

Figure-12. Remesh tool

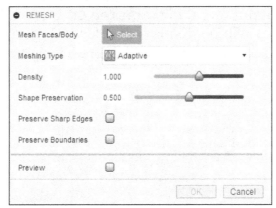

Figure-13. REMESH dialog box

- The **Select** button of **Mesh Faces/Body** section is active by default. Click on the faces to select. You can also drag the cursor to select multiple faces.
- Select **Uniform** option from **Meshing Type** drop-down for creating the similar size face on the entire selection. This option is used for keeping the face sizes even.
- Select **Adaptive** option from **Meshing Type** drop-down for smaller faces in the region of high detail and larger faces in the region of low detail. This option is used for preserving details on the selected model.
- Click in the **Density** edit box and enter desired value of density. You can also specify the value of density by moving the **Density** slider from **REMESH** dialog box; refer to Figure-14. It controls the number of faces created.

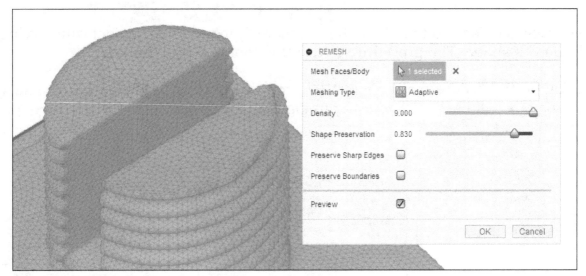

Figure-14. Specifying density

- Click in the **Shape Preservation** edit box and enter the value. You can also specify the value by moving the **Shape Preservation** slider. The value of **Shape Preservation** lies between 0 to 1. Note that this option will be available if you have selected **Adaptive** option from the **Meshing Type** drop-down in dialog box.
- Select the **Preserve Sharp Edges** check box from **REMESH** dialog box to preserve the sharp edges from the input mesh.
- Select the **Preserve Boundaries** check box from **REMESH** dialog box to make sure that any open boundaries of the selected model do not change shape. This option is useful if you have two separate bodies meeting at open boundaries that you wish to merge later.
- Select the **Preview** check box to view the mesh preview, before it is created.
- After specifying the parameters, click on the **OK** button from **REMESH** dialog box. The created mesh body will be displayed; refer to Figure-15.

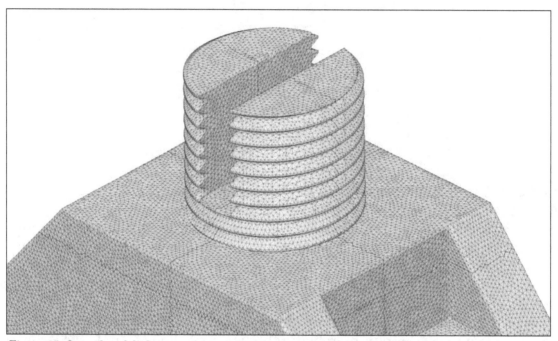

Figure-15. Created mesh body

Reduce

The **Reduce** tool is used to reduce the number of faces on your model while trying to maintain its shape. The procedure to use this tool is discussed next.

- Click on the **Reduce** tool from **MODIFY** drop-down; refer to Figure-16. The **REDUCE** dialog box will be displayed; refer to Figure-17.

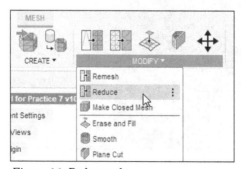

Figure-16. Reduce tool

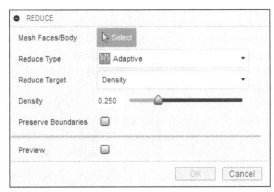

Figure-17. REDUCE dialog box

- The **Select** button of **Mesh Faces/Body** section is active by default. Click on the mesh body from the **BROWSER** or select faces of mesh.
- Select desired option from the **Reduce Target** drop-down to define which parameter is to be reduced.
- Move the slider to increase or decrease the value of parameter selected in **Reduce Target** drop-down; refer to Figure-18.

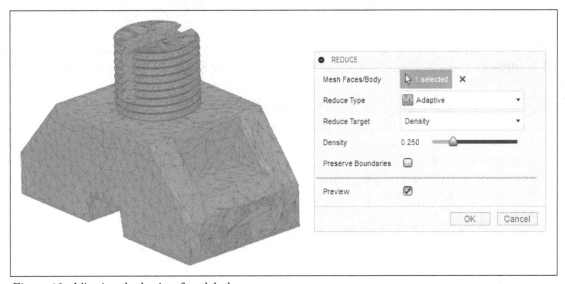

Figure-18. Adjusting the density of mesh body

- Select the **Preserve Boundaries** check box to retain the existing boundaries of body after modification.
- Select the **Preview** check box to view the mesh preview, before it is created.
- After specifying the parameter, click on the **OK** button from **REDUCE** dialog box. The mesh body will be created; refer to Figure-19.

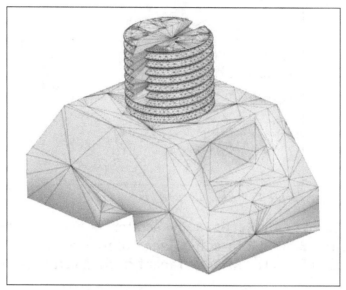

Figure-19. Created mesh body after Reduce tool

Make Closed Mesh

The **Make Closed Mesh** tool is used for rebuilding the selected mesh body as a new closed mesh. If there is any gap in your mesh part then it will be filled automatically to form a closed mesh. The procedure to use this tool is discussed next.

- Click on the **Make Closed Mesh** tool from **MODIFY** drop-down; refer to Figure-20. The **MAKE CLOSED MESH** dialog box will be displayed; refer to Figure-21.

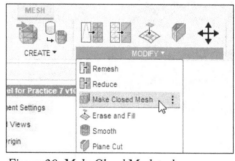

Figure-20. Make Closed Mesh tool

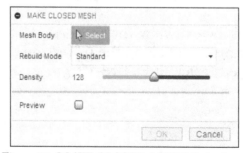

Figure-21. MAKE CLOSED MESH dialog box

- The **Select** button of **Mesh Body** section is active by default. Click on the body to be selected; refer to Figure-22. Note that if there is a single body then it will get selected automatically.

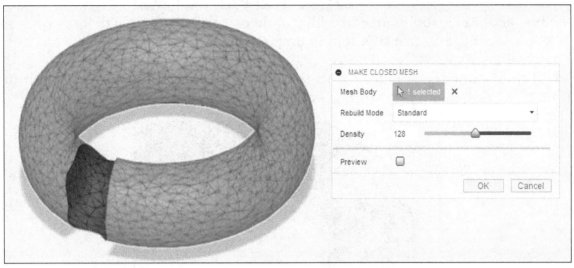

Figure-22. Selecting mesh body for creating closed mesh

- Select **Standard** option from **Rebuild Mode** drop-down to rebuild the mesh with default behavior. This option provides a good balance of speed and accuracy, but sharp edges will become soft.
- Select **Preserve Sharp Edges** option from **Rebuild Mode** drop-down to rebuild the mesh similar to the **Standard** option but it will preserve sharp edges. In this option, the mesh density is higher near the edge.
- Select **Accurate** option from **Rebuild Mode** drop-down to create a closed mesh of the selected body with high accuracy. The performance will be slower than standard but accuracy may be improved.
- Select **Blocky** option from **Rebuild Mode** drop-down to rebuild the model as simple cubes. This option does not provide an accurate approximation of the input shape, this is just intended to give your model a bulky aesthetic.
- Click in the **Density** edit box and enter desired value. You can also specify the value of density by moving the **Density** slider.
- Select the **Preview** check box to view the mesh preview before it is created.
- After specifying the parameter, click on the **OK** button from **MAKE CLOSED MESH** dialog box. The mesh body will be created; refer to Figure-23.

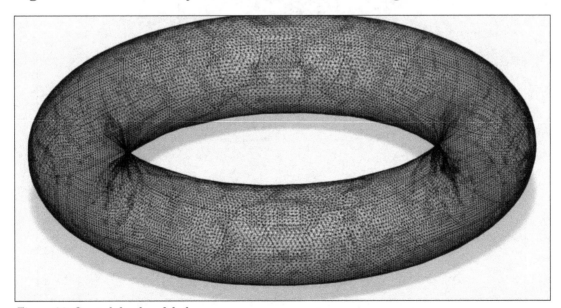

Figure-23. Created closed mesh body

Erase and Fill

The **Erase and Fill** tool is used to fill a hole or heal regions or defects on a mesh body. The procedure to use this tool is discussed next.

* Select the boundary of hole/cut in the mesh or select the complete mesh model to be filled or healed; refer to Figure-24.

Figure-24. Boundary of cut selected for healing

* Click on the **Erase and Fill** tool from **MODIFY** drop-down in the **Toolbar**; refer to Figure-25. The **ERASE AND FILL** dialog box will be displayed along with preview of fill feature; refer to Figure-26.

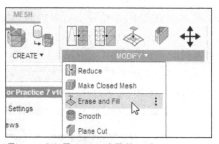

Figure-25. Erase and Fill tool

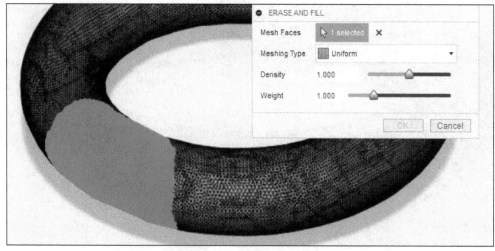

Figure-26. ERASE AND FILL dialog box

- Select **Uniform** option from **Meshing Type** drop-down to fill the selected region with regular shaped triangles. This option gives the smoothest and most reliable results.
- Select **Minimal** option from **Meshing Type** drop-down to use the minimum number of faces to fill the selected hole.
- (For **Uniform** option) Click in the **Density** edit box and enter desired value. You can also specify the value of density by moving the **Density** slider.
- (For **Uniform** option) Click in the **Weight** edit box and enter the weight mesh after filling. You can also specify the weight by moving the **Weight** slider. Note that weight defines the maximum deviation of new fill face from surrounding faces.
- After specifying the parameters, click on the **OK** button from **ERASE AND FILL** dialog box. The filled hole will be displayed; refer to Figure-27.

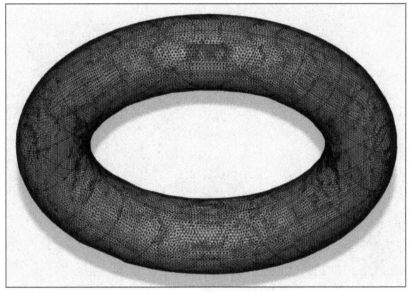

Figure-27. Filled hole

Smooth

The **Smooth** tool is used to smooth out uneven regions on the mesh. The procedure to use this tool is discussed next.

- Click on the **Smooth** tool from **MODIFY** drop-down; refer to Figure-28. The **SMOOTH** dialog box will be displayed; refer to Figure-29.

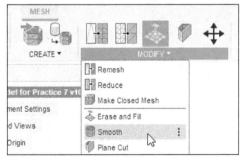

Figure-28. Smooth tool

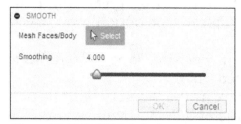

Figure-29. SMOOTH dialog box

- The **Select** button of **Mesh Faces/Body** section is active by default. Click on the body to select; refer to Figure-30.

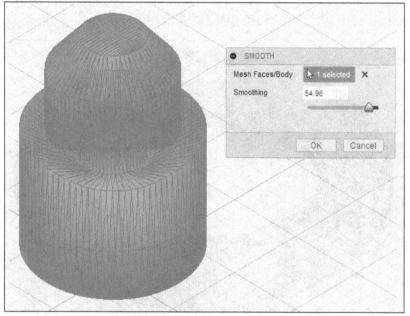

Figure-30. Selecting Mesh body for smooth

- Click in the **Smoothing** edit box and enter desired value for smoothing of mesh body. You can also move the **Smoothing** slider to specify the smoothing value.
- After specifying the parameters, click on the **OK** button from **SMOOTH** dialog box. The mesh body will be displayed; refer to Figure-31.

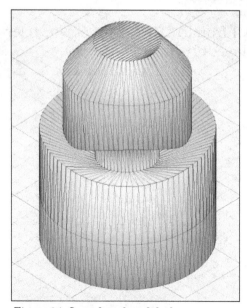

Figure-31. Smoothened mesh body

Plane Cut

The **Plane Cut** tool is used to cut the selected mesh body using a plane/face/surface. The procedure to use this tool is discussed next.

* Click on the **Plane Cut** tool from **MODIFY** drop-down; refer to Figure-32. The **PLANE CUT** dialog box will be displayed; refer to Figure-33.

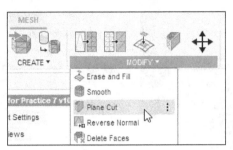

Figure-32. Plane Cut tool

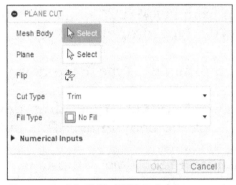

Figure-33. PLANE CUT dialog box

* The **Select** button of **Mesh Body** section is active by default. Click on the body to select. The manipulator will be displayed on the selected body; refer to Figure-34.

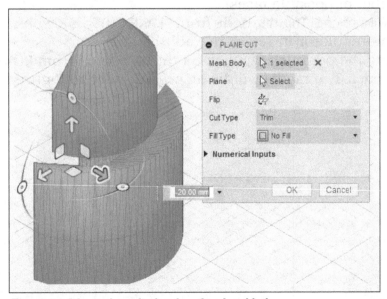

Figure-34. Manipulator displayed on the selected body

* Move the manipulator to split the body. To select desired cutting plane, click on the **Select** button of **Plane** section and select required plane; refer to Figure-35. The preview of model cut by selected plane will be displayed.

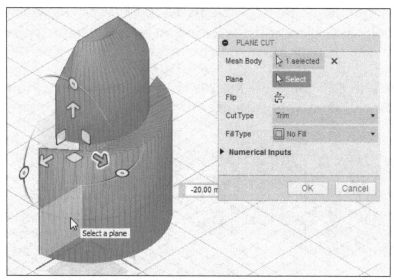

Figure-35. Selecting plane to split

- Click on the **Flip** button from **PLANE CUT** dialog box to flip the direction of split.
- Select **Trim** option from **Cut Type** drop-down to split the body into two sides and removes one of the sides.
- Select **Split Body** option from **Cut Type** drop-down to split the body to create two separate mesh body.
- Select **Split Faces** option from **Cut Type** drop-down to split the faces that intersect the plane but keep the body intact. This option will create a new face group on one side of the split.
- Select **No Fill** option from **Fill Type** drop-down if you want to leave open boundary at the cut.
- Select **Uniform** option from **Fill Type** drop-down if you want to fill the hole with new faces of regular shape.
- Select **Minimal** option from **Fill Type** drop-down if you want to fill the hole with minimal number of possible faces.
- Click on the **Numerical Inputs** node from **PLANE CUT** dialog box to manually enter the value of manipulator in respective edit box.
- After specifying the parameters, click on the **OK** button from **PLANE CUT** dialog box to split the model. The model will be displayed; refer to Figure-36.

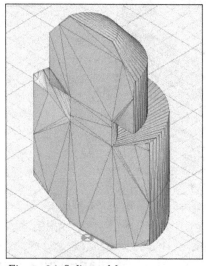

Figure-36. Split model

Reverse Normal

The **Reverse Normal** tool is used to flip the normal direction of the selected face. The procedure to use this tool is discussed next.

- Click on the **Reverse Normal** tool from **MODIFY** drop-down; refer to Figure-37. The **REVERSE NORMAL** dialog box will be displayed; refer to Figure-38.

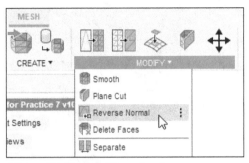

Figure-37. Reverse Normal tool

Figure-38. REVERSE NORMAL dialog box

- The **Select** button of **Mesh Faces** section is active by default. Click on the face of mesh body to select; refer to Figure-39.

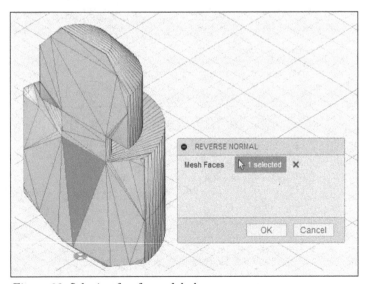

Figure-39. Selecting face for mesh body

- You can also select multiple faces by clicking on them.
- After selecting required faces, click on the **OK** button from **REVERSE NORMAL** dialog box. The normal direction of selected face will be flipped; refer to Figure-40.

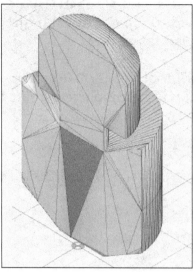

Figure-40. Applied reverse normal on selected face

Delete Faces

The **Delete Faces** tool is used to remove selected face from the body. The procedure to use this tool is discussed next.

- Click on the **Delete Faces** tool from **MODIFY** drop-down; refer to Figure-41. The **DELETE FACES** dialog box will be displayed; refer to Figure-42.

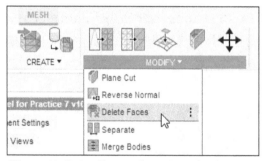

Figure-41. Delete Faces tool

Figure-42. DELETE FACES dialog box

- The **Select** button of **Mesh Faces** section is active by default. Click on desired faces to select; refer to Figure-43.

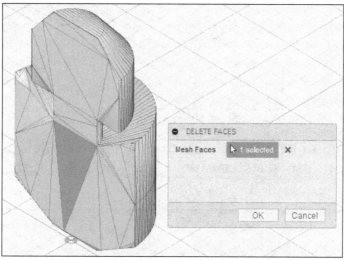

Figure–43. Selecting face to delete

*   Click on the **OK** button from **DELETE FACES** dialog box to delete the selected faces. The faces will be deleted; refer to Figure-44.

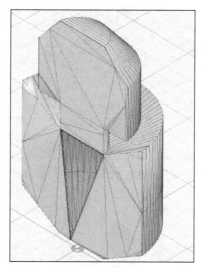

Figure–44. Deleted Face

Separate

The **Separate** tool is used to create a new mesh body from selected set of faces. The procedure to use this tool is discussed next.

*   Click on the **Separate** tool from **MODIFY** drop-down; refer to Figure-45. The **SEPARATE** dialog box will be displayed; refer to Figure-46.

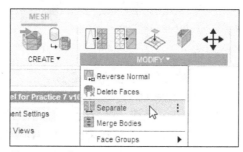

Figure–45. Separate tool

Figure-46. SEPARATE dialog box

- The **Select** button of **Mesh Faces** section is active by default. Click on the face to select. You can also select multiple faces by holding the **CTRL** key; refer to Figure-47.

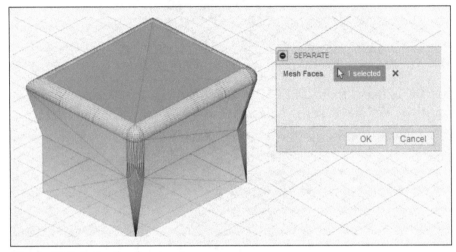

Figure-47. Selecting face for separate

- After selection of required faces, click on the **OK** button from **SEPARATE** dialog box. The selected face will be separated; refer to Figure-48.

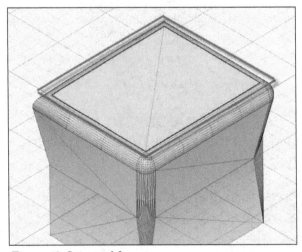

Figure-48. Separated face

Merge Bodies

The **Merge Bodies** tool is used to create a single body from multiple input bodies. If the input bodies have touching boundary edges then these edges will be stitched together. The procedure to use this tool is discussed next.

- Click on the **Merge Bodies** tool from **MODIFY** drop-down; refer to Figure-49. The **MERGE BODIES** dialog box will be displayed; refer to Figure-50.

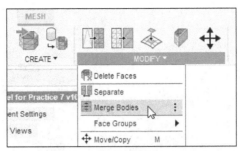

Figure-49. Merge Bodies tool

Figure-50. MERGE BODIES dialog box

- The **Select** button of **Mesh Bodies** section is active by default. Click on the bodies to be merged; refer to Figure-51.

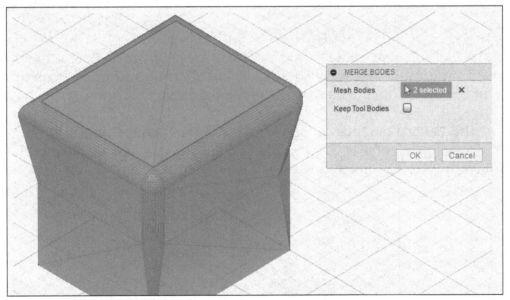

Figure-51. Selecting bodies to merge

- Select the **Keep Tool Bodies** check box from **MERGE BODIES** dialog box to keep the input bodies after merging operation as separate bodies.
- After specifying the parameters, click on the **OK** button from **MERGE BODIES** dialog box. The selected bodies will be merged; refer to Figure-52.

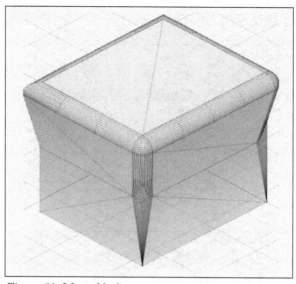
Figure-52. Merged bodies

Face Groups

The tools of **Face Groups** cascading menu are used to segment a mesh body into logical regions. The tools in this cascading menu are used to generate, clear, or create the face groups. Note that you can select the whole face group by double-clicking on one of its face.

Generate Face Groups

The **Generate Face Groups** tool is used to divide faces of the selected mesh body into groups based on normal angles of their faces or the origination of faces in the mesh body. The procedure to use this tool is discussed next.

• Click on the **Generate Face Groups** tool of **Face Groups** cascading menu from **MODIFY** drop-down; refer to Figure-53. The **GENERATE FACE GROUPS** dialog box will be displayed; refer to Figure-54.

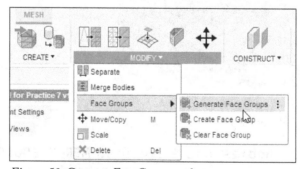
Figure-53. Generate Face Groups tool

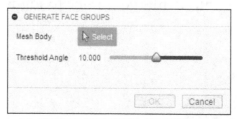
Figure-54. GENERATE FACE GROUPS dialog box

- The **Select** button of **Mesh Body** section is active by default. Click on the mesh body to select; refer to Figure-55.

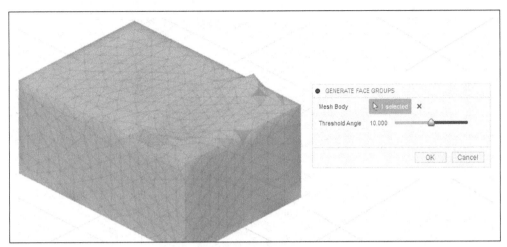

Figure-55. Selected mesh body for generating face groups

- Click in the **Threshold Angle** edit box and enter the value of angle for faces at which they will be counter in different groups. Example: suppose, there are 20 faces which have angle of inclination from 0 to 10 then they will be counted in one type of group. There are another group of faces which have angle value from 11 to 20 then they will become another face group.
- After specifying the parameters, click on the **OK** button from **GENERATE FACE GROUPS** dialog box. The generated face groups will be displayed; refer to Figure-56.

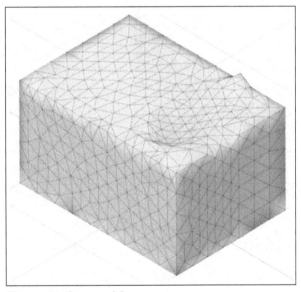

Figure-56. Generated face groups

Create Face Group

The **Create Face Group** tool is used to create a new face group from a selected set of faces. The procedure to use this tool is discussed next.

- Click on the **Create Face Group** tool of **Face Groups** cascading menu from **MODIFY** drop-down; refer to Figure-57. The **CREATE FACE GROUP** dialog box will be displayed; refer to Figure-58.

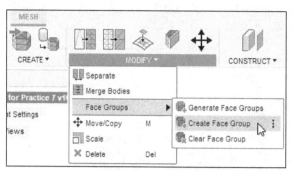

Figure-57. Create Face Group tool

Figure-58. CREATE FACE GROUP dialog box

- The **Select** button of **Mesh Faces** section is active by default. Click on desired face to select; refer to Figure-59.

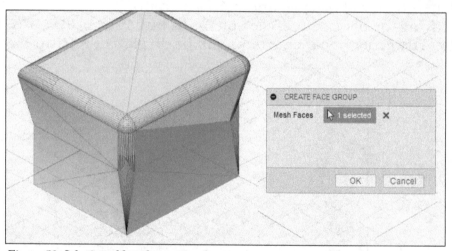

Figure-59. Selection of faces for creating face group

- After selection of required faces, click on the **OK** button from **CREATE FACE GROUP** dialog box. The created face group will be displayed; refer to Figure-60.

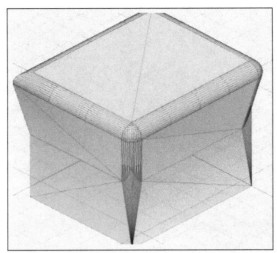

Figure-60. Created face group

Clear Face Group

The **Clear Face Group** tool is used to clear any existing face group from selected body. The procedure to use this tool is discussed next.

• Click on the **Clear Face Group** tool of **Face Groups** cascading menu from **MODIFY** drop-down; refer to Figure-61. The **CLEAR FACE GROUP** dialog box will be displayed; refer to Figure-62.

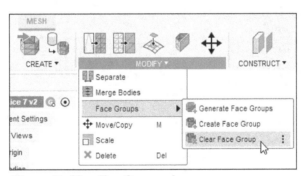

Figure-61. Clear Face Group tool

Figure-62. CLEAR FACE GROUP dialog box

• The **Select** button of **Mesh Faces/Body** section is active by default. Click on the existing face groups to be cleared or select the whole mesh body; refer to Figure-63.

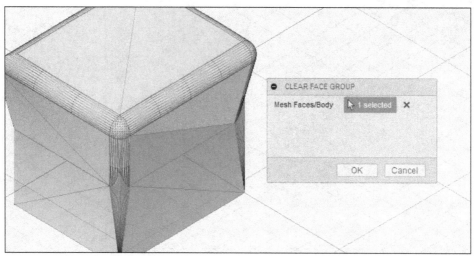

Figure-63. Selected faces to clear

- After selection of faces, click on the **OK** button from **CLEAR FACE GROUP** dialog box. The selected face will be cleared; refer to Figure-64.

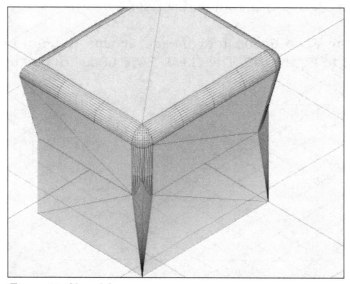

Figure-64. Cleared face group

- Other tools of **MODIFY** drop-down are same as discussed earlier in this book.
- After creating or modifying the mesh body, click on the **FINISH MESH** button from **Toolbar** to exit the **MESH Workspace**.

PRACTICAL

Create the mesh model as shown in Figure-65.

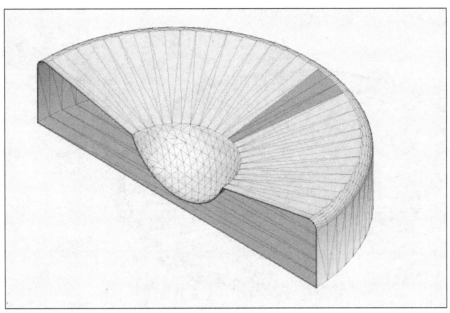

Figure-65. Final model

Converting model into mesh

Before modifying the mesh file, we need to convert the model into mesh file.

- Create or open the model of this practical in the **MODEL Workspace**. The file is available in the resource kit of this book.
- Click on the **Create Mesh** tool from **CREATE** drop-down in **SOLID** tab of **Toolbar**. The **MESH Workspace** will be displayed along with the model; refer to Figure-66.

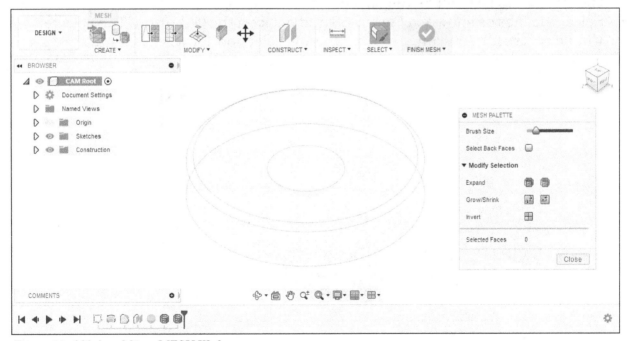

Figure-66. Added model into MESH Workspace

- Click on the **BRep to Mesh** tool of **CREATE** drop-down from **Toolbar**. The **BREP TO MESH** dialog box will be displayed.
- The **Select** button of **Body** section is active by default. Select the transparent model to convert it into mesh model and specify the parameters as shown in Figure-67.

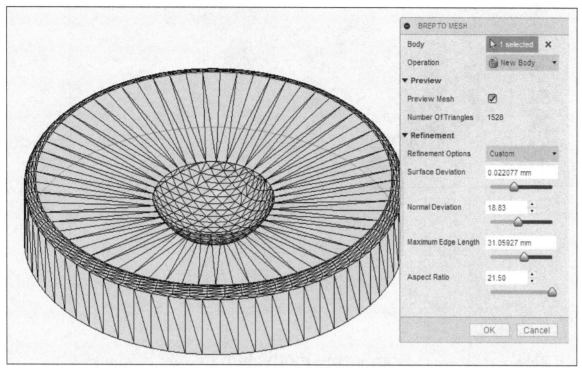

Figure-67. Specifying parameters for creating mesh file

- After specifying the parameters, click on the **OK** button from **BREP TO MESH** dialog box. The mesh body will be created and displayed on the **MESH** workspace.

Reverse the face

- Click on the **Reverse Normal** tool of **MODIFY** drop-down from **Toolbar**. The **REVERSE NORMAL** dialog box will be displayed.
- Select the face as shown in Figure-68 and click on the **OK** button. The selected face will be reversed. You can also use window selection for selecting multiple faces.

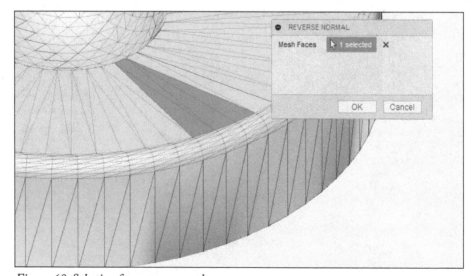

Figure-68. Selection for reverse normal

Plane Cut

- Click on the **Plane Cut** tool of **MODIFY** drop-down from **Toolbar**. The **PLANE CUT** dialog box will be displayed.

- The **Select** button of **Mesh Body** section is active by default. You need to select the model for plane cut. Click on the model to select.
- Click on the **Select** button of **Plane** section and select the **YZ** plane to cut the body; refer to Figure-69.

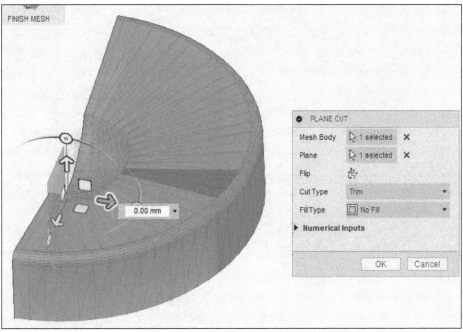

Figure-69. Selecting plane for plane cut

- Specify the parameters as shown in above figure and click on the **OK** button from **PLANE CUT** dialog box.
- After following all the steps discussed above, the model will be displayed as shown in Figure-70.

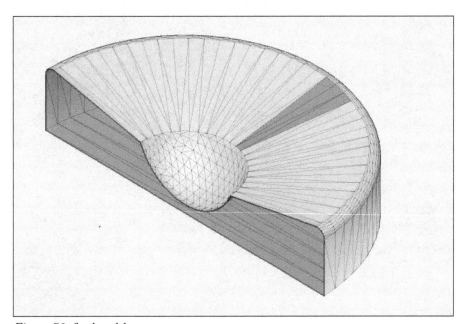

Figure-70. final model

PRACTICE

Create the model as displayed in the Figure-71.

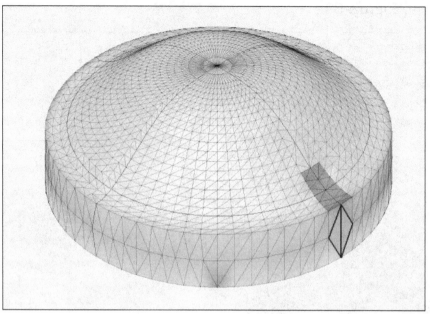

Figure-71. Practice 1

SELF ASSESSMENT

Q1. Which of the following are included in a mesh model?

a. Vertices   b. Edges
c. Faces   d. Surfaces
e. Hole features   f. Sketches

Q2. In Mesh workspace, selection brush is used instead of window selection to select elements. (T/F)

Q3. Which of the following is not an object created by BRep method?

a. Sheetmetal model
b. Surface model
c. Solid model
d. Mesh model

Q4. In Autodesk Fusion 360, mesh model is combination of only triangulated faces. (T/F)

Q5. Which of the following tool is used to automatically fill all the holes and cuts in the selected mesh body?

a. Erase and Fill
b. Make Closed Mesh
c. Reduce
d. Remesh

FOR STUDENTS NOTES

Chapter 15

Manufacturing

Topics Covered

The major topics covered in this chapter are:

- *New Setup*
- *Milling Machine Setup*
- *Turning Machine Setup*
- *Milling and Turning Tools*
- *Creating new Mill and Turning tool*

INTRODUCTION

CAM stands for Computer Aided Manufacturing. CAM is a mode where you can convert the 3D model into a machine readable program codes used for the manufacturing process (usually G code). Some of the common manufacturing processes that are studied under CAM are Milling, Turning, Drilling, and Laser Cutting. In the CAM workspace of Autodesk Fusion, you will be able to generate high-quality toolpaths within minutes. Depending on the Fusion 360 version, you can create high quality 2D, 3D, 5-Axis milling, and turning toolpaths for high speed machining (HSM). You can also generate toolpaths for laser cutting machines and additive manufacturing.

STARTING WITH MANUFACTURE WORKSPACE

The tools in **MANUFACTURE** workspace are used to generate toolpaths for all types of manufacturing processes. The procedure to start **MANUFACTURE** workspace is discussed next.

• Click on the **MANUFACTURE** option from the **Workspace** drop-down while in any workspace; refer to Figure-1. The **MANUFACTURE** workspace will be displayed; refer to Figure-2.

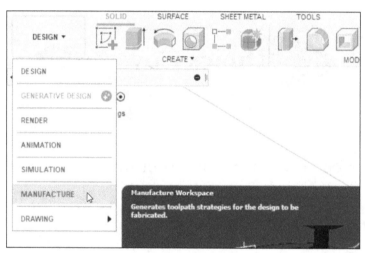

Figure-1. MANUFACTURE workspace option

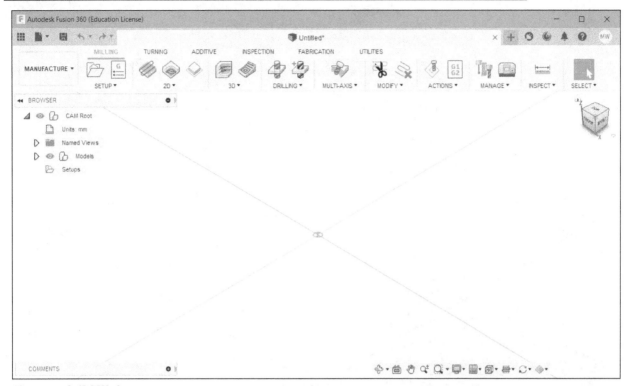

Figure-2. CAM Workspace

JOB SETUP

In this section, we will discuss the procedure of setting up the workpiece and create the required material stock for machining process. Job setup lets you define your stock for machining and machine type to be used like Milling or Turning. Stock is the workpiece (piece of raw material) out of which the final product will be produced after machining. The shape and size of stock depends upon the final model which is to be created by machining.

New Setup

Before starting any machining project, you need to define a job setup to tell Autodesk Fusion 360 that the toolpaths will be generated for Milling machine, Turning machine, or any other supported machine. You also need to set zero location to be used as machining coordinates origin. A machining coordinate origin acts as 0,0,0 coordinate for defining other locations of the part. This location is manually set in the CNC machines by moving cutting tool to desired location. You can also define any fixture component for machining. A fixture is generally used to hold the workpiece in desired orientation. The procedure to setup the workpiece is discussed next.

- Click on the **New Setup** tool from **SETUP** drop-down; refer to Figure-3. The **SETUP** dialog box will be displayed along with the workpiece; refer to Figure-4.

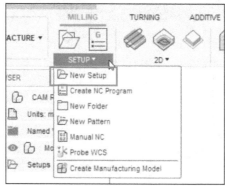

Figure-3. New Setup tool

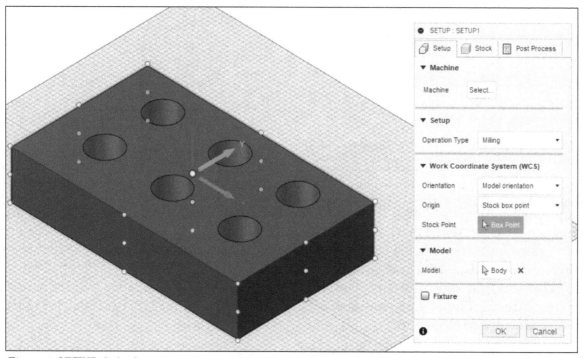

Figure-4. SETUP dialog box

Setting a Milling Machine

The options in **Machine** node of **SETUP** dialog box deals with set up of machines. Depending on your requirement, you can select a machine or you can add a new entry of your machine in dialog box. First we will discuss the procedure of setting a milling machine and later, we will discuss the procedure of setting a turning machine.

Setup

- Click on the **Select** button in the **Machine** section of the dialog box to select predefined machine or specify machine related parameters. The **Machine Library** dialog box will be displayed; refer to Figure-5.

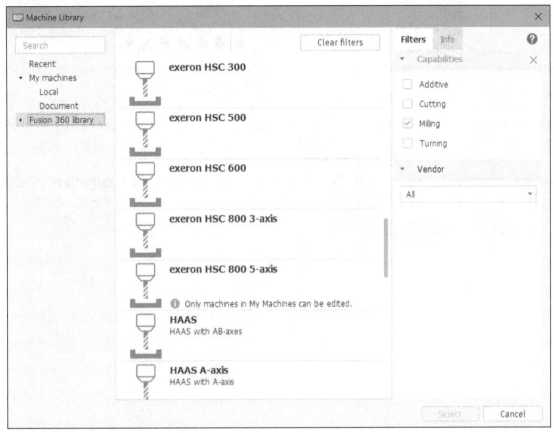

Figure-5. Machines dialog box

- Select desired check box(es) from the **Capabilities** node in the right of dialog box to define which type of machine you want to use. By default, the **Milling** check box is selected so predefined set of milling machines is displayed.
- If you want to search desired machine using name of vendor (maker of machine) then select desired manufacturer from drop-down in the **Vendor** node.
- Select desired machine from the dialog box. The information about machine will be displayed in the **Info** tab of dialog box; refer to Figure-6.

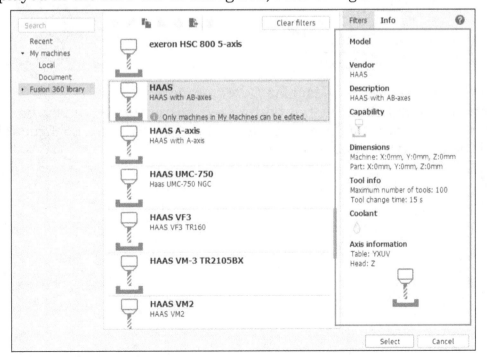

Figure-6. Info about machine

- If you want to edit a machine then select the machine and click on the **Edit** button from top of toolbar in dialog box to modify the parameters of the selected machine. The **Machine Configuration** dialog box will be displayed; refer to Figure-7. Note that you can only modify machine definitions which are created by user. You cannot modify machine definitions predefined by Autodesk.

- To create a user defined machine definition, click on the **Create New** button from the top in the dialog box after selecting **Local** location in **My machines** node at the left in the dialog box. A list box will be displayed to define the type of machine to be created; refer to Figure-8. Select desired option from the list. In our case, we have selected **Milling** option. On doing so, the **Machine Configuration** dialog box will be displayed. Various options of this dialog box are discussed next.

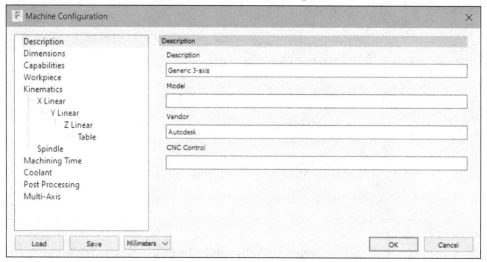

Figure-7. *Machine Configuration dialog box*

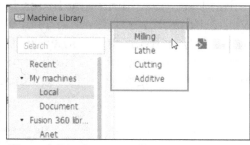

Figure-8. *List for new machine types*

Machine Configuration

- Click on the **Load** button at the bottom left corner of the dialog box to set the parameters as per pre-defined machines. The **Open Machine Configuration** dialog box will be displayed; refer to Figure-9.

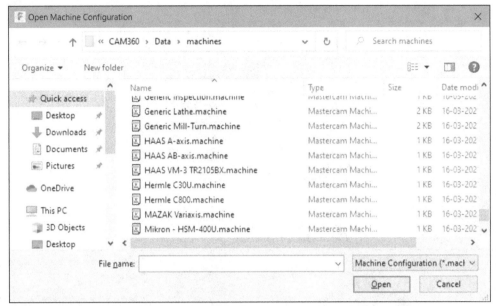

Figure-9. Open Machine Configuration dialog box

- Select desired configuration file and click on the **Open** button from the dialog box. The parameters will be updated automatically in the **Machine Configuration** dialog box.

- If you want to define each of the machine configuration parameter manually, then specify the desired parameters in the **Machine Configuration** dialog box and click on the **Save** button from the dialog box. The **Save Machine Configuration** dialog box will be displayed. Save the machine configuration at desired location and then click on the **OK** button from the **Machine Configuration** dialog box.

Note: The options of **Machine Configuration** dialog box have been discussed later in this book.

Operation Type

- Select the **Milling** option from the **Operation Type** drop-down in **Setup** section of **Setup** tab for setting up a milling operation. Note that if you have not defined machine then you need to select desired option from **Operation Type** drop-down but if you have selected a machine in the **Machine** section of the dialog box then this option will be selected automatically in the **Operation Type** drop-down based on type of the machine.

Setting Work Coordinate System

- Select **Model Orientation** option in **Orientation** section from **Work Coordinate System (WCS)** to set the orientation of coordinate system based on key points of workpiece for machining.

- Select the **Select Z axis/plane & X axis** option from the **Orientation** section of **Work Coordinate System (WCS)** node to select the Z axis and X axis for setting the orientation of workpiece; refer to Figure-10. The updated **WCS** section will be displayed along with axis or face selection on part; refer to Figure-11. By default, **Z Axis** selection button is active and you are asked to select a reference to define direction of Z axis. Select desired face or edge. You will be asked to select direction reference for X axis. Select desired face or edge. If you want to flip direction of axes then select the **Flip Z Axis** and/or **Flip X Axis** check boxes.

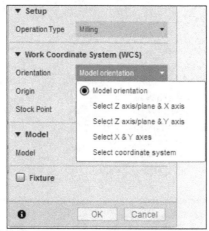

Figure-10. Orientation

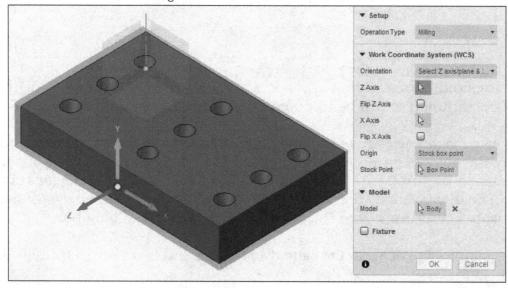

Figure-11. Updated WCS section of dialog box

- Select the **Select Z axis/plane & Y axis** option from the **Orientation** section of the **Work Coordinate System (WCS)** node to select references for defining orientation of Z axis and Y axis of WCS (Work Coordinate System). Click on the **Z Axis** button from **Work Coordinate System (WCS)** section and click on the desired axis or plane from Origin node in the **Browser** to define the Z axis. The Z axis should be perpendicular to machining plane. Select the **Flip Z Axis** check box to flip the selected direction of Z axis at 180 degree. The **X Axis** button is active by default. Click on the axis or plane to define the X axis; refer to Figure-12. The X axis will be defined perpendicular to the selected plane. Select the **Flip X Axis** check box to flip the selected direction of X axis at 180 degree.

- Select the **Select X & Y axes** option from the **Orientation** section of **Work Coordinate System (WCS)** node to select references for defining orientation of X axis and Y axis of WCS. Select desired faces or edges as discussed earlier.

- Select the **Select Coordinate System** option from **Orientation** section of **Work Coordinate System (WCS)** node to use a user defined coordinate system in the model to set the orientation of WCS.

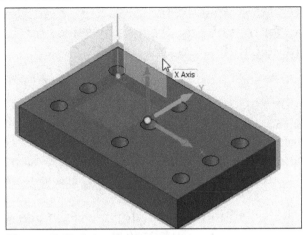

Figure-12. Defining X axis

- Select **Model Origin** option from the **Origin** drop-down of the **Work Coordinate System (WCS)** node to use World Coordinate System (WCS) origin of the current model as Work Coordinate System origin. Note that this drop-down is not available when the **Select coordinate system** option is selected in the **Orientation** drop-down of the dialog box.

- Select the **Selected point** option from the **Origin** drop-down to select a vertex or key point of model for defining WCS origin. Click on the desired vertex or key point of an edge to define WCS origin.

- Select the **Model box point** option of the **Origin** drop-down from **Work Coordinate System (WCS)** section to define a point for WCS origin by selecting a point on the model bounding box. The **Model Point** selection button for **Stock Point** section is active by default. Click at desired box point of model to define WCS origin.

- Select the **Stock box point** option of the **Origin** drop-down from the **Work Coordinate System (WCS)** section to define WCS origin by selecting a point on the stock bounding box. The **Stock Point** button of **Origin** section will be activated. Click on the stock point from workpiece to define WCS origin; refer to Figure-13.

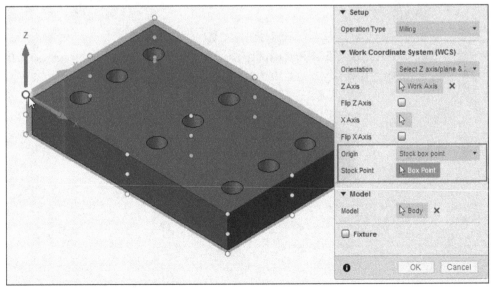

Figure-13. Defining stock box point

Model Body Selection

- If there is only one model in the drawing area then it will be considered as machining model for generating toolpaths and will be active by default in the **Model** section of dialog box.

- If there are multiple solid bodies in the drawing area then it is recommended to select required model for machining process by selecting the **Body** selection button from **Model** node and select desired body.

Fixture Selection

- Select the **Fixture** check box from the **SETUP** dialog box to define fixture for the workpiece. A fixture is used to hold workpiece in desired orientation.
- The **Fixture** selection button is active by default on selecting the **Fixture** check box. You need to select the component/body to be defined as fixture; refer to Figure-14. Note that components defined as fixture will be avoided by tool while cutting.
- Similarly, click on the selection button for the **Fixture Attachment** section in **Fixture** node and select desired attachment body.

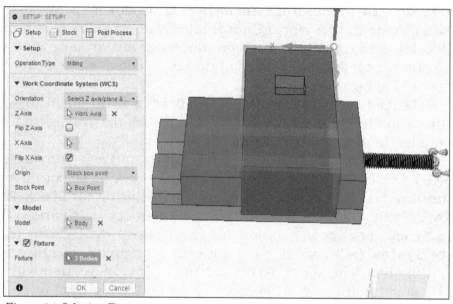

Figure-14. Selecting Fixture

Stock tab

- Click on the **Stock** tab of **SETUP** dialog box to define the workpiece dimensions. The **Stock** tab will be displayed; refer to Figure-15.

Figure-15. Setup tab

Fixed size box

- Select the **Fixed size box** option from the **Mode** drop-down in the **Stock** tab to create a rectangular stock body of defined parameter. The updated **Stock** tab will be displayed; refer to Figure-16.

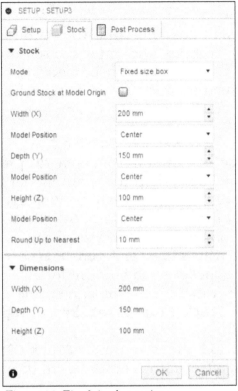

Figure-16. Fixed size box options

- Select the **Ground Stock at Model Origin** check box to place the origin of stock at origin of machining model. (Machining model is the final part design to be achieved after performing machining). Note that on selecting this check box, you will be able to define only offset distances along X, Y, and Z axes for positioning stock on the model.
- Click in the **Width (X)** edit box and specify the width of stock body.
- Select the **Offset from left side (-X)** option of **Model Position** drop-down from **Stock** section to offset the stock to the left side of model along X axis of WCS.
- Click in the **Offset** edit box and enter desired value; refer to Figure-17.

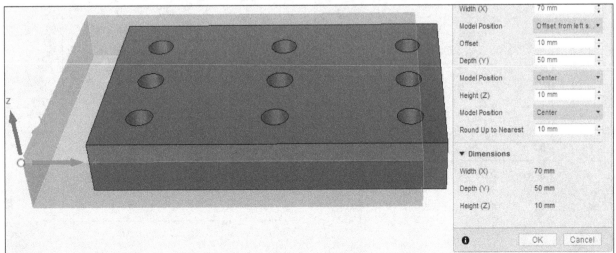

Figure-17. Offset from left side

- Select **Center** option of **Model Position** drop-down from **Stock** section to place the stock on the center of model.
- Select the **Offset from right side (+X)** option of **Model Position** drop-down from **Stock** section to offset the stock to the right side of model.
- Click in the **Offset** edit box and enter the required value. In our case we are selecting the **Center** option.
- Click in the **Depth (Y)** edit box and enter the required depth of the stock.
- Click in the **Height (Z)** edit box and enter the required height of stock.
- Click in the **Round Up to Nearest** edit box and enter the value of increment/decrement to be performed in stock size when you click on the spinner buttons to increase or decrease stock size in the dialog box.

Relative Size Box

- Select the **Relative size box** option from **Mode** drop-down in **Stock** tab to create a rectangular stock body larger then the main model by specified offset distance values. The updated **SETUP** dialog box will be displayed; refer to Figure-18.
- Select the **No additional stock** option of **Stock Offset Mode** drop-down from **Stock** section to create a stock equal to size of the model.
- Select the **Add stock to the sides and top-bottom** option of **Stock Offset Mode** drop-down to create stock of specified amount for all sides of model except top and bottom of model. For Top and Bottom, you can specify different values. Specify the respective values in the **Stock Side Offset**, **Stock Top Offset**, and **Stock Bottom Offset** edit boxes.
- Select the **Add stock to all sides** option of **Stock Offset Mode** drop-down if you want to specify different offset values for different directions of model for creating stock; refer to Figure-19. Click in the offset edit boxes and specify the values as desired.

Figure-18. Relative size box option

Figure-19. Add stock to all sides mode

Fixed size cylinder

- Select the **Fixed size cylinder** option of **Mode** drop-down from **Stock** section to create a fixed size cylinder stock body. The updated **SETUP** dialog box will be displayed; refer to Figure-20.
- The **Axis** button of **Setup** section is active by default. Select the axis from model to be used as center axis of cylindrical stock; refer to Figure-20.

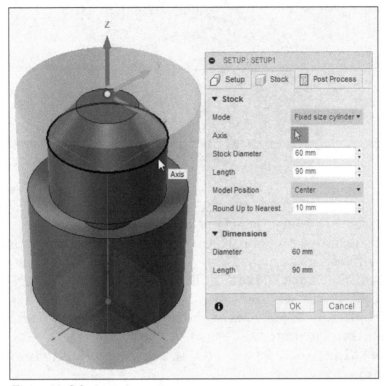

Figure-20. Selecting axis

- Click in the **Stock Diameter** edit box from the **Stock** section and enter desired diameter of stock.
- Click in the **Length** edit box and enter desired length value for stock.

Relative size cylinder

- Select the **Relative size cylinder** option from **Mode** drop-down of **Stock** section to create cylindrical stock body of specified offset value. Here, offset values will act as thickness over the main model. The updated **SETUP** dialog box will be displayed; refer to Figure-21.

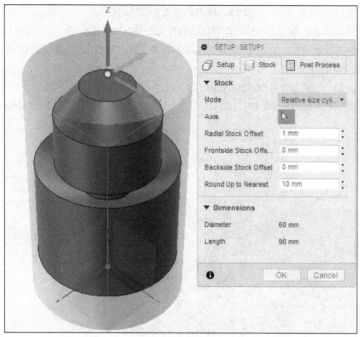

Figure-21. Relative size cylinder option

- The **Axis** button is active by default. Click on the axis from model to use as center axis of stock cylinder.
- Click in the **Radial Stock Offset** edit box and specify desired value to define thickness of stock in radial direction.
- Click in the **Frontside Stock Offset** edit box and specify thickness of stock on the top side of cylindrical stock.
- Click in the **Backside Stock Offset** edit box and specify thickness of stock on the bottom side of cylindrical stock.

Fixed size tube

- Select the **Fixed size tube** option of **Mode** drop-down from **Stock** section to create a tube stock body of fixed size. A tube is similar to cylinder with a through hole. The updated **SETUP** dialog box will be displayed; refer to Figure-22.

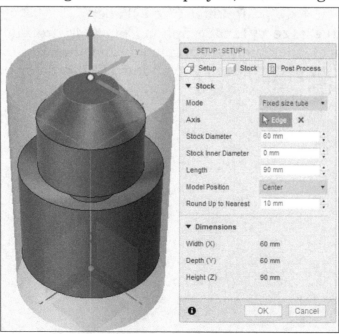

Figure-22. Fixed size tube option

- The **Axis** button is active by default. Click on the axis from model to select as center axis of tube.
- Click in the **Stock Diameter** edit box from **Stock** section and specify desired value for diameter of stock.
- Click in the **Stock Inner Diameter** edit box and specify value for inner diameter (hole diameter) of stock; refer to Figure-23.

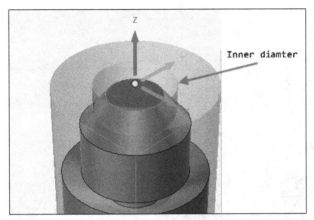

Figure-23. Inner diameter

- Click in the **Length** edit box and specify desired length of stock tube.
- Select the **Offset from front** option from **Model Position** drop-down of **Stock** section to move the stock to top side of model by value specified in **Offset** edit box; refer to Figure-24.

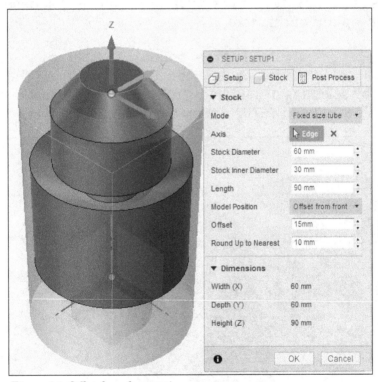

Figure-24. Offset from front option

- Click in the **Offset** edit box and specify desired value to move stock.
- Select the **Center** option of **Model Position** drop-down from **Stock** section to place the stock at the center of model.
- Select the **Offset from back** option from the **Model Position** drop-down of the **Stock** section to move the stock to bottom side of model.

Relative size tube

- Select the **Relative size tube** option from the **Mode** drop-down in the **Stock** section to create stock of specified thickness with respect to main model. The updated **SETUP** dialog box will be displayed; refer to Figure-25.

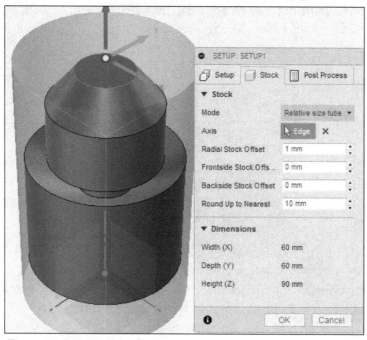

Figure-25. Relative size tube

- The edit boxes of the **Relative size tube** are same as discussed in the **Relative size cylinder** option of this tool.

From solid

- Select the **From solid** option of the **Mode** drop-down from the **Stock** section to create a stock by selecting a solid body from multi-body part or from a part file in an assembly. The updated **SETUP** dialog box will be displayed; refer to Figure-26.

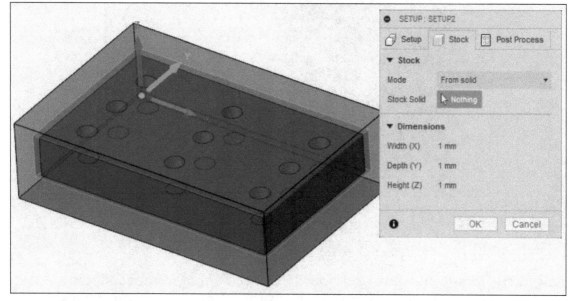

Figure-26. From solid option

- Click on the selection button of **Stock Solid** section and click on the body to be used as stock; refer to Figure-27.

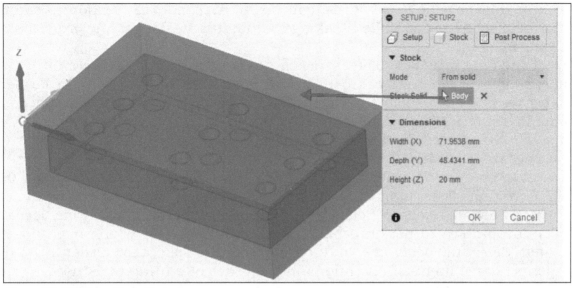

Figure-27. Selection of body for stock

- Click on the **Dimensions** node of **Stock** tab from **SETUP** dialog box to check the exact dimensions of the stock.

Post Process tab

- Click on the **Post Process** tab from **SETUP** dialog box to define post processing parameters. The **Post Process** tab will be displayed; refer to Figure-28. Post processing is the step at which CNC program generated by software is modified to suit specific machine.

Figure-28. Post Process tab

- Click in the **Program Name/Number** edit box of the **Post Process** tab from the **Setup** dialog box to define NC program name/number. This number is output at the start of NC program at the "0" number line. It is also used as storage name on the CNC controller storage.
- Click in the **Program Comment** edit box and enter desired comment to be added with the program. The comment text will not be read by CNC control but it will be displayed in program and machine display.
- Click in the **WCS Offset** edit box of the **Machine WCS** section from **Post Process** tab to move WCS by specified value for providing tool compensation. The output in the NC program is generally produced by G54 through G59 codes, but will vary between various NC Controls/Machines. Zero (0) in WCS Offset will output the first available fixture offset and 1 will output the 2nd available offset.

- Click in the **Probe WCS override** edit box of **Machine WCS** section to enter the value of offset for probe while checking coordinates in CMM or any probe installed in your machine.

- Select the **Multiple WCS Offsets** check box if you want to create multiple WCS offset values for tool wear compensations. Specify desired value in the **Number of Instances** edit box to define number of WCS offsets to be created. Specify desired value in the **WCS Offset Increment** edit box to define number of duplicate WCS Offsets to be created. Note that you can still change the values.

- The **Operation Order** drop-down below the **Multiple WCS Offsets** check box is used to specify the order of individual operations. Select the **Preserve Order** option from the **Operation Order** drop-down to machine operations by the order in which they are selected. Select the **Order by Operation** option from **Operation Order** drop-down to perform machining by the order in which operations are created. Select the **Order by tool** option from the **Operation Order** drop-down to perform machining of operations in the order of cutting tools used by them. So, if there are two cutting tools used in overall machining then first operations with tool number 1 will be performed then operation with tool number 2 will be performed.

- After specifying desired parameters, click on the **OK** button from the **SETUP** dialog box. The **Setup1** will be added in **Browser**; refer to Figure-29.

- If you want to edit the earlier created stock or setup parameter then right-click on respective **Setup** option from the **Setups** node of **BROWSER** and click on **Edit** button from marking menu/shortcut menu; refer to Figure-30. The **SETUP** dialog box will be displayed as discussed earlier.

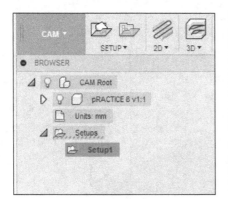

Figure-29. Setup1 in BROWSER

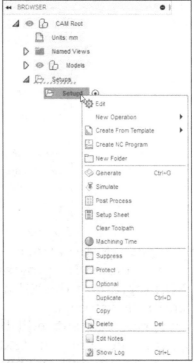

Figure-30. Edit stock

Turning Machine Setup

In this section, we will discuss about the procedure of setting machine for **Turning** operation. The procedure is discussed next.

- Click on the **Turning or mill/turn** option from **Operation Type** drop-down in **SETUP** dialog box. The options used in turning process will be displayed along with the model; refer to Figure-31. You can also select a turning machine by using **Select** button in **Machine** section of this dialog box. The procedure is same as discussed for milling machine.

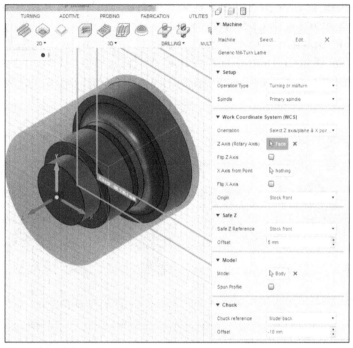

Figure-31. Turning or mill turn option

- Select **Primary spindle** or **Secondary spindle** option from **Spindle** drop-down in **Setup** section to specify the spindle to be used if your machine has two spindles.
- Click on the **Z Axis (Rotary Axis)** button of **Work Coordinate System (WCS)** section from **SETUP** dialog box and select a reference for Z axis as desired; refer to Figure-32.

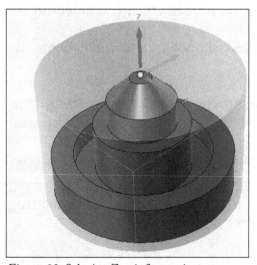

Figure-32. Selecting Z axis for turning

- Select the **Flip Z Axis** check box to flip selected direction of Z axis by 180 degree.
- The **X Axis** button is active by default. Click on the axis or plane to define the X axis.
- Select the **Flip X Axis** check box to flip selected direction of X axis by 180 degree.

- The **Origin** drop-down will define where zero position will be located on the part. Click on the **Origin** drop-down from **Work Coordinate System (WCS)**section and select the desired option to set origin point of WCS.
- Click in the **Safe Z Reference** drop-down and select desired reference for safe Z position where tool should move after making cutting passes.
- Click in the **Offset** edit box of **Safe Z** node in the dialog box and specify the distance of safe Z location from selected reference.
- Select the **Spun Profile** check box of **Model** section from **Setup** tab to generate profile for turning. If you are going to mill-turn a part which has irregular surface then you should select this check box to avoid accident; refer to Figure-33.

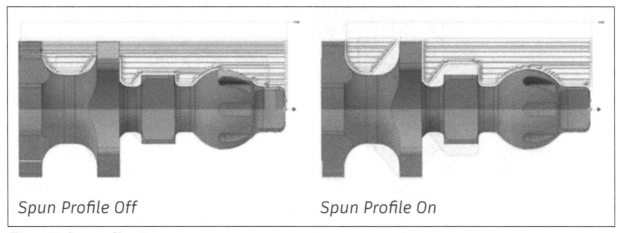

Spun Profile Off   Spun Profile On

Figure-33. Spun profile option

- Click in the **Spun Profile Tolerance** edit box of **Model** section and specify desired value of tolerance up to which profile can deviate from the model boundaries.
- Select the **Spun Profile Smoothing** check box to smoothen the profile at curves.
- Select desired reference for defining position of chuck from the **Chuck reference** drop-down.
- Click in the **Offset** edit box of the **Chuck** node and specify desired value of distance at which chuck should be placed from the selected reference.

The tools of **Stock** and **Post Process** tab have been discussed earlier in **Milling** section.

- After specifying the parameters, click on the **OK** button from the **SETUP** dialog box to complete the process of creating stock and defining machine parameters.

Cutting Machine Setup

In this section, we will discuss the procedure of setting up a cutting machine. This machine can be water jet machine, laser/plasma cutting machine, and so on. The procedure is discussed next.

- Select the **Cutting** option from **Operation Type** drop-down of the **SETUP** dialog box. The options used in cutting process will be displayed; refer to Figure-34.

Figure-34. SETUP dialog box for
Cutting operation

- Set the orientation and zero point for cutting operation in the **SETUP** dialog box.
- The other options in the dialog box have been discussed earlier.
- After specifying the parameters, click on the **OK** button from the **SETUP** dialog box.

Additive Manufacturing Machine Setup

In this section, we will setup a 3D printing machine. The procedure is given next.

- Select an additive manufacturing machine by using the **Select** button from the **Machine** node in the **SETUP** dialog box (like Aconity3D). The options will be displayed as shown in Figure-35.

Figure-35. SETUP dialog box for
additive manufacturing

- Make sure the **Additive** option is selected in the **Operation Type** drop-down of Setup node.
- Select the **Automatic** check box from the **Arrangement** node to automatically place all the parts on the machine bed. Other options in this dialog box have been discussed earlier.
- After specifying the parameters, click on the **OK** button from **SETUP** dialog box. The setup will be created.

TOOL SELECTION

Before proceeding towards the tools used to generate 2D and 3D path, you need to know various types of tools used in machining and their selection criteria. For this we need to discuss the **Select Tool** dialog box; refer to Figure-36. To display this dialog box, click on the **2D Adaptive Clearing** tool from the **2D** drop-down in the Toolbar. A dialog box will be displayed. Click on the **Select** button for **Tool** section in the **Tool** node of the dialog box. The **Select Tool** dialog box will be displayed. You can also display a similar dialog box by clicking on the **Tool Library** tool from the **MANAGE** drop-down of **MILLING** tab in the **Ribbon**; refer to Figure-37. Note that in this chapter, we will discuss about cutting tools only. In next chapter, we will discuss the toolpaths.

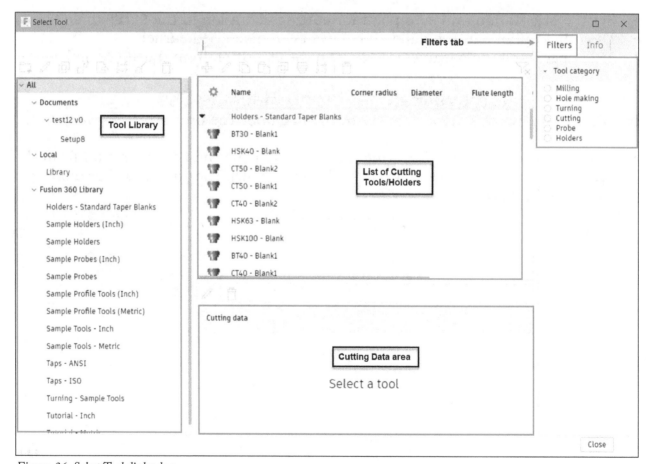

Figure-36. Select Tool dialog box

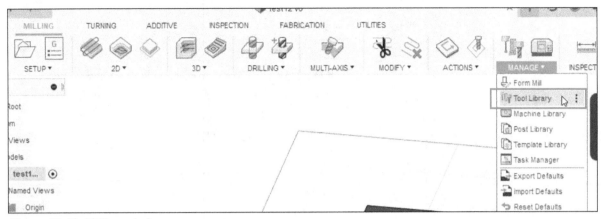

Figure-37. Tool Library tool

- Select desired category of tool from the **Tool Library** area at the left in the dialog box. List of cutting tools will be displayed.
- Select desired radio button from the **Filters** tab at the right in the dialog box to filter list of cutting tools; refer to Figure-38. On selecting a radio button, various options to further filter the list of tools will be displayed in the **Filters** tab.

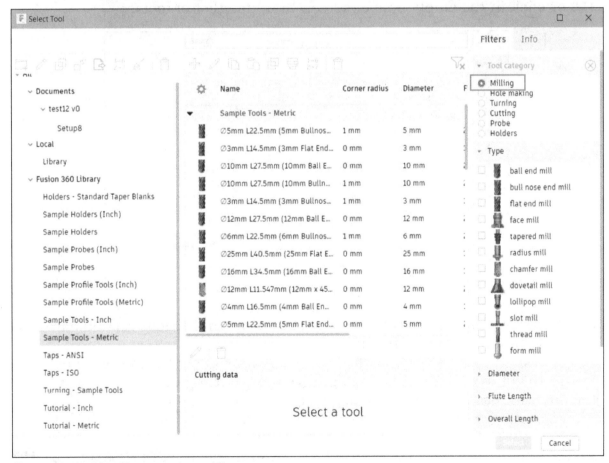

Figure-38. Applying filters to cutting tool list

- Select desired check box from **Type** rollout in the **Filters** tab to define type of cutting tool. For example, select the **ball end mill** check box to display only ball end mill cutting tools in the list. You can select multiple check box to display tools of multiple types.
- Expand the **Diameter** node and select desired operator to be used for specifying diameter range for cutting tools; refer to Figure-39.

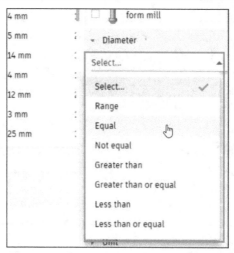

Figure-39. Operators for defining diameter range

- Select desired operator and specify related value in the edit box below it. For example, we have selected **Greater than** operator and specified value as **8 mm** so all the cutting tools which have diameter greater than 8 will be displayed; refer to Figure-40. Note that after specifying value in the edit box, you need to press **TAB** or click in the empty area of screen to update list of tools.

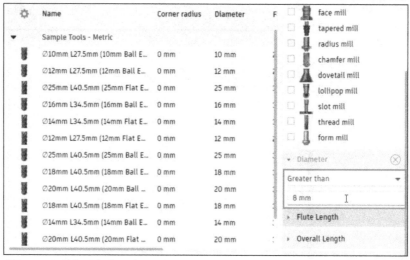

Figure-40. Filtering list of cutting tools

- Similarly, expand other rollouts in the **Filters** tab and specify related parameters to further filter list of cutting tools.
- Select desired cutting tool from the list. Information about the tool will be displayed in the **Info** tab of the dialog box and available cutting data will be displayed in the **Cutting data** area of the dialog box; refer to Figure-41.
- After selecting desired cutting tool, click on the **Select** button from the dialog box.

To check properties of selected tool, click on the **View** tool from **Toolbar** in the cutting tools list area of dialog box; refer to Figure-42. The **Select Tool** dialog box will be displayed with information of selected cutting tool; refer to Figure-43.

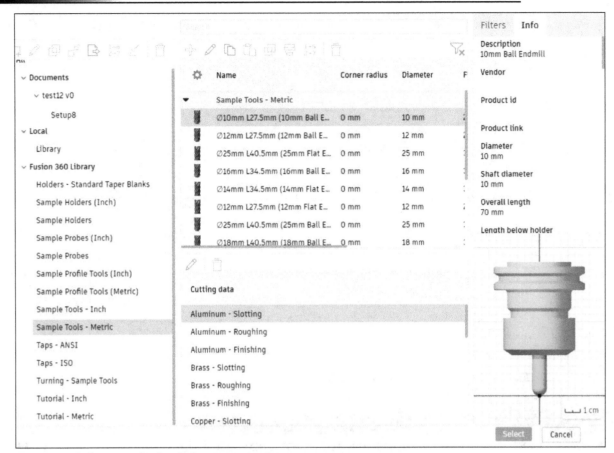

Figure-41. Information displayed about cutting tool

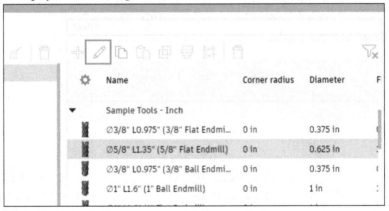

Figure-42. View tool

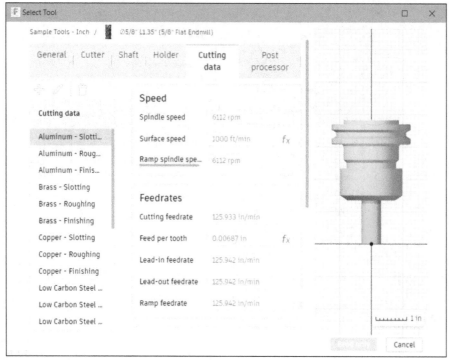

Figure-43. Information about cutting tool

TOOLS USED IN CNC MILLING AND TURNING

The tools used in CNC machines are made of cemented carbide, High Speed Steel, Tungsten Alloys, Ceramics, and many other hard materials. The shapes and sizes of tools used in Milling machines and Lathe machines are different from each other. These tools are discussed next.

Milling Tools

There are various type of milling tools for different applications. These tools are discussed next.

End Mill

End mills are used for producing precision shapes and holes on a Milling or Turning machine. The correct selection and use of end milling cutters is paramount with either machining centers or lathes. End mills are available in a variety of design styles and materials; refer to Figure-44.

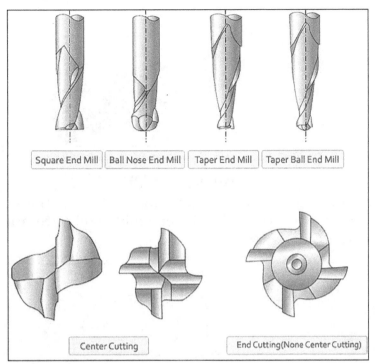

Figure-44. End Mill tool types

Titanium coated end mills are available for extended tool life requirements. The successful application of end milling depends on how well the tool is held (supported) by the tool holder. To achieve best results an end mill must be mounted concentric in a tool holder. The end mill can be selected for the following basic processes:

FACE MILLING - For small face areas, of relatively shallow depth of cut. The surface finish produced can be 'scratchy".
KEYWAY PRODUCTION - Normally two separate end mills are required to produce a quality keyway.
WOODRUFF KEYWAYS - Normally produced with a single cutter, in a straight plunge operation.
SPECIALTY CUTTING - Includes milling of tapered surfaces, "T" shaped slots & dovetail production.
FINISH PROFILING - To finish the inside/outside shape on a part with a parallel side wall.
CAVITY DIE WORK - Generally involves plunging and finish cutting of pockets in die steel. Cavity work requires the production of three dimensional shapes. A Ball type end mill is used for the finishing cutter with this application.

Roughing End Mills, also known as ripping cutters or hoggers, are designed to remove large amounts of metal quickly and more efficiently than standard end mills; refer to Figure-45. Coarse tooth roughing end mills remove large chips for heavy cuts, deep slotting and rapid stock removal on low to medium carbon steel and alloy steel prior to a finishing application. Fine tooth roughing end mills remove less material but the pressure is distributed over many more teeth, for longer tool life and a smoother finish on high temperature alloys and stainless steel.

Figure-45. Roughing End Mill

Bull Nose Mill

Bull nose mill look alike end mill but they have radius at the corners. Using this tool, you can cut round corners in the die or mold steels. Shape of bull nose mill tool is given in Figure-46.

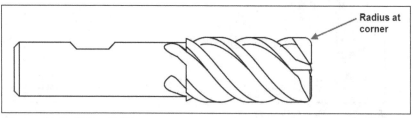

Figure-46. Bull Nose Mill cutter

Ball Nose Mill

Ball nose cutters or ball end mills has the end shape hemispherical; refer to Figure-44. They are ideal for machining 3-dimensional contoured shapes in machining centres, for example in moulds and dies. They are sometimes called ball mills in shop-floor slang. They are also used to add a radius between perpendicular faces to reduce stress concentrations.

Face Mill

The Face mill tool or face mill cutter is used to remove material from the face of workpiece and make it plane; refer to Figure-47.

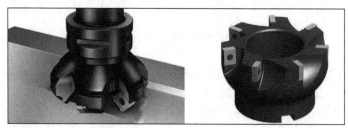

Figure-47. Face milling tool

Radius Mill and Chamfer Mill

The Radius mill tool is used to apply round (fillet) at the edges of the part. The Chamfer mill tool is used to apply chamfer at the edges of the part. Figure-48 shows the radius mill tool and chamfer mill tool.

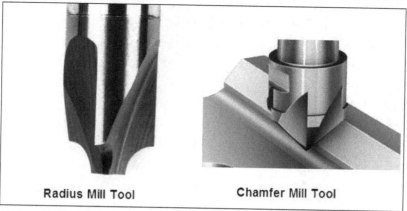

Radius Mill Tool   Chamfer Mill Tool

Figure-48. Radius mill and Chamfer mill tool

Slot Mill

The Slot mill tool is used to create slot or groove in the part metal. Figure-49 shows the shape of slot mill tool.

Figure-49. Slot mill tool

Taper Mill

In CNC machining, taper end mills are used in many industries for a large number of applications, such as walls with draft or clearance angle, tool and die work, mold work, even for reaming holes to make them conical. There are mainly two types of taper mills, Taper End Mill and Taper Ball Mill; refer to Figure-44.

Dove Mill

Dove mill or Dovetail cutters are designed for cutting dovetails in a wide variety of materials. Dovetail cutters can also be used for chamfering or milling angles on the bottom surface of a part. Dovetail cutters are available in a wide variety of diameters and in 15, 45, and 60 degree angles; refer to Figure-50.

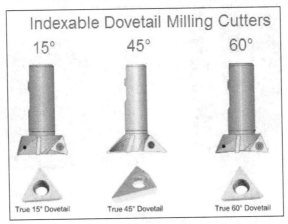

Figure-50. Dovetail milling cutters

Lollipop Mill

The Lollipop mill tool is used to cut round slot or undercuts in workpiece. Some tool suppliers use a name Undercut mill tool in place of Lollipop mill in their catalog. The shape of lollipop mill tool is given in Figure-51.

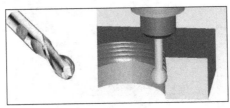

Figure-51. Lollipop mill tool

Engrave Mill

The Engrave mill tool is used to perform engraving on the surface of workpiece. Engraving has always been an art and it is also true for CNC machinist. You can find various shapes of engraving tool that are single flute or multi-flute; refer to Figure-52. You can use ball mill/end mill for engraving or you can use specialized engrave mill tool for engraving. This all depends on your requirement. If you want to perform engraving on softer materials or plastics then it is better to use ball end mill but if you want an artistic shade on the surface then use the respective engrave mill tool. Keep a note of maximum depth and spindle speed mentioned by your engrave mill tool supplier.

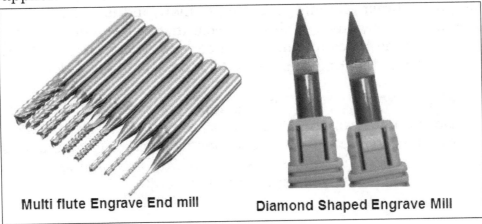

Figure-52. Engrave mill tools

Thread Mill

The Thread mill tool is used to generate internal or external threads in the workpiece. The most common question here is if we have Taps to create thread then why is there need of Thread mill tool. The answer is less machining time on CNC, tool cost saving, more parts per tool, and better thread finish. Now, you will ask why to use tapping. The answer is low machine cost. Figure-53 shows thread mill tools.

Figure-53. Thread Mill

Barrel Mill

Barrel Mill tool is the tool recently being highly used in machining turbine/impeller blades and other 5-axis milling operations. Barrel Mill has conical shape with radius at its end; refer toFigure-54. Note that earlier Ball mill tools were used for irregular surface contouring but Barrel Mill tools give much better surface finish so they are highly in demand for 5-axis milling now a days.

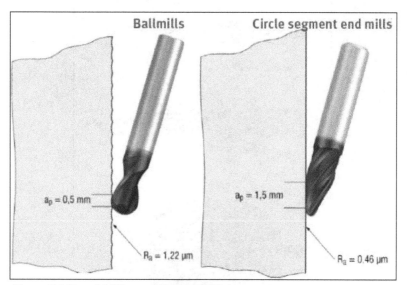

Figure-54. Barrel Mill versus Ball Mill Tool

Drill Bit

Drill bit is used to make a hole in the workpiece. The hole shape depends on the shape of drill bit. Drill bits for various purposes are shown in Figure-55. Note that drill is the machine or holder in which drill bit is installed to make cylindrical holes. There are mainly four categories of drill bit; Twist drill bit, Step drill bit, Unibit (or conical bit), and Hole Saw bit (Refer to Figure-56). Twist drill bits are used for drilling holes in wood, metal, plastic and other materials. For soft materials the point angle is 90 degree, for hard materials the point angle is 150 degree and general purpose twist drill bits have angle of 150 degree at end point. The Step drill bits are used to make counter bore or countersunk holes. The Unibits are generally used for drilling holes in sheetmetal but they can also be used for drilling plastic, plywood, aluminium and thin steel sheets. One unibit can give holes of different sizes. The Hole saw bit is used to cut a large hole from the workpiece. They remove material only from the edge of the hole, cutting out an intact disc of material, unlike many drills which remove all material in the interior of the hole. They can be used to make large holes in wood, sheet metal, and other materials.

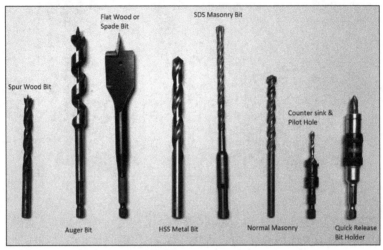

Figure-55. Drill Bits for different purposes

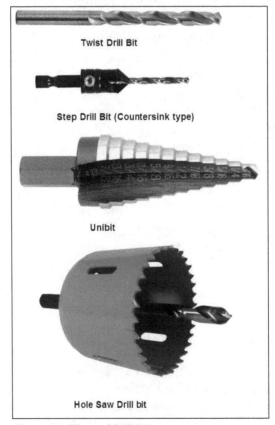

Figure-56. Types of drill bits

Reamer

Reamer is a tool similar to drill bit but its purpose is to finish the hole or increase the size of hole precisely. Figure-57 shows the shape of a reamer.

Bore Bar

Bore Bar or Boring Bar is used to increase the size of hole; refer to Figure-58. One common question is why to use bore bar if we can perform reaming or why to perform reaming when we have bore bar. The answer is accuracy. A reamer does not give tight tolerance in location but gives good finish in hole diameter. A bore bar gives tight tolerance in location but takes more time to machine hole as compared to reamer. The decision to choose the process is on machinist. If you need a highly accurate hole then perform drilling, then boring and then reaming to get best result.

Figure-58. Boring Bar

Figure-57. Reamer tool

Lathe Tools or Turning Tools

The tools used in CNC lathe machines use a different nomenclature. In CNC lathe machines, we use insert for cutting material. The Insert Holder and Inserts have a special nomenclature scheme to define their shapes. First we will discuss the nomenclature of Insert holder and then we will discuss the nomenclature of Inserts.

Insert Holders

Turning holder names follow an ISO nomenclature standard. If you are working on a CNC shop floor with lathes, knowing the ISO nomenclature is a must. The name looks complicated, but is actually very easy to interpret; refer to Figure-59.

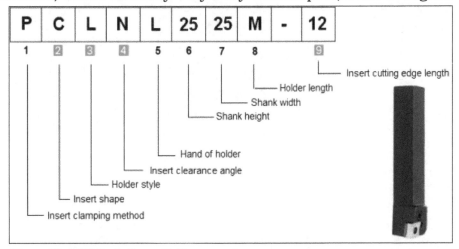

Figure-59. CNC Lathe Insert Holder nomenclature

When selecting a holder for an application, you mainly have to concentrate on the numbers marked in red in the above nomenclature. The others are decided automatically (e.g., the shank width and height are decided by the machine), or require less effort. In Figure-60, the rows with the question mark indicate the parameters that require the decision by machinist based on job.

	Parameter		How is this decided ?
1	Insert clamping method		Select based on cutting forces. Top clamping is the most sturdy, screw clamping the least.
2	Insert shape	?	Decided by the contour that you want to turn.
3	Holder style	?	Decided by the contour that you want to turn.
4	Insert clearance angle	?	Positive / Negative, based on application.
5	Hand of holder		Decided based on whether you want to cut towards the chuck or away from the chuck, and on turret position - turret front / rear
6	Shank height		Decided by holder size.
7	Shank width		Decided by machine.
8	Holder length		Decided by machine.
9	Insert cutting edge length	?	Decide based on depth of cut you want to use.

Figure-60. CNC Lathe Insert Holder nomenclature parameters

Figure-61 and Figure-62 show the options available for each of the parameters.

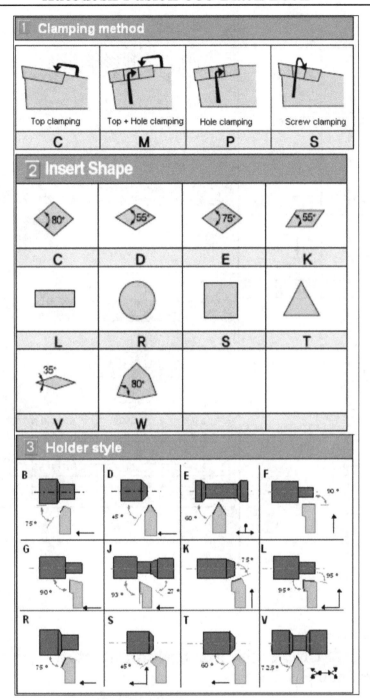

Figure-61. Clamping Method, Insert Shapes, and Holder Style

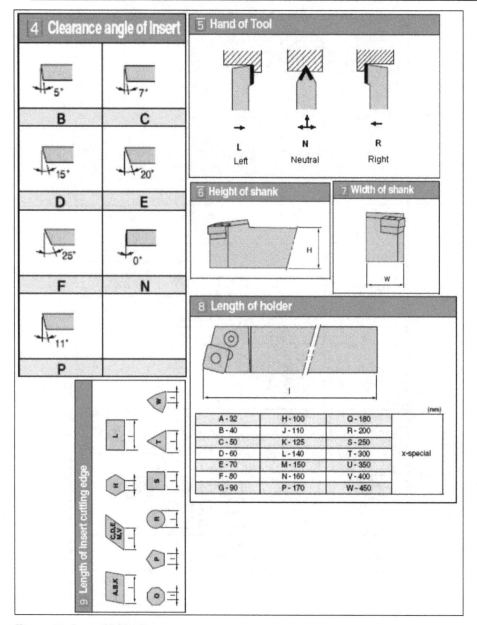

Figure-62. Insert Holder Parameters

CNC Lathe Insert Nomenclature

General CNC Insert name is given as

Meaning of each box in nomenclature is given next.

1 = Turning Insert Shape

The first letter in general turning insert nomenclature tells us about the general turning insert shape, turning inserts shape codes are like C, D, K, R, S, T, V, W. Most of these codes surely express the turning insert shape like

C = C Shape Turning Insert
D = D Shape Turning Insert
K = K Shape Turning Insert
R = Round Turning Insert
S = Square Turning Insert
T = Triangle Turning Insert
V = V Shape Turning Insert
W = W Shape Turning Insert

Figure-63 shows the turning inserts shapes.

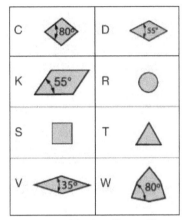

Figure-63. Turning Insert Shapes

The general turning insert shape play a very important role when we choose an insert for machining. Not every turning insert with one shape can be replaced with the other for a machining operation. As C, D, W type turning inserts are normally used for roughing or rough machining.

2 = Turning Insert Clearance Angle

The second letter in general turning insert nomenclature tells us about the turning insert clearance angle.

The clearance angle for a turning insert is shown in Figure-64.

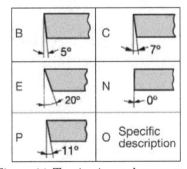

Figure-64. Turning insert clearance angle

Turning insert clearance angle plays a big role while choosing an insert for internal machining or boring small components, because if not properly chosen the insert bottom corner might rub with the component which will give poor machining. On the other hand a turning insert with 0° clearance angle is mostly used for rough machining.

3 = Turning Insert Tolerances

The third letter of general turning insert nomenclature tells us about the turning insert tolerances. Figure-65 shows the tolerance chart.

Code Letter	Cornerpoint (inches)	Thickness (inches)	Inscribed Circle (in)	Cornerpoint (mm)	Thickness (mm)	Inscribed Circle (mm)
A	.0002"	.001"	.001"	.005mm	.025mm	.025mm
C	.0005"	.001"	.001"	.013mm	.025mm	.025mm
E	.001"	.001"	.001"	.025mm	.025mm	.025mm
F	.0002"	.001"	.0005"	.005mm	.025mm	.013mm
G	.001"	.005"	.001"	.025mm	.13mm	.025mm
H	.0005"	.001"	.0005"	.013mm	.025mm	.013mm
J	.002"	.001"	.002-.005"	.005mm	.025mm	.05-.13mm
K	.0005"	.001"	.002-.005"	.013mm	.025mm	.05-.13mm
L	.001"	.001"	.002-.005"	.025mm	.025mm	.05-.13mm
M	.002-.005"	.005"	.002-.005"	.05-.13mm	.13mm	.05-.15mm
U	.005-.012"	.005"	.005-.010"	.06-.25mm	.13mm	.08-.25mm

Figure-65. Insert tolerance chart

4 = Turning Insert Type

The fourth letter of general turning insert nomenclature tells us about the turning insert hole shape and chip breaker type; refer to Figure-66.

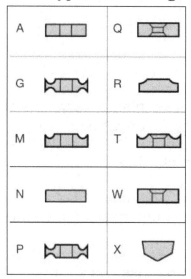

Figure-66. Turning Insert hole shape and chip breaker

5 = Turning Insert Size

This numeric value of general turning insert tells us the cutting edge length of the turning insert; refer to Figure-67.

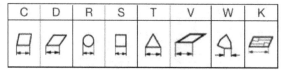

Figure-67. Turning Insert Cutting Edge Length

6 = Turning Insert Thickness

This numeric value of general turning insert tells us about the thickness of the turning insert.

7 = Turning Insert Nose Radius

This numeric value of general turning insert tells us about the nose radius of the turning insert.

Code	=	Radius Value
04	=	0.4
08	=	0.8
12	=	1.2
16	=	1.6

You can learn more about machining tools from your tool supplier manual.

TOOL LIBRARY

Tool Library is used to create and manage cutting tools used in Fusion 360. Click on the **Tool Library** tool from **MANAGE** drop-down in **MILLING** tab of the **Ribbon**. The **Tool Library** dialog box will be displayed; refer to Figure-68. The process to create a new tool is discussed next.

Creating a New Mill Tool

You have learned to use an already available tool in the library but what if the tool is not available in library. The procedure to create a new mill tool is discussed next.

- Select the **Library** option from the **Local** node in the **Tool Library** area at the left in the dialog box and click on the **New** tool from **Toolbar** in **Cutting Tools list** area of the dialog box. The **New tool** page will be displayed as shown in Figure-69.

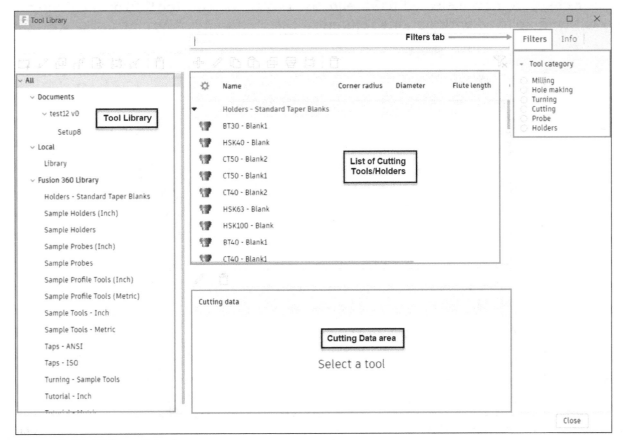

Figure-68. Tool Library dialog box

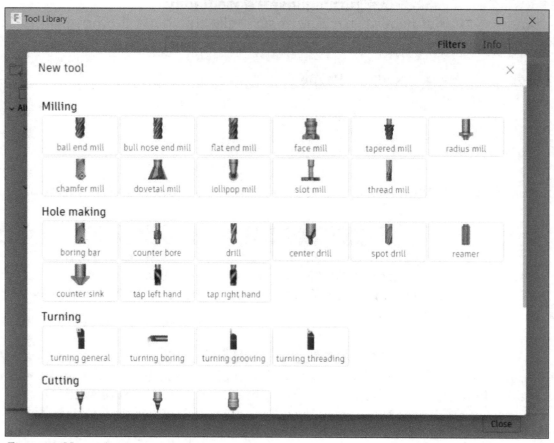

Figure-69. New tool page

- Select desired button from the page to create respective type of cutting tool. In our case, we have selected **ball end mill** button. The **Tool Library** dialog box will be displayed with options related to selected tool type; refer to Figure-70

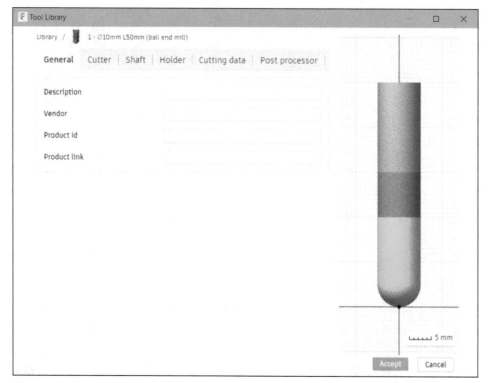

Figure-70. Library dialog box

- Specify desired parameters in the edit boxes of **General** tab in the dialog box like description of tool, name of vendor, and so on.
- Click on the **Cutter** tab in the dialog box. The options will be displayed as shown in Figure-71.

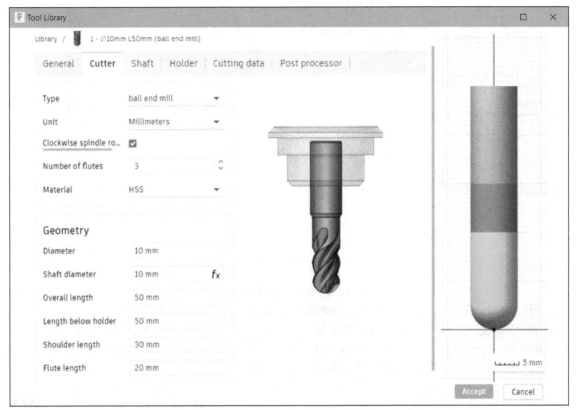

Figure-71. Cutter tab in Tool Library dialog box

- Click on the **Type** drop-down of **Cutter** tab from **Library** dialog box and select desired type of cutting tool if you want to change the tool type. Preview of the selected tool type will be displayed in graphics section of dialog box.
- Click on the **Unit** drop-down from the **Cutter** tab and select desired unit type.
- Select the **Clockwise spindle rotation** check box to set the rotation of spindle to clockwise direction.
- Click in the **Number of flutes** edit box and specify desired value to define number of cutting flutes in the tool.
- Click on the **Material** drop-down and select desired material used for cutting tool.
- Click in the **Diameter** edit box of the **Geometry** section and specify the diameter of cutting tool.
- Click in the **Shaft diameter** edit box of the **Geometry** section and specify the diameter of shaft.
- Click in the **Overall length** edit box to specify total length of cutting tool.
- Similarly, click in the other edit boxes and specify desired parameters in **Geometry** section of **Cutter** tab.
- Click on the **Shaft** tab from the **Tool Library** dialog box. The **Shaft** tab will be displayed; refer to Figure-72.

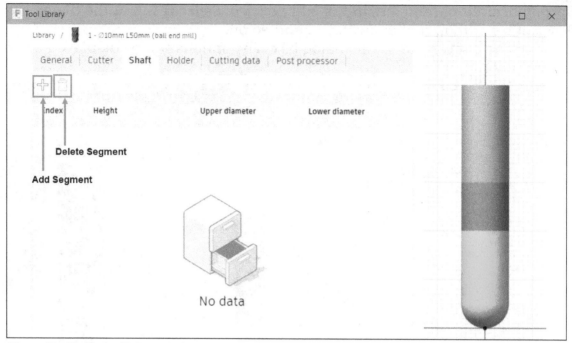

Figure-72. Shaft tab of Library dialog Box

- To add the shaft, click on the **Add segment** button and if you want to delete the created shaft then click on the **Remove** button from **Shaft** tab.
- After clicking on the **Add segment** button, double-click in the edit boxes for shaft parameter to set the shaft shape and size. Preview of the shaft will be displayed on the right in the dialog box; refer to Figure-73.

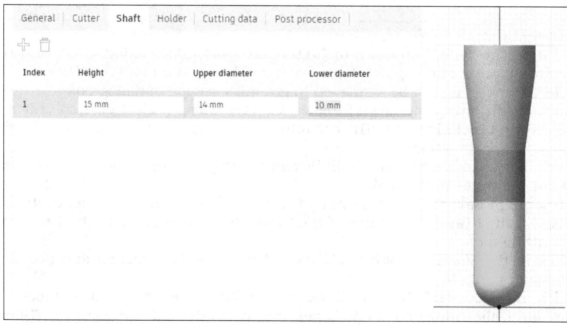

Figure-73. Editing the value of shaft

- If you want to delete a segment of shaft then select the shaft from list and click on the **Delete segment** button.
- Click on the **Holder** tab from the **Library** dialog box to select desired tool holder. The **Holder** tab will be displayed; refer to Figure-74.

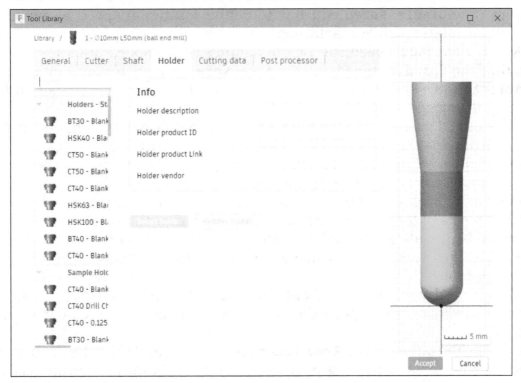

Figure-74. Holder tab

- Select desired holder from the list at the left in the dialog box and click on the **Select holder** button from the **Holder** tab to use selected holder. Preview of tool holder will be displayed. Specify desired parameters in the **Info** area of the dialog box.
- Click on the **Cutting data** tab of the **Tool Library** dialog box to set the feed and speed of cutting tool. Options in the **Feed & Speed** tab will be displayed as shown in Figure-75. You can use multiple cutting data presets for different materials using the **Add preset** button at the left in the dialog box.

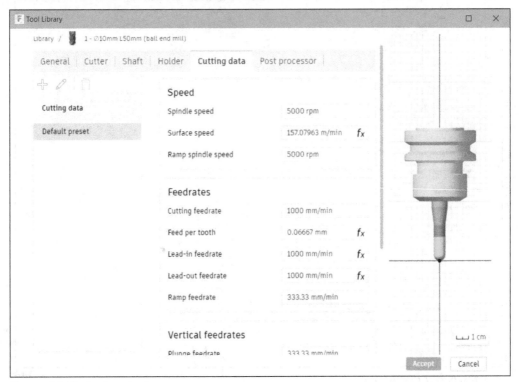

Figure-75. Feed & Speed tab

- Click in the **Spindle speed** edit box to define rotational speed at which cutting tool will be rotating while machining.
- Click in the **Surface speed** edit box to define surface speed at tools cutting edge. Note that the **Spindle speed** and **Surface speed** parameters are inter-related by formula "Surface speed = Π x D x RPM". Here, D is diameter of tool and RPM is the Spindle speed.
- Specify desired value in the **Ramp spindle speed** edit box to define spindle speed when cutting tool is moving in ramp motion while cutting.
- Specify desired value in the **Cutting feedrate** edit box to define speed at which tool will move on the path while cutting material.
- Specify desired value in the **Feed per tooth** edit box if you want to define cutting speed in terms of tool movement per tooth. Note that the value in this edit box is generated automatically based on value specified in **Cutting feedrate** edit box by relation "Feed per tooth = Cutting feedrate/(Spindle speed x Number of flutes in cutting tool).
- Specify desired values in the **Lead-in feedrate** and **Lead-out feedrate** edit boxes to specify at what speed the cutting tool will enter and exit the workpiece, respectively.
- Specify desired value in the **Ramp feedrate** edit box to define the cutting feedrate when cutting tools is moving in a ramp motion. Ramp motion is generated in toolpath when cutting tool moves up or down in the workpiece while cutting.
- Specify desired value in the **Plunge feedrate** edit box of **Vertical feedrates** section in the dialog box to define vertical movement speed of tool when entering the workpiece. Similarly, you can specify value in **Feed per revolution** edit box to define feed rate with respect to revolution of cutting tool.
- Click in the **Post Processor** tab of **Library** dialog box to specify the post processor related values of NC machining. The **Post Processor** tab will be displayed; refer to Figure-76.

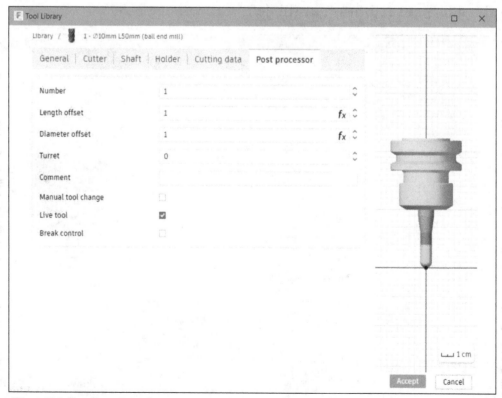

Figure-76. Post Processor tab

- The parameters in this tab are used to define compensation for tool wear while machining. Note that due to repetitive cutting cycles, the cutting edges/tips of tools wear and cause lesser depth of cuts over the period of time for same toolpaths. To compensate that we add values specified in offsets of post processor to cutting operation movement for extending depths.
- Click in the **Number** edit box and specify the tool number for which you want to define parameters.
- Click in the **Length offset** edit box and define length offset number to be associated with the current tool number.
- Click in the **Diameter offset** edit box and define diameter offset number to be associated with the current tool number.
- Click in the **Turret** edit box and specify the turret number to be used for defining parameters related to cutting tools. A turret is generally a big round mechanism that holds multiple different cutting tools in the machine indexed by tool numbers. Some machines can have multiple turrets but commonly you will find a single turret in the machine.
- Click in the **Comment** edit box to associate a text comment with current selected tool. This comment tells the properties of current tool in output file. Note that the comments do not take any role in cutting operations and are provided only for information of users.
- Select **Manual tool change** check box to force manual change of cutting tool by user after each operation even in an automatic machine.
- Select the **Live tool** check box to tell the machine that current tool is a live (active) cutting tool. If this check box is not selected then current tool will not be used for cutting operations and will be considered as empty slot in turret.
- Select the **Break control** check box to check whether cutting tool is in good condition or not. Make sure your machine and post processor support this function before selecting this check box.
- After specifying all the parameters, click on the **Accept** button. The mill tool will be created and added in the tools list.

Creating New Tool Library

A tool library is an organized collection cutting tools. You can create a new tool library specific to your products in your industry. To create a new library, select the **Local** node from left area in the dialog box and click on the **New library** tool from **Toolbar** in **Tool Library** dialog box. You will be asked to specify name for new library; refer to Figure-77. Specify desired name for new library and press **ENTER**. The library will be created. Note that if you select a library and create a new tool then new tool will be part of selected library.

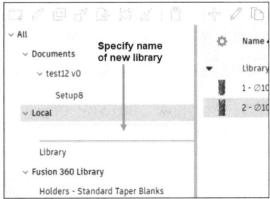

Figure-77. New library added

You can use the **Rename** tool in **Toolbar** of the dialog box after selecting a user defined library in the dialog box to change the name of the library. Using the **Duplicate library** tool, you can create a duplicate copy of selected user defined library.

Importing Libraries

The **Import libraries** tool is used to include one or more tool libraries from selected data files. The procedure to use this tool is given next.

• Click on the **Import libraries** tool ⬚ from the **Toolbar** in the **Tool Library** dialog box after selecting a user defined library. The **Open File(s)** dialog box will be displayed; refer to Figure-78.

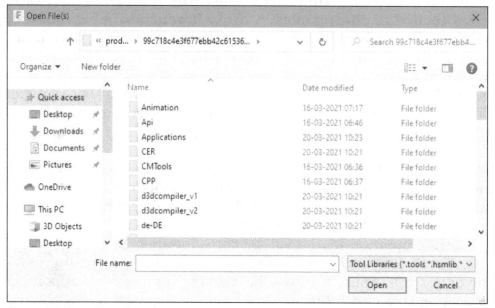

Figure-78. Open Files dialog box

• Select desired files to import library data stored in them. You can select multiple files while holding the **CTRL** key. Make sure the files are in supported formats (*.tools, *.hsmlib, *.json, and *.tsv).
• After selecting files, click on the **Open** button. The imported libraries will be added in the dialog box.

Exporting Tool Library

The **Export library** tool is used to export selected tool library data in a file. The procedure to use this tool is given next.

• Click on the **Export library** tool from the **Toolbar** after selecting a library from left area in the dialog box. The **Save File** dialog box will be displayed as shown in Figure-79.
• Specify desired name of file in the **File name** edit box at the bottom in the dialog box.
• Click in the **Save as type** drop-down and select desired format in which the file will be saved.

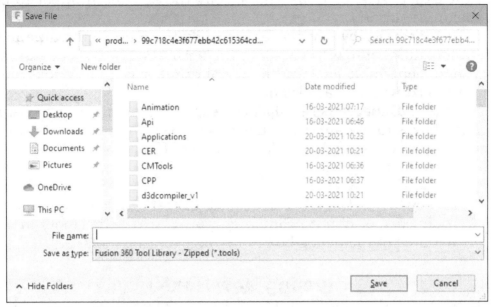

Figure-79. Save File dialog box1

- You can move to desired directory by using common Operation System functions. After setting desired parameters, click on the **Save** button. The file will be exported to specified directory.

Renumbering Tools of a Library

The **Renumber tools** tool is used to change the numbering of cutting tools in selected library by specified increment value. The procedure to use this tool is given next.

- Select the user defined library whose tool numbers are to be changed from the left area in the dialog box and click on the **Renumber tools** tool ⊞ from the **Toolbar** in the **Tool Library** dialog box. The **Renumber tools** page will be displayed in the dialog box; refer to Figure-80.

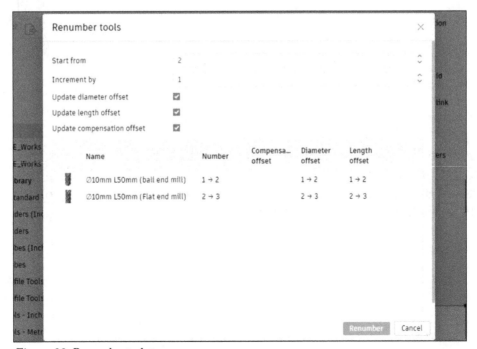

Figure-80. Renumber tools page

- Click in the **Start from** edit box and specify the starting number to be applied to first tool in the library.
- Click in the **Increment by** edit box and specify the increment value to be added in current tool number to get next tool number in the library. For example, if you have set increment value as 2 and first tool number is 1 then next tool number will be 3, up next will be 5 and so on.
- Select the **Update diameter offset**, **Update length offset**, and **Update compensation offset** check boxes to update respective offsets based on new tool numbers.
- After setting desired parameters, click on the **Renumber** button from the page. The tool numbers will change accordingly.

The **Delete library** tool in the **Toolbar** of **Tool Library** dialog box is used to delete selected user defined tool library. Similarly, you can use the tools in **Toolbar** of Cutting tools list area of the dialog box to manage cutting tools.

Creating New Holder

In this section, we will discuss about the procedure of creating new tool holder. The procedure is discussed next.

- Select the library in which you want to add new tool holder and click on the **New tool** button ⊞ from the **Toolbar** in the **Tool Library** dialog box. The **New tool** page will be displayed in the dialog box as discussed earlier.
- Click on the **holder** button from the **Holders** section of the page. The **Tool Library** dialog box will be displayed; refer to Figure-57.

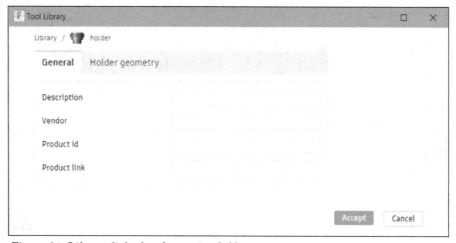

Figure-81. Library dialog box for creating holder

- Click in the edit boxes of **General** tab from **Library** dialog box and specify the information of holder as desired.
- Click on the **Holder Geometry** tab from **Library** dialog box to view or change the geometry of holder; refer to Figure-82.

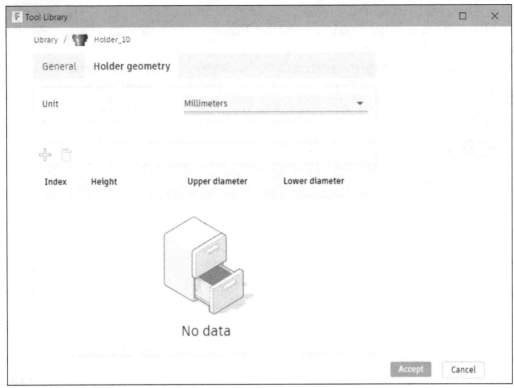

Figure-82. Holder Geometry tab of Library dialog box

- Click on the **Add segment** button from the dialog box to add desired segment in tool holder. The procedure is same as discussed earlier while creating new tool.
- To change a parameter, click in the respective edit box and specify desired value. The holder in graphic window will be updated; refer to Figure-83.

Figure-83. Preview of tool holder

- After specifying the parameter, click on the **Accept** button from **Library** dialog box. The created tool holder will be displayed in **Tool Library** dialog box.

Creating New Turning tool

In this section, we will discuss about the procedure of creating a new turning tool. The procedure is discussed next.

- Select the library in which you want to add new turning tool and click on the **New tool** button ⊕ from the **Toolbar** in the **Tool Library** dialog box. The **New tool** page will be displayed in the dialog box as discussed earlier.

- Click on desired button from the **Turning** section of the page. Respective options will be displayed in the **Tool Library** dialog box; refer to Figure-84 (in case of turning general button selected).
- Specify desired parameters in the **General** tab of the dialog box.
- Click on the **Insert** tab from the dialog box to specify parameters related to cutting tool insert used in turning tool. The dialog box will be displayed as shown in Figure-85.

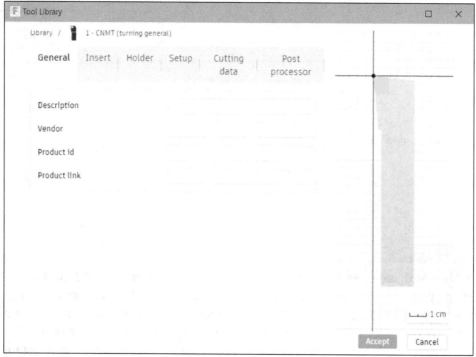

Figure-84. General tab in Library dialog box

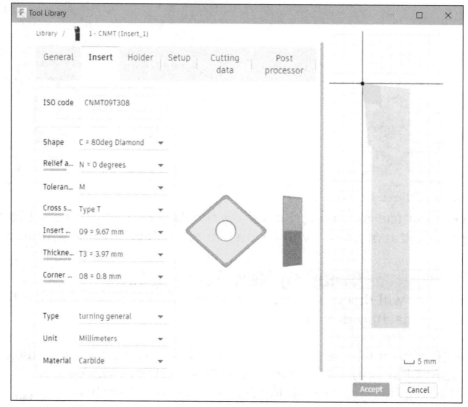

Figure-85. Library dialog box for turning tool

- Click in the **ISO code** edit box to specify desired ISO code of cutting tool insert from your Tool Manufacturer's manual. All the parameters in the dialog box will be modified accordingly. You can also specify parameters in the other edit boxes of dialog box and value in ISO code edit box will be modified accordingly.
- Click in the **Shape** drop-down and select desired shape for tool insert. Preview of the cutting tool insert will be displayed; refer to Figure-86.

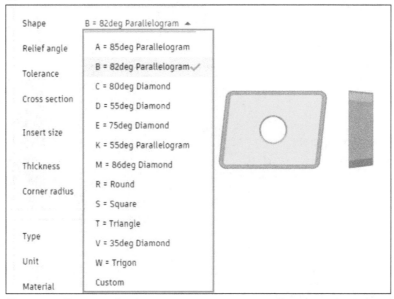

Figure-86. Preview of tool insert

- Click in the **Relief angle** drop-down and specify the slope angle to be applied on the side face of insert for breaking chips while cutting.
- Click in the **Tolerance** drop-down and select desired option to define tolerance in tool insert size.
- Click in the **Cross section** drop-down to define cross-sectional shape of the cutting tool inserts.
- Click in the **Insert size** drop-down to define size of cutting tool insert.
- Click in the **Thickness** drop-down to define thickness of cutting tool insert.
- Click in the **Corner radius** drop-down to define radius at the corners of tool insert.
- Select desired option from the Type drop-down to define the type of cutting insert to be created.
- Similarly, set the other parameters for turning tool in **Insert** tab.
- Specify desired parameter in the **Holder** tab of **Tool Library** dialog box as discussed earlier.
- Click in the **Setup** tab from the **Tool Library** dialog box to setup the tool. The **Setup** tab will be displayed; refer to Figure-87.

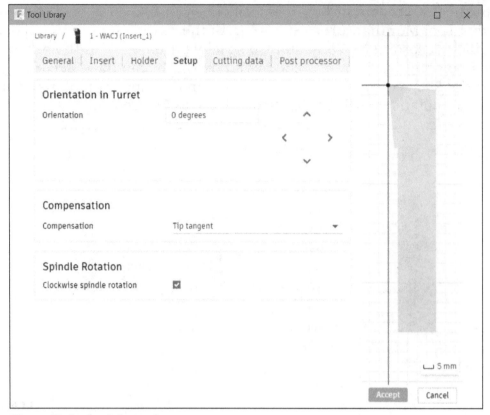

Figure-87. Setup tab for turning

- The four arrows keys are used to define the orientation of the cutting tool in turret. This orientation defines which side of tool holder will be used for cutting. Click on the upper key to orient the tool's tip upward, left key to orient the tool to left side, right key to orient the tool to right side, and down key to orient the tool's tip downward. If you want to orient the tool at specified angle then click in the **Orientation** edit box and specify the angle.
- Click in the **Compensation** drop-down from **Setup** tab and select the desired option to define where tool wear compensation will be applied.
- Select the **Clockwise spindle rotation** check box from **Spindle Rotation** section to set the rotation direction of spindle as clockwise.
- Set the parameters in other tabs of **Tool Library** dialog box as discussed earlier and click on the **Accept** button. The new turning tool will be created and displayed in the **Tool Library** dialog box.

Creating new tool for Waterjet/ Plasma Cutter/ Laser Cutter

In this section, we will discuss about the procedure of creating the Waterjet/Plasma Cutter/Laser Cutter tool. The procedure is discussed next.

- Select the library in which you want to add new turning tool and click on the **New tool** button ⊞ from the **Toolbar** in the **Tool Library** dialog box. The **New tool** page will be displayed in the dialog box as discussed earlier.
- Click on desired button from the **Cutting** section of the page. Respective options will be displayed in the **Tool Library** dialog box; refer to Figure-88 (In case of waterjet).

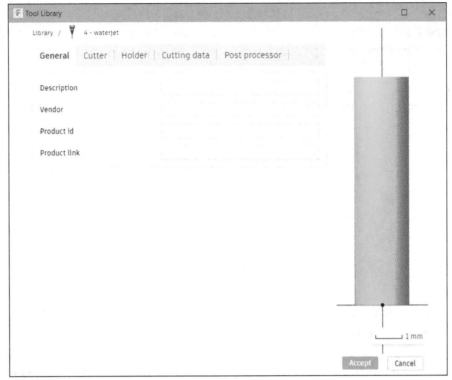

Figure-88. Library dialog box for water jet

- The options of the **Tool Library** dialog box are similar as discussed earlier in **Creating New Mill tool** section of this chapter.
- After specifying the parameters, click on the **OK** button from the **Tool Library** dialog box. The created tool will be displayed in **Tool Library** dialog box.
- After creating desired tools, click on the **Close** button from the dialog box.

MACHINE LIBRARY

The **Machine Library** tool is used to create and manage machines used in Fusion 360 CAM. Click on the **Machine Library** tool from the **MANAGE** drop-down in the **MILLING** tab of the **Ribbon**. The **Machine Library** dialog box will be displayed; refer to Figure-89.

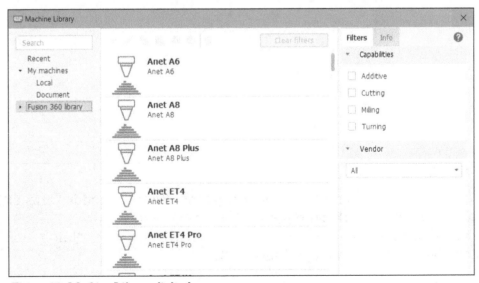

Figure-89. Machine Library dialog box

Creating a New Machine

The **Create new** tool in the **Toolbar** of the **Machine Library** dialog box to create a new machine of desired type. The procedure to use this tool is given next.

- Select the **Local** option in the **My machines** node from left area of the **Machine Library** dialog box and click on the **Create new** tool from the **Toolbar**. A flyout will be displayed with list of machines that can be created; refer to Figure-90.

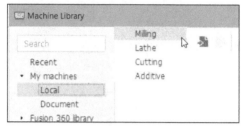

Figure-90. Create new flyout

- Select desired option from the flyout (**Milling** option in our case). The **Machine Configuration** dialog box will be displayed; refer to Figure-91.

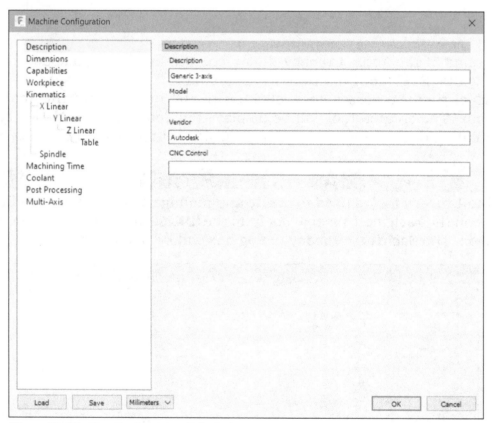

Figure-91. Machine Configuration dialog box

- Specify desired parameters in the **Description**, **Model**, **Vendor**, and **CNC Control** edit boxes of the **Description** page in the dialog box.
- Select the **Dimensions** option from the left area in the dialog box to define dimensions of machine; refer to Figure-92.

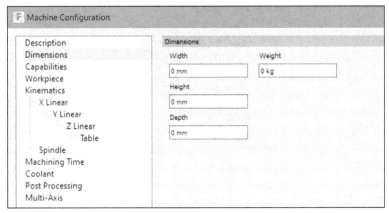

Figure-92. Dimensions page

- Specify desired values in **Width**, **Height**, and **Depth** edit boxes to define width, height, and depth of machine, respectively.
- Specify desired value in the **Weight** edit box to define total weight of machine.
- Select the **Capabilities** option from the left area in the dialog box. The **Capabilities** page will be displayed; refer to Figure-93.

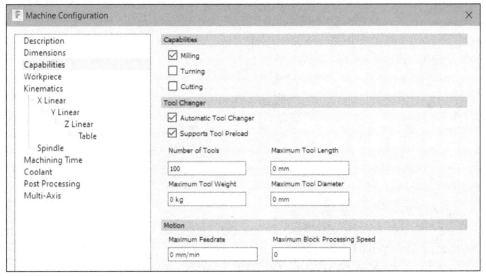

Figure-93. Capabilities page

- Select check boxes from the **Capabilities** section of the dialog box to define what are the capabilities of current machine. You can create a machine which can perform milling, turning, and cutting as well.
- Select the **Automatic Tool Changer** check box to define that machine can automatically change cutting tool.
- Select the **Supports Tool Preload** check box if your machine supports preloading of tool spindle. Proper preloading extends the life of spindle bearing.
- Select the **Workpiece** option from the left area in the dialog box to define parameters related to maximum size of workpiece allowed on machine. The **Workpiece** page will be displayed; refer to Figure-94. Specify the size of workpiece along X, Y, and Z axes of machine in respective edit boxes. You can also specify maximum weight capacity that machine can hold for workpiece in the **Weight Capacity** edit box.

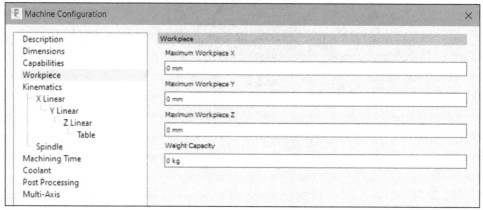

Figure-94. Workpiece page

- Select the **X Linear** option from the left in **Kinematics** node to define parameters related to linear motion of tool and workpiece along X axis. The options will be displayed as shown in Figure-95.

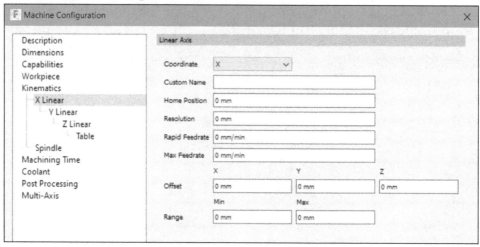

Figure-95. X Linear Kinematics options

- Specify desired text in the **Custom Name** edit box to apply a custom name to the X axis. For example, you can write a custom name as "large base side" to identify x axis where larger side of model will be placed on machine bed,
- Click in the **Home Position** edit box to specify start position for cutting tool along X axis.
- Specify desired value in the **Resolution** edit box to define maximum allowed deviation (tolerance) in tool movement along X axis. Note that when a tool moves from 0 to 10 mm then it is not exact 10 mm position and there is some amount of error in that position like 10.004 or 9.9995. Using the value defined for resolution, we notify the user about accuracy of movements along X axis.
- Specify desired value in the **Rapid Feedrate** edit box to define the maximum speed up to which cutting tool can move for non-cutting passes toolpath.
- Specify desired value in the **Max Feedrate** edit box to define the maximum speed up to which cutting tool can move for cutting passes in toolpath.
- Specify desired values in the **Offset** edit boxes to move the X axis at respective location. For example, if you specify 10 in X, Y, and Z edit boxes for Offset then X axis will move to 10 mm in X direction, 10 mm in Y direction and 10 mm in Z direction from its original position.
- Specify desired values in the **Min** and **Max** edit boxes for **Range** to define the minimum and maximum distance from origin within which the cutting tool can move in machine.

- Similarly, you can specify parameters in **Y Linear** and **Z Linear** pages of the dialog box. Note that you can drag and drop options in the **Kinematics** node to desired location for defining order of respective components in the machine.

- If you want to add a rotary axis in the machine then right-click on any option under **Kinematics** node and select the **Add rotary axis** option from the shortcut menu; refer to Figure-96. A new rotary axis will be added in the **Kinematics** node after selected option; refer to Figure-97. These option are similar to linear axis options discussed earlier with only difference being degree in place of mm. Figure-98 shows how rotary axes are generally placed in a 5 axis milling machine.

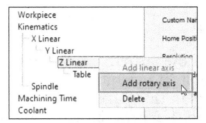

Figure-96. Add rotaty axis option

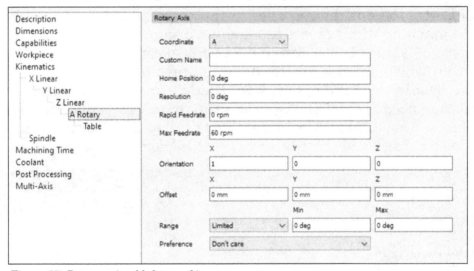

Figure-97. Rotary axis added to machine

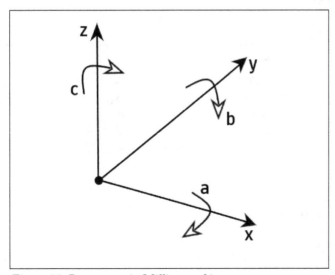

Figure-98. Rotary axes in Milling machine

- Select the **Spindle** option from left area in the dialog box to define parameters related to spindle of machine. The options will be displayed as shown in Figure-99.

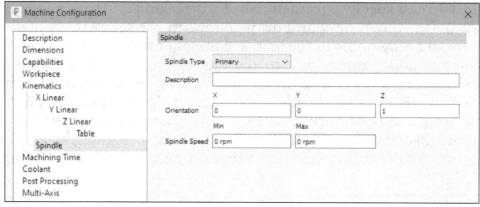

Figure-99. Spindle page

- Select desired option from the **Spindle Type** drop-down to specify whether you want to use primary spindle of machine or secondary spindle of machine. Note that rest of the parameters in this page will be defined for selected spindle type.

- Specify desired text in the **Description** edit box to define user description of spindle.

- In the **Orientation** edit boxes, specify the coordinates for defining location of spindle in machine with respect to machine origin.

- Specify desired values in the **Min** and **Max** edit boxes for **Spindle Speed** to define range within which spindle can rotate in the machine.

- Select the **Machining Time** option from the left area in the dialog box to define standard time need for tool change and ratio up to which maximum feedrate can be used for safe machining. The options will be displayed as shown in Figure-100. Specify desired values in the edit boxes of this page.

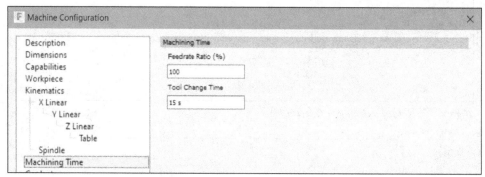

Figure-100. Machining Time page

- Select the **Coolant** option from the left area in the dialog box to define what type of coolant functions are supported by machine; refer to Figure-101.

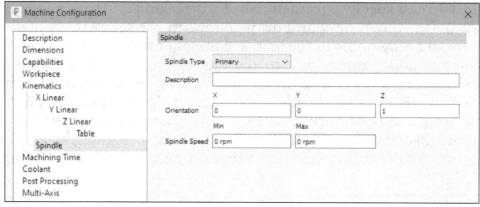

Figure-101. Coolant page

- Select desired check boxes from this page to define coolant capabilities of machine.
- Select the **Post Processing** option from the left area in the dialog box to define parameters related to post processor of machine; refer to Figure-102.

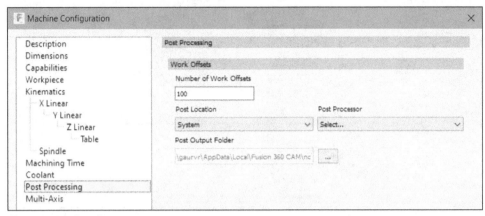

Figure-102. Post Processing page

- Specify desired value in the **Number of Work Offsets** edit box to define maximum limit of creating work offsets in machine.
- Select desired option from the **Post Location** drop-down to define where Autodesk Fusion 360 should search for post processors of machines. A post processor is a piece of application which converts the generic machine codes generated by Autodesk Fusion 360 CAM to a machine specific NC program. After setting desired location in the drop-down, select a post processor specific to your machine from **Post Processor** drop-down.
- Click on the **Browse** button [...] for **Post Output Folder** in this page to define location where NC program will be saved after post processing of CAM data. The **Post Directory** dialog box will be displayed; refer to Figure-103. Browse to desired location and click on the **Select Folder** button to use respected location as post directory.

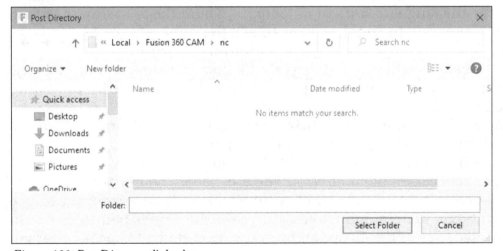

Figure-103. Post Directory dialog box

- Select the **Multi-Axis** option from the left area in the dialog box to specify parameters for multi-axis milling of the machine; refer to Figure-104.

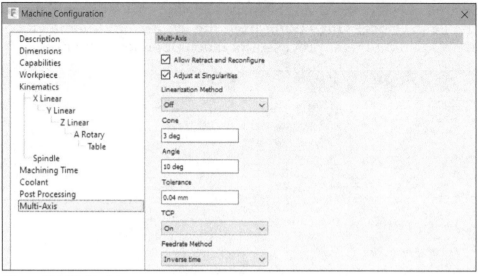

Figure-104. Multi-Axis page

- Select the **Allow Retract and Reconfigure** check box from the page to allow backward movement of cutting tool during multi-axis cutting operation.
- Select the **Adjust at Singularities** check box to adjust axis of cutting tool during simultaneous multi-axis operation for minimum deviation from main toolpath. If this check box is not selected then cutting tool will have to move a larger rotation for performing cuts which are perpendicular to current tool axis. In simple words, selecting this check box will allow tilting of cutting tool for performing multi-axis operations.
- Select the **Linear** option from the **Linearization Method** drop-down to convert simultaneous multi-axis movements to linear movements within specified tolerance value specified in **Linearization Tolerance** edit box. Select the **Rotary** option from the **Linearization Method** drop-down to convert rotary multi-axis movements to curves following the circumference of cylindrical/revolved features.
- Specify desired value in the **Cone** edit box to define angular range for the cutting tool axis near the singularity point of toolpath where the adjustments of singularities will be performed. In simple words, this is the range within which the cutting tool will tilt for performing simultaneous multi-axis operations.
- Specify desired value in the **Angle** edit box to define minimum angular delta movement that cutting tool should move before singularity adjustment will be applied in toolpath.
- Specify desired value in the **Tolerance** edit box to define maximum deviation allowed from path for singularity adjustments.
- Select desired option from the **TCP** drop-down to specify whether you machine supports tool center point control. The tool center point control system is used to automatically locate the center of cutting tool while performing cutting operations in a multiaxis system and makes sure that the center point lies on the toolpath as expected. Select the **On** option from the drop-down to specify that your custom machine supports tool center point control. Select the **Per axis** option from the drop-down if tool center point control is available in your machine separately for each axis.
- Select desired option from the **Feedrate Method** drop-down to define how feedrate will be applied for multiaxis toolpaths. Select the **Inverse time** option from the drop-down to specify cutting feedrate in terms of time required to perform the operation. Note that you need to specify adequate time for operation to avoid over speeding of cutting tool which can cause accident. Select the **IPM** option from the

drop-down to apply feedrate in Inch Per Minute value. Select the **DPM** option from the drop-down to apply feedrate in Degree Per Minute value. Select the **DPM-IPM** option to use combination of both Degree Per Minute and Inch Per Minute values for defining feedrate. Select the **TCP** option from the drop-down to use tool center point programming.

*   Click on the **OK** button from the dialog box to create machine with specified configuration.

Editing A User-Defined Machine

To edit a user-defined machine, select the machine from the list of machines and click on the **Edit selected** button from **Toolbar** in the dialog box. The **Machine Configuration** dialog box will be displayed. The options of this dialog box have been discussed earlier.

You can use the other buttons in **Toolbar** of the **Machine Library** dialog box as discussed earlier. After setting desired parameters, click on the **Close** button from the dialog box to exit.

PRACTICAL

Create the stock of the given model Figure-105 and create the required milling tool.

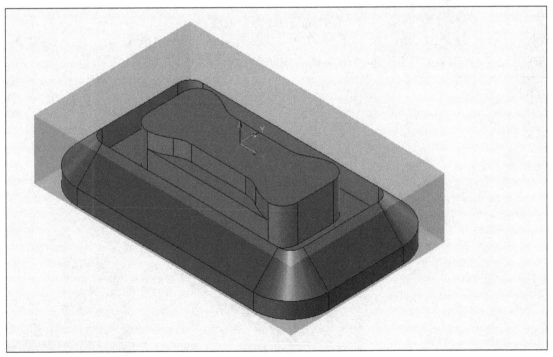

Figure-105. Practical 1

Opening Model in CAM

*   Create and save the part in **DESIGN** workspace for this practical. The part file is also available in Chapter 15 folder of **Autodesk Fusion 360** resource kit so you can open it from there instead of creating.
*   Click on the **MANUFACTURE** option from **Workspace** drop-down. The model will be displayed in the **MANUFACTURE** workspace; refer to Figure-106.

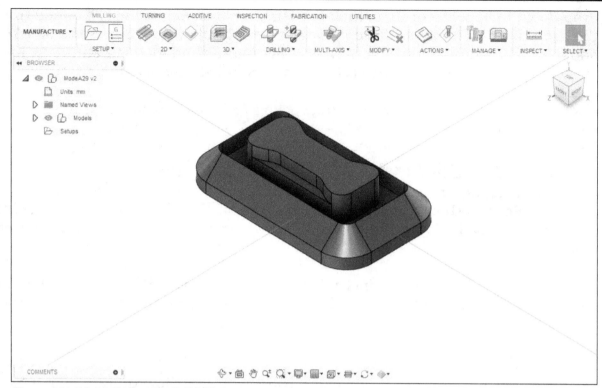

Figure-106. Model displayed in CAM Workspace

Creating Setup

- Click on the **New Setup** tool of **SETUP** drop-down from **Toolbar**. The **SETUP** dialog box will be displayed along with the setup; refer to Figure-107.

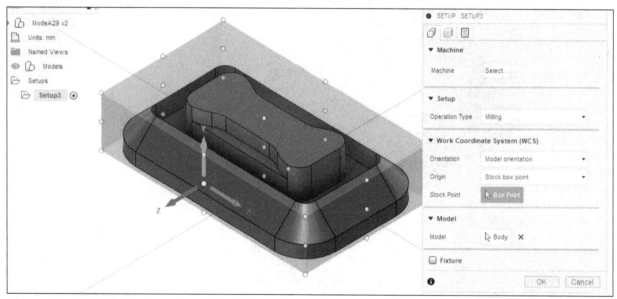

Figure-107. SETUP dialog box along with model

- Select the **Milling** option from the **Operation Type** drop-down in the **Setup** tab to specify the type of operation as milling.
- Select the **Select Z axis/plane & X axis** option from the **Orientation** drop-down of **Setup** tab and select the Z-axis for milling; refer to Figure-108. Note that you can select any desired face/edge to define Z axis for setup.

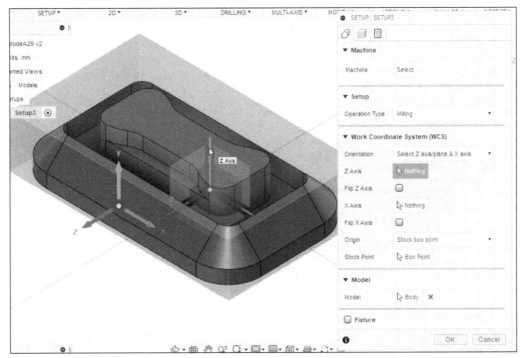

Figure-108. Selecting Z axis for milling

- The selection button of **X Axis** option is active by default. Select the X axis of WCS to select as X axis or select desired edge of model to define X axis direction.
- Click on the **Box Point** selection button for **Stock Point** option in **SETUP** tab and select the box point as displayed in the Figure-109.

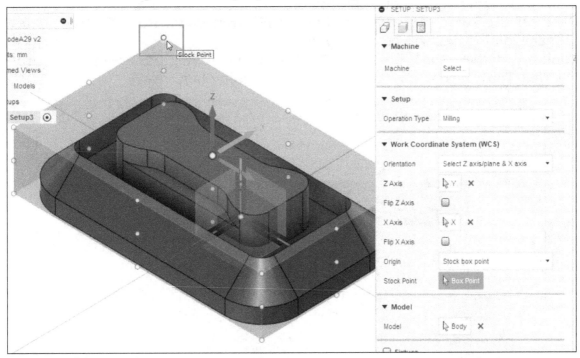

Figure-109. Selecting stock point

- Select the **Stock** tab at the top in the dialog box to define stock related parameters.
- Select the **Relative Size Box** option from the **Mode** drop-down in **Stock** tab and enter the parameters as shown in Figure-110.

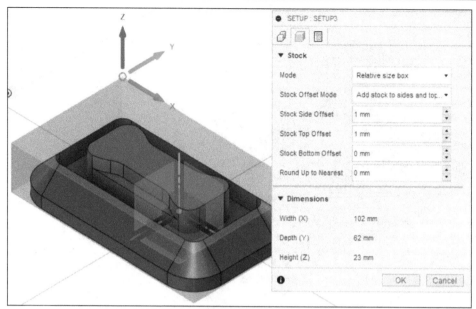

Figure-110. Specifying the parameters for creating stock

- After specifying the parameters in **SETUP** dialog box, click on the **OK** button from **SETUP** dialog box. The stock of model will be created.

Creating a new Milling tool

- Before creating a new tool, you need to know the inner and outer dimension of the model like height of pocket, radius, etc. Use the **Measure** tool in **INSPECT** drop-down of **Toolbar** to find out parameters.
- Click on the **Tool Library** tool of **MANAGE** drop-down from **Toolbar**. The **Tool Library** dialog box will be displayed.
- Click on the **New Tool** button of **Tool Library** dialog box; refer to Figure-111. The **New Tool** page will be displayed in the dialog box.

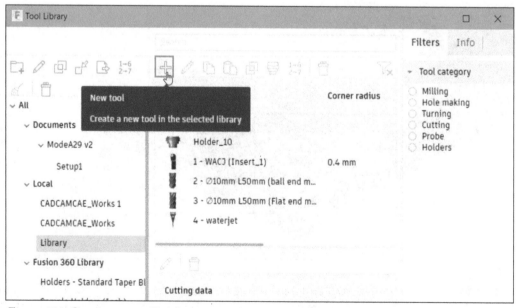

Figure-111. Creating new mill tool

- Select the **flat end mill** button from the page. The Tool Library dialog box will be displayed.
- Specify desired parameters in the **General** and **Cutter** tab of **Library** dialog box; refer to Figure-112.

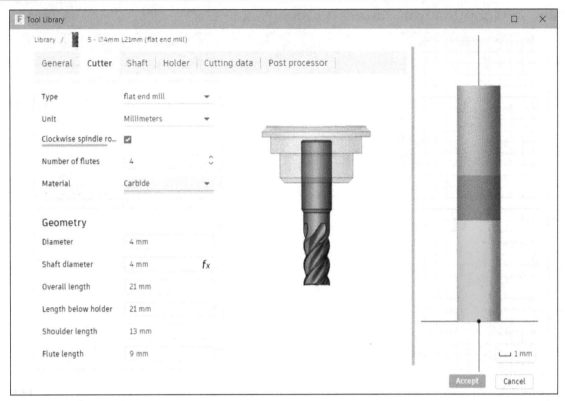

Figure-112. Cutter tab

- Select the **Shaft** tab of the **Library** dialog box and specify the parameters as shown in Figure-113.

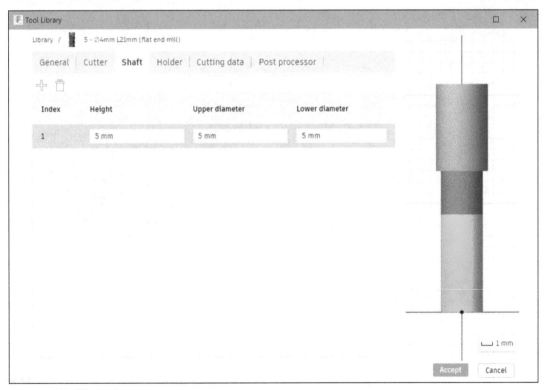

Figure-113. Shaft tab

- Select the **Holder** tab of **Library** dialog box to select desired holder.
- Select desired tool holder from the list box at the left in the dialog box and click on the **Select Holder** button from the **Holder** tab. The selected holder will be displayed as shown in Figure-114.

- Click on **OK** button from **Select Holder** dialog box.

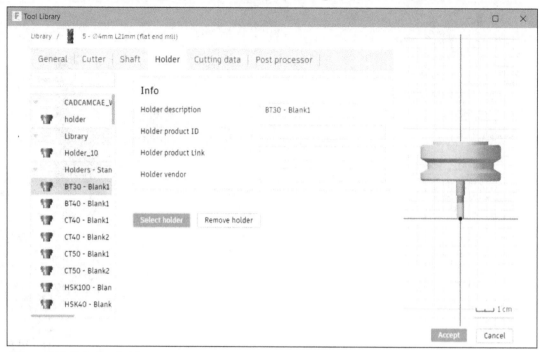

Figure-114. Selecting holder

- Click on the **Cutting data** tab of the **Library** dialog box and specify the parameters as shown in Figure-115.

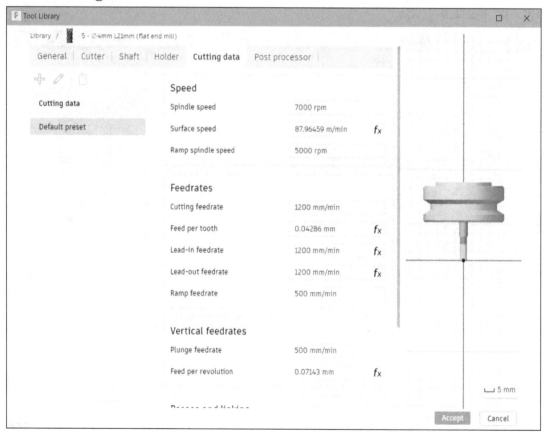

Figure-115. Cutting data tab

- After specifying the parameters, click on the **Accept** button from **Tool Library** dialog box. The tool will be created and displayed in the tool library. Click on the **Close** button to exit dialog box.

PRACTICAL 2

Create the stock for lathe machining as shown in Figure-116.

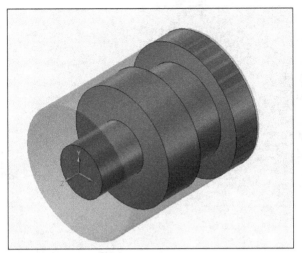

Figure-116. Practical 2

Adding model in CAM Workspace

- Create and save desired part in **DESIGN** workspace. The part file is also available in the current chapter folder of **Autodesk Fusion 360 Resources**.
- Click on the **MANUFACTURE** option from **Workspace** drop-down. The model will be displayed in the **MANUFACTURE** workspace; refer to Figure-117.

Figure-117. Model in CAM Workspace

Creating Setup

- Click on the **New Setup** tool of **SETUP** drop-down from **Toolbar**. The **SETUP** dialog box will be displayed along with the stock of model; refer to Figure-118.

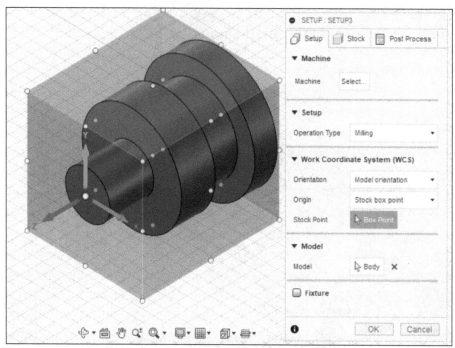

Figure-118. Model along with SETUP dialog box

- Select the **Turning or mill/turn** option from the **Operation Type** drop-down in the **Setup** tab to specify the machining type.
- The **Selection** button of **Z Axis(Rotary Axis)** option is active by default. Select the edge as displayed in Figure-119.

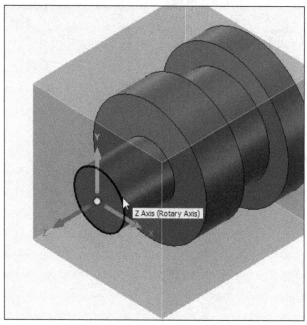

Figure-119. Selecting Z axis

- Click on the **Stock** tab of **SETUP** dialog box and specify the parameters as shown in Figure-120.

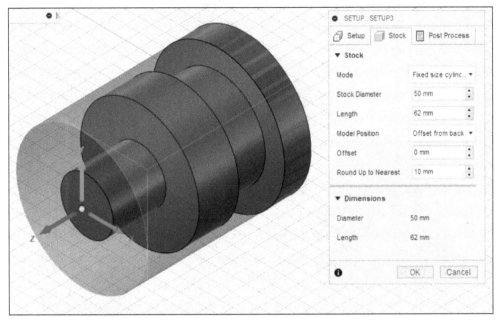

Figure-120. Specifying parameters in stock tab

- Click in the **Post Process** tab of **SETUP** dialog box and specify the parameters as shown in Figure-121.

Figure-121. Post process tab of turning

- After specifying the parameters, click on the **OK** button from **SETUP** dialog box. The stock will be created and displayed with the model; refer to Figure-122.

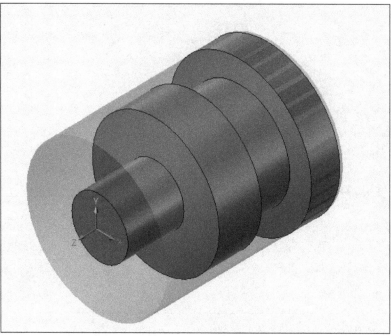

Figure-122. Practical 2

Creating new Turning tool

- Before creating a new tool, you need to know the dimensions of slot, radius of part, length of part, etc. Use the **Measure** tool in **INSPECT** drop-down to find out various parameters.
- Click on the **Tool Library** tool of **MANAGE** drop-down from **Toolbar**. The **Tool Library** dialog box will be displayed.
- Select the Library option from the left area in the dialog box and click on the **New Tool** button from the **Toolbar** in the **Tool Library** dialog box. The New Tool page will be displayed in the dialog box.
- Select the **turning general** button from the **Turning** section of the page. The **Tool Library** dialog box will be displayed for specifying parameters of new tool.
- Specify the **Description** of tool as Roughing tool in the **General** tab of dialog box.
- Specify the parameters in **Insert** tab from **Library** dialog box as shown in Figure-123.

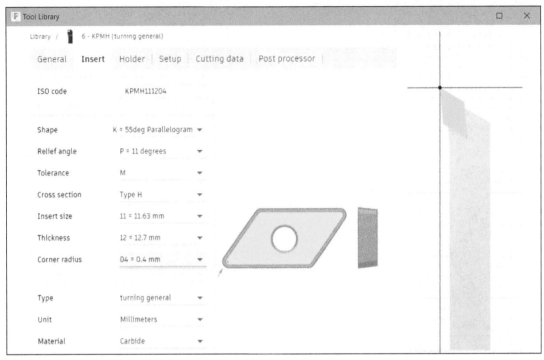

Figure-123. Insert tab

- Click on the **Holder** tab of **Library** dialog box and specify the parameters as shown in Figure-124.

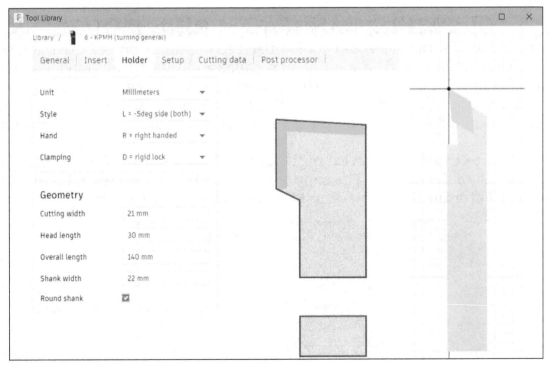

Figure-124. Holder tab

- Click on the **Cutting data** tab of **Library** dialog box and specify the parameters as shown in Figure-125.

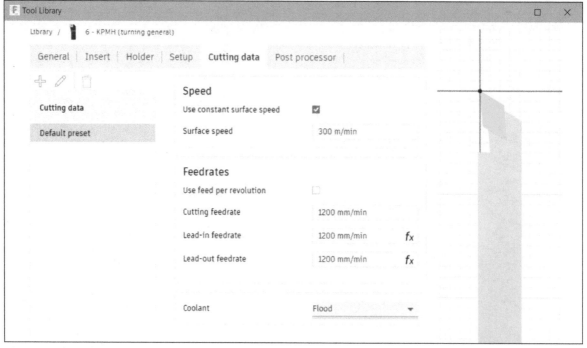

Figure-125. Feed & Speed tab

- After specifying the parameters, click on the **Accept** button from **Tool Library** dialog box. The tool will be created and displayed in the tool library. Click on the **Close** button to exit the dialog box.

CREATING MANUFACTURING MODEL

The **Create Manufacturing Model** tool is used to create a copy of the main model. You can use edit this copy of main model called manufacturing model as desired without affecting the main model. There is generally a need of removing features from the model to be 3D printed like holes, chamfers etc. Using this tool, ensures that your main model is not modified. The procedure to use this tool is given next.

- Click on the **Create Manufacturing Model** tool from the **SETUP** drop-down in the **MILLING** tab of the **Ribbon**. The copy of main model will be created and added in the **Manufacturing Models** node in **BROWSER**; refer to Figure-126.

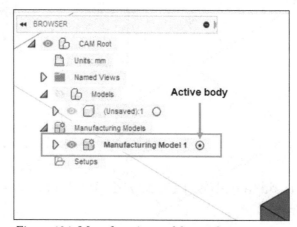

Figure-126. Manufacturing model created

SELF ASSESSMENT

Q1. Which of the following workspace is used to generate toolpaths for NC programs?

a. DESIGN   b. ANIMATION
c. SIMULATION   d. MANUFACTURE

Q2. The **New Setup** tool in **SETUP** panel can be used to prepare model for additive manufacturing. (T/F)

Q3. If the workpiece provided for machining your part has an irregular shape then which of the following stock mode should be used?

a. Relative size cylinder   b. Fixed size box
c. From Part   d. Fixed size tube

Q4. Generally in NC programming G54 through G59 codes are used to define work coordinate system offsets. (T/F)

Q5. Which of the following cutting tools can be used for pocket milling?

a. Spot Drill   b. Thread
c. Dovetail   d. Radius

Q6. Engrave mill tool is used to apply chamfer at the edges of part. (T/F)

PRACTICE

Create the stock of given part; refer to Figure-127.

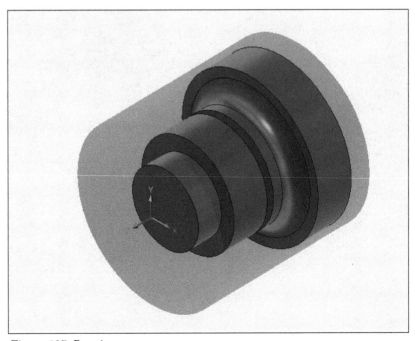

Figure-127. Practice

FOR STUDENT NOTES

Chapter 16

Generating Milling Toolpaths - 1

Topics Covered

The major topics covered in this chapter are:

- *2D Pocket Toolpath*
- *2D Contour Toolpath*
- *Trace Toolpath*
- *Bore Toolpath*
- *Engrave Toolpath*
- *Adaptive Clearing Toolpath*
- *Parallel Toolpath*
- *Contour Toolpath*
- *Horizontal Toolpath*
- *Scallop Toolpath*

GENERATING 2D TOOLPATHS

In previous chapter, you have learned the procedure of creating stock for a given model and selecting/creating cutting tools to perform machining operations. In this chapter, you will learn to create 2D Milling toolpaths using tools available in the **2D** drop-down of **MILLING** tab in the **Toolbar**.

2D Adaptive Clearing Toolpath

The **2D Adaptive Clearing** tool is used to create a machining operation on the part which uses the adaptive path as per the parts curvature to avoid abrupt direction changes. This tool path is generally used as roughing toolpath for removing large amount of stock from the workpiece. The procedure to use this tool is discussed next.

* Click on the **2D Adaptive Clearing** tool of **2D** drop-down from **Toolbar**; refer to Figure-1. The **2D ADAPTIVE** dialog box will be displayed; refer to Figure-2.

Figure-1. 2D Adaptive Clearing tool

Figure-2. 2D ADAPTIVE dialog box

* Click on the **Select** button for **Tool** section from the **2D ADAPTIVE** dialog box. The **Select Tool** dialog box will be displayed.
* Select the **Sample Tools - Metric** option from the **Fusion 360 Library** node at the left in the dialog box and select 10 mm flat Endmill cutting tool from the dialog box; refer to Figure-3. Click on the **Select** button from the dialog box to use this tool for machining.

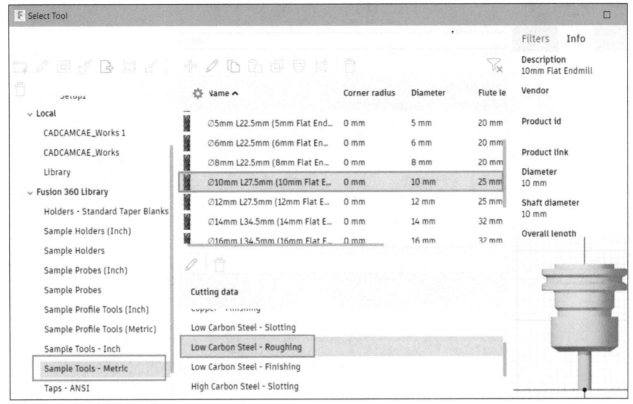

Figure-3. Tool selected with cutting strategy

- Select desired option from the **Coolant** drop-down to define how coolant will be sprayed on the workpiece while performing cutting operations.
- Select desired option from the **Preset** drop-down to specify cutting parameters related to respective material. For example, if you want to rough machine a low carbon steel workpiece then select the **Low Carbon Steel - Roughing** option from the **Preset** drop-down; refer to Figure-4.

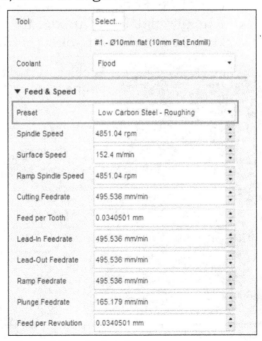

Figure-4. Preset selected

- You can specify desired parameters in the edit boxes of the **Feed & Speed** section if you do not want to use presets.

Geometry Tab

The options in the **Geometry** tab are used to define the geometry to be machined by cutting tool. The procedure is discussed next.

- Click on the **Geometry** tab from the **2D ADAPTIVE** dialog box. The options will be displayed as shown in Figure-5.

Figure-5. Geometry tab for 2D ADAPTIVE dialog box

- The **Nothing** selection button of **Pocket Selections** section in **Geometry** node is active by default. Click on the geometries of model to be machined as shown in Figure-6. Note that selected geometries are pockets with closed chain of curves.

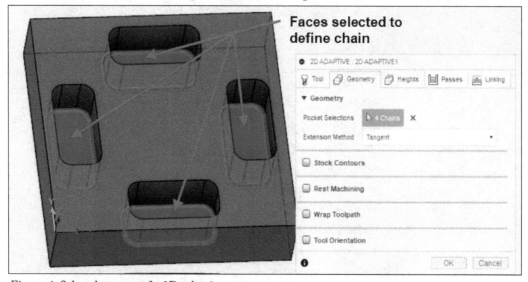

Figure-6. Selected geometry for 2D adaptive

- Select desired option from the **Extension Method** drop-down to define method to be used for extending profile of selected pocket geometries. Select the **Tangent** option from the drop-down to extend geometry tangentially if possible/needed.

Select the **Closest boundary** option if you want to extend selected geometries in direction closest to boundary of model/workpiece. Select the **Parallel** option from the drop-down to extend selected geometry chains in such a way that the extended length of geometries are parallel to each other till they intersect with boundary; refer to Figure-7. Note that extension of geometries is possible only for open pocket sections.

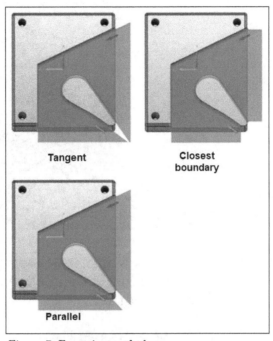

Figure-7. Extension methods

- Select the **Stock Contours** check box of **Geometry** tab from **2D ADAPTIVE** dialog box to specify the boundaries of stock. After selecting this check box, select the edges to be defined as boundaries of the stock, the cutting tool will not perform cutting moves outside this boundary although it can move in/out of the boundaries for non-cutting moves. This option is not useful if you are machining closed pockets. Use this option to define boundaries for an open pocket; refer to Figure-8.

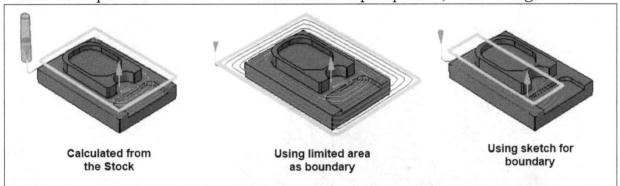

Figure-8. Stock contour selections

- Select the **Rest Machining** check box of the **Geometry** tab from the dialog box to machine only left over material from previous tool. Specify the tool diameter and corner radius of previous tool in the respective edit boxes of **Rest Machining** node for rest machining.
- Select the **Wrap Toolpath** check box from the **Geometry** tab of the **2D ADAPTIVE** dialog box to create a wrap toolpath around selected cylindrical object. After

selecting this check box, select the cylindrical face; refer to Figure-9. Specify desired value in the **Offset** edit box to increase/decrease the diameter of wrap cylinder for generating toolpath.

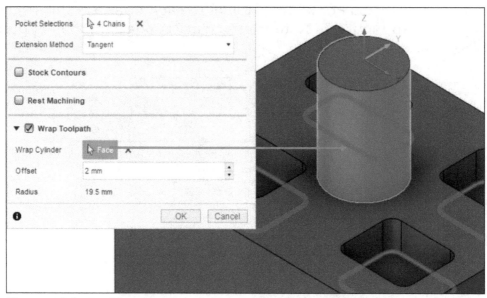

Figure-9. Cylindrical face selected

• Select the **Tool Orientation** check box from the **Geometry** tab in the **2D ADAPTIVE** dialog box to override the orientation of tool earlier defined in setup. The procedure to orient the tool has been discussed earlier.

Heights

The **Heights** tab of the **2D ADAPTIVE** dialog box is used to set height of different planes for standard tool movement. The procedure to define these heights are discussed next.

• Click on the **Heights** tab of the **2D ADAPTIVE** dialog box. The **Heights** tab will be displayed along with preview of planes in modeling area; refer to Figure-10.

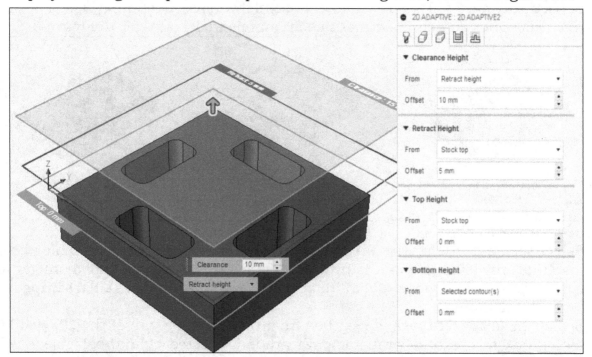

Figure-10. Heights tab

- The Clearance height is the distance between tool and workpiece at which the tool rapidly moves before the start of machining process. Click in the **From** drop-down of **Clearance Height** section in **Heights** tab and select desired reference from which clearance height will be measured.

- Click in the **Offset** edit box of **Clearance Height** section and specify the value of distance by which clearance will be away from the selected reference.

- The Retract height is the distance at which cutting tool will move upward after a cutting pass and before performing next cutting pass. The Retract height should be set above the Feed Height and Top of workpiece.

- Click in the **From** drop-down from the **Retract Height** section of the **Heights** tab and select desired reference type to be used for defining retract plane.

- Click in the **Offset** edit box of the **Retract Height** section and specify the distance value of retract height from selected reference.

- The Top height is the term used to define the top level of the cut. Click in the **From** drop-down from the **Top Height** section of the **Heights** tab and select desired option.

- Click in the **Offset** edit box from the **Top Height** section and specify height from selected reference.

- The Bottom Height is the term used to define the depth of the cut in a workpiece. If it is not defined in the dialog box then system will automatically assume the stock depth.

- Click in the **From** drop-down from the **Bottom Height** section in **Heights** tab and select desired option.

- Click in the **Offset** edit box from the **Bottom Height** section and specify the distance from selected reference for bottom height.

Passes

The options in **Passes** tab are used to define various parameters related to cutting and non-cutting passes of tool. If you are getting chunks of material left after machining or there is stock left in some area of the workpiece then options in this tab are responsible for the problem. These options are discussed next.

Passes Options

- Click in the **Passes** tab of **2D ADAPTIVE** dialog box. The options of the tab will be displayed as shown in Figure-11.

Figure-11. Passes tab

- Click in the **Tolerance** edit box of the **Passes** section from the **Passes** tab and enter desired value of deviation allowed between linear representation of curve and their original form while cutting. A very low value of tolerance can increase processing time or make the tool path impossible to cut by your tool if you are using a roughing tool.
- Click in the **Optimal Load** edit box of the **Passes** section from the **Passes** tab and enter desired value to specify the amount of engagement the adaptive strategies should maintain while cutting. This is the amount of material being removed in one cutting pass while tool is in same XY plane. The Optimal load parameter is directly linked with tool diameter and strength.
- Select the **Both Ways** check box to allow cutting during both forward and return motion of tool. Note that using this option can increase the load on tool and workpiece so you should select it cautiously and only for soft materials. On selecting this check box, the **Optimal Load Other Way** and **Other Way Feedrate** edit boxes will be displayed below the check box. Specify lower values of optimal load and feedrate in these edit boxes to reduce chances of accident.
- Click in the **Minimum Cutting Radius** edit box and specify desired value of minimum cutting radius that can be achieved while machining. If a corner on workpiece has fillet value lower than specified value then it will be ignored while machining.
- Select the **Use Slot Clearing** check box to start cutting from the middle of slot with plunge entry and continue a spiral motion for rest of the pocket; refer to Figure-12.

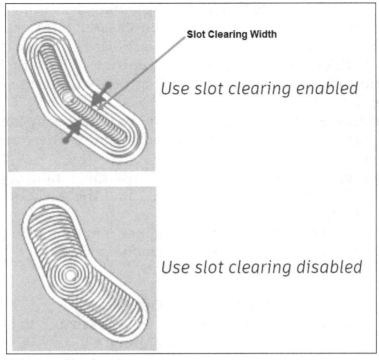

Figure-12. Slot clearing

- Click in the **Slot Clearing Width** edit box and specify desired value for width of slot. Once tool has performed machining in the slot, it will return to spiral motion of cutting near the walls of pockets.
- Select the **Conventional** option from the **Direction** drop-down of the **Passes** section to set the material removal process by conventional milling. In **Conventional** milling, the cutting process is performed in opposite direction of tool movement which causes the tool to scoop up the material; refer to Figure-13. The cutting process starts at zero thickness and increases up to maximum.
- Select the **Climb** option from the **Direction** drop-down in the **Passes** section to specify the material removal process by climb milling. In climb milling, the cutting process starts by machining maximum thickness and then decreases to zero.

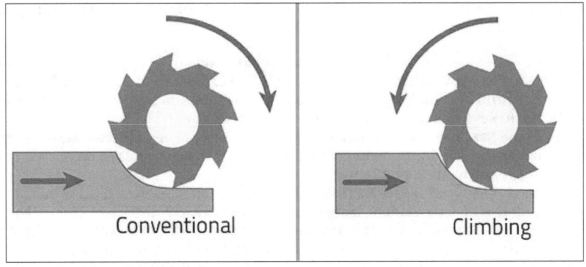

Figure-13. Conventional and climb machining

Multiple Depth Options

- Select the **Multiple Depths** check box from **Passes** tab to cut the material in multiple depths instead of cutting in one go. Note that every cutting tool has a length range by which tool can not perform cutting. When the **Multiple Depths** check box is not selected then the tool will cut only up to the depth allowed by its cutting length.
- Click in the **Maximum Roughing Stepdown** edit box and specify the maximum value of step-down distance between two roughing depths. Stepdown is the vertical distance between two cutting passes.
- Select the **Order By Depth** check box to sequence machining of pockets based on their depths. If this check box is selected then all the pockets will be machined at same level simultaneously.
- Select the **Order By Area** check box to sequence machining by pockets based on their areas. If this check box is selected then one pocket will be machined before moving to the next.

Stock To Leave Options

- Select the **Stock to leave** check box from the **Passes** tab to define the amount of stock to leave on the workpiece after roughing. This stock will be removed in finishing pass.
- Click in the **Radial Stock to Leave** edit box of the **Stock to Leave** section and specify the value of stock to remain around cylindrical faces of the part or side walls of the pockets after current operation.
- Click in the **Axial Stock to Leave** edit box of the **Stock to Leave** section and specify the value of stock to remain at the bottom flat surface/face of pocket after current operation.

Smoothing Options

- Select the **Smoothing** check box from the **Passes** tab to smoothen the toolpath by removing excessive points and fitting arcs within specified tolerance. Selecting this check box can reduce the size of G-codes as it will convert multiple connected lines into arcs wherever possible and similarly, multiple consecutive points will be converted into a line.
- Click in the **Smoothing Tolerance** edit box of the **Smoothing** section and specify desired value of tolerance.

Feed Optimization Options

- Select the **Feed Optimization** check box of the **Passes** tab to specify reduced feed at corners and curves. Note that the feed optimization options will not be available if the **Both Ways** check box is selected in the dialog box.
- Click in the **Maximum Directional Change** check box of **Feed Optimization** section and specify the maximum value of angle change allowed toolpath before feed rate is changed automatically to lower value.
- Click in the **Reduced Feed Radius** edit box and specify the value of minimum radius allowed for movement before the feed is reduced.
- Click in the **Reduced Feed Distance** edit box and specify the value for distance when approaching to a corner at which reduced feed rate will be applied.
- Click in the **Reduced Feedrate** edit box and specify the value of slower feed rate to be used when the tool is moving in X direction while cutting material.

- Select the **Only Inner Corners** check box from the **Feed Optimization** section to reduce the feed rate at inner corners.

Linking

The options of the **Linking** tab are used to define, how toolpath passes should be linked in different directions. The procedure to use these options is discussed next.

- Click on the **Linking** tab from the **2D ADAPTIVE** dialog box. The options of the **Linking** tab will be displayed; refer to Figure-14.

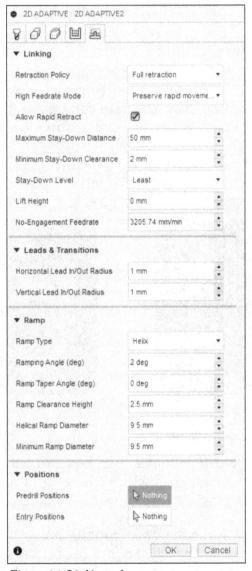

Figure-14. Linking tab

Linking Options

- Select the **Full retraction** option from the **Retraction Policy** drop-down to retract the tool up to retract height at the end of cutting pass before moving to the start of next machining pass.
- Select **Minimum retraction** option of **Retraction Policy** drop-down to move the tool up to the minimum retraction height from one cutting pass to other pass where the tool clears the workpiece.

- The **High Feedrate Mode** drop-down is used to specify the tool rapid movements which should be output as true rapid movement. Click on the **High Feedrate Mode** drop-down and select the desired option to define in which direction the rapid movement of tool will be retained during operation. Note that rapid movements are those in which tool does not perform any cutting operation and moves at full high feed rate to desired location.

- Select the **Allow Rapid Retract** check box from the **Linking** tab to perform the retracts as rapid movements. Clear the **Allow Rapid Retract** check box to force retract at lead-out feedrate.

- Click in the **Maximum Stay-Down Distance** edit box and specify maximum value of distance between different cutting passes up to which the tool stays down and does not retract. If there are more retractions in toolpath than expected then you should increase this value.

- Click in the **Minimum Stay-Down Clearance** edit box and specify minimum radial clearance distance value up to which cutting tool stay down.

- Click in the **Stay-Down Level** drop-down of the **Linking** section and select desired option to control how much the tool will try to stay down around obstacles during cutting process.

- Click in the **Lift Height** edit box and specify value of lift distance during reposition moves of the tool. You should specify this value to avoid cutting tool colliding with leftover stock on workpiece after cutting pass; refer to Figure-15.

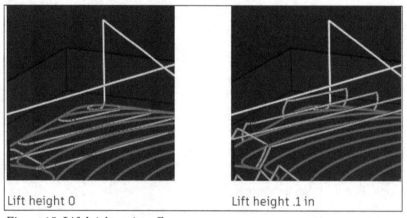

Lift height 0   Lift height .1 in

Figure-15. Lift height option effect

- Click in the **No-Engagement Feedrate** edit box of **Linking** section to specify the feedrate used for rapid movements where the tool is not in engagement with the material of workpiece during cutting pass.

Leads and Transitions

- The options of the **Leads & Transitions** section of the **Linking** tab is used to specify how leads and transitions should be generated at the time of tool entering or exiting the material.

- Click in the **Horizontal Lead In/Out Radius** edit box and specify the value of radius required for horizontal entry/exit.

- Click in the **Vertical Lead In/Out Radius** edit box and specify the value of radius required for vertical entry/exit.

Ramp Options

- The options in the **Ramps** section of **Linking** tab are used to define how tool will enter into the material during its first cutting pass.

- Hover the cursor on the options of the **Ramp Type** drop-down, the explanation of these options will be displayed on the screen. Select desired option from the drop-down and specify related parameters.

Position Options

- The options of the **Positions** section in the **Linking** tab are used to specify the tool entry point and predrill positions if required.
- The **Nothing** selection button of **Predrill Positions** section is active by default. Click on the point in workpiece where drill has been done earlier so that the cutting tool can enter the material.
- Click on the **Nothing** selection button of **Entry Positions** section and click on the geometry near the location of workpiece where you want to enter the tool.
- After specifying the parameter for **2D ADAPTIVE** dialog box, click on the **OK** button. The toolpath will be created and displayed on the workpiece; refer to Figure-16.

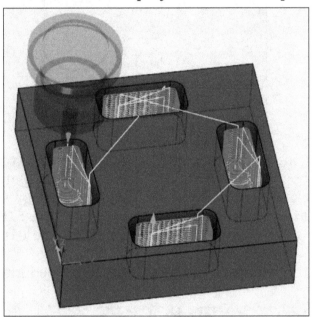

Figure-16. Toolpath created by adaptive clearing

- The option of the created toolpath will be added in **BROWSER**. Right-click on the toolpath. A shortcut menu will be displayed; refer to Figure-17.

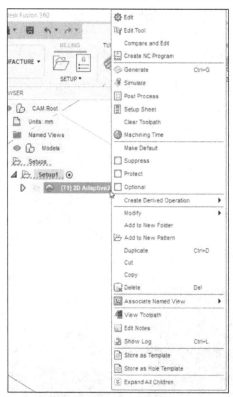

Figure-17. Shortcut menu

- If you want to edit the parameters then click on the **Edit** button from shortcut menu. The dialog box will be displayed for selected toolpath.
- If you want to check the animation of material cutting by selected toolpath then click on the **Simulate** button from shortcut menu. The **SIMULATE** dialog box will be displayed along with simulation keys. Click on the **Play** or **Pause** button as required.
- The other tools of this shortcut menu will be discussed later.

2D POCKET

The **2D Pocket** tool is used to remove material from pockets in the model. The procedure to use this tool is discussed next.

- Click on the **2D Pocket** tool from the **2D** drop-down in **MILLING** tab of **Toolbar**; refer to Figure-18. The **2D POCKET** dialog box will be displayed; refer to Figure-19.
- Click on the **Select** button of **Tool** section. The **Select Tool** dialog box will be displayed.
- Click on the required tool from the dialog box and click on the **OK** button from **Select Tool** dialog box. The tool will be selected and displayed in **Tool** section of **2D POCKET** dialog box.
- Click on the **Geometry** tab of **2D POCKET** dialog box. The **Geometry** tab will be displayed.
- The **Nothing** button of **Pocket Selections** section from **Geometry** tab is active by default. Click on the surface/edge of the model to select; refer to Figure-20.

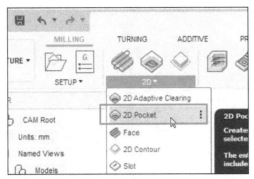

Figure-18. 2D Pocket tool

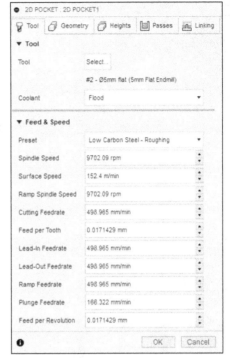

Figure-19. 2D POCKET dialog box

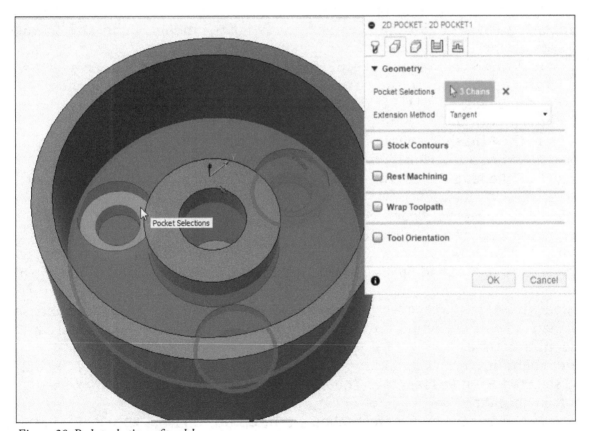

Figure-20. Pocket selections of model

- Select desired option from the **Sideways Compensation** drop-down of the **Passes** section in **Passes** tab to define the motion of tool with respect to walls.
- Select the **Finishing Passes** check box if you want to perform finishing pass. The options related to finishing passes will be displayed in the dialog box; refer to Figure-21.

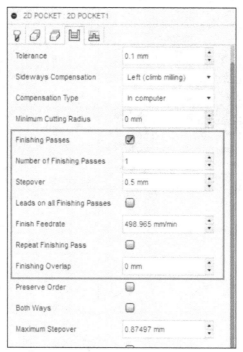

Figure-21. Finishing pass options

- Specify the total number of finishing passes to be performed in the **Number of Finishing Passes** edit box to get better finish of the model. Generally, 2 finishing passes are performed to get better finish.
- Specify desired value in the **Stepover** edit box to define the distance by which tool will move left or right in the cutting plane to perform next cutting pass.
- Select the **Leads on all Finishing Passes** check box to apply full length of lead in and lead out to finishing passes.
- Click in the **Finish Feedrate** edit box to specify movement speed by which cutting tool will move for performing finishing operation.
- Select the **Repeat Finishing Pass** check box to repeat the finish operation once more when 0 stock left.
- Select the **Preserve Order** check box if you want to machine the part in same sequence as they were selected.
- Select the **Both Ways** check box to machine both forward and backward.
- Select the **Use Morphed Spiral Machining** check box from the dialog box to perform smoother machining.
- Select the **Allow Stepover Cusps** check box if you want to allow cusps being formed at corners while cutting material. Selecting this option decreases the machining time.
- For machining multiple depth pockets, the **Multiple Depths** check box should be selected from **Passes** tab. The other parameters have been discussed earlier in this chapter.
- After specifying the parameters, click on the **OK** button from **2D POCKET** dialog box. The toolpath will be generated and displayed; refer to Figure-22.

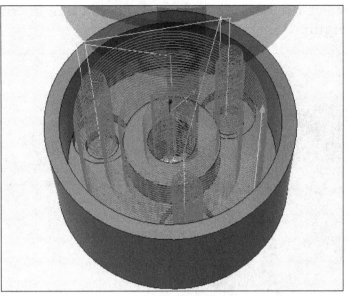

Figure-22. Created toolpath for 2D Pocket

Face

The **Face** tool is used to remove material from the top face of the workpiece. In any machining sequence, this is generally the first toolpath to be generated for flat head workpiece. The procedure to use this tool is discussed next.

*   Click on the **Face** tool of **2D** drop-down from **Toolbar**; refer to Figure-23. The **FACE** dialog box will be displayed; refer to Figure-24.

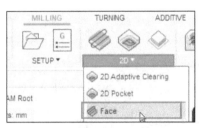

Figure-23. Face tool

Figure-24. FACE dialog box

*   Click on the **Select** button from **Tool** tab and select desired face mill tool for facing operation according to your model from **Select Tool** dialog box. Generally, we use flat end mill or face mill cutting tools for this operation.
*   Click on the **Geometry** tab of **FACE** dialog box. The **Geometry** tab will be displayed and **Nothing** selection button of **Stock Selections** section will be active by default

asking you to select the boundary of workpiece. Click on the outer edge of the model; refer to Figure-25.

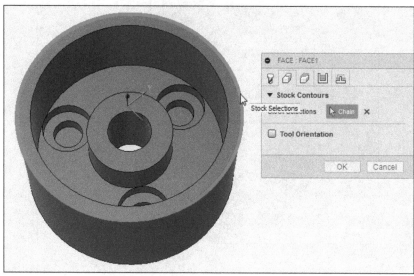

Figure-25. Selection for FACE dialog box

- Select the **Use Chip Thinning** check box from **Passes** tab of dialog box to make tool roll while cutting so that chips formed are thin.
- The other options of the dialog box have been discussed earlier. After specifying the parameters, click on the **OK** button from the **FACE** dialog box. The toolpath will be generated and displayed on workpiece; refer to Figure-26.

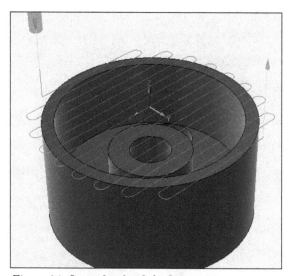

Figure-26. Created toolpath for facing

- Right-click on the recently created **Face** toolpath and click on the **Simulate** button to check the animation of facing operation.

2D Contour

The **2D Contour** tool is used to remove material by following the contour of the model. This tool is generally used to remove the material from outer/inner walls of the selected workpiece. You can use this toolpath for path roughing and finishing. The procedure to use this tool is discussed next.

- Click on the **2D Contour** tool from the **2D** drop-down of the **Toolbar**; refer to Figure-27. The **2D CONTOUR** dialog box will be displayed; refer to Figure-28.

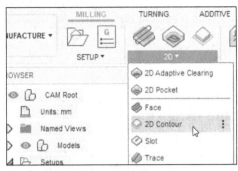

Figure-27. 2D Contour tool

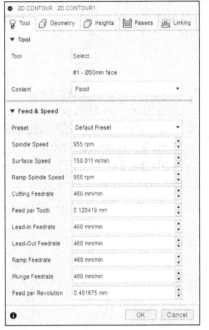

Figure-28. 2D CONTOUR dialog box

- Click on the **Select** button from the **Tool** tab of the **2D CONTOUR** dialog box and select the required tool according to your model from **Select Tool** dialog box.
- Click on the **Geometry** tab of **2D CONTOUR** dialog box. The **Geometry** tab will be displayed and the **Nothing** button of **Contour Selection** section will be active by default. Click on the boundary edges of the model to select; refer to Figure-29.
- Click in the **Tangential Extension Distance** edit box of **Geometry** tab and enter the value of distance to extend the open contours tangentially.
- Select the **Separate Tangentially End Extension** check box to enable tangential extension at the end of cutting pass.

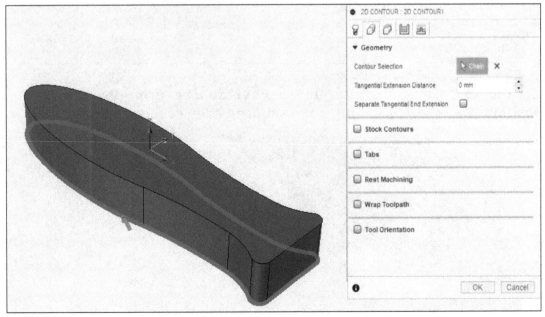

Figure-29. Selection of edge for contour

Tabs

The **Tabs** check box is used to define parameters for holding the workpiece when you are cutting a part from sheet. This option is generally used when you are working on a sheet and after operation the part will be detached from this sheet. When you use this option, cutting tool will leave small amount of stock in the form of tabs to hold the workpiece after machining. The procedure is discussed next.

*   Click on the **Tabs** check box to enable options related to tab.
*   Select desired shape of tab from **Tab Shape** drop-down. You can select rectangular or triangular shapes for creating tab.
*   Click in the **Tab Width** edit box and specify the value of width of tab.
*   Click in the **Tab Height** edit box and specify the value of height of tab.
*   Select the **By distance** option from the **Tab Positioning** drop-down of **Tabs** section to position the tab around the workpiece by specified distance; refer to Figure-30.

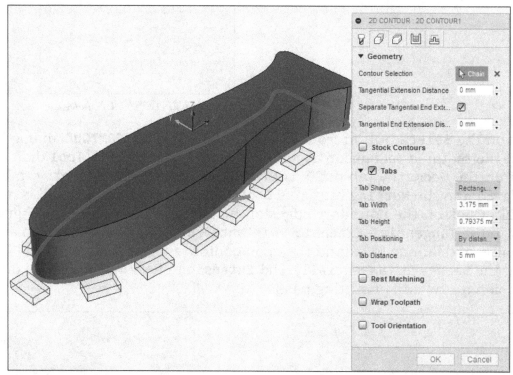

Figure-30. Tab positioning by distance

*   Select the **At points** option from **Tab Positioning** drop-down in **Tabs** section to position the tab by selecting points on the edge of model; refer to Figure-31.

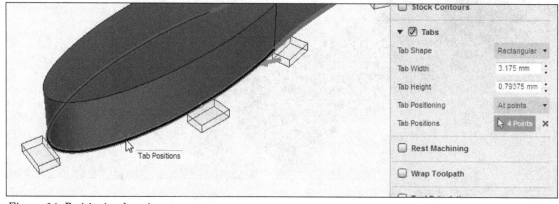

Figure-31. Positioning by points

- The other options of this dialog box are same as discussed earlier.
- After specifying the parameters, click on the **OK** button from the **2D CONTOUR** dialog box. The toolpath will be generated and displayed on the model; refer to Figure-32.

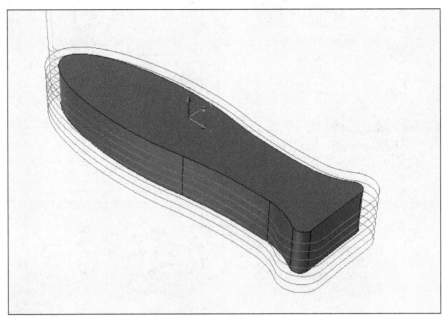

Figure-32. Created toolpath for solid

Slot

The **Slot** tool is used to remove material from model in slot pattern. Note that this slot milling is governed in 2D plane so the depth of cut need to be specified explicitly. Note that the Slot toolpath follows centerline/curve of the slot feature for cutting passes. The slot features must be created using the Slot creation tools in Design workspace of Autodesk Fusion. The procedure to use this tool is discussed next.

- Click on the **Slot** tool from the **2D** drop-down of **Toolbar**; refer to Figure-33. The **SLOT** dialog box will be displayed; refer to Figure-34.
- Click on the **Select** button from **Tool** tab of the **SLOT** dialog box. The **Select Tool** dialog box will be displayed. Select the tool as required.
- Click on the **Geometry** tab of **2D CONTOUR** dialog box. The **Geometry** tab will be displayed and the **Nothing** selection button for **Pocket Selections** option will be active by default.
- Select the curve chains for the slot; refer to Figure-35.

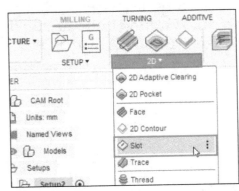

Figure-33. Slot Tool

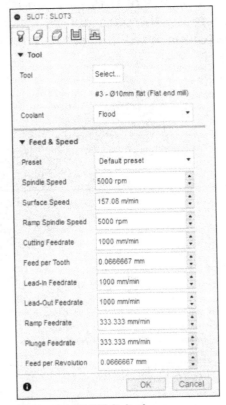

Figure-34. SLOT dialog box

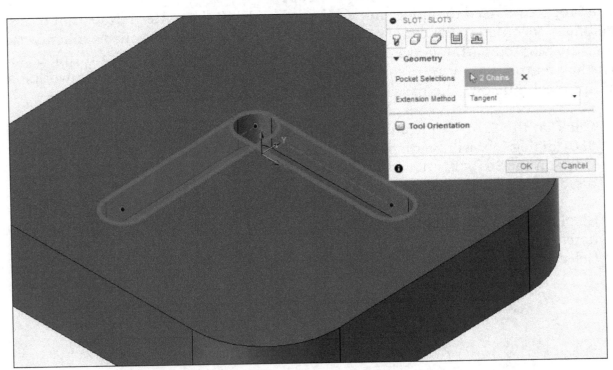

Figure-35. Selection for slot

- Select the desired option from the **Ramp Type** drop-down in Linking tab of the dialog box. In our case, we have applied **Plunge** option of **Ramp Type** drop-down from **Linking** tab for better generation of toolpath as tool entry is not available.
- Specify the desired parameters for ramp if asked based on selected option. The other options of this dialog box have been discussed earlier.

- After specifying the parameters, click on the **OK** button from the **SLOT** dialog box. The toolpath will be generated and displayed on the model; refer to Figure-36.

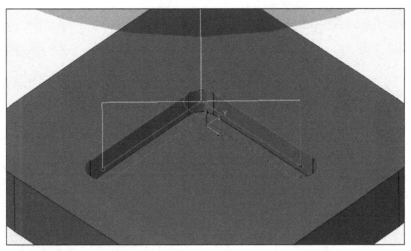

Figure-36. Generated toolpath for slot

Note that the main difference between pocket toolpaths and slot toolpaths in Autodesk Fusion 360 is that Pocket toolpaths can machine any regular and irregular slots whereas the Slot toolpaths can machine only regular slots created in Autodesk Fusion. A pocket toolpath is fine when you are using end mill or other similar cutting tools but if you have a slot cutter with size equal to width of slot.

Trace

The **Trace** tool is used to create toolpath which follows selected curves. If you want to machine a sketch curve up to specified depth then this toolpath is the right selection. Note that there is not need of cut feature to exist in the model when using this toolpath, as depth of cut is explicitly specified. The procedure to use this tool is discussed next.

- Click on the **Trace** tool of **2D** drop-down from **Toolbar**; refer to Figure-37. The **TRACE** dialog box will be displayed; refer to Figure-38.
- Click on the **Select** button of **Tool** tab from **TRACE** dialog box. The **Select Tool** dialog box will be displayed. Select the tool as required.
- Click on the **Geometry** tab of **TRACE** dialog box. The **Geometry** tab will be displayed and you will be asked to select curve chains to be machined.

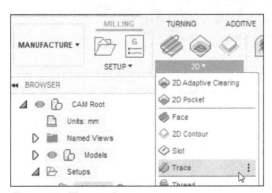

Figure-37. Trace tool

Figure-38. TRACE dialog box

- The **Nothing** button of **Curve Selections** section of **Geometry** tab is active by default. Select the geometry for trace; refer to Figure-39. You can also select sketch curves for this toolpath.

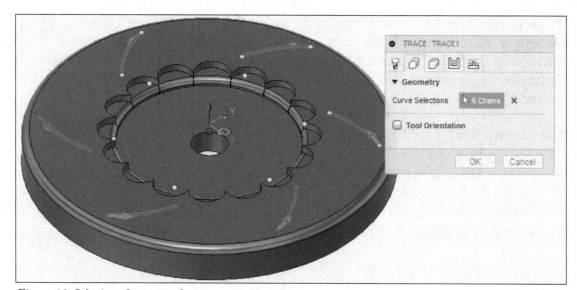

Figure-39. Selection of geometry for trace

- Click on the **Passes** tab of **TRACE** dialog box. The **Passes** tab will be displayed; refer to Figure-40.

Figure-40. Passes tab of TRACE dialog box

- Click in the **Pass Extension** edit box and enter the value of distance by which the toolpath along length will be extended.
- Select the **Repeat Passes** check box from the **Passes** tab to perform an additional finishing pass with zero stock for better finish.
- Select the **Preserve Order** check box from the **Passes** tab to specify that the geometry machined in the order in which they are selected.
- Select the **Both Ways** check box from the **Passes** tab to use the climb and conventional machining for open profile of the selected geometry.
- Click in the **Maximum Angle (deg)** edit box and specify maximum plunge angle for machining. Plunge angle is the angle at which cutting tool will enter workpiece while cutting.
- Click in the **Axial Offset** edit box and specify the value of axial distance by which the toolpath will be moved up/down along the spindle axis.
- Select the **Axial Offset Passes** check box from the **Passes** tab to enable multiple cutting passes with specified depth distance between two passes along Z axis.
- Click in the **Up/Down Milling** drop-down and select the desired option from the drop-down. Select the **Both** option to perform cutting in both upward and downward directions. Select the **Up milling** option to perform cutting operation in upward direction only. Select the **Down milling** option to perform cutting operation in downward direction only. If the **Up milling** or **Down milling** option is selected then specify desired angle value in the **Up/Down Shallow Angle** edit box at which cutting tool will move upward or downward.
- Select the **Chamfer** check box of **Passes** tab from **TRACE** dialog box to perform a chamfer operation on selected geometry. This option is available if you have selected chamfer mill as cutting tool.

- Click in the **Chamfer Width** edit box of the **Chamfer** section from the **Passes** tab and specify the value of width for machining of chamfer.
- Click in the **Chamfer Depth** box of **Chamfer** section from **Passes** tab and specify the value of depth of chamfer for machining.
- The other options of this dialog box are same as discussed earlier. After specifying the parameters, click on the **OK** button from **TRACE** dialog box. The toolpath will be generated and displayed on the model; refer to Figure-41.

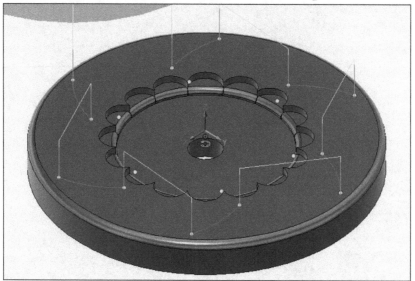

Figure-41. Created toolpath for trace

Thread

The **Thread** tool is used to create internal and external threads on the selected geometry. The procedure to use this tool is discussed next.

- Click on the **Thread** tool of **2D** drop-down from **Toolbar**; refer to Figure-42. The **THREAD** dialog box will be displayed; refer to Figure-43.
- Click on the **Select** button of **Tool** tab from **THREAD** dialog box. The **Select Tool** dialog box will be displayed. Select the tool as required. Note that you can use threading tool or tap for performing threading operations.
- Click on the **Geometry** tab of **THREAD** dialog box. The **Nothing** button of **Circular Face Selections** option in **Geometry** tab will be active by default and you will be asked to select circular faces.
- Select the circular faces on which you want to machine threads; refer to Figure-44
- If you want to select all the faces which have same diameter for creating threads then select the **Select Same Diameter** check box and then select the round face of model. To further filter your selection, select the **Only Same Hole Depth** check to select only those holes which have same depth as the selected one. Select the **Only Same Z Top Height** check box to select only those holes/boss features which have same height.
- Click on the selection button for **Containment Boundary** option and select desired boundary chain within which the holes/boss features should be selected automatically.
- Set the other options as discussed earlier in the **Geometry** tab of dialog box.

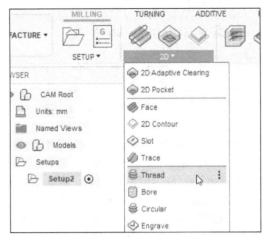

Figure-42. *Thread tool*

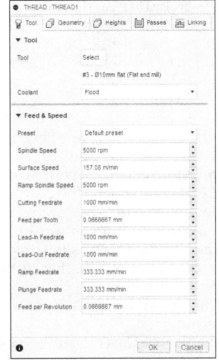

Figure-43. *THREAD dialog box*

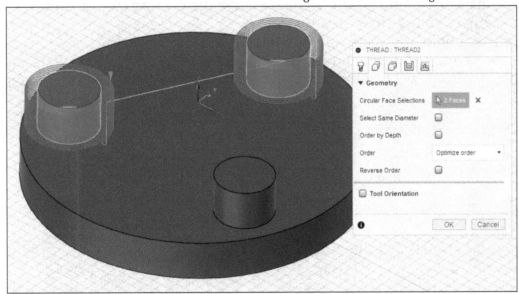

Figure-44. *Selecting Circular faces*

- Click on the **Passes** tab from the **THREAD** dialog box. The **Passes** tab will be displayed as shown in Figure-45.
- Specify desired value of thread entry angle in the **Start Angle** edit box.
- Select the **Right handed** option of **Threading Hand** drop-down from **Passes** tab to create a right handed thread direction. Select the **Left handed** option of **Threading Hand** drop-down from **Passes** tab to create a left handed thread direction. Thread direction defines how screws (clockwise or counter clockwise) are fastened in holes to which threads have been applied.
- Click in the **Thread Pitch** edit box of the **Passes** tab to specify value of distance between two thread cuts.
- Click in the **Pitch Diameter Offset** edit box of **Passes** tab to enter the value of difference between major and minor thread diameter. This value defines depth of thread.

- Select the **Do Multiple Threads** check box from the **Passes** tab to create multiple start threads and specify number of threads value in **Number of Threads** edit box. This will create a multi-start threading.

- Select desired option from **Compensation Type** drop-down to define the compensation type. Select the **In computer** option from the drop-down if you want to create the path which already accommodates compensation based on tool and toolpath. Select the **In control** option to insert G41/42 codes in NC program so that the operator can specify compensation while working on machine. Select the **Wear** option if you want to use benefit of both **In computer** and **In control** which means the compensation is provided in the toolpath itself and G41/42 codes are also provided in NC program for manual changes. Note that the wear compensation needs to be entered as negative if **Wear** option is selected. Select the **Inverse wear** option if you want to specify wear compensation as positive value.

- Select the **Multiple Passes** check box of the **Passes** tab to specify value for multiple depth cuts while thread milling. This option is used when you want to create deeper threads.

- Select the **Repeat Passes** option if you want to repeat cutting passes once again for better finish.

- Select the **Use Helical Leads** check box to create lead in and lead out paths as helical curves.

- Select the **Lead To Center** check box to move the cutting tool at center of selected chain for performing threading operation.

- The other options of this dialog box have been discussed earlier.

- After specifying the parameters, click on the **OK** button from the **THREAD** dialog box. The toolpath will be generated and displayed on the model; refer to Figure-46.

Figure-45. Passes tab of THREAD dialog box

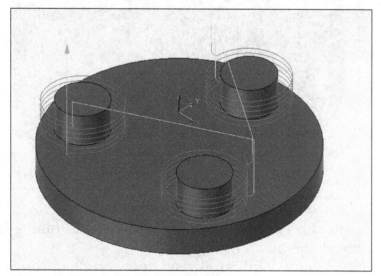

Figure-46. Generated Toolpath for thread

Bore

The **Bore** tool is used to increase the diameter of hole in tight tolerance. You can also finish cylindrical islands in the model by using this tool. The procedure to use this tool is discussed next.

- Click on the **Bore** tool from the **2D** drop-down in the **Toolbar**; refer to Figure-47. The **BORE** dialog box will be displayed; refer to Figure-48.

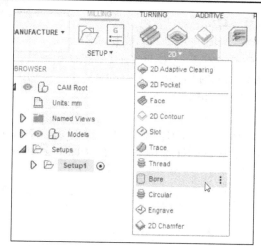

Figure-47. Bore tool

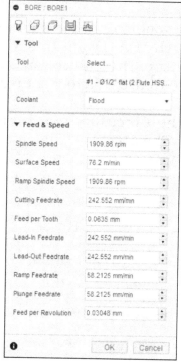

Figure-48. BORE dialog box

- Click on the **Select** button from the **Tool** tab in the **BORE** dialog box. The **Select Tool** dialog box will be displayed. Select the tool as required.
- Click on the **Geometry** tab in **BORE** dialog box. The **Geometry** tab will be displayed and the **Nothing** button for **Circular Face Selections** option in **Geometry** tab will be active by default. You need to click on the circular faces of holes which were drilled earlier and you want to perform finish operation on those holes; refer to Figure-49.

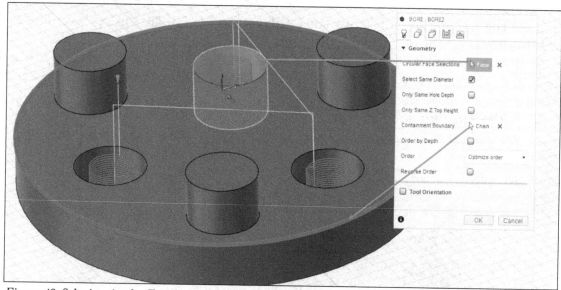

Figure-49. Selecting circular Face for bore

- The other options of the dialog box have been discussed earlier.
- After specifying the parameter, click on the **OK** button from the **BORE** dialog box. The toolpath will be generated and displayed on the model.

Circular

The **Circular** tool is used for milling circular or round pocket/boss features. This tool is useful when you want to derived height and depth of toolpath from selected cylindrical feature. The feature can be a circular pocket as well as a cylindrical island. The procedure to use this tool is discussed next.

- Click on the **Circular** tool from the **2D** drop-down in the **Toolbar**; refer to Figure-50. The **CIRCULAR** dialog box will be displayed; refer to Figure-51.

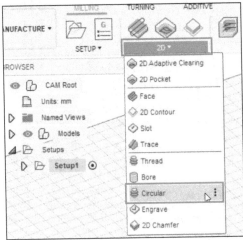

Figure-50. Circular tool

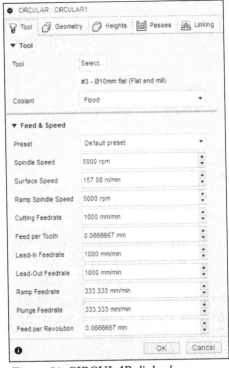

Figure-51. CIRCULAR dialog box

- Click on the **Select** button from **Tool** tab in the **CIRCULAR** dialog box. The **Select Tool** dialog box will be displayed. Select the cutting tool as required.
- Click on the **Geometry** tab in the **CIRCULAR** dialog box. The **Geometry** tab will be displayed. The **Nothing** button for **Circular Face Selections** option in **Geometry** tab will be active by default.
- Click on the circular face in the model to select it; refer to Figure-52.

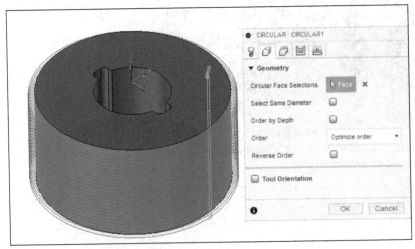

Figure-52. Selecting face for Circular tool

- The other options of the dialog box have been discussed earlier. After specifying the parameters, click on the **OK** button from the **CIRCULAR** dialog box. The toolpath will be generated and displayed on the model.

Engrave

The **Engrave** tool is used to perform artistic machining on the workpiece like creating logos. This tool is also used to print text on the press dies. The procedure to use this tool is discussed next.

- Click on the **Engrave** tool from the **2D** drop-down in the **Toolbar**; refer to Figure-53. The **ENGRAVE** dialog box will be displayed; refer to Figure-54.

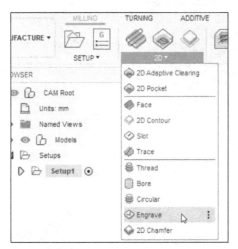

Figure-53. Engrave tool

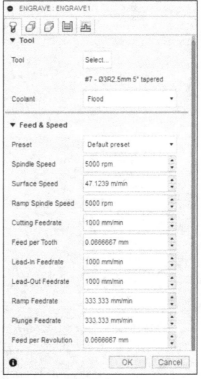

Figure-54. ENGRAVE dialog box

- Click on the **Select** button from **Tool** tab in **ENGRAVE** dialog box. The **Select Tool** dialog box will be displayed. Select the tool as required preferably a tapered cutting tool.
- Click on the **Geometry** tab of the **ENGRAVE** dialog box. The options of **Geometry** tab will be displayed and the **Nothing** button for **Contour Selection** option in **Geometry** tab will be active by default. You need to select the curves to be engraved; refer to Figure-55.

Note that sometimes people confuse in engrave and emboss features. Engraving is achieved when pockets of desired shape are created in the workpiece whereas embossing is achieved when islands of desired shapes are created in the workpiece.

- Specify desired value in the **Sharp Corner Angle** edit box to define the maximum value of angle between edges at corner up to which corners will be cleared during machining. Note that in general, cutting tool creates sharp corners in engraving toolpaths.

- The other options of the dialog box are same as discussed earlier. After specifying the parameters, click on the **OK** button from **ENGRAVE** dialog box to complete the process. The toolpath will be generated and displayed on the model; refer to Figure-56.

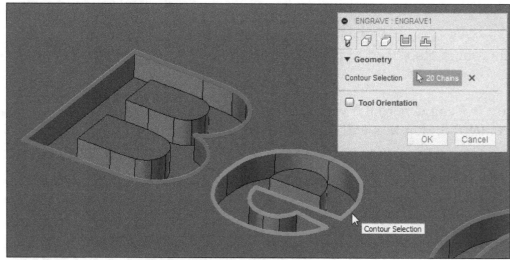

Figure-55. Contour selection for engraving

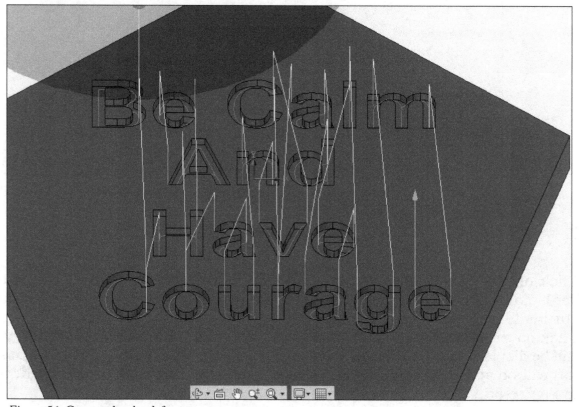

Figure-56. Generated toolpath for engrave

2D Chamfer

The **2D Chamfer** tool is used to create chamfer profile on the edge of model. The procedure to use this tool is discussed next.

- Click on the **2D Chamfer** tool from **2D** drop-down in **Toolbar**; refer to Figure-57. The **2D CHAMFER** dialog box will be displayed; refer to Figure-58.
- Click on the **Select** button of **Tool** tab from **2D CHAMFER** dialog box. The **Select Tool** dialog box will be displayed. Select the tool as required.

- Click on the **Geometry** tab of **2D CHAMFER** dialog box. The **Geometry** tab will be displayed.
- The **Nothing** button of **Contour Selection** section from **Geometry** tab is active by default. You need to select the edge for creating chamfer; refer to Figure-59.

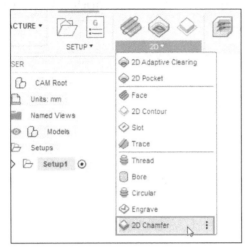

Figure-57. 2D Chamfer tool

Figure-58. CHAMFER dialog box

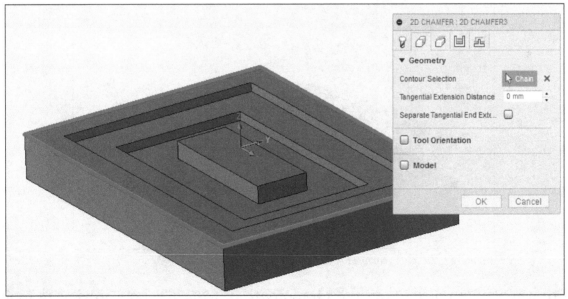

Figure-59. Selection of edge for chamfer

- Click in the **Passes** tab from **2D CHAMFER** dialog box. The options of **Passes** tab will be displayed; refer to Figure-60.
- Click in the **Chamfer Width** edit box of **Chamfer** section from **Passes** tab and specify the value of width of chamfer.
- Click in the **Chamfer Tip Offset** edit box of **Chamfer** section from **Passes** tab and specify the value of tip offset so that cut is not made from the tip but from the approximate middle of cutting edge of tool.

- Click in the **Chamfer Clearance** edit box of **Chamfer** section and enter the value of clearance from near by walls to avoid collision.
- The other options of the dialog box are same as discussed earlier.
- After specifying the parameters, click on the **OK** button from the **2D CHAMFER** dialog box. The toolpath will be generated and displayed on the model; refer to Figure-61.

Figure-60. Passes tab 2D CHAMFER dialog box

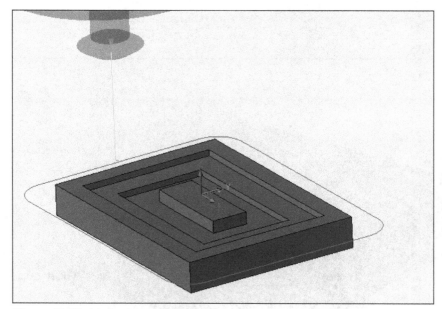

Figure-61. Toolpath generated for chamfer

Note that when you hover cursor on a toolpath in **BROWSER** then preview of toolpath is displayed in modeling area; refer to Figure-62 but when you select a toolpath from **BROWSER** then preview of material removed from stock is displayed; refer to Figure-63.

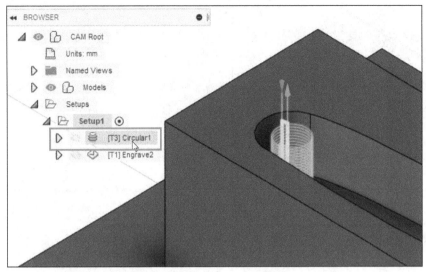

Figure-62. Hovering cursor on toolpath

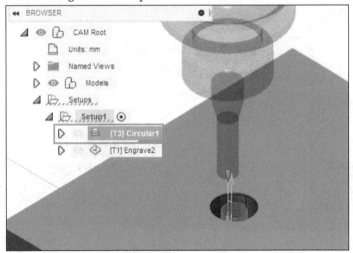

Figure-63. On selecting a toolpath

GENERATING 3D TOOLPATH

Till now, we have discussed the procedure of generating 2D Toolpaths. In this section, we will discuss the tools used for generating 3D Toolpaths. 3D toolpaths enable simultaneous movement of tool in horizontal and vertical direction whereas in 2D toolpaths, the cutting tool performs all the cutting operations in current XY plane and then moves up/down in Z direction.

Adaptive Clearing

The **Adaptive Clearing** tool is used to remove the material in bulk from workpiece. It is generally a roughing process. It uses an advanced strategy of milling motion so that there is less load on the tool. This tool is similar to **2D Adaptive Clearing** tool but with operation scope in 3D. The procedure to use this tool is discussed next.

* Click on the **Adaptive Clearing** tool from the **3D** drop-down of **MILLING** tab in the **Toolbar**; refer to Figure-64. The **ADAPTIVE** dialog box will be displayed; refer to Figure-65.

Figure-64. Adaptive Clearing tool

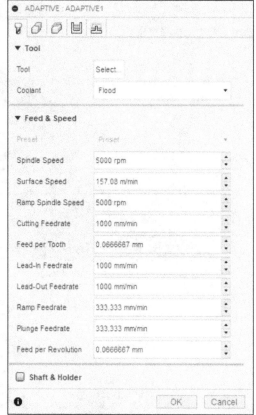

Figure-65. ADAPTIVE dialog box

- Click on the **Select** button of **Tool** tab from **ADAPTIVE** dialog box. The **Select Tool** dialog box will be displayed. Select the tool as required generally an end mill or ball/bull mill.

- Select the **Shaft & Holder** check box of **Tool** tab from **ADAPTIVE** dialog box to specify how the shaft and holder of the tool will avoid possible collisions with stock or workpiece.

- Click in the **Shaft and Holder Mode** drop-down of **Shaft and Holder** section from **Tool** tab and select the desired option to define motion for avoiding collision. Select the **Pull away** option from the drop-down to create a toolpath away from the workpiece by specified shaft clearance distance. On selecting this option, the **Use Shaft** check box and **Shaft Clearance** edit boxes will become available. Select the **Use Shaft** check box to use length of tool shaft as safe distance away from workpiece while creating the toolpath. Specify desired value in the **Shaft Clearance** edit box to define extended length after shaft length to be avoided from workpiece surfaces. Select the **Detect tool length** option from the drop-down to use total length of cutting tool as safe distance of tool holder from workpiece for creating toolpath. Select the **Fail on collision** option from the drop-down to abort machining toolpath at the event of possible collision between tool holder and workpiece. Select the **Trimmed** option from the drop-down to created trimmed toolpaths at the points of possible collision of tool holder with workpiece. Figure-66 shows various available strategies.

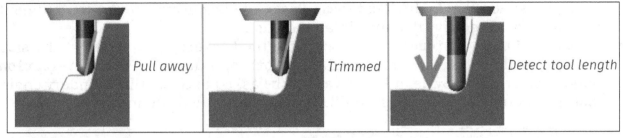

Figure-66. Holder avoidance strategies

- Select the **Use Holder** check box of the **Shaft & Holder** node to use holder of selected cutting tool in the toolpath calculation to avoid collisions along with shaft length.
- Click in the **Holder Clearance** edit box of **Shaft & Holder** node and specify desired value of tool clearance to keep the holder away from the part.

Geometry

- Click on the **Geometry** tab of **ADAPTIVE** dialog box. The options will be displayed as shown in Figure-67.

Figure-67. Geometry tab of ADAPTIVE dialog box

- Select the **None** option in the **Machining Boundary** drop-down of **Geometry** tab to machine all the stock without limitation. Note that the **None** option is not available for all machining strategies.
- Select the **Bounding Box** option in the **Machining Boundary** drop-down from **Geometry** tab to generate toolpath within boundaries of the part viewed from the WCS.
- Select the **Silhouette** option in the **Machining Boundary** drop-down from **Geometry** tab to machine the toolpath within boundary defined by projection of part on machining plane. Note that on selecting this option, the curvature of part will also be taken into account while calculation stock for machining.
- Select the **Selection** option in the **Machining Boundary** drop-down from **Geometry** tab to machine the toolpath within a region bound by selected boundary curves. Note that on selecting this option, you will be asked to select curve chains.
- You can specify the position of tool with respect to boundary chain while cutting by using the options in **Tool Containment** drop-down.

- Specify desired value in the **Additional Offset** edit box to expand the containment boundary by respective distance in all directions.
- Select the **Stock Contours** check box from the **Geometry** tab to select the side walls of model to be face machined. The **Nothing** button for **Stock Selection** option will be active by default and you will be asked to select the boundary chain (perimeter curves); refer to Figure-68. Select desired curve chains from the model.

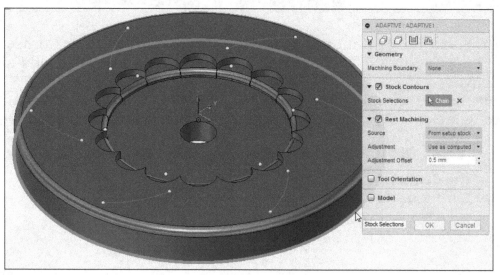

Figure-68. Selection of geometry

- Select the **Rest Machining** check box of **Geometry** tab to limit the operation to removal of the material left by the tools in previous operations or marked by specified bodies.
- Click in the **Source** drop-down of **Rest Machining** section and select the required option to define source to be used for calculating remaining stock. Select the From previous operation(s) option from the **Source** drop-down to calculate remaining stock from all the previous operations and use it as stock for rest machining. Select the **From file** option from the drop-down to use specified model file as source of stock for rest machining. On selecting this option, the **File** edit box below the drop-down will be displayed and you need to specify the path of model file. Select the **From bodies** option in the drop-down to use selected bodies as source for rest machining operations. On selecting this option, the selection button for **Bodies** option will be active and you need to select desired bodies from modeling area. Select the **From setup stock** option from the drop-down to use remaining amount of stock as source for rest machining operation.
- Click in the **Adjustment** drop-down of the **Rest Machining** section and select the required option. If you want to machine small cusps left by tool in previous operation then select **Machine Cusps** option. If you do not want to machine them then select the **Ignore Cusps** option. Selecting the **Use as computed** option allows the system to automatically decide whether to remove cusps or not based on the value specified in **Adjustment Offset** edit box.
- Click in the **Adjustment Offset** edit box of **Rest Machining** check box and enter the value of stock to be removed, depending on the rest material adjustment setting.

Passes

- Click on the **Passes** tab of the **ADAPTIVE** dialog box. The options of **Passes** tab will be displayed; refer to Figure-69.

Figure-69. Passes tab of ADAPTIVE dialog box

- Click in the **Tolerance** edit box of the **Passes** tab and specify the value of tolerance for curves of model.
- Select the **Machine Shallow Areas** check box of the **Passes** tab to remove excessive cusps from the shallow areas of Z-level. After selecting check box, specify desired parameters in the **Minimum Shallow Stepdown** and **Maximum Shallow Stepover** edit boxes to define minimum vertical downward movement and maximum horizontal (left/right) movement while cutting shallow areas, respectively.
- Click in the **Optimal Load** edit box of the **Passes** tab and specify the value of engagement (stepover) between cutting tool and stock which is allowed by adaptive strategies. Note that stepover can increase/decrease based on optimum load on cutting tool.
- Select the **Both Ways** check box from the drop-down to machine stock when cutting tool is moving upward as well as downward. Set the related parameters as discussed earlier.
- Click in the **Minimum Cutting Radius** edit box of the **Passes** tab and specify the value of minimum radius that can be cut using this toolpath.
- Select the **Machine Cavities** check box of **Passes** tab to machine the pockets of the model.
- Select the **Use Slot Clearing** check box of **Passes** tab to start pocket clearing with a slot at its middle before continuing with a spiral motion towards the pocket walls.

- Select the **Fillets** check box of the **Passes** tab and specify the value of fillet radius.
- The other options of the dialog box are same as discussed earlier.
- After specifying the parameters, click on the **OK** button from **ADAPTIVE** dialog box. The toolpath will be generated and displayed on the model; refer to Figure-70.

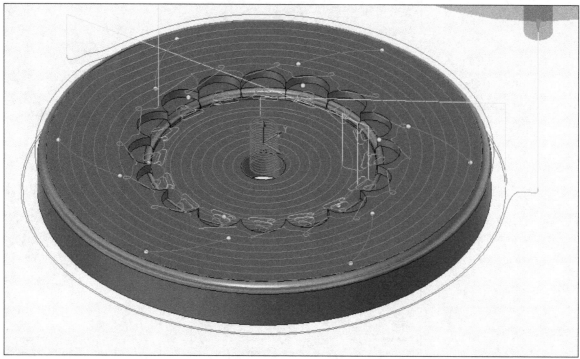

Figure-70. Generated toolpath of adaptive clearing

Pocket Clearing

The **Pocket Clearing** tool is mainly used for clearing large quantity of stock material from pockets of the model. Note that in this toolpath, the cutting tool can move simultaneously along all three axes in 3D space. The procedure to use this tool is discussed next.

- Click on the **Pocket Clearing** tool from the **3D** drop-down in the **Toolbar**; refer to Figure-71. The **POCKET** dialog box will be displayed; refer to Figure-72
- Click on the **Select** button of **Tool** tab from **POCKET** dialog box. The **Select Tool** dialog box will be displayed. Select the tool as required.
- The options of this dialog box are same as discussed earlier.
- After specifying the parameters, click on the **OK** button from **POCKET** dialog box. The toolpath will be generated and displayed on the model; refer to Figure-73.

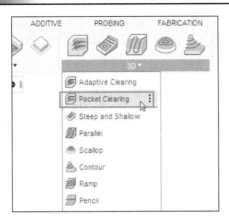

Figure-71. Pocket Clearing tool

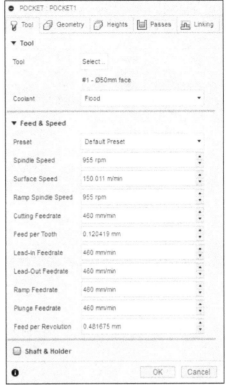

Figure-72. POCKET dialog Box

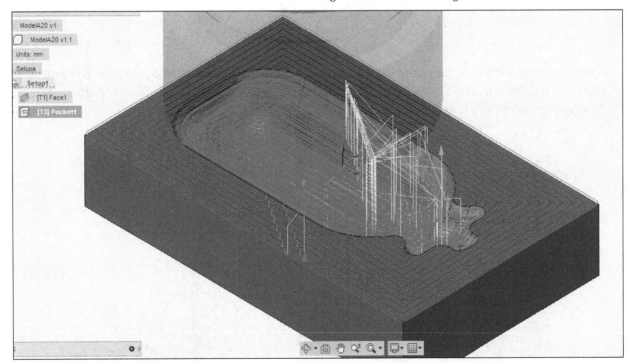

Figure-73. Generated toolpath for pocket

Steep and Shallow

As the name suggests, this tool is used to create toolpath for machining steep and shallow areas of part based on specified angle threshold. When the area to be machined is steep then contour passes are used and when the area is shallow then parallel/scallop passes are used for machining. The procedure to use this tool is given next.

- Click on the **Steep and Shallow** tool from the **3D** drop-down in the **Toolbar**; refer to Figure-74. The **STEEP AND SHALLOW** dialog box will be displayed refer to Figure-75.

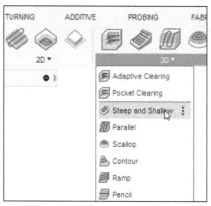

Figure-74. Steep and Shallow tool

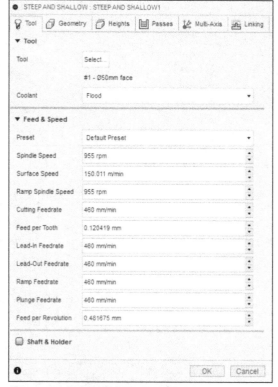

Figure-75. STEEP AND SHALLOW dialog box

- Specify the desired parameters in **Tool**, **Geometry**, **Heights**, and **Linking** tabs of dialog box as discussed earlier. Click on the **Passes** tab. The options will be displayed as shown in Figure-76.

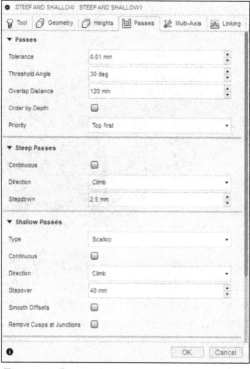

Figure-76. Passes tab

- Specify desired value of angle in **Threshold Angle** edit box. The areas which have slope angle larger than this value with respect to horizontal line will be counted as shallow areas and below this value will be counted as steep areas.
- The value specified in **Overlap Distance** edit box is distance past the threshold angle line which will act as transitional area for blending steep and shallow areas.
- Select the **Continuous** check box from the **Steep Passes** section to create continuous spiral toolpath for steep sections of the model; refer to Figure-77.

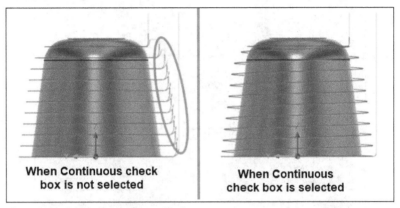

Figure-77. Steep continuous toolpaths

- Select desired option from the **Type** drop-down of **Shallow Passes** section to define whether you want to use parallel or scallop toolpaths for shallow areas.
- Select the **Smooth Offsets** check box to replace sharp corners in the rounds in the steep corners of the model.
- Select the **Remove Cusps at Junctions** check box to create centerline toolpaths at corners to machine areas generally left untouched by cutting tool.
- Click on the **Multi-Axis** tab in the dialog box to define general parameters for using Multi-axis machining tool. The options will be displayed as shown in Figure-78.
- Select desired option from the **Primary Mode** drop-down to define which axes are to be used for multi-axis machining.
- Specify desired value in the **Smoothing Distance** edit box to define minimum distance at which the cutting tool will tilt in Multi-Axis machine to creating smooth surface. Similarly, you can specify the maximum tilt angle allowed for smoothening surface during machining in **Smoothing Angle** edit box.

Figure-78. Multi-Axis tab

- Select the **Collision Avoidance** check box to allow tilting of cutting tool for avoiding collision with stock.
- Select the **Tool Axis Limits** check box to define maximum tilting limits for cutting tool in Multi-axis milling machines.
- Set the other parameters as discussed earlier and click on the **OK** button from the dialog box. The toolpath will be generated; refer to Figure-79.

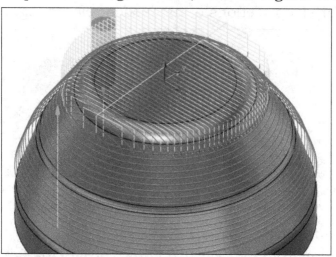

Figure-79. Steep and Shallow toolpath

Parallel

The **Parallel** tool is used mainly as finishing strategies. This toolpath is used when parallel cutting passes are needed to machine the workpiece. This toolpath is useful for machining shallow areas. The procedure to use this tool is discussed next.

- Click on the **Parallel** tool from the **3D** drop-down in the **Toolbar**; refer to Figure-80. The **PARALLEL** dialog box will be displayed; refer to Figure-81.

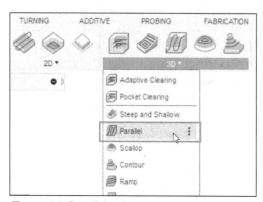

Figure-80. Parallel tool

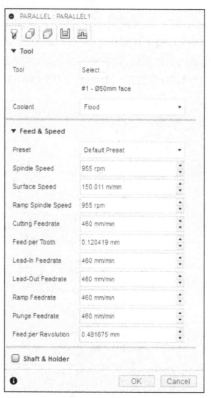

Figure-81. PARALLEL dialog box

- Click on the **Select** button in the **Tool** tab from the **PARALLEL** dialog box. The **Select Tool** dialog box will be displayed. Select the tool as required.
- Click on the **Geometry** tab of the **PARALLEL** dialog box. The **Geometry** tab will be displayed; refer to Figure-82.

Figure-82. Geometry tab of PARALLEL dialog box

- Select the **Contact Point Boundary** check box of **Geometry** section to extend the boundary limits of toolpath where the tool actually touches the surface of model rather than the tool center location while cutting.
- Select the **Contact Only** check box of **Geometry** section to not generate the toolpath on the model where the tool is not in contact with the machining surface. You should select this check box to avoid creating toolpaths over openings (like holes) in the model.
- Select the **Slope** check box of **Geometry** tab to specify the range of slope within which the faces of part will be considered for machining.
- Click in the **From Slope Angle** edit box of **Slope** check box and specify the angle value greater than 0 degree to define starting angle of slope confinement. Only area equal to or more than this values will be machined.
- Click in the **To Slope Angle** edit box of **Slope** check box and specify the value less than 90 degree to define end angle for slope confinement. Only area equal to or less than this values will be machined.
- Select the **Avoid/Touch Surface** check box of the **Geometry** tab to define surfaces which will be avoided or touched while machining within a specified distance. The **Nothing** button of **Avoid/Touch Surfaces** node will be active by default. You need to click on the surfaces to be avoided.
- Click in the **Avoid/Touch Surface Clearance** edit box of the **Avoid/Touch Surface** section and specify desired value of clearance near avoidance surfaces.

- Select the **Touch Surfaces** check box of **Avoid/Touch Surfaces** section to invert the meaning of **Avoid/Touch Check Surface** check box. Now, selected surfaces will be touched and other surfaces will be avoided.
- The other options of the dialog box are same as discussed earlier.
- After specifying the parameters, click on the **OK** button from **PARALLEL** dialog box. The toolpath will be generated and displayed on the model; refer to Figure-83.

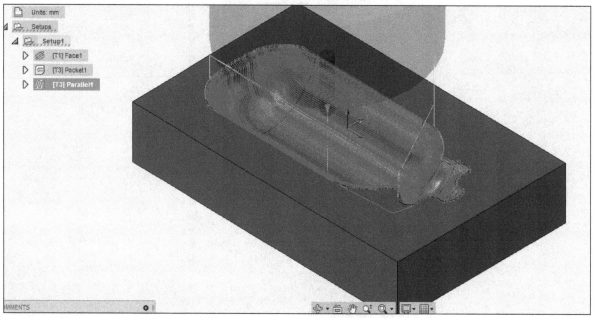

Figure-83. The generated toolpath for parallel

Scallop

The **Scallop** tool is used to finish round faces of the model with slopes. This tool can also be used to machine fillets. Although, you can use the scallop toolpath for performing roughing operations, this toolpath is generally used after Contour and Parallel toolpaths for performing finishing operation. The procedure to use this tool is discussed next.

- Click on the **Scallop** tool from the **3D** drop-down in the **Toolbar**; refer to Figure-84. The **SCALLOP** dialog box will be displayed; refer to Figure-85.

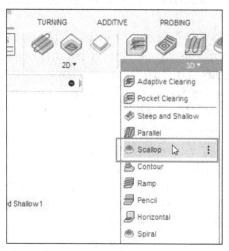

Figure-84. Scallop tool

- Click on the **Select** button of **Tool** tab from **SCALLOP** dialog box. The **Select Tool** dialog box will be displayed. Select the tool as desired.
- Click on the **Passes** tab of **SCALLOP** dialog box. The **Passes** tab will be displayed; refer to Figure-86.

Figure-85. SCALLOP dialog box

Figure-86. Passes tab of SCALLOP dialog box

- Select the **Link from Inside to Outside** check box of the **SCALLOP** dialog box to specify that linking should be done by moving from inside passes to outside passes in toolpath.
- Click in the **Inside/Outside Direction** drop-down of **Passes** tab and select desired option to define how cutting will progress on surface.
- Select the **Limit Number of Stepovers** check box in **Passes** tab to limit the number of steps that can be created to finish surface of model for given stepover distance.
- The options of the dialog box are same as discussed earlier.
- After specifying the parameters, click on the **OK** button from the **SCALLOP** dialog box. The generated toolpath will be displayed on the model; refer Figure-87.

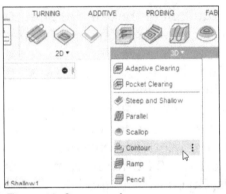

Figure-87. Generated toolpath of scallop tool

Contour

The **Contour** tool is generally used for finishing steep walls. This tool can be used for machining vertical walls. The procedure to use this tool is discussed next.

- Click on the **Contour** tool from the **3D** drop-down in the **Toolbar**; refer to Figure-88. The **CONTOUR** dialog box will be displayed; refer to Figure-89.

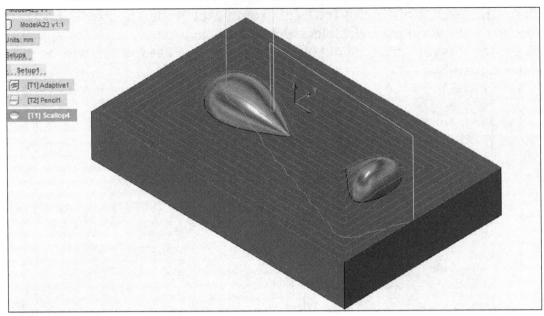

Figure-88. Contour tool

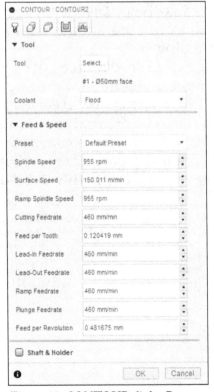

Figure-89. CONTOUR dialog Box

- Click on the **Select** button of **Tool** tab in the **POCKET** dialog box. The **Select Tool** dialog box will be displayed. Select the tool as required, generally a ball end mill.

- Click on the **Passes** tab of the **CONTOUR** dialog box. The **Passes** tab will be displayed; refer to Figure-90

- Select the **Multi-Axis Tilting** check box of **Passes** tab from **CONTOUR** dialog box to enable multi axis tilting to avoid collision of model with tool holder when using short tools or machining complex areas.

- Click in the **Maximum Tilt** edit box of **Multi-Axis Tilting** check box and specify the maximum allowed tilt with respect to operation tool axis.

- Click in the **Maximum Segment Length** edit box of **Multi-Axis Tilting** check box and specify the length of single segment for the generated toolpath. This value is similar to stepover over distance.

- Click in the **Maximum Tool Axis Sweep** edit box of **Multi-Axis Tilting** section and specify the value of maximum angle change for a single tool axis sweep in the generated toolpath.

- The other options of the dialog box are same as discussed earlier.

- After specifying the parameters, click on the **OK** button from **CONTOUR** dialog box. The toolpath will be generated and displayed on the model; refer to Figure-91

Figure-90. Passes tab of CONTOUR dialog box

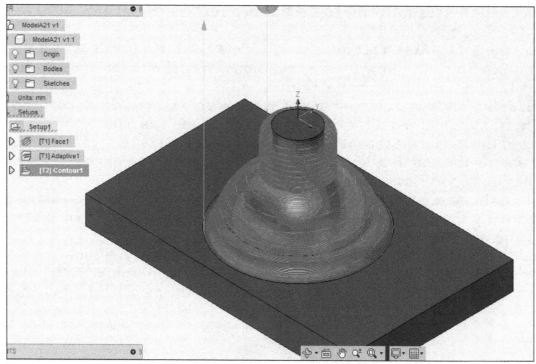

Figure-91. The generated toolpath for Contour tool

Ramp

The **Ramp** tool is used to create a finishing operation meant for machining step areas similar to Contour tool. The procedure to use this tool is discussed next.

- Click on the **Ramp** tool of **3D** drop-down from **Toolbar**; refer to Figure-92. The **RAMP** dialog box will be displayed; refer to Figure-93.

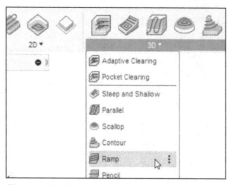

Figure-92. Ramp tool

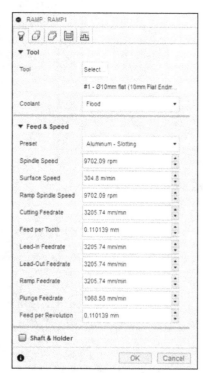

Figure-93. RAMP dialog box

- Click on the **Select** button of **Tool** tab from **RAMP** dialog box. The **Select Tool** dialog box will be displayed. Select the tool as required generally a ball end mill.
- Click on the **Passes** tab in the dialog box and select the **Flat Area Detection** check box to detect flat areas in the selected model faces and respond with suitable cutting strategy.
- Select the **Order Bottom-Up** check box in **Passes** tab to generate contour cutting passes for the ramp starting from minimum Z value to maximum Z value of the selected model.
- The other options of the dialog box have been discussed earlier.
- After specifying the parameters, click on the **OK** button from the **RAMP** dialog box. The generated toolpath will be displayed on the model; refer to Figure-94.

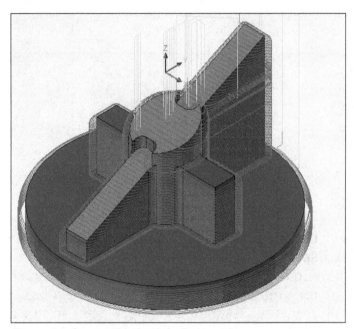

Figure-94. Generated toolpath for Ramp tool

Pencil

The **Pencil** tool is used to create toolpaths along sharp internal corners with small radii tool. This tool is generally used to remove material where no other toolpath can work. After performing finishing operation, create this toolpath for cleaning. The procedure to use this tool is discussed next.

- Click on the **Pencil** tool of **3D** drop-down from **Toolbar**; refer to Figure-95. The **PENCIL** dialog box will be displayed; refer to Figure-96.

Figure-95. Pencil tool

Figure-96. PENCIL dialog box

- Click on the **Select** button in **Tool** tab from **PENCIL** dialog box. The **Select Tool** dialog box will be displayed. Select the tool as required.
- Specify desired value in the **Overthickness** edit box to define additional thickness to be applied to the cutting tool for machining corner fillets.
- Specify desired value in the **Bitangency Angle** edit box to define angular span in which cutting tool touches the workpiece while cutting. A higher value means more tool area for cutting.
- The other options of the dialog box have been discussed earlier.
- After specifying the parameters, click on the **OK** button from **PENCIL** dialog box. The generated toolpath will be displayed on the model; refer to Figure-97.

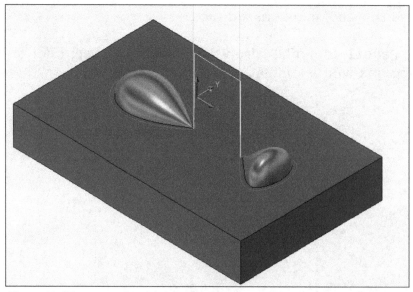

Figure-97. The generated toolpath for Pencil tool

Note: The Pencil toolpaths and Trace toolpaths look similar in operation but the different between the two is their field of operation. Trace toolpaths can be assumed as 2D version of pencil toolpaths. While pencil toolpaths generate simultaneous 3 axis movements, the Trace toolpaths generate XY movements for removing material on one plane and then move upward/downward for next cutting plane.

Horizontal

The **Horizontal** tool is used for machining flat surfaces of model which are surrounded by other features. Horizontal toolpaths are generally used in collaboration with ramp toolpaths for finishing models having multiple steep walls. The procedure to use this tool is discussed next.

- Click on the **Horizontal** tool from the **3D** drop-down in the **Toolbar**; refer to Figure-98. The **HORIZONTAL** dialog box will be displayed; refer to Figure-99

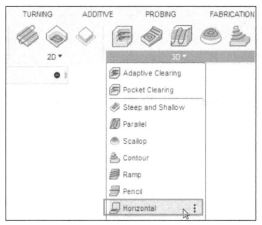

Figure-98. Horizontal tool

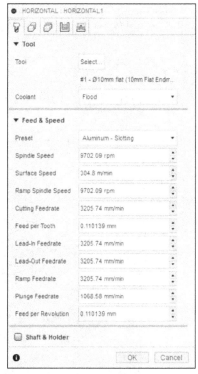

Figure-99. HORIZONTAL dialog box

- Click on the **Select** button from **Tool** tab in the **HORIZONTAL** dialog box. The **Select Tool** dialog box will be displayed. Select the tool as required.
- Select the **Use Morphed Spiral Machining** check box in the **Passes** tab of the dialog box to generate constant spiral move toolpaths when machining flat pockets. Using this option keeps load on cutting tool in check.
- The other options of the dialog box are same as discussed earlier. After specifying the parameters, click on the **OK** button from the **HORIZONTAL** dialog box. The generated toolpath will be displayed on the model; refer to Figure-100.

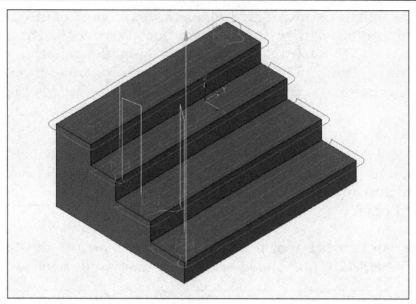

Figure-100. The generated toolpath for Horizontal tool

PRACTICAL

Generate the toolpath of the given model shown in Figure-101 using 2D Toolpath tools.

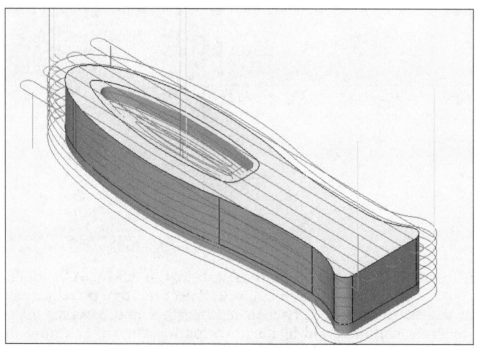

Figure-101. Practical

Adding Model to CAM

*   Create and save the part in **DESIGN** workspace. The part file is available in the respective chapter folder of **Autodesk Fusion 360 Black Book Resources**.
*   Select the **MANUFACTURE** option from the **Workspace** drop-down. The model will be displayed in the **MANUFACTURE** workspace; refer to Figure-102.

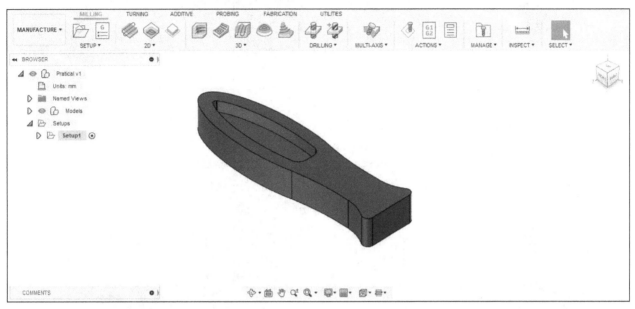

Figure-102. Practical 1

Creating Stock

*   Click on the **New Setup** tool from the **SETUP** drop-down in the **Toolbar**. The **SETUP** dialog box will be displayed along with the stock of model.
*   Click on the **Setup** tab of the **SETUP** dialog box and specify the parameters as displayed in Figure-103.

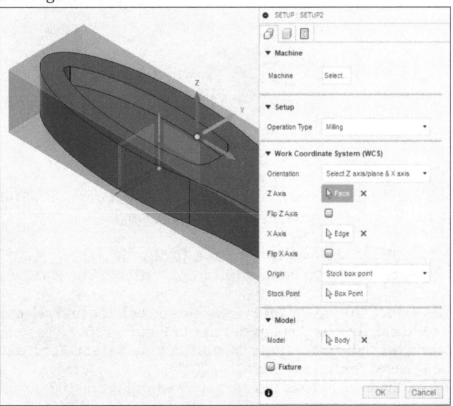

Figure-103. Setup tab of SETUP dialog

*   Click on the **Stock** tab of **SETUP** dialog box and enter the parameters as displayed in Figure-104.

Figure-104. Stock tab of practical

*   Click on the **Post Process** tab of **SETUP** dialog box and specify the parameters as displayed in Figure-105.

Figure-105. Post Process tab of SETUP dialog box

*   After specifying the parameters, click on the **OK** button from the **SETUP** dialog box. The stock will be created and displayed on the model.

Generating Face toolpath

*   Click on the **Face** tool of **2D** drop-down from **Toolbar**. The **FACE** dialog box will be displayed.
*   Click on the **Select** button for **Tool** option from **Tool** tab and select the tool from **Select Tool** dialog box as displayed; refer to Figure-106.
*   After selecting the tool, click on the **OK** button from **Select Tool** dialog box. The tool will be selected for machining.
*   Specify the parameters of **Tool** tab as displayed in Figure-107.

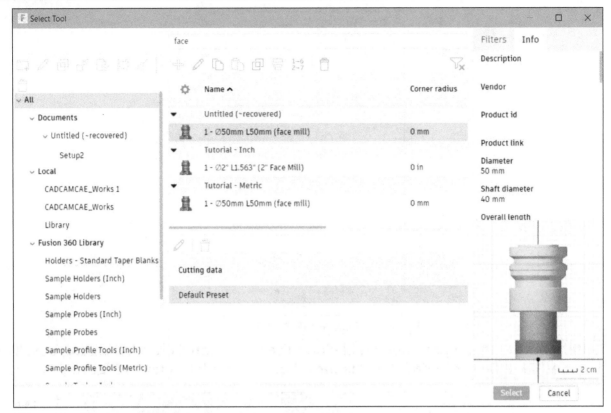

Figure-106. Selecting tool for facing

- Click in the **Passes** tab of the **FACE** dialog box and specify the parameters as displayed in Figure-108.

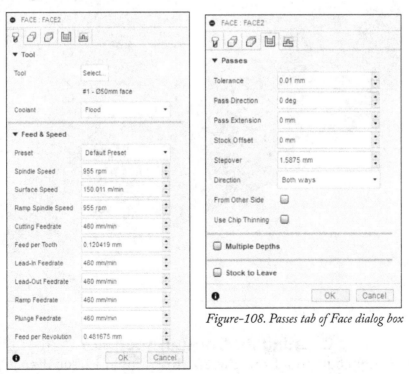

Figure-107. Tool tab

Figure-108. Passes tab of Face dialog box

- Click on the **Linking** tab of **FACE** dialog box and specify the parameters as displayed in Figure-109.

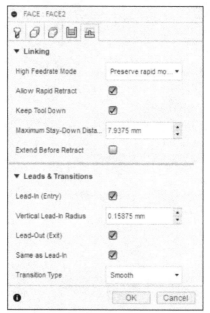

Figure-109. Linking Tab of FACE dialog box

- After specifying the parameters, click on the **OK** button from **FACE** dialog box. The toolpath will be generated and displayed on the model; refer to Figure-110.

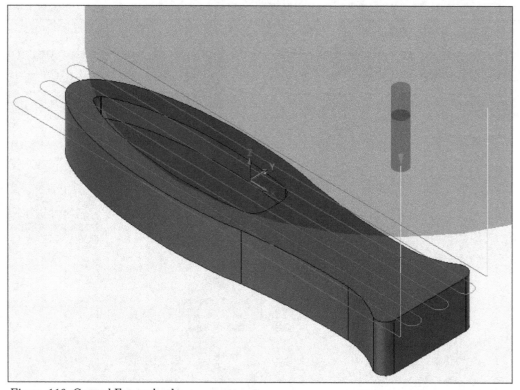

Figure-110. Created Face toolpath

Creating 2D Contour toolpath

- Click on the **2D Contour** tool from the **2D** drop-down in **Toolbar**. The **2D CONTOUR** dialog box will be displayed.
- Click on the **Select** button of **Tool** option from **Tool** tab and select the tool from **Select Tool** dialog box as displayed in Figure-111.

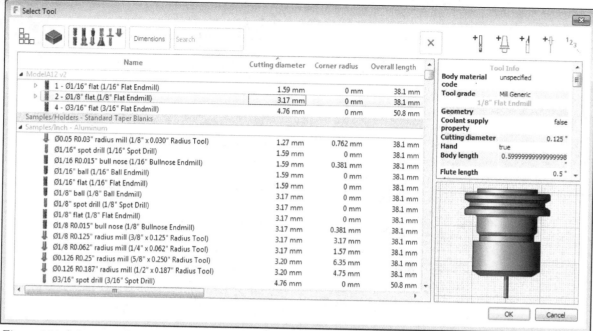

Figure-111. Selecting tool for 2D Contour

- After selecting desired cutting tool, click on the **OK** button from **Select Tool** dialog box.
- Click on the **Geometry** tab of the **2D CONTOUR** dialog box and specify the parameters as displayed in Figure-112.

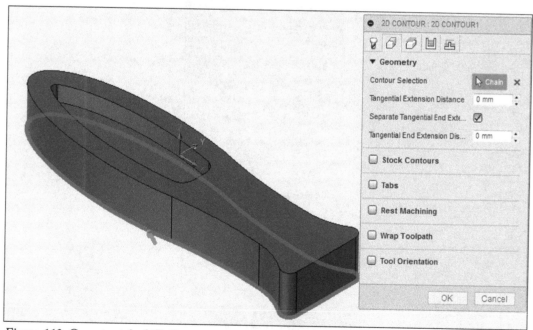

Figure-112. Geometry tab of 2D Contour

- Click in the **Passes** tab of **2D CONTOUR** dialog box and specify the parameters of **Passes** and **Multiple Depths** sections as displayed in Figure-113.
- Click on the **Linking** tab of **2D CONTOUR** dialog box and specify the parameters as displayed in Figure-114.

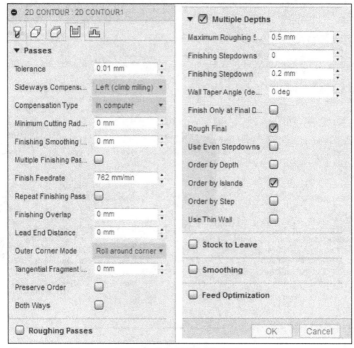

Figure-113. Passes tab of 2D CONTOUR dialog box

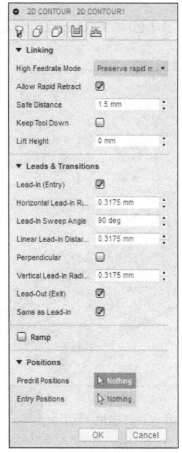

Figure-114. Linking tab of 2D CONTOUR dialog box

• After specifying the parameters, click on the **OK** button from **2D CONTOUR** dialog box. The toolpath will be generated and displayed on the model; refer to Figure-115.

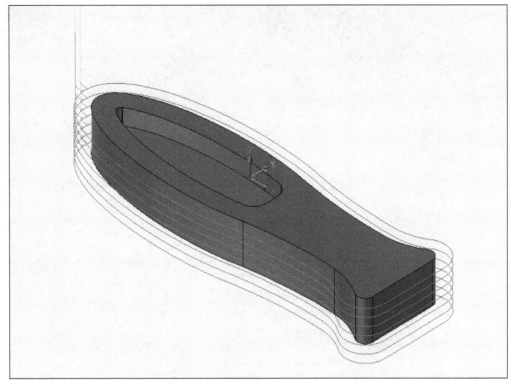

Figure-115. Generated toolpath for 2D Contour tool

Generating Pocket toolpath

- Click on the **2D Pocket** tool of **2D** drop-down from **Toolbar**. The **2D POCKET** dialog box will be displayed.
- Click on the **Select** button of **Tool** option from **Tool** tab and select the tool from **Select Tool** dialog box as displayed in Figure-116.

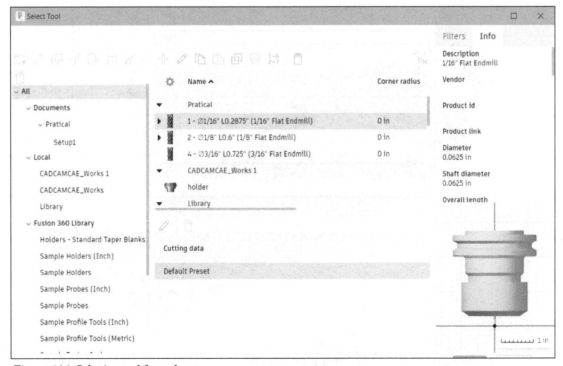

Figure-116. Selecting tool for pocket

- Specify the parameters of **Tools** tab as shown in Figure-117.

Figure-117. Tools tab of 2D POCKET dialog box

- Click on the **Geometry** tab of **2D POCKET** dialog box and specify the parameters as shown in Figure-118.

Figure-118. Geometry tab of 2D POCKET dialog box

- Click in the **Passes** tab of **2D POCKET** dialog box and specify the parameters as shown in Figure-119.
- Click in the **Linking** tab of **2D POCKET** dialog box and specify the parameters as shown in Figure-120.

*Figure-119. Passes tab of 2D
POCKET dialog box*

*Figure-120. Linking Tab of 2D
POCKET dialog box*

- After specifying the parameters, click on the **OK** button from **2D POCKET** dialog box. The toolpath will be generated and displayed on the model; refer to Figure-121.
- After creating the required operations for a model, we need to simulate the generated toolpath.

Simulating the toolpath

- Select the **Setup1** option from **BROWSER** and right-click on it. A shortcut menu will be displayed.
- Click on the **Simulate** tool from shortcut menu. The **SIMULATE** dialog box will be displayed along with simulation keys; refer to Figure-122.

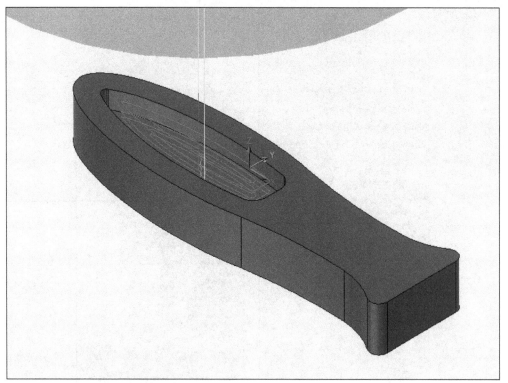

Figure-121. Generated toolpath of pocket

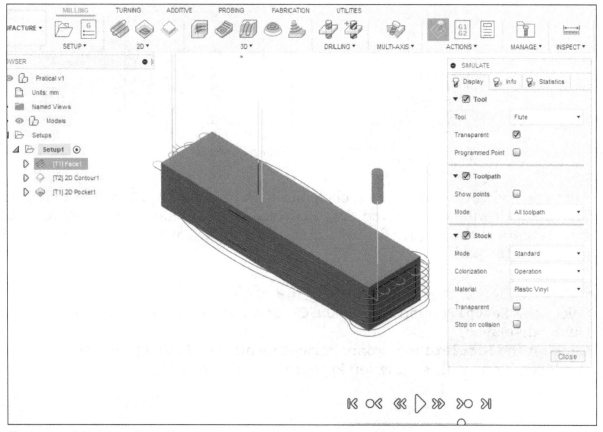

Figure-122. Simulating the model

- Click on the **Play** button to view the simulation process. One by one all generated toolpath will be applied to the model and at the end of simulation the part will display along with toolpaths on the model; refer to Figure-123. Note that till now only roughing toolpaths have been created. Create the finishing toolpaths as discussed earlier.

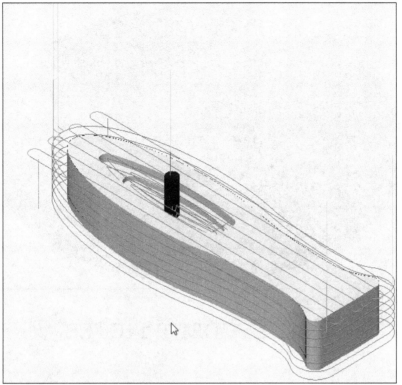

Figure-123. Running simulation process

PRACTICE 1

Machine the stock of diameter as 55 mm and length as 12 mm to create the part as shown in Figure-124. The pat file of this model is available in the respective folder of **Autodesk Fusion 360 Black Book Resources**.

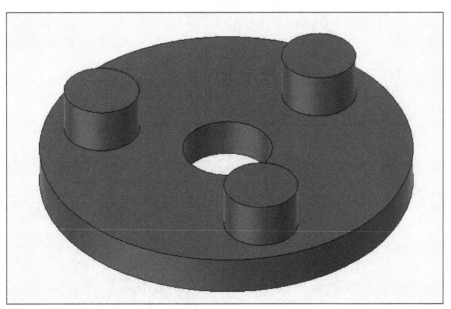

Figure-124. Practice 1

PRACTICE 2

Machine the stock of diameter as 55 mm and length as 8 mm to create the part as shown in Figure-125.

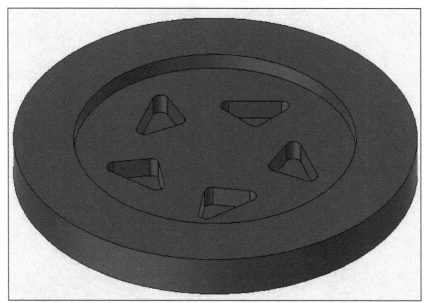

Figure-125. Practice 2

FOR STUDENT NOTES

Chapter 17

Generating Milling
Toolpaths - 2

Topics Covered

The major topics covered in this chapter are:

- *Spiral Toolpath*
- *Radial Toolpath*
- *Morphed Spiral Toolpath*
- *Project Toolpath*
- *Swarf Toolpath*
- *Multi-Axis Contour Toolpath*
- *Drilling Toolpath*

3D TOOLPATHS

In previous chapter, we have worked on 2D toolpaths and some 3D toolpaths available in **Toolbar**. In this chapter, we will discuss rest of the 3D toolpaths. Later, we will also work on multi-axis toolpaths and drilling operations.

Spiral

The spiral toolpath is used to cut material in spiral fashion. This toolpath is useful when you need to finish faces of cylindrical objects like shown in Figure-1. The procedure to use this tool is discussed next.

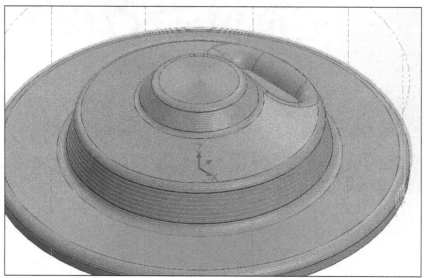

Figure-1. Spiral Toolpath

- Click on the **Spiral** tool in **3D** drop-down from **Toolbar**; refer to Figure-2. The **SPIRAL** dialog box will be displayed; refer to Figure-3.

Figure-2. Spiral tool

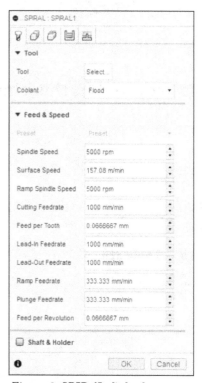

Figure-3. SPIRAL dialog box

- Click on the **Select** button of **Tool** section from **Tool** tab. The **Select Tool** dialog box will be displayed. Select the desired tool generally an end mill or ball mill.
- Click on the **Geometry** tab in the dialog box and select desired point to be used as center for toolpath.
- Click on the **Passes** tab of **SPIRAL** dialog box. The options will be displayed as shown in Figure-4.

Figure-4. Passes tab of SPIRAL dialog box

- Select the **Clockwise** check box of **Passes** tab to set the direction of spiral to clockwise.
- Select the **Spiral** option from **Spiral** drop-down to create a spiral toolpath which starts from the center and ends at the outermost boundary.
- Select the **Spiral with circles** option from **Spiral** drop-down to create a circular toolpath at the minimum and maximum radius. This toolpath will only be created if the minimum radius is larger than zero and maximum radius is smaller than the radius of regulation boundary.
- Select the **Concentric circles** option from **Spiral** drop-down to create a concentric circle toolpath; refer to Figure-5.

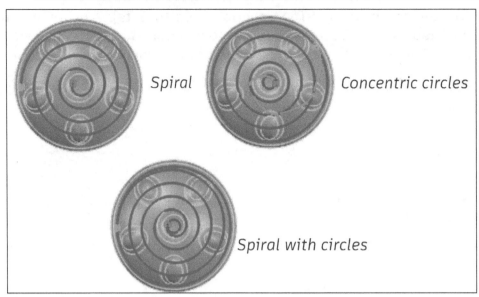

Figure-5. Spiral modes

- Click in the **Inner Limit** edit box of **Passes** tab to set the minimum inner radius. If no inner limit is specified then spiral toolpath will start from center of the selected circular face.
- Click in the **Outer Limit** edit box of **Passes** tab to set the maximum outer radius. If outer limit is not specified then spiral toolpath will extend up to the outer edge of selected circular face.
- Click in the **Stepover** edit box of **Passes** tab and specify distance between two consecutive cutting passes.
- The other options of the dialog box are same as discussed earlier. After specifying the parameters, click on the **OK** button from **SPIRAL** dialog box. The toolpath will be generated and displayed on the model; refer to Figure-6

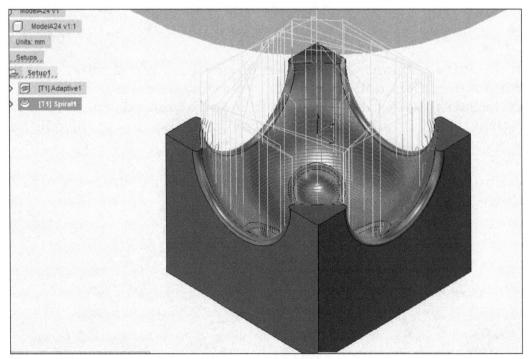

Figure-6. Toolpath generated for Spiral tool

Radial

The **Radial** tool is used to create toolpath along the radii of an arc. The toolpath created by radial is similar to the spokes of wheel which are then projected down on the surface. Note that Radial and Spiral both the toolpaths are used to machine round shallow parts. The different between two toolpaths is their direction of cutting. In case of Radial toolpaths, the cutting passes will be straight lines connecting the center with outer edge. When we go deep in metallurgy of parts, these two toolpaths can generate different grain patterns on surface of the same part The procedure to use Radial tool is discussed next.

*   Click on the **Radial** tool of **3D** drop-down from **Toolbar**; refer to Figure-7. The **RADIAL** dialog box will be displayed; refer to Figure-8.
*   Click on the **Select** button of **Tool** section from **Tool** tab. The **Select Tool** dialog box will be displayed. Select the required tool.
*   Click on the **Geometry** tab in the dialog box and select desired point to be used as center for toolpath.
*   Click on the **Passes** tab of **RADIAL** dialog box. The options will be displayed as shown in Figure-9.
*   Click in the **Angular Step** edit box of **Passes** tab and specify the angular value of distance between two consecutive radial passes.
*   Click in the **Angle From** edit box of **Passes** tab and specify the value of radial angle measured from the X-axis from where first cutting pass will be generated.
*   Click in the **Angle To** edit box of **Passes** tab and specify the radial angle measured from the X-axis up to which the cutting passes can be generated.
*   The other options of the dialog box are same as discussed earlier.
*   After specifying the parameters click on the **OK** button from **RADIAL** dialog box. The toolpath will be generated and displayed on the model; refer to Figure-10.

Figure-7. Radial tool

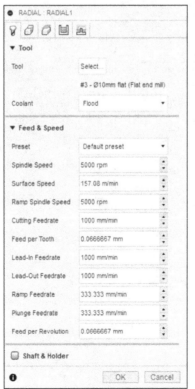

Figure-8. RADIAL dialog box

Figure-9. Passes tab of RADIAL dialog box

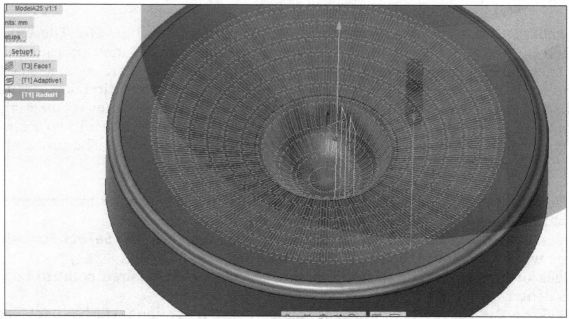

Figure-10. Toolpath generated for Radial tool

Morphed Spiral

The **Morphed Spiral** tool is mainly used to create toolpath of free-form surfaces. In this toolpath, spiral cutting passes are generated following the surface curvature within specified boundaries. The procedure to use this tool is discussed next.

• Click on the **Morphed Spiral** tool of **3D** drop-down from **Toolbar**; refer to Figure-11. The **MORPHED SPIRAL** dialog box will be displayed; refer to Figure-12.

Figure-11. Morphed Spiral tool

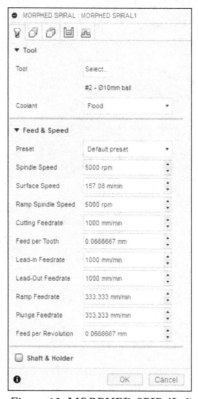

Figure-12. MORPHED SPIRAL dialog box

• Click on the **Select** button of **Tool** section from **Tool** tab. The **Select Tool** dialog box will be displayed. Select the required tool, generally a ball end mill.

- Click on the **Geometry** tab and select the boundary curve to be used as containment boundary for toolpaths. If no chain is selected then by default, bounding box of stock is used as boundary of toolpaths.
- The other options of the dialog box are same as discussed earlier. After specifying the parameters, click on the **OK** button from **MORPHED Spiral** dialog box. The toolpath will be created and displayed on the model; refer to Figure-13.

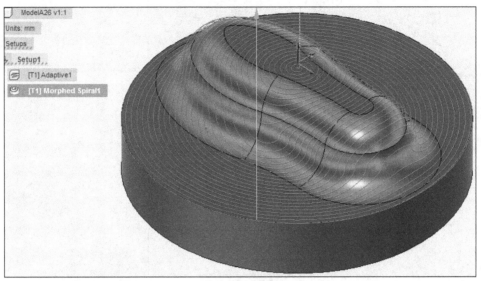

Figure-13. Generated toolpath for Morphed Spiral tool

Project

The **Project** tool is used to create a toolpath along the selected contour. The contour will be machined with the center of the tool. Note that Project toolpaths work similar to Engrave toolpaths discussed earlier. The only different is number of dimensions in which toolpaths are generated. Engrave toolpaths are 2 dimensional whereas project toolpaths are 3 dimensional which means if you have an irregular surface on which engraving of selected curves is to be performed then you should use Project toolpaths. The procedure to use this tool is discussed next.

- Click on the **Project** tool of **3D** drop-down from **Toolbar**; refer to Figure-14. The **PROJECT** dialog box will be displayed; refer to Figure-15.

Figure-14. Project tool

Figure-15. PROJECT dialog box

- Click on the **Select** button of **Tool** section from **Tool** tab. The **Select Tool** dialog box will be displayed. Select the desired tool generally a tapered mill.
- Click on desired curve from the model which are to be engraved. Note that there is no need of curves to be placed directly on the face of model. Make sure that projection of curves fall on the face of model where curves are to be engraved; refer to Figure-16.

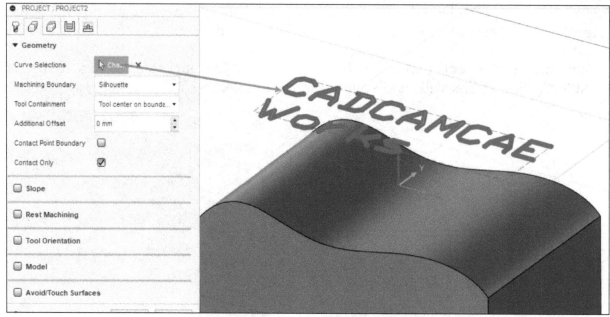

Figure-16. Curve chain selected

- The other options of the dialog box are same as discussed earlier. After specifying the parameter, click on the **OK** button from **PROJECT** dialog box. The toolpath will be generated and displayed on the model; refer to Figure-17.

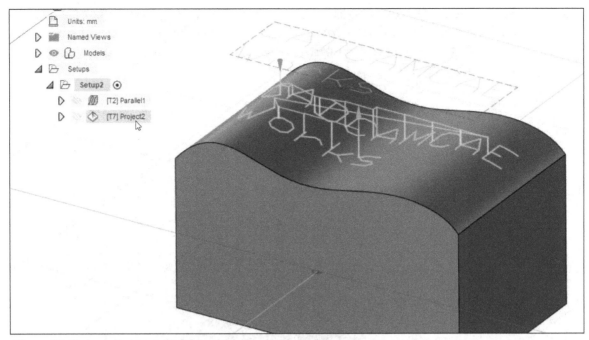

Figure-17. Generated toolpath for Project tool

Morph

The **Morph** tool is used to create the toolpath for machining shallow areas between selected contour. The morph toolpaths look similar to flowlines or parallel toolpaths but the main difference is containment boundaries. In morph toolpaths, cutting tool enters from one of the open side and exits from the other side while avoiding selected contours. The procedure to use this tool is discussed next.

- Click on the **Morph** tool of **3D** drop-down from **Toolbar**; refer to Figure-18. The **MORPH** dialog box will be displayed; refer to Figure-19.

Figure-18. Morph tool

Figure-19. MORPH dialog box

- Click on the **Select** button of **Tool** section from **Tool** tab. The **Select Tool** dialog box will be displayed. Select the required tool generally a ball or bull nose end mill.
- Select the curve to be used as containment boundaries for toolpath from the model; refer to Figure-20.

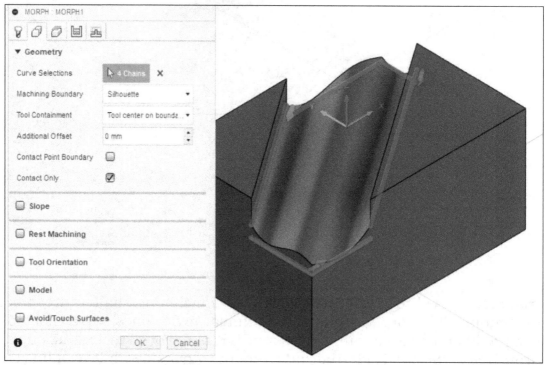

Figure-20. Curves selected for morph toolpath

- Select desired option from the **Morph Mode** drop-down in the **Passes** tab of dialog box to define how direction of toolpath will be decided. Select the **Simple** option to create linear movements of tool from one curve chain side to another. Select the **Closest** option from drop-down to make cutting tool jump to nearest point in another side while cutting. Note that sometimes selecting **Closest** option can cause unmachined areas of model. Use this option when all the chain sides are equal.
- The options of the dialog box are same as discussed earlier.
- After specifying the various parameters, click on the **OK** button from the **MORPH** dialog box. The toolpath will be generated and displayed on the model; refer to Figure-21.

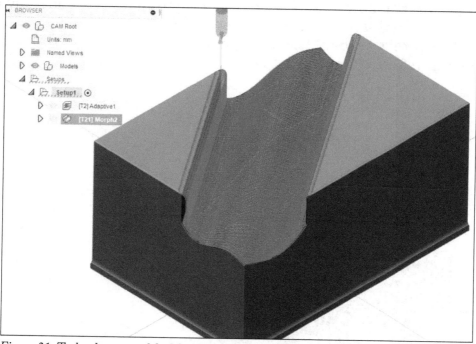

Figure-21. Toolpath generated for Morph tool

Creating Flow Toolpath

The Flow Toolpath is used to machine irregular surfaces of the model by following isocurves of the model. By default, the flow toolpaths are 3D toolpaths but you can also activate the multi-axes mode if tilting of tool axis is required. Using this toolpath offers a better finish of 3D surfaces. The procedure to create this toolpath is given next.

- Click on the **Flow** tool from the **3D** drop-down in the **Ribbon**. The **Flow** dialog box will be displayed; refer to Figure-22.

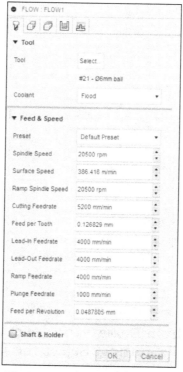

Figure-22. Flow dialog box

- Click on the **Select** button and select the desired fillet tool, generally a ball end mill. Specify desired feed and speed parameters.
- Click on the **Geometry** tab and select the round faces on which you want to apply flow toolpath; refer to Figure-23.

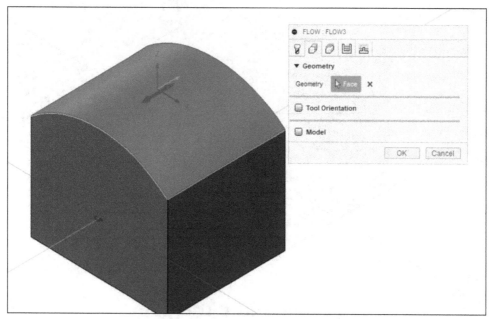

Figure-23. Faces selected for flow toolpath

- Click on the **Passes** tab in the dialog box. The options will be displayed as shown in Figure-24.

Figure-24. Passes tab for Flow dialog box

- Select the **Along u** option from **Isometric Direction** drop-down if you want the flow lines along X axis while cutting. Select the **Along v** option from the drop-down if you want the flow lines along Y axis while cutting. Note that the selected flow direction will be displayed on the model by two headed arrow on the face of model; refer to Figure-23.

- Specify the desired value of **Number of Stepovers** edit box to increase the finishing steps of cut. Note that cutting tool will perform only specified number of cutting passes so make sure to input a high value for better finish.
- Select the **Use Multi-Axis** check box to allow tilting of tool/workpiece bed to create multi-axis toolpath.
- Specify the other parameters as discussed earlier and click on the **OK** button. The toolpath will be created; refer to Figure-25.

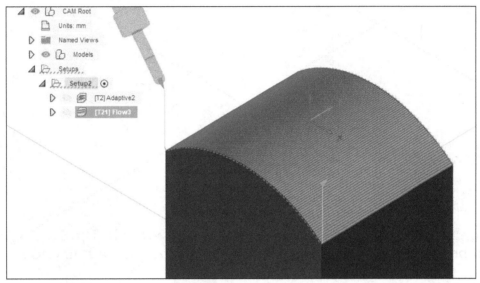

Figure-25. Flow toolpath created

MULTI-AXIS TOOLPATH

Till now, we have discussed the procedure of creating 2D and 3D toolpath. In this section, we will discuss the procedure of creating multi-axis toolpaths which are used to machine complex 3D shapes.

Creating Swarf Toolpath

Swarf is a multi-axis toolpath which is used for side cutting of model. It is used for milling the beveled edges and tapered walls. The procedure to create this toolpath is discussed next.

- Click on the **Swarf** tool of **MULTI-AXIS** drop-down from **Toolbar**; refer to Figure-26. The **SWARF** dialog box will be displayed; refer to Figure-27.

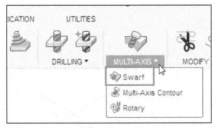

Figure-26. Swarf tool

Figure-27. SWARF dialog box

- Click on the **Select** button of **Tool** section from **Tool** tab. The **Select Tool** dialog box will be displayed. Select the required tool generally, a ball end mill.

Geometry

- Click on the **Geometry** tab of **SWARF** dialog box. The **Geometry** tab will be displayed; refer to Figure-28.

Figure-28. Geometry tab of SWARF dialog box

- Select the **Contours** option from **Drive Mode** drop-down to use boundary chains of selected face of model as driving element for cutting toolpaths.
- Select the **Surface** option from **Drive Mode** drop-down to use selected surfaces of the model as driving element for cutting toolpaths.
- Select **Faces** option from **Selection Mode** drop-down to select continuous faces from model which are to be machined.

- Select **Contour pairs** option from **Selection Mode** drop-down to select the upper and lower edges chain from the tapered face which will be machined.
- Select **Manual** option from **Selection Mode** drop-down to select the individual surfaces from the model to be machined.
- The **Nothing** button of **Surfaces** section is active by default. You need to select the face, edges, or contours as desired.
- Select the **Machine other Side** check box of **Geometry** section to force the toolpath to be created on the other side of selected contour.
- Select the **Flip Top-Bottom** check box of **Geometry** section to change the direction of tool by switching the lower and upper contour. This option will only be applied when **Contours pairs** option is not selected in the **Select Mode** drop-down.
- Click in the **Synchronization Mode** drop-down from **Geometry** tab and select the desired option if **Contours** option is selected in the **Drive Mode** drop-down. The option specified in this drop-down defines how two consecutive cutting passes are related to each other.

Passes

- Click on the **Passes** tab of the **SWARF** dialog box. The options of **Passes** tab will be displayed; refer to Figure-29.

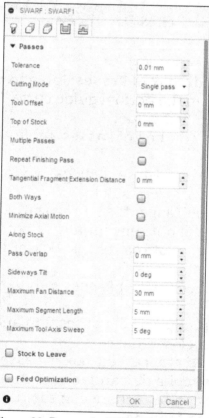

Figure-29. Passes tab of SWARF dialog box

- Click on the **Cutting Mode** drop-down of **Passes** tab and select the required option to specify the pattern in which cutting passes will be created; refer to Figure-30.

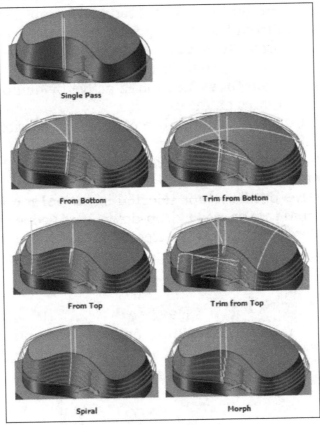

Figure-30. Cutting mode options

- Click in the **Tool Offset** edit box of **Passes** tab and enter the value of tool offset along the tool axis relative to the bottom guide curve. The cutting tool will be away from face of model by specified value.
- Click in the **Top of Stock** edit box of **Passes** tab to specify the overall thickness of the stock.
- Select the **Repeat Finishing Passes** check box of **Passes** tab to perform an additional finishing pass with zero stock.
- Specify desired value in the **Tangential Fragment Extension Distance** edit box of **Passes** tab to extends the cut tangentially at both ends of the pass.
- Select the **Minimize axial Motion** check box of **Passes** tab to minimize the motion of tool.
- Select the **Along Stock** check box of **Passes** tab to move the tool along stock for machining.
- Click in the **Pass Overlap** edit box of **Passes** tab and enter the value of distance to extend machining for a closed pass.
- Click in the **Maximum Fan Distance** edit box of **Passes** tab and specify maximum distance up to which the tool axis can fan out. As the toolpath moves from one surface to another, there can be a change in the ruling direction causing the tool axis to move. This is called fanning.
- Click in the **Sideways tilt** edit box of **Passes** tab and specify the value in degrees at which the cutting tool should be tilted sideways.
- Click in the **Maximum Segment Length** edit box of **Passes** tab and specify the value of maximum length for a single segment for the generated for toolpath.
- Click in the **Maximum Tool Axis Sweep** edit box of **Passes** tab and specify the maximum angle up to which tool can deviate about its axis while cutting.

- The other options of the dialog box are same as discussed earlier. After specifying desired parameters, click on the **OK** button from **SWARF** dialog box. The toolpath will be generated and displayed on the model; refer to Figure-31.

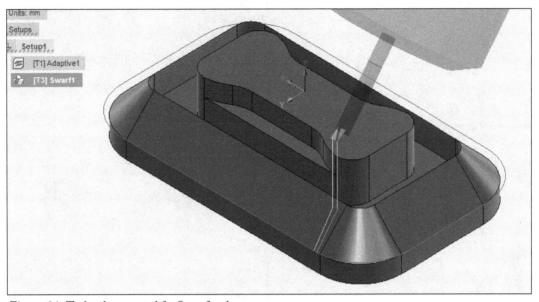

Figure-31. Toolpath generated for Swarf tool

Creating Multi-Axis Contour Toolpath

The **Multi-Axis Contour** tool is used for machining curves with the help of 5-Axis machining. These curves lie on the face of a model forming different 3D curvatures. The procedure to use this tool is discussed next.

- Click on the **Multi-Axis Contour** tool of the **Multi-Axis** drop-down from the **Toolbar**; refer to Figure-32. The **MULTI-AXIS CONTOUR** dialog box will be displayed; refer to Figure-33.

Figure-32. Multi-Axis Contour tool

- Click on the **Select** button of the **Tool** section from the **Tool** tab. The **Select Tool** dialog box will be displayed. Select the required tool generally a ball end mill.
- Click on the curve from the model to select for machining. You can also select more than 1 curves for machining.
- Click on the **Passes** tab of **MULTI-AXIS CONTOUR** dialog box. The **Passes** tab will be displayed; refer to Figure-34.

Figure-33. *MULTI-AXIS CONTOUR dialog box*

Figure-34. *Passes tab of MULTI-AXIS CONTOUR dialog box*

- Click on the **Cutting Mode** drop-down from **MULTI-AXIS CONTOUR** dialog box and select required option to specify the method for machining along a specific contact path.
- Click on the **Sideways Compensation** drop-down and select the required option.
- Click in the **Forward Tilt** edit box of **MULTI-AXIS CONTOUR** dialog box and enter the angular value by which tool should be tilted forward.
- Click in the **Minimum Tilt** edit box of **MULTI-AXIS CONTOUR** dialog box and specify the minimum value of tilting allowed from the selected operation tool axis.
- Click in the **Maximum Tilt** edit box of **MULTI-AXIS CONTOUR** dialog box and specify the maximum value of tilting allowed from the selected operation tool axis.
- The other options of the dialog box are same as discussed earlier. After specifying the parameter, click on the **OK** button from **MULTI-AXIS CONTOUR** dialog box. The toolpath will be generated and displayed on the model; refer to Figure-35.

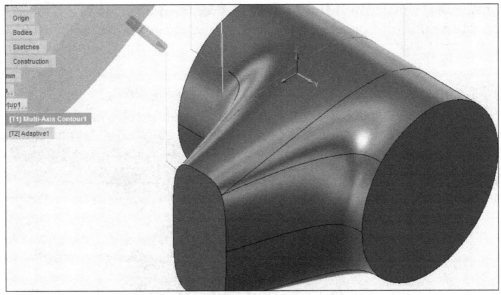

Figure-35. Generated toolpath for Multi-Axis Contour tool

Creating Multi-Axis Rotary Toolpath

The multi-axis rotary toolpath is created to machine objects with rotary axis. The procedure to create this toolpath is given next.

- Click on the **Rotary** tool from the **MULTI-AXIS** drop-down in the **MILLING** tab of the **Ribbon**. The **ROTARY** dialog box will be displayed; refer to Figure-36.

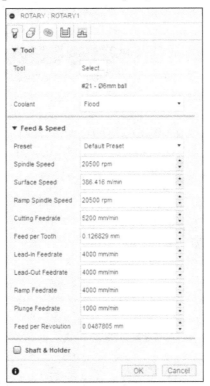

Figure-36. ROTARY dialog box

- Select desired cutting tool and specify cutting feed parameters in the **Tool** tab of dialog box as discussed earlier.
- Click on the **Passes** tab and select desired option from the **Style** drop-down to define pattern for rotary cutting passes. Select the **Spiral** option to general spiral

cutting passes, select the **Line** option to create linear cutting passes, or select the **Circular** option to create concentric circular toolpaths; refer to Figure-37.

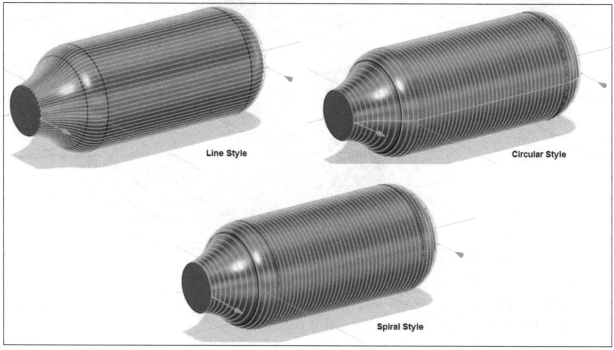

Figure-37. Rotary toolpath styles

• Specify other parameters as discussed earlier in the dialog box and click on the **OK** button. The multi-axis toolpath will be generated.

CREATING DRILL TOOLPATH

The **Drill** tool is used to perform drilling at the specified locations. The procedure to use this tool is discussed next.

• Click on the **Drill** tool from **DRILLING** panel in the **Toolbar**; refer to Figure-38. The **DRILL** dialog box will be displayed; refer to Figure-39.

Figure-38. Drilling tool

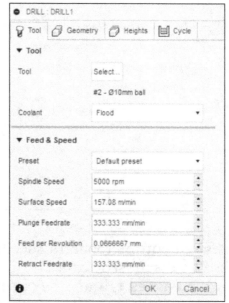

Figure-39. DRILL dialog box

- Click on the **Select** button of **Tool** section from **Tool** tab. The **Select Tool** dialog box will be displayed. Select required drill tool for drilling. Note that you can also select boring, threading, tapping, reaming, and probe tool if you are going to perform related operation.

Geometry

- Click on the **Geometry** tab of **DRILL** dialog box. The **Geometry** tab will be displayed; refer to Figure-40.

Figure-40. Geometry tab of DRILL dialog box

- Select the **Selected faces** option from **Hole Mode** drop-down to select faces of hole from model for drilling. Select the **Selected points** option from **Hole Mode** drop-down to select points of holes for drilling. Select the **Diameter range** option from drop-down if you want to select all the holes inside specified maximum & minimum diameter range.
- The **Nothing** button for **Hole Faces/Hole Points/Containment Boundary** option (based on your selection in **Hole Mode** drop-down) in **Geometry** tab is active by default. Click on desired geometries from the model; refer to Figure-41.

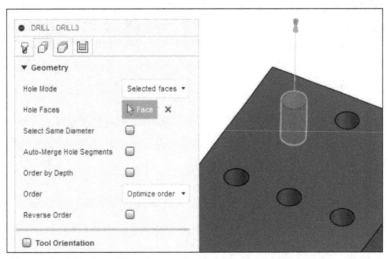

Figure-41. Selection of face for creating hole

- Select the **Select Same Diameter** check box of **Geometry** tab to select the hole of same diameter.

- Select the **Auto-Merge Hole Segments** check box of **Geometry** tab to include the neighboring segments automatically while drilling a hole with multiple segments.
- Select the **Optimize Order** check box of **Geometry** tab to minimize the ordering of hole by ordering the holes.
- Select desired option from the **Order** drop-down to define order of drilling holes.

Cycle

- Click on the **Cycle** tab of **DRILL** dialog box. The options will be displayed; refer to Figure-42.

Figure-42. Cycle tab of DRILL dialog box

- Click in the **Cycle Type** drop-down of **Cycle** tab and select desired option.
- The other options of the dialog box are same as discussed earlier.
- After specifying the parameters, click on the **OK** button from **DRILL** dialog box. The toolpath will be generated and displayed on the model; refer to Figure-43.

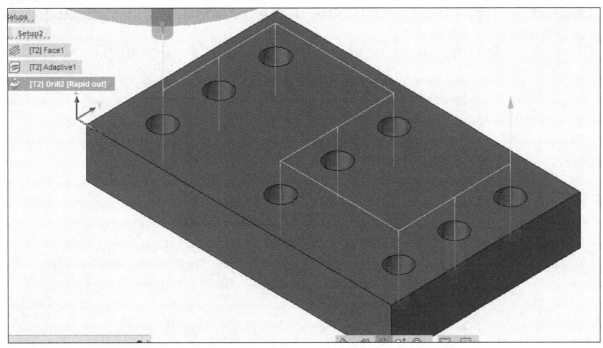

Figure–43. Toolpath generated for drill

TOOLPATH BY HOLE RECOGNITION

The **Hole Recognition** tool is used to create toolpath for drilling holes automatically based on recognized holes. The procedure to create toolpath is given next.

- Click on the **Hole Recognition** tool from the **DRILLING** panel in the **MILLING** tab of **Toolbar**. The **Hole Recognition** dialog box will be displayed with all the holes selected; refer to Figure-44.

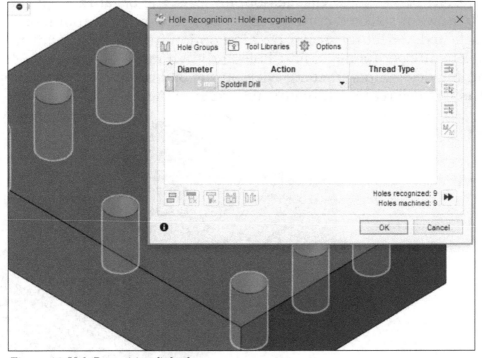

Figure–44. Hole Recognition dialog box

- Select desired tool type from the drop-down in the **Action** column of **Hole Groups** tab in the dialog box like Simple Drill, Spot Drill, and so on.

- Click on the **Tool Libraries** tab in the dialog box and select check boxes for tools you want to use for drilling.
- Click on the **Options** tab in the dialog box. The options will be displayed as shown in Figure-45.

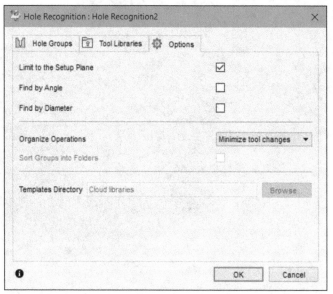

Figure-45. Options tab in Hole Recognition dialog box

- Select the **Limit to the Setup Plane** check box to limit search for holes to current active plane.
- Select the **Find by Angle** check box and specify the minimum-maximum angle limits recognizing holes.
- Select the **Find by Diameter** check box and specify the maximum diameter up to which you want to search for the holes.
- Select the **Minimize tool changes** option from the **Organize Operations** drop-down if you want selected cutting tool to complete all operations before the next tool is selected. Select the **Group by size** option from the drop-down if you want to complete all operations on first hole and then move to another hole.
- After setting desired parameters, click on the **OK** button from the dialog box. The toolpaths will be created automatically; refer to Figure-46.

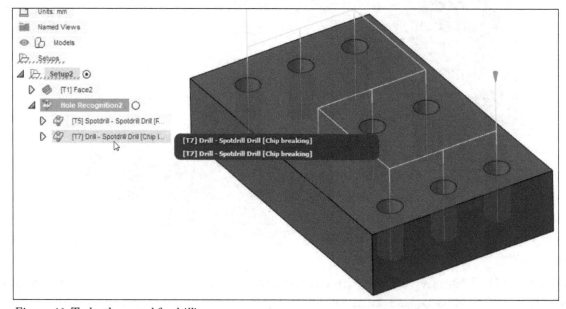

Figure-46. Toolpaths created for drilling

PRACTICAL

Create CAM program of the given model using 3D and 2D toolpaths; refer to Figure-47.

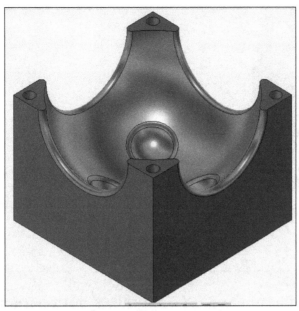

Figure-47. Practical 1

Adding Model to MANUFACTURE

- Create and save the part in **MANUFACTURE** workspace. The part file is available in the respective chapter folder of **Autodesk Fusion 360 Black Book** resources.
- Click on the **MANUFACTURE** workspace from **Workspace** drop-down. The model will be displayed in the **MANUFACTURE** workspace.

Creating Stock

- Click on the **New Setup** tool of **SETUP** drop-down from **Toolbar**. The **SETUP** dialog box will be displayed along with the stock of model.
- Specify the parameters of **Setup** tab, **Stock** tab and **Post Process** tab of **SETUP** dialog box as shown in Figure-48.

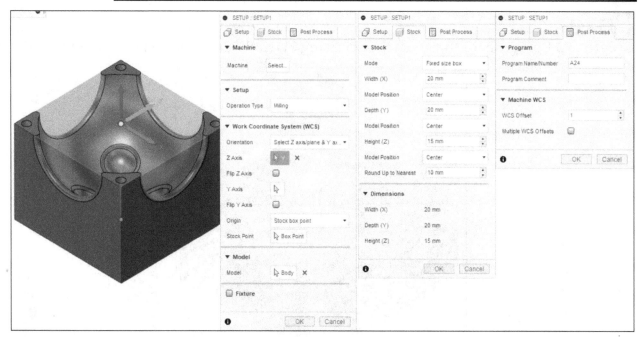

Figure–48. Specifying parameters for stock

- After specifying the parameters, click on the **OK** button from **SETUP** dialog box. The stock will be created and displayed on the model.

Generating Adaptive Clearing Toolpath

- Click on the **Adaptive Clearing** tool of **3D** drop-down from **Toolbar**. The **ADAPTIVE** dialog box will be displayed.
- Click on the **Select** button of **Tool** tab and select the tool from the record list of **Select Tool** dialog box as shown in Figure-49.
- If the tool is not available in the list then create a new ball mill tool. After selecting the tool, click on the **OK** button from **Select Tool** dialog box. The tool will be added in the **ADAPTIVE** dialog box.

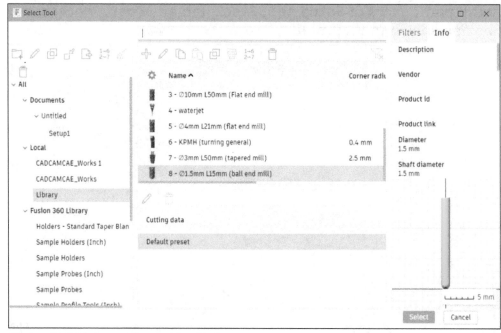

Figure–49. Selecting tool for adaptive clearing

- Specify the parameters of **Geometry** tab, **Passes** tab, and **Linking** tab of **ADAPTIVE** dialog box; refer to Figure-50.

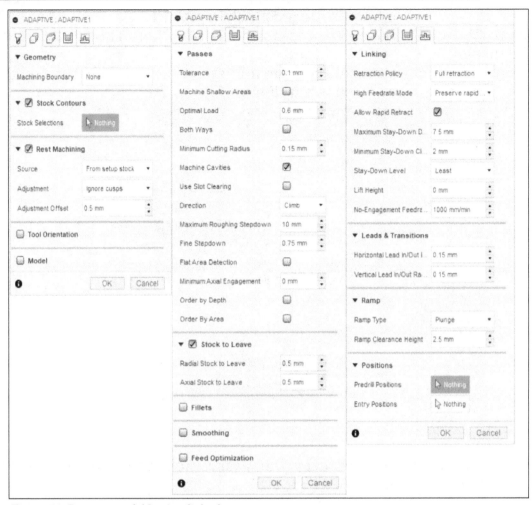

Figure-50. Parameters of Adaptive dialog box

- After specifying the parameters, click on the **OK** button from **ADAPTIVE** dialog box. The toolpath will be generated and displayed on the model; refer to Figure-51.

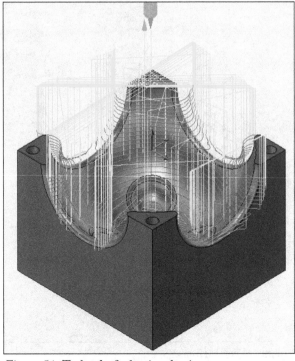

Figure-51. Toolpath of adaptive clearing

Generating Spiral Toolpath

- Click on the **Spiral** tool of **3D** drop-down from **Toolbar**. The **SPIRAL** dialog box will be displayed.
- Click on the **Select** button of **Tool** tab and select the tool from the record list of **Select Tool** dialog box as shown in Figure-52.

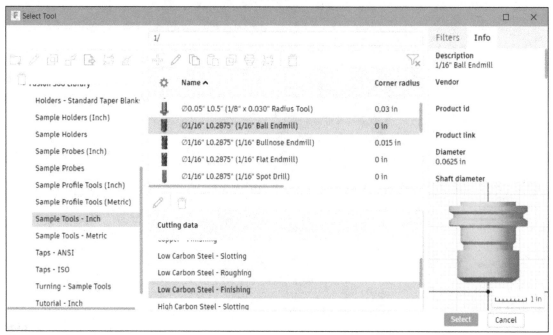

Figure-52. Selecting tool for spiral

- Specify the parameters of **Geometry** tab, **Passes** tab, and **Linking** tab of **SPIRAL** dialog box; refer to Figure-53.

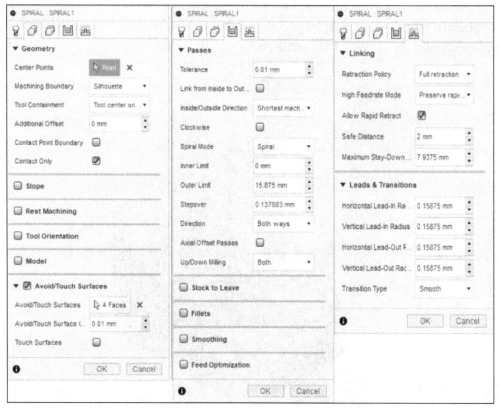

Figure-53. Specifying parameters for spiral

Generating Scallop Toolpath

- Click on the **Scallop** tool of **3D** drop-down from **Toolbar**. The **SCALLOP** dialog box will be displayed.
- Click on the **Select** button of **Tool** section and select the tool which is selected in last operation from **Select Tool** dialog box.
- Specify the parameters of **Tool** tab, **Geometry** tab, **Passes** tab, and **Linking** tab as shown in Figure-54.

Figure-54. Parameters for Scallop tool

- After specifying the parameters, click on the **OK** button from the **SCALLOP** dialog box. The toolpath will be created and displayed on the model; refer to Figure-55.

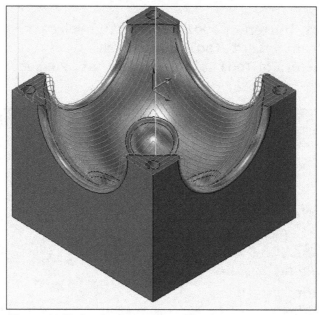

Figure-55. Toolpath generated for scallop

Generating Radial Toolpath

- Click on the **Radial** tool of **3D** drop-down from **Toolbar**. The **RADIAL** dialog box will be displayed.
- Click on the **Select** button of **Tool** section and select the tool which is selected in last operation from **Select Tool** dialog box.
- Click on the **Geometry** tab of the **RADIAL** dialog box. The **Geometry** tab will be displayed.
- Click on the **Machine Boundary** drop-down and select the **Selection** option.
- The **Nothing** button of **Machine Boundary** option is active by default. You need to select the chain for machining; refer to Figure-56.

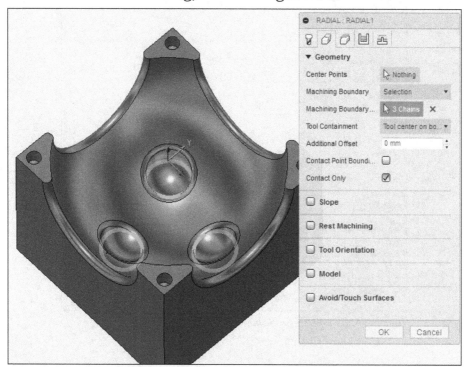

Figure-56. Selecting chains for machining

- Specify the parameters of **Passes** tab and **Geometry** tab as displayed in Figure-57.

Figure-57. Parameters for Radial dialog box

- After specifying the parameters, click on the **OK** button from **RADIAL** dialog box. The toolpath will be created and displayed on the model; refer to Figure-58.

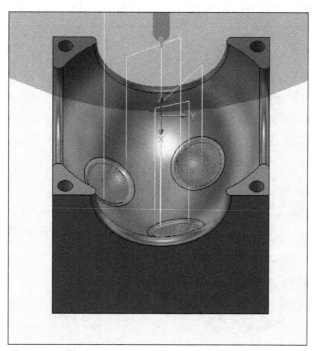

Figure-58. Toolpath generated for Radial tool

Generating Drill Toolpath

- Click on the **Drill** tool from **Toolbar**. The **DRILL** dialog box will be displayed.
- Click on the **Select** button of **Tool** section and select the tool from **Select Tool** dialog box as displayed in Figure-59.

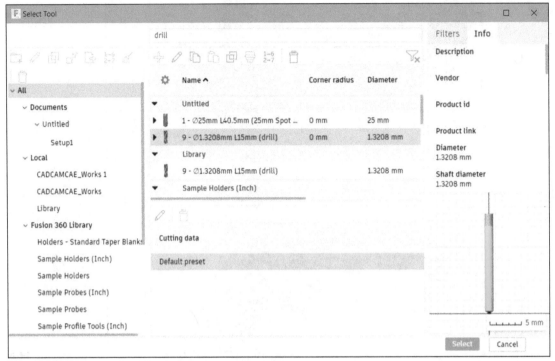

Figure-59. Selection of tool for drill

- After selecting the tool for drill, click on the **OK** button from **Select Tool** dialog box. The tool will be added in the **DRILL** dialog box.
- Specify the parameters of **Geometry** tab from **DRILL** dialog box as displayed in Figure-60. Select the round face of one of the hole.

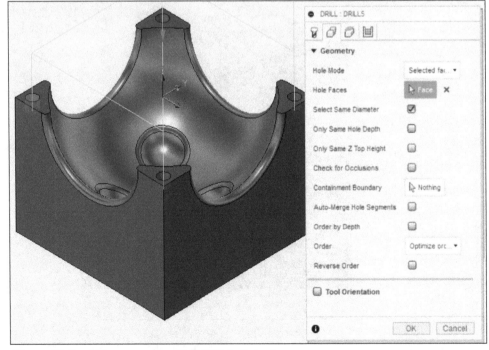

Figure-60. Specifying geometry tab of DRILL dialog box

- After specifying the parameters, click on the **OK** button from **DRILL** dialog box. The toolpath will be generated and displayed on the model; refer to Figure-61.

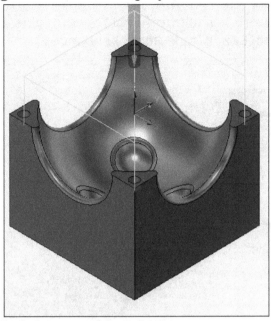

Figure-61. Generated toolpath for Drill tool

- Till now, we have applied the tools to create the required toolpath. Now, we will simulate the whole process to check whether the generated toolpath is working properly or there is a collision of tool with stock.
- Click on the **Simulate** button from shortcut menu of **Setup1** option in the **BROWSER**. The **SIMULATE** dialog box will be displayed.
- Specify the parameters of **SIMULATE** dialog box as required and click on the **Play** button. The simulation of machining process will be displayed; refer to Figure-62. Make sure to create finishing toolpaths as discussed earlier.

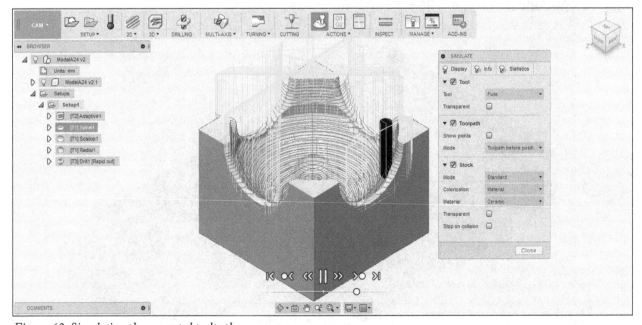

Figure-62. Simulating the generated toolpath

PRACTICE 1

Machine the stock of length as 148 mm, width as 52 mm, height as 21 mm to create the part as shown in Figure-63. The part file of this model is available in the respective folder of **Autodesk Fusion 360 Black Book Resources**.

Figure-63. Practice 1

PRACTICE 2

Machine the stock of length as 102 mm, width as 62 mm, height as 23 mm to create the part as shown in Figure-64. The part file of this model is available in the respective folder of **Autodesk Fusion 360 Black Book Resources**.

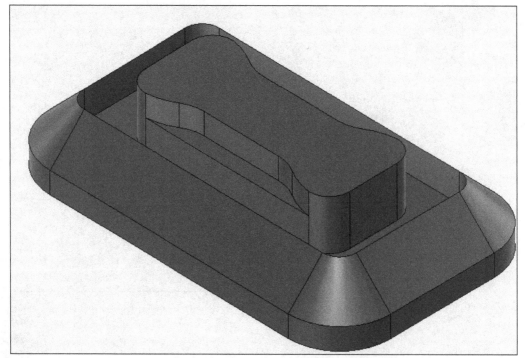

Figure-64. Practice 2

PRACTICE 3

Machine the stock of stock diameter as 80 mm, thickness as 7 mm, and align stock to the center of the part as shown in Figure-65. The part file of this model is available in the respective folder of **Autodesk Fusion 360 Black Book Resources**.

Figure-65. Practice 3

FOR STUDENT NOTES

Chapter 18

Generating Turning and Cutting Toolpaths

Topics Covered

The major topics covered in this chapter are:

- *Turning Profile*
- *Turning Face*
- *Turning Chamfer*
- *Turning Thread*
- *Turning Groove*
- *Cutting Toolpaths*

GENERATING TURNING TOOLPATH

Till now, we have discussed the procedure of creating milling toolpaths with the help of various tools. In this section, we will discuss the tools used for creating Turning toolpaths. You need to open a cylindrical part for turning in Autodesk Fusion and create a machine setup of Turning machine as discussed earlier in Chapter 15 before working on this chapter. The tools to generate turning toolpaths are available in the **TURNING** tab in the **Ribbon**; refer to Figure-1.

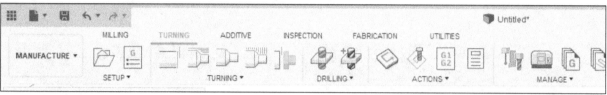

Figure-1. TURNING tab

Turning Face

The **Turning Face** tool is used for machining the front face of the part. The turning face toolpath is generally the first toolpath created for machining on CNC Turn machine. The procedure to use this tool is discussed next.

- Click on the **Turning Face** tool of **TURNING** drop-down from **Toolbar**; refer to Figure-2. The **FACE** dialog box will be displayed; refer to Figure-3.

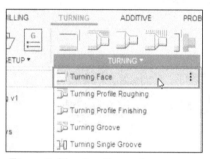

Figure-2. Turning Face tool

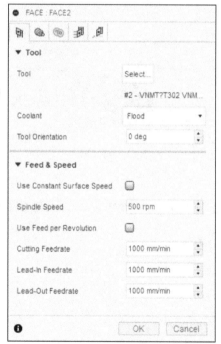

Figure-3. FACE dialog box

- Click on the **Select** selection button from **Tool** section. The **Select Tool** dialog box will be displayed. Select desired tool and click on **OK** button. The tool will be added in the **FACE** dialog box.
- Click in the **Tool Orientation** edit box of the **Tool** tab and enter the angular value to orient the tool as required on the tool post of machine.
- The other options of this dialog box have been discussed in next topic of this chapter.
- After specifying the parameters, click on the **OK** button from **FACE** dialog box. The toolpath will be generated and displayed on the model; refer to Figure-4.

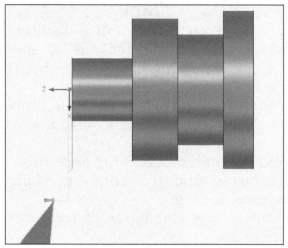

Figure-4. Toolpath generated for facing

Turning Profile Roughing

The **Turning Profile Roughing** tool is used for roughing the profile of model using various tools. The procedure to use this tool is discussed next.

- Click on the **Turning Profile Roughing** tool of **TURNING** drop-down from **Toolbar**; refer to Figure-5. The **PROFILE ROUGHING** dialog box will be displayed; refer to Figure-6.

Figure-5. Turning Profile tool

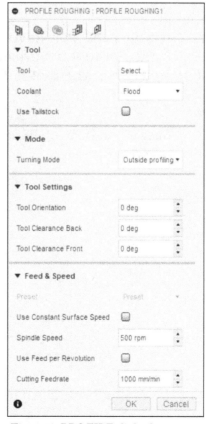

Figure-6. PROFILE dialog box

- Click on the **Select** button of **Tool** option from **Tool** tab. The **Select Tool** dialog box will be displayed. Select the desired tool for profiling and click on the **OK** button. Note that the shape and size of cutting tool depends on shape and size of model. The tool will be added in the **PROFILE ROUGHING** dialog box.

Tool

- Click on the **Coolant** drop-down of **Tool** tab and select desired option.
- Select the **Use Tailstock** check box of **Tool** tab to apply support to the part to be machined. When you have a long part for turning then you should use tail stock to support the other end of part.
- Select the **Outside profiling** option of **Turning Mode** drop-down from **Mode** section to retract the tool outside of the stock and machine axially depending on the direction setting.
- Select the **Inside Profiling** option of **Turning Mode** drop-down from **Mode** section to approach or retract the tool from the center-line of part and machines radially depending on the direction setting.
- Set desired angle values in the edit boxes of **Tool Settings** section to define orientation of the cutting tool.
- Select the **Use Constant Surface Speed** check box of **Feed & Speed** section from **Tool** tab to automatically adjust the spindle speed for maintaining a constant surface speed between tool and the workpiece.
- Click in the **Surface Speed** edit box of **Feed & Speed** section to specify the value of spindle speed expressed as the speed of the tool on the surface.
- Click in the **Maximum Spindle Speed** edit box of **Feed & Speed** section and specify the value of maximum allowed spindle speed when using constant surface speed.
- Select the **Use Feed Per Revolution** check box of **Feed & Speed** section to automatically adjust the feed rate based on the RPM of the spindle to maintain a constant chip load.
- Click in the **Cutting Feed Per Revolution** edit box of **Feed & Speed** section and specify the distance of tool advancement into the material for each 360 degree rotation of spindle when the tool is fully engaged.

Geometry

- Click on the **Geometry** tab of **PROFILE** dialog box. The **Geometry** tab will be displayed; refer to Figure-7.

Figure-7. Geometry tab of PROFILE dialog box

- Select the **Model** check box of **Geometry** tab to select a model contour for machining. The **Nothing** button of **Model Contour** option will be active by default. You need to click on the contour of model to be used as profile for cutting.

- Select desired option from **Front Mode** drop-down and define the reference for front confinement region. Confinement region is an imaginary space within which machining of part will be done.
- Click in the **Offset** edit box of **Front** section in **Geometry** tab and specify the value to define offset distance for front confinement plane from selected reference in the **Front Mode** drop-down. You can specify positive as well as negative values.
- Specify the value of extension in **Tangential Extension** edit box if you want the open profiles to be tangentially extended.
- Similarly, specify parameters for back plane of confinement region.
- Select the **Rest Machining** check box of **Geometry** tab to specify that only stock left after the previous operations should be machined.
- Click in the **Source** drop-down of **Rest Machining** section and select the desired option to specify the source from which the rest machining stock should be calculated.

Radii

Click on the **Radii** tab of **PROFILE** dialog box. The options of **Radii** tab will be displayed for defining clearance radius, outer radius and inner radius of part; refer to Figure-8.

Figure-8. Radii tab of PROFILE dialog box

- Click i the **From** drop-down of **Clearance** section and select the required reference for clearance location.
- Click in the **Offset** edit box of **Clearance** section and specify desired value to increase clearance radius with respect to selected reference.
- Click in the **From** drop-down of **Outer Radius** section and select desired reference for defining outer diameter of part.
- Click in the **Offset** edit box of **Outer Radius** section and specify desired value to increase outer radius with respect to selected reference.
- Click in the **From** drop-down of **Inner Radius** section and select desired reference for inner diameter of part.
- Click in the **Offset** edit box of **Inner Radius** section and enter the desired value.

- Select desired option from the **Tool Limit** drop-down to define the reference to be used as limiting point for creating toolpath.
- Click in the **Distance to Cut Below Inner Radius** edit box and specify the value of distance up to which the cutting tool should continue cutting when the tool has reached centerline of part during parting or facing operations.

Passes

Click on the **Passes** tab of **PROFILE** dialog box to specify number of cutting passes to be created and their direction. The options of **Passes** tab will be displayed as shown in Figure-9.

Figure-9. Passes tab of PROFILE
ROUGHING dialog box

- Select desired option from the **Cycle** drop-down to define how tool will retract and come back for next cutting pass. Select the **Horizontal passes**, **Vertical passes**, or **Back cutting** option to create cutting passes oriented in respective direction.
- Select the **Front to back** option of **Direction** drop-down from **Cycle and Direction** section to cut from the front side of the stock towards the back side when **Horizontal passes** or **Vertical passes** option is selected in **Cycle** drop-down.
- Select the **Back to Front** option of **Direction** drop-down from **Cycle and Direction**

section to cut the material of stock from back side towards the front side when **Horizontal passes** or **Vertical passes** option is selected in **Cycle** drop-down.

- Select the **Both Ways** option of **Direction** drop-down from **Cycle and Direction** section to cut the stock in both forward and backward direction.
- Click in the **Grooving** drop-down from **Cycle and Direction** section and select desired option to define which type of grooving is allowed while cutting.
- Click in the **Tolerance** edit box of **Passes** tab and enter the value of tolerance for each cutting pass.
- Specify the maximum depth of cut allowed for each cutting pass in the **Maximum Depth of Cut** edit box.
- Select the **Make Sharp Corners** check box of **Passes** tab to specify that sharp corners must be created at the corners and edges if radius is not specified.
- Select the **Use Pecking** check box to use multiple retractions along its path when tool is cutting material so that long chips are not formed while cutting. Specify the related depth and retraction value for pecking in respective edit boxes below the **Use Pecking** check box.
- Specify the desired values of material to be left for finishing toolpath in the edit boxes of **Stock to Leave** section of the dialog box.
- Select the **Extend to Stock** check box to extend toolpaths beyond the part surface until it reaches the boundaries of stock.

Linking

Click on the **Linking** tab of **PROFILE ROUGHING** dialog box to define how cutting passes will be connected in the toolpath. The options of **Linking** tab will be displayed; refer to Figure-10.

- Select desired option from the **High Feedrate Mode** drop-down to define when to use G01 code of rapid feed move and when to use G00 code of rapid retraction move if there is not cutting operation involved in the current step. Select the **Always use high feed** option when there are chances of collision at maximum rapid move speed and specify the desired speed in the **High Feedrate** edit box.
- Set desired options in **Approach Z** and **Retract Z** drop-downs to define the respective safe Z locations. After cutting passes tool will move to this location.
- Select the **Override Setup Safe Z** check box if you want to override default reference for Safe Z location and specify related offset distance.
- Specify desired values in the **Z Clearance** and **X Clearance** edit boxes of **Clearance** section to define distance in Z and X direction from the part up to which the cutting tool can reach after performing first cutting pass.
- Specify desired value of retraction distance for cutting pass in the **Retract Distance** edit box.
- Select the **Angled Entry** check box if you want the cutting tool to enter stock from specified angle with respect to positive Z axis. After selecting the check box, specify desired entry angle value in the **Entry Angle** edit box. Click in the **Entry Clearance** edit box to define distance from the stock at which cutting tool will start moving in cutting feed mode. Click in the **Entry Feedrate** edit box and specify the feed rate at which cutting tool will move while entering the stock.
- The other options of the dialog box are same as discussed earlier in the book.
- After specifying the parameter, click on the **OK** button from the **PROFILE ROUGHING** dialog box. The toolpath will be generated and displayed on the part; refer to Figure-11.

Figure-10. Linking Tab of PROFILE ROUGHING dialog box

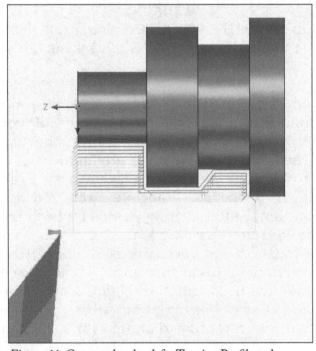

Figure-11. Generated toolpath for Turning Profile tool

Turning Profile Finishing

The **Turning Profile Finishing** tool is used generate toolpath for performing finishing operation on the stock left by previous roughing operations. The procedure to use this tool is discussed next.

- Click on the **Turning Profile Finishing** tool from the **TURNING** drop-down in the **TURNING** tab of **Toolbar**. The **PROFILE FINISHING** dialog box will be displayed as shown in Figure-12.

Figure-12. PROFILE FINISHING dialog box

Most of the options in this dialog box are same as discussed for **PROFILE ROUGHING** dialog box. The options which have not been discussed earlier are discussed next.

- Click in the **Tool Orientation** edit box of **Tool Settings** section and specify the angular value by which cutting tool will be oriented on the tool post. This option is used when your lathe turret has a programmable B-Axis. This option is useful for cutting the stock corners of part.
- Click in the **Tool Clearance Back** edit box of **Tool Settings** section and specify the angular value which is added to the tool angle for providing clearance behind the cutting edge.
- Click in the **Tool Clearance Front** edit box of **Tool Settings** section and specify the angular value which is added to the tool angle for providing clearance in front of the cutting edge. Depending on the condition of your cutting tool insert, you can increase this value to use more portion of cutting tool insert for cutting.
- Select desired option from the **Preset** drop-down to use predefined cutting parameters.
- Select the **Use Feed per Revolution** check box to specify feedrate in mm per revolution of spindle. Note that on selecting this check box, the **Cutting Feedrate**, **Lead-in Feedrate**, and **Lead-Out Feedrate** edit boxes will be converted to **Cutting Feed per Revolution**, **Lead-In Feed per Revolution** and **Lead-Out Feed per Revolution** edit boxes, respectively.

- Click in the **Lead-In feed per Revolution** edit box of **Feed & Speed** section and specify the feed rate of tool in cutting mode before the tool approaches and initially enters the material.
- Click in the **Lead-Out feed per Revolution** edit box of **Feed & Speed** section and specify the feedrate of tool in cutting mode when the tool exists the material.
- Select the **Spring Pass** check box from the **Passes** section of **Passes** tab in the dialog box to create additional finishing pass with zero stock thickness for better finish.
- Select the **No Dragging** check box from the **Passes** section of **Passes** tab in the dialog box to allow changing of tool position when cutting direction is switched between axial and radial. In this way, less drag (stress) is applied on the tool while cutting.
- Specify desired values for no drag parameters: **No Drag Limit**, **No Drag Clearance**, **No Drag Overlap** in their respective edit boxes. The **No Drag Limit** parameter is used to specify the minimum angle after which no drag mode will activate. The **No Drag Clearance** parameter is used to specify the distance from model face at which cutting tool will retract and switch to no drag mode. The **No Drag Overlap** parameter is used to specify overlapping distance of last cut before retracting to activate no drag mode; refer to Figure-13.

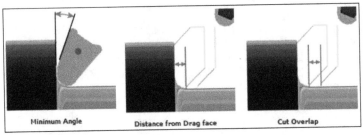

Figure-13. No Drag parameters

- Specify desired parameters in **Leads & Transitions** section of **Linking** tab in the dialog box to define how cutting tool will enter the workpiece in first cutting pass and how tool will exit in last cutting pass.
- Specify the other parameters as discussed earlier and click on the **OK** button from the dialog box to generate the toolpath.

Turning Groove

The **Turning Groove** tool is used for roughing and finishing strategy to create groove cuts in the model. The procedure to use this tool is discussed next.

- Click on the **Turning Groove** tool of **TURNING** drop-down from **Toolbar**; refer to Figure-14. The **GROOVE** dialog box will be displayed; refer to Figure-15.

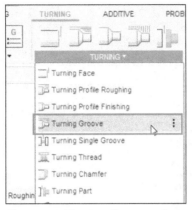

Figure-14. Turning Groove tool

Figure-15. The GROOVE dialog box

- Click on the **Select** button of **Tool** section. The **Select Tool** dialog box will be displayed. Select desired tool, generally an ID groove turning or OD groove turning tool and click on **OK** button. The tool will be added in the **GROOVE** dialog box.

- Click in the **Turning Mode** drop-down of **Mode & Direction** section from the **TOOL** tab and select desired option according to your turning strategy. If you want to create groove on outer diameter of part then select the **Outside grooving** option and if you want to create groove in the inner diameter of part then select the **Inside grooving** option.

- Click in the **Direction** drop-down of **Mode & Direction** section and select desired option to define in which direction the cutting passes will be generated.

- Select desired option from the **Preset** drop-down to set cutting speed and feed rates predefined in tool library for selected cutting tool.

- Click on the **Geometry** tab in the dialog box and set the containment region of groove as shown in Figure-16.

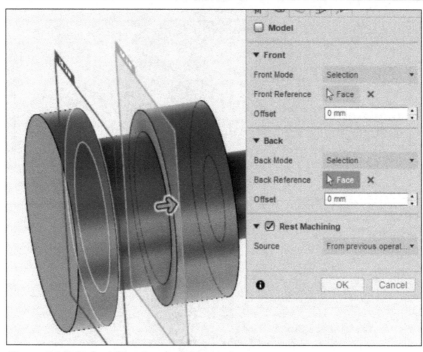

Figure-16. Selection of face for grooving

Passes tab

- Click on the **Passes** tab of the **GROOVE** dialog box. The **Passes** tab will be displayed; refer to Figure-17.

Figure-17. Passes tab of GROOVE dialog box

- Click in the **Tolerance** edit box of **Passes** tab and specify the value of tolerance allowed for linearizing toolpath curves.

- Select the **Use Reduced Feedrate** check box of **Passes** tab to reduce the feedrate when creating groove cuts along the X-Axis.

- Click in the **Up/Down Direction** drop-down of **Passes** tab and select desired option to control the direction of finishing passes.

- Click in the **Pass Overlap** edit box of **Passes** tab to specify the overlap distance used for finishing pass split due to only up and only down machining directions.

- Click in the **Compensation Type** drop-down of **Passes** tab and select desired compensation type.

- Click in the **Backoff Distance** edit box of the **Passes** tab and specify the distance to move back at cutting feed rate from the stock before retracting at rapid rate.

- Select the **Finishing Passes** check box of the **Passes** tab to remove the material from stock using secondary finishing operation. This option is only available when **Roughing pass** check box is selected in this tab.

- Click in the **Number of Stepovers** check box of **Passes** tab and specify the value of number of finishing stepovers to be applied.

- Click in the **Stepover** edit box of **Passes** tab and specify the value of distance between two consecutive cutting passes.

- Click in the **Finish Feedrate** edit box of **Passes** tab and specify the value of feed rate used for final finishing pass.

- Select the **Repeat Finishing Pass** check box of **Passes** tab to perform an additional finishing pass with zero stock.

- Select **Roughing Passes** check box to enable roughing of the part before running finishing passes.

- Select the **Full stepdown** option of **Grooving Pattern** drop-down from **Passes** tab to remove the stock material radially, before moving along the spindle axis to remove material.

- Select the **Partial stepdown** option of **Grooving Pattern** drop-down to remove all the material along the spindle axis before starting to remove the material at the next depth which is specified by the **Maximum Groove Stepdown** value. The **Maximum Groove Stepdown** edit box is displayed when **Partial stepdown** option is selected in the **Grooving Pattern** drop-down.

- Select the **Sideways with Partial Stepdown** option of **Grooving Pattern** drop-down to remove the material from groove by moving cutting tool side by side. Due to this strategy the tool life is increased and it results in smaller chips formation.

- Select the **Use Pecking** check box of **Passes** tab to cut the material by specified pecking depth using pecking force with the help of cutting tool. After performing one cutting pass, the tool then retracts along its path by the specified pecking retract distance.

- The other tools of the dialog box are same as discussed earlier. After specifying the parameters, click on the **OK** button from the **GROOVE** dialog box. The toolpath will be generated and displayed on the model; refer to Figure-18.

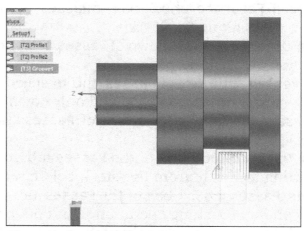

Figure-18. Generated toolpath Groove tool

Turning Single Groove

The **Turning Single Groove** tool is used for machining groove at selected position only. This tool will create a groove equal to the width of the tool. This is perfect for making a clearance groove behind the thread. The procedure to use this tool is discussed next.

• Click on the **Turning Single Groove** tool of **TURNING** drop-down from **Toolbar**; refer to Figure-19. The **SINGLE GROOVE** dialog box will be displayed; refer to Figure-20

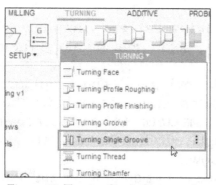

Figure-19. Turning Single Groove tool

Figure-20. SINGLE GROOVE dialog box

• Click on the **Select** button of **Tool** section. The **Select Tool** dialog box will be displayed. Select desired tool, generally a grooving tool and click on **OK** button. The tool will be added in the **SINGLE GROOVE** dialog box.

Geometry

- Click on the **Geometry** tab of **SINGLE GROOVE** dialog box. The options tab will be displayed; refer to Figure-21. The **Nothing** button of **Groove Positions** option of **Geometry** tab will be active by default and you will be asked select edge of the groove to be machined.

Figure-21. Geometry tab of
SINGLE GROOVE dialog box

- Select desired edge of the model for groove machining.
- Click on the **Groove Side Alignment** drop-down of **Geometry** section and select desired option for alignment of tools side edge.
- Click on the **Groove Tip Alignment** drop-down of **Geometry** section and select desired option for alignment of tool tip.
- Select the **Dwell Before Retract** check box to give a pause of specified seconds before grooving tool comes out of the workpiece after cutting. Specify desired dwell time (in seconds) in the **Dwelling Period** edit box.
- Select the **Allow Rapid Retract** check box if you want to use G00 code for retraction of tool. By default, rapid feed rate is used for retraction.
- The other options of the dialog box are same as discussed earlier. After specifying various parameters, click on the **OK** button from the **SINGLE GROOVE** dialog box. The toolpath will be generated and displayed on the model; refer to Figure-22.

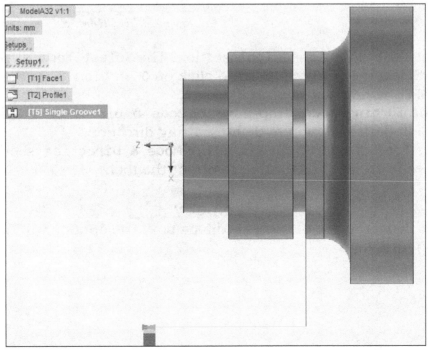

Figure-22. Toolpath generated for Single Groove tool

Turning Chamfer

The **Turning Chamfer** tool is used to create toolpath for chamfering sharp corners of the model that have not been created in the design. The procedure to use this tool is discussed next.

- Click on the **Turning Chamfer** tool of **TURNING** drop-down from the **Toolbar**; refer to Figure-23. The **CHAMFER** dialog box will be displayed; refer to Figure-24.

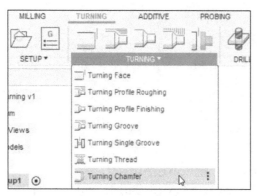

Figure-23. Turning Chamfer tool

Figure-24. The CHAMFER dialog box

- Click on the **Select** button of **Tool** section. The **Select Tool** dialog box will be displayed. Select the required tool and click on **OK** button. The tool will be added in the **CHAMFER** dialog box.
- Click on the **Turning Mode** drop-down of **Mode & Direction** section from **Tool** tab and select desired option to define cutting direction.
- Click in the **Tool Orientation** edit box of **Mode & Direction** section from **Tool** tab and specify desired angle value to orient the tool.

Geometry

- Click in the **Geometry** tab of **CHAMFER** dialog box. The options will be displayed as shown in Figure-25.

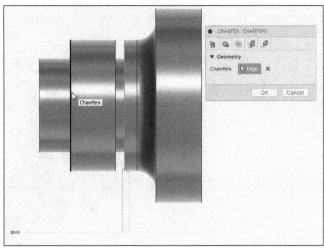

Figure-25. Geometry tab of CHAMFER dialog box

- The **Nothing** button of **Chamfers** option is active by default. Select the edges from model on which you want to apply chamfer.

Passes

- Click on the **Passes** tab of **CHAMFER** dialog box. The **Passes** tab will be displayed; refer to Figure-26.

Figure-26. Passes tab of CHAMFER dialog box

- Click in the **Chamfer Width** edit box of **Passes** tab and specify the value of width of chamfer.
- Click in the **Chamfer Extension** edit box of **Passes** tab and specify the value by which to extend the chamfer cutting pass.
- Click in the **Chamfer Angle** edit box of **Passes** tab and specify the angle value of chamfer measured from the Z-Axis.
- The other options of the dialog box are same as discussed earlier in the book. After specifying the parameters, click on the **OK** button from **CHAMFER** dialog box. The toolpath will be generated and displayed on the model; refer to Figure-27.

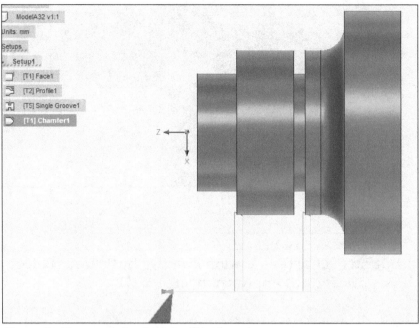

Figure-27. The generated toolpath for Chamfer tool

Turning Part

The **Turning Part** tool is used to cut off the part from the bar. This tool is also known as cut off operation. The procedure to use this tool is discussed next.

- Click on the **Turning Part** tool of **TURNING** drop-down from **Toolbar**; refer to Figure-28. The **PART** dialog box will be displayed; refer to Figure-29.

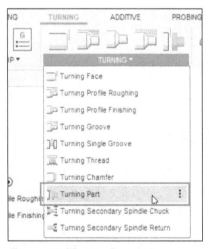

Figure-28. Turning Part tool

Figure-29. PART dialog box

- Click on the **Select** button from **Tool** section. The **Select Tool** dialog box will be displayed. Select desired tool and click on **OK** button. The tool will be added in the **PART** dialog box.
- Select the **Edge Break** check box from the **Geometry** tab in the dialog box to define how edges will be conditioned while parting; refer to Figure-30.

- Select the **Chamfer** option from the **Edge Break Type** drop-down to create chamfer at the outer edge before cut off. Select the **Fillet** option from the **Edge Break Type** drop-down to create round at the outer edge before cut off. Specify the parameters related to the selected option in the edit boxes of **Edge Break** section.

Figure-30. Edge Break options

- The other options of the dialog box are same as discussed earlier. After specifying the parameters, click on the **OK** button from the **PART** dialog box. The toolpath will be generated and displayed on the model; refer to Figure-31.

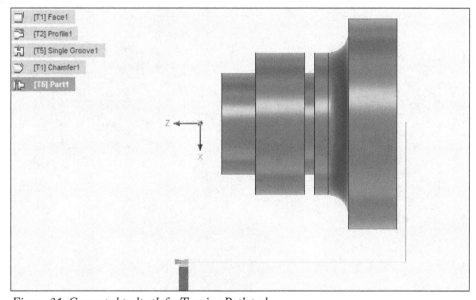

Figure-31. Generated toolpath for Turning Path tool

Turning Thread

The **Turning Thread** tool is used to cut threads on cylindrical and conical surfaces. The procedure to use this tool is discussed next.

- Click on the **Turning Thread** tool of **TURNING** drop-down from **Toolbar**; refer to Figure-32. The **THREAD** dialog box will be displayed; refer to Figure-33.

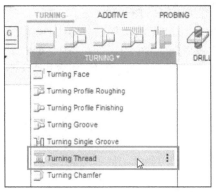

Figure-32. Turning Thread tool

Figure-33. THREAD dialog box

- Click on the **Select** button of **Tool** section. The **Select Tool** dialog box will be displayed. Select the desired tool and click on **OK** button. The tool will be added in the **THREAD** dialog box.
- Click on the **Nothing** button for **Thread Faces** option from **Geometry** tab and select the round faces on which you want to machine threads; refer to Figure-34.

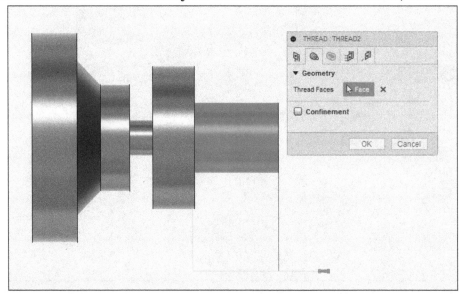

Figure-34. Selection of Face for thread

Passes

- Click on the **Passes** tab of **THREAD** dialog box. The **Passes** tab will be displayed; refer to Figure-35.

Figure-35. Passes tab of THREAD dialog box

- Click on the **Threading Hand** drop-down of **Passes** tab and select desired threading direction.
- Click in the **Thread Depth** edit box of **Passes** tab and specify desired value to define the depth of thread.
- Click in the **Number of Stepdowns** edit box of **Passes** tab and specify desired number of step downs to be performed for achieving desired depth.
- Click in the **Thread Pitch** edit box of **Passes** tab and specify the value of thread pitch (distance between two consecutive strands of thread).
- Select the **Do Multiple Threads** check box of **Passes** tab to create multi-start threads and specify number of threads to be created.
- Click in the **Infeed Mode** drop-down of **Passes** tab and select the required option.
- Click in the **Infeed Angle** edit box of **Passes** tab and specify angular infeed value at which cutting tool will move while cutting.
- Select the **Fade Thread End** check box of **Passes** tab to fade out the thread at the end. This fading helps easy assembly of fasteners.
- Select the **Spring Pass** check box of **Passes** tab to perform the final finishing pass twice to remove stock left due to tool deflection.
- Select the **Use cycle** check box of **Passes** tab to request output as canned cycle.
- The other options of the dialog box were discussed earlier.
- After specifying the parameters, click on the **OK** button from the **THREAD** dialog box. The toolpath will be generated and displayed on the model; refer to Figure-36.

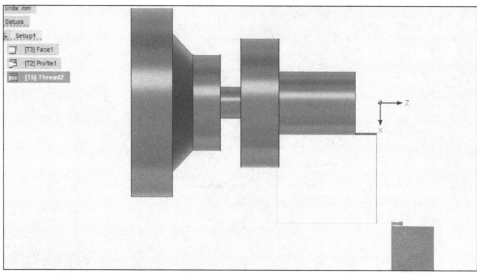

Figure-36. Toolpath generated for thread tool

- After simulating the thread toolpath, the model will be displayed as shown in Figure-37.

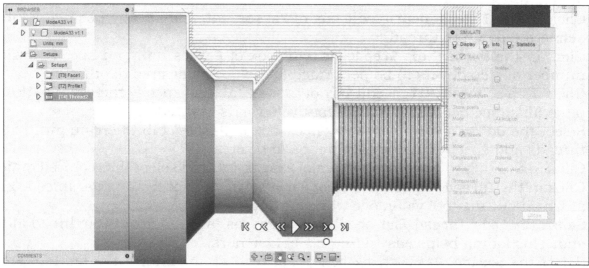

Figure-37. Simulating thread toolpath

Turning Secondary Spindle Chuck

The **Turning Secondary Spindle Chuck** tool is used to transfer stock from one chuck to another (Sub-spindle). This tool is useful when you want to machine back side of part after machining the front side. The procedure to use this tool is discussed next.

- Click on the **Turning Secondary Spindle Chuck** tool of **TURNING** drop-down from **Toolbar**; refer to Figure-38. The **SECONDARY SPINDLE CHUCK** dialog box will be displayed; refer to Figure-39

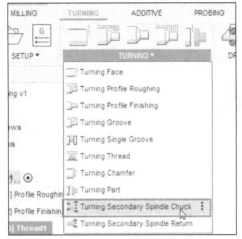

Figure-38. *Turning Secondary Spindle Chuck tool*

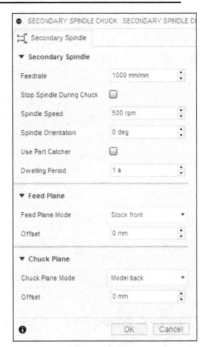

Figure-39. *SECONDARY SPINDLE CHUCK dialog box*

- Click in the **Feedrate** edit box of **Secondary Spindle** section from **SECONDARY SPINDLE CHUCK** dialog box and specify the value of feed rate at which chuck will move for stock transfer.
- Select the **Stop Spindle During Chuck** check box of **Secondary Spindle** section to keep spindle stopped during stock transfer operations. If this check box is not selected then you can specify rotation speed of spindle during stock transfer.
- Click in the **Spindle Orientation** edit box of **Secondary Spindle** section and specify the angular value between axes of primary and secondary chucks.
- Select the **Use Part Catcher** check box of **Secondary Spindle** section to activate part catcher when available.
- Click in the **Dwelling Period** edit box of **Secondary Spindle** section and enter a time for the operation to dwell.
- Click in the **Feed Plane Mode** drop-down of **Feed Plane** section and select desired option to specify the feed plane at which transfer will occur.
- Click in the **Offset** edit box of **Feed Plane** section and specify the offset distance from feed plane.
- Click in the **Chuck Plane Mode** drop-down from **Chuck Plane** section and select the desired option to specify the chuck plane.
- Click in the **Offset** edit box of **Chuck Plane** section and specify the offset distance from selected Chuck Plane.
- Click on the **OK** button from the dialog box. The operation will be created and displayed in the **BROWSER**. Note that there is no simulation for this operation. Only NC codes are generated for this operation based on selected machine.

Turning Secondary Spindle Return

The **Turning Secondary Spindle Return** tool is used for automatic stock return from secondary chuck to main chuck. No toolpath is associated with the strategy. The procedure to use this tool is discussed next.

- Click on the **Turning Secondary Spindle Return** tool of **TURNING** drop-down from **Toolbar**; refer to Figure-40. The **SECONDARY SPINDLE RETURN** dialog box will be displayed; refer to Figure-41.

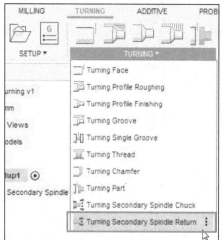

Figure-41. SECONDARY SPINDLE RETURN dialog box

Figure-40. Turning Secondary Spindle Return tool

- Click in the **Feedrate** edit box of the **Secondary Spindle** section from **SECONDARY SPINDLE RETURN** dialog box and specify desired value of feedrate.
- Select the **Stop Spindle During Return** check box of **Secondary Spindle** section to keep spindle stopped during operations.
- Click in the **Spindle Speed** edit box of **Secondary Spindle** section and specify the value of spindle speed in rpm for each machining operation.
- Click in the **Dwelling Period** edit box of **Secondary Spindle** section and specify time for the operation to pause.
- Click on the **Open Spindle Chucks** drop-down of **Secondary Spindle** section and select desired option to choose which spindle chucks to open before returning the secondary spindle to original position.
- Click on the **Feed Plane Mode** drop-down from the **Feed Plane** section and select desired option to specify the feed plane.
- Click in the **Offset** edit box of the **Feed Plane** section and specify offset distance from selected reference for feed plane; refer to Figure-42.

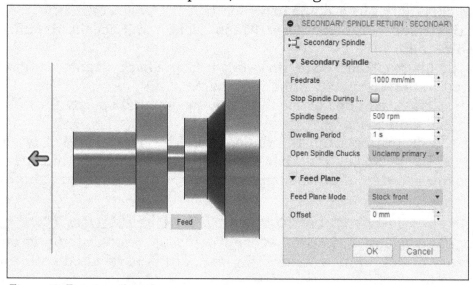

Figure-42. Entering offset value

- After specifying the parameters, click on the **OK** button from the **SECONDARY SPINDLE RETURN** dialog box. The operation will be added in the **BROWSER**.

GENERATING CUTTING TOOLPATHS

Till now, we have discussed the procedure to use turning toolpaths and milling toolpaths. In this section, we will discuss the procedure of generating toolpaths for 2D profile cutting using machines like water jet, laser, and plasma cutters.

Cutting

The **2D Profile** tool is used to create toolpaths for cutting 2D Profile of the stock with the help of Laser machine, Plasma cutters, and Water jet machines. The procedure to use this tool is discussed next.

- Click on the **2D Profile** tool from the **CUTTING** panel in the **FABRICATION** tab of **Toolbar**; refer to Figure-43. The **2D PROFILE** dialog box will be displayed; refer to Figure-44.

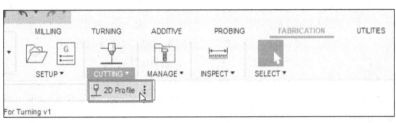

Figure-43. 2D Profile tool

Figure-44. The 2D PROFILE dialog box

- Click on the selection button for **Tool** section. The **Select Tool** dialog box will be displayed. Select the desired tool for cutting (waterjet, laser cutting, or plasma cutter) and click on **OK** button. The tool will be added in the **2D PROFILE** dialog box.
- Click on the **Cutting Mode** drop-down of **Tool** section and select desired option to set the appropriate cutting quality. In some machines, there are internal quality tables to set the value.

Geometry

- Click on the **Geometry** tab of **2D PROFILE** dialog box. The options will be displayed as shown in Figure-45.

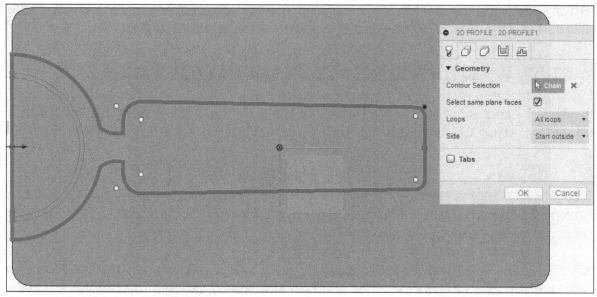

Figure-45. Geometry Tab of 2D PROFILE dialog box

- The **Nothing** button of **Contour Selection** option is active by default. You need to click on the curve contours from the model or sketches from **BROWSER** to be cut.
- Select the **Select same plane faces** check box of **Geometry** tab to select all the faces created on the same plane, from all parts within the stock boundary to automatically detect internal curve chains for cutting.
- Click on the **Loops** drop-down of **Geometry** tab and select desired option to filter outer or inner chain as needed.
- Click on the **Side** drop-down of **Geometry** tab and select desired option to offset selected edges or sketches.
- Select the **Tabs** check box and provide support tabs as discussed earlier.

Linking

- Click on the **Linking** tab of **2D PROFILE** dialog box. The options will be displayed as shown in Figure-46.

Figure-46. Linking Tab of 2D PROFILE dialog box

- Select the **Keep Nozzle Down** check box of **Linking** tab to avoid retracts and avoid previously cut areas. Set desired parameters in **Linking** section which are displayed on selecting the **Keep Nozzle Down** check box as discussed earlier.
- Select the **Lead-In (Entry)** check box of **Leads** section to enable a contour blend while entering workpiece.
- Click in the **Lead-In Radius** edit box of **Leads** section and specify the value to define the radius for lead in moves.
- Click in the **Lead-In Sweep Angle** edit box of **Leads** section and specify the value to define the sweep angle of the lead-in arc.
- Click in the **Lead-In Distance** edit box of **Leads** section and specify the value to define length of the linear lead-in move.
- Select the **Lead-Out (Exit)** check box of **Linking** tab to enable a contour blend while exiting the workpiece after operation.
- Select the **Same as Lead-In** check box of **Linking** tab to set the lead-out values identical to lead-in values as specified earlier.
- Click in the **Pierce Clearance** edit box of **Piercing** section and specify the value to distance away from the part edge to start the cut.
- The **Nothing** button for **Entry Positions** option from **Positions** section is active by default. You need to click on the point from model to define tool entry location.
- The other options of the dialog box are same as discussed earlier. After specifying the parameters, click on the **OK** button from the **2D PROFILE** dialog box. The toolpath will be generated and displayed on the model; refer to Figure-47.

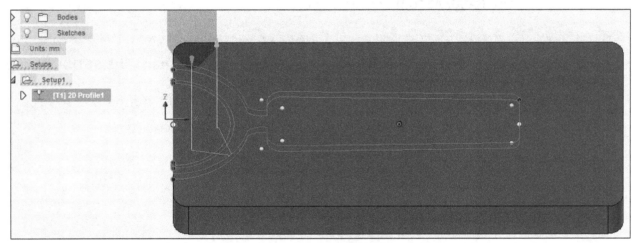

Figure-47. Generated toolpath for Cutting tool

PRACTICAL

Generate the toolpaths of the given model shown in Figure-48 using Turning tools.

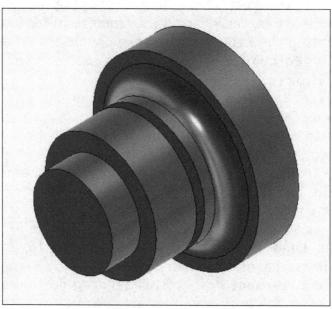

Figure-48. Practical

Adding Model to CAM

- Create and save the part in **DESIGN** workspace. The part file is available in respective chapter folder of **Autodesk Fusion 360 Black Book Resources** folder.
- Select the **MANUFACTURE** option from **Workspace** drop-down. The model will be displayed in the **MANUFACTURE** workspace.

Creating Stock

- Click on the **New Setup** tool of **SETUP** drop-down from **Toolbar**. The **SETUP** dialog box will be displayed along with the stock of model.
- Specify the parameters of **Setup** tab and **Stock** tab of **SETUP** dialog box to create the stock; refer to Figure-49.

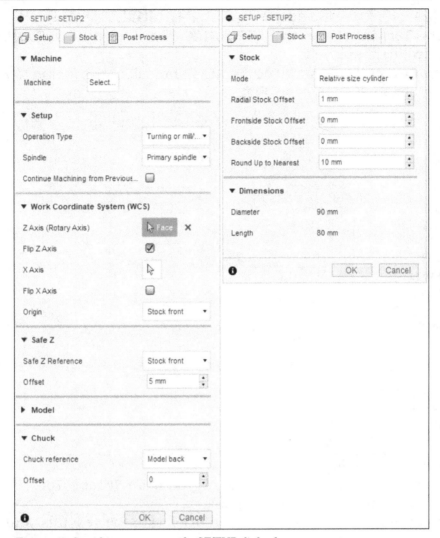

Figure-49. Specifying parameters for SETUP dialog box

- After specifying the parameters of **SETUP** dialog box, click on the **OK** button. The stock will be created and displayed on the model; refer to Figure-50.

Figure-50. Created stock for practical

Generating Face Toolpath

- Click on the **Turning Face** tool of **TURNING** drop-down from **Toolbar**. The **FACE** dialog box will be displayed.
- Click on the **Select** button for **Tool** option and select the facing tool from **Select Tool** dialog box as displayed in Figure-51.

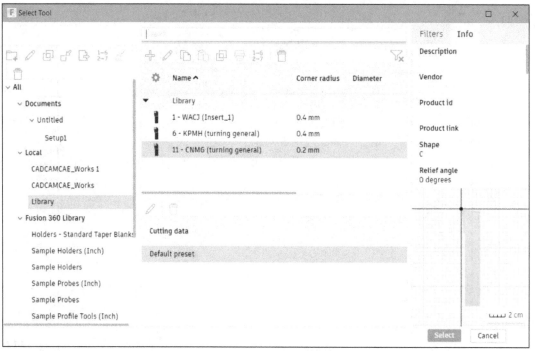

Figure-51. Selecting tool for turning

- After selecting the tool, click on the **OK** button from **Select Tool** dialog box. The tool will be added in the **Face** dialog box.
- Specify the parameters of **Tool** tab, **Geometry** tab, **Passes** tab, and **Linking** tab of **FACE** dialog box as shown in Figure-52.

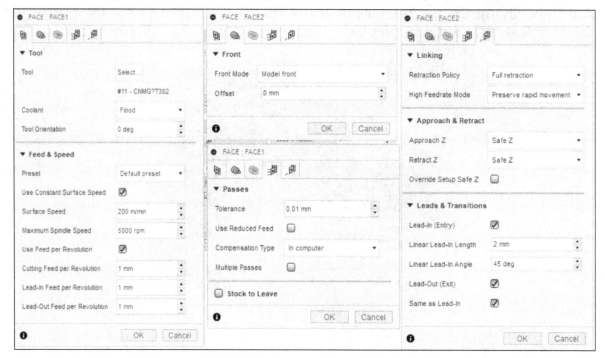

Figure-52. Specifying the parameters of FACE dialog box

- After specifying the parameters, click on the **OK** button from the **FACE** dialog box to generate the toolpath. The toolpath will be generated and displayed on the model.

Generating Turning Profile Roughing Toolpath

- Click on the **Turning Profile Roughing** tool of **TURNING** drop-down from **Toolbar**. The **PROFILE ROUGHING** dialog box will be displayed.
- Click on the **Select** button of **Tool** option and select the required tool from **Select Tool** dialog box.
- Specify the parameters of **Tool** tab, **Geometry** tab, **Passes** tab, and **Linking** tab of **PROFILE** dialog box as shown in Figure-53.

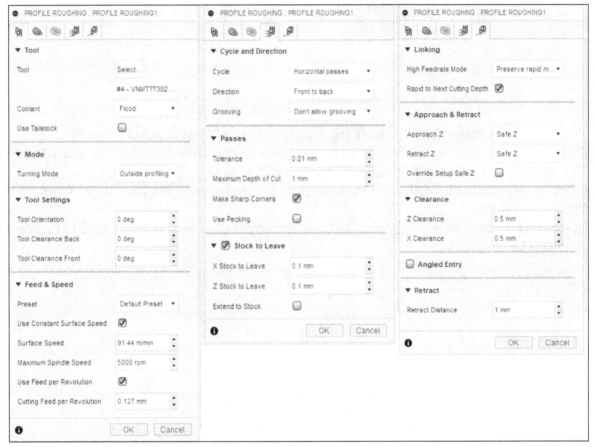

Figure-53. Specifying parameters for PROFILE dialog box

- After specifying the parameters, click on the **OK** button from **PROFILE** dialog box. The toolpath will be generated and displayed on the model; refer to Figure-54.

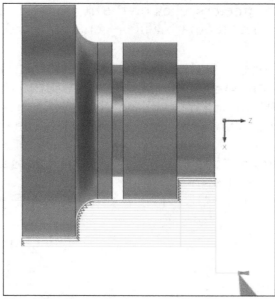

Figure-54. Generated toolpath of profile tool

Generating Turning Profile Finishing Toolpath

- Click on the **Turning Profile Finishing** tool of **TURNING** drop-down from **Toolbar**. The **PROFILE FINISHING** dialog box will be displayed.
- Click on the **Select** button of **Tool** option and select the required tool from **Select Tool** dialog box.
- Specify the parameters of **Tool** tab, **Geometry** tab, **Passes** tab, and **Linking** tab of **PROFILE** dialog box as shown in Figure-55.

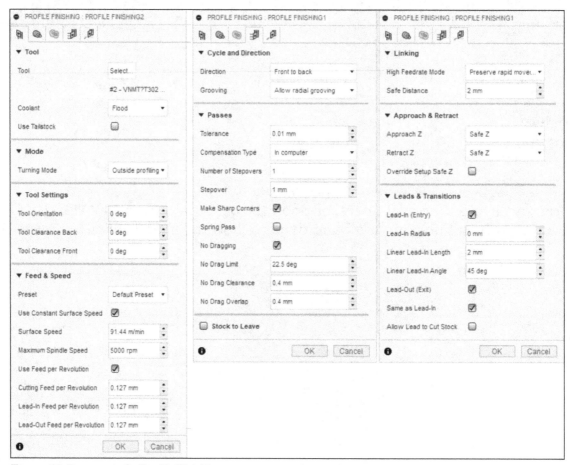

Figure-55. Parameters for Profile Finishing

- After setting the parameters, click on the **OK** button from the dialog box. The finishing toolpath for profile will be created.

Generating Turning Single Groove Toolpath

- Click on the **Turning Single Groove** tool of **TURNING** drop-down from **Toolbar**. The **SINGLE GROOVE** dialog box will be displayed.
- Click on the **Select** button of **Tool** option and select an OD groove square tool of insert thickness 4 mm from **Select Tool** dialog box.
- Specify the parameters of **Tool** tab, **Geometry** tab, **Passes** tab and **Linking** tab of **PROFILE** dialog box as shown in Figure-56.

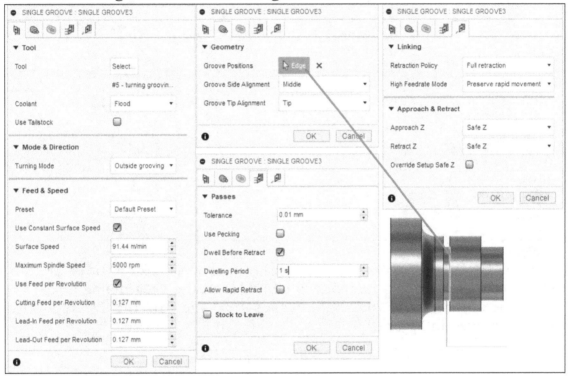

Figure-56. Specifying parameters for SINGLE GROOVE dialog box

- After specifying the parameters, click on the **OK** button from **SINGLE GROOVE** dialog box. The toolpath will be generated and displayed on the model; refer to Figure-57

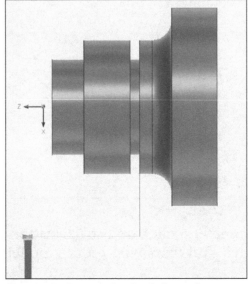

Figure-57. Generated toolpath for single groove

Generating Turning Part toolpath

- Click on the **Turning Part** tool from **TURNING** drop-down in **Toolbar**. The **PART** dialog box will be displayed.
- Click on the **Select** button of **Tool** option from **PART** dialog box and select the grooving tool of parameters shown in Figure-58.

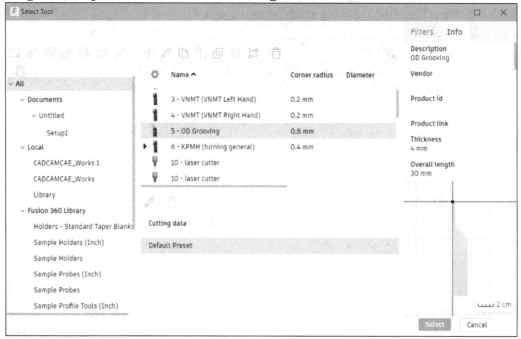

Figure-58. Selecting tool for part

- After selecting the tool of parameters displayed above, click on the **OK** button from **Select Tool** dialog box. The tool will be added in the **PART** dialog box.
- Specify the parameter of **Tool** tab, **Passes** tab, and **Linking** tab of **PART** dialog box as displayed in Figure-59.

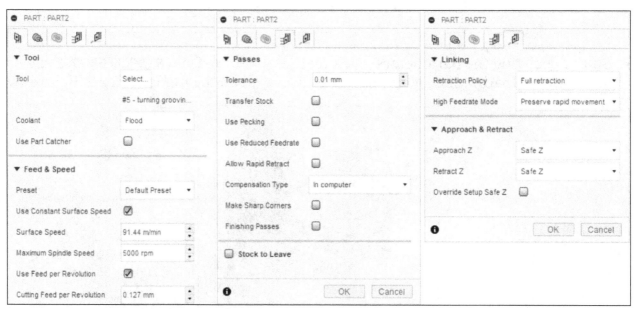

Figure-59. Specifying the parameters of PART dialog box

- After specifying the parameters, click on the **OK** button from **PART** dialog box. The toolpath will be generated and displayed on the model; refer to Figure-60.

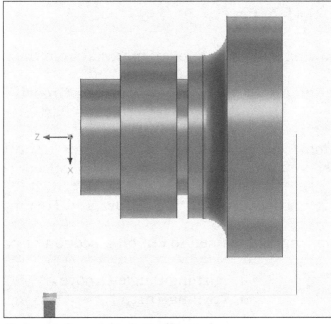

Figure-60. Generated toolpath of Part tool

Simulating the Toopath

- Right-click on the **Setup1** option from **BROWSER** and click on the **Simulate** tool from the shortcut menu. The **SIMULATE** dialog box will be displayed along with the model.
- Use the simulation keys to play the machining animation to check the errors occurred while machining process; refer to Figure-61.

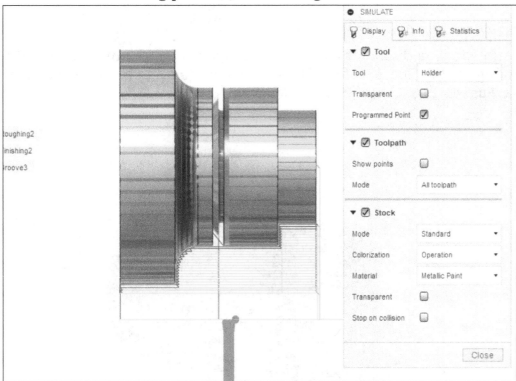

Figure-61. Simulation process

SELF ASSESSMENT

Q1. Which of the following tool is used to cut material from the profile of round part?

a. Turning Profile Roughing　　　　b. Turning Groove
c. Turning Chamfer　　　　　　　　d. Turning Part

Q2. You can specify maximum spindle speed parameter for cutting tool in Autodesk Fusion 360 only when you are using constant surface speed. (T/F)

Q3. What is the difference between Turning Groove and Turning Single Groove tool?

Q4. Which of the following tool is used to cut off a section of part?

a. Turning Part　　　　　　　　　b. Turning Single Groove
c. Turning Chamfer　　　　　　　　d. Turning Thread

Q5. Which of the following tool is used to generate NC program for cutting material using water jet machine?

a. Turning Part　　　　　　　　　b. 2D Adaptive Clearing
c. 2D Profile　　　　　　　　　　　d. Trace

PRACTICE 1

Machine the stock of diameter as 50 mm and length as 62 mm to create the part as shown in Figure-62. The part file of this model is available in the respective folder of **Autodesk Fusion 360 Black Book Resources**.

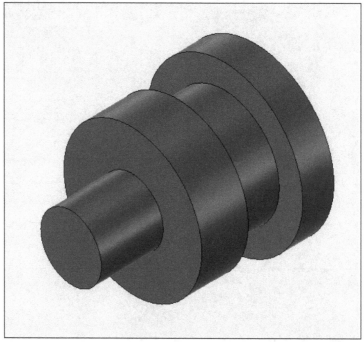

Figure-62. Practice 1

PRACTICE 2

Machine the stock of diameter as 80 mm and length as 130 mm to create the part as shown in Figure-63. The part file of this model is available in the respective folder of **Autodesk Fusion 360 Black Book Resources**.

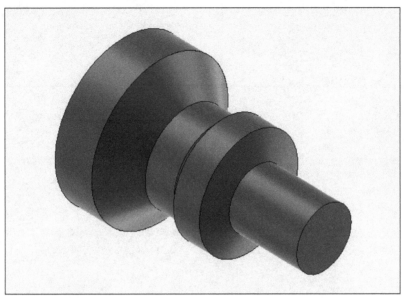

Figure-63. Practice

FOR STUDENT NOTES

Chapter 19

Probing, Additive Manufacturing, and Miscellaneous CAM Tools

Topics Covered

The major topics covered in this chapter are:

- *New Folder*
- *New Pattern*
- *Probing*
- *Inspecting Surface*
- *Simulating*
- *Generate Toolpath and Clear Toolpath*
- *Machining Time, Tool Library, and Task Manager*
- *Post Process*
- *Creating Form Mill tool*

INTRODUCTION

Till now, we have learned the procedure of generating Milling and Turning toolpath. In this chapter, we will discuss some of the other tools used to organize and manipulate the toolpaths.

NEW FOLDER

The **New Folder** tool is used for creating a folder to combine similar group operation. The procedure to use this tool is discussed next.

- Select the operations from **BROWSER** by holding **CTRL** key and right-click on any of the selected operations. A shortcut menu will be displayed; refer to Figure-1.

Figure-1. Add to new folder tool

- Click on the **Add to New Folder** button from the menu. The selected operations will be added in a new folder. The folder will be displayed in the **BROWSER** with the name as **Folder**; refer to Figure-2.

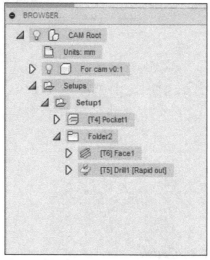

Figure-2. Moved operations

- If you want to rename the newly created folder then double-click on the folder with a pause between the clicks and specify desired name.

There is an another method for creating the folder which is discussed next.

- Click on the **New Folder** tool from **SETUP** drop-down in **Toolbar**; refer to Figure-3. The folder will be created in the **BROWSER**.

Figure-3. New Folder tool

- Select the operations while holding the **CTRL** key from **BROWSER** and drag the operations into newly created folder. The operations will be added in the created folder.

NEW PATTERN

The **New Pattern** tool is used for duplicating generated toolpath on the same model in **Linear**, **Circular**, **Mirror**, **Component**, and **Duplicate** pattern. The use of **New Pattern** tool can speed up your entire programming process since all changes to a pattern take effect immediately and no toolpath has to be updated. The procedure to use this tool is discussed next.

- Right-click on the operation(s) that you want to pattern. The shortcut menu will be displayed; refer to Figure-4.

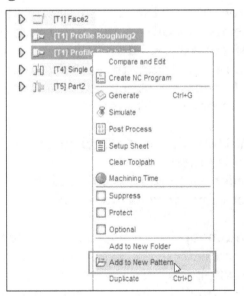

Figure-4. The Add to New Pattern tool

- Click on the **Add to New Pattern** tool from the displayed menu. The **FOLDER : PATTERN** dialog box will be displayed; refer to Figure-5.

Figure-5. FOLDER PATTERN dialog box

Linear Pattern

In this section, we will discuss the procedure of creating linear pattern.

- Click on the **Liner Pattern** option of **Pattern Type** drop-down from **FOLDER : PATTERN** dialog box for creating linear pattern.
- The selection button for **Direction 1** option of **Pattern** tab is active by default. You need to click on the edge or face from model to define the direction in which instances of pattern will be created.
- Select the **Flip Direction 1** check box of **Pattern** tab to flip the direction of pattern along selected edge or face.
- Click in the **Spacing for Direction 1** edit box of **Pattern** tab and specify the value of distance between two consecutive instances of the pattern.
- Click in the **Number of instances 1** edit box of **Pattern** tab and specify the number of instances to be created in the pattern along first direction.
- Select the **Additional Direction** check box of **Pattern** tab to use an additional direction for creating the pattern. The options to create instances in direction 2 will be displayed.
- Select the **Keep Original** check box of **Pattern** tab to keep the original toolpaths as well after creating new pattern.
- Select the **Preserve order** option from **Operation Order** drop-down in **Pattern** tab to machine all operations in each instance of the pattern before moving to the next instance.
- Select **Order by operation** option of **Operation Order** drop-down from **Pattern** tab to machine all occurrences of same operation in all instances of the pattern before moving to the next operation.
- Select **Order by tool** option of **Operation Order** drop-down from **Pattern** tab to machines all operations in the pattern that use the current tool before changing tools.
- Select the **Reverse** check box to reverse the order of cutting operations in pattern.
- The other tools of the drop-down are same as discussed earlier in this book. After specifying the parameters, click on the **OK** button from **FOLDER : PATTERN** dialog box; refer to Figure-6. The pattern will be created and displayed in the **BROWSER**.

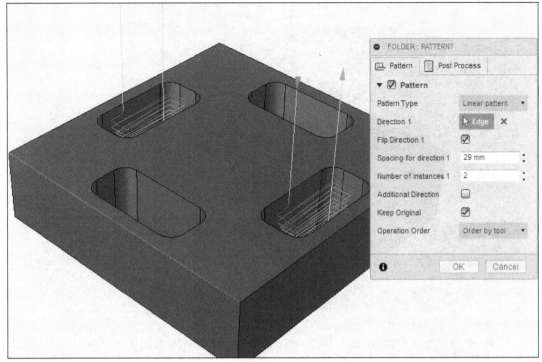

Figure-6. Applied linear pattern

Circular Pattern

In this section, we will discuss the procedure of creating the circular patter.

• Select the **Circular Pattern** option of **Pattern Type** drop-down from **FOLDER :
   PATTERN** dialog box for creating circular pattern; refer to Figure-7.

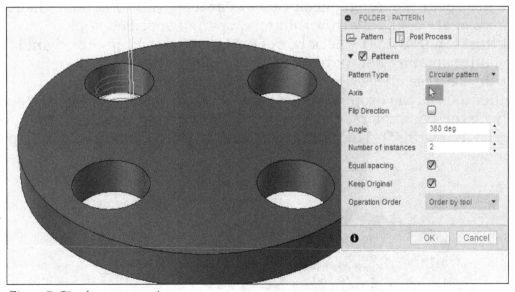

Figure-7. Circular pattern option

• The selection button of **Axis** section in **Pattern** tab is active by default. You need
   to click on the axis or edge from model to define center axis of pattern; refer to
   Figure-8.

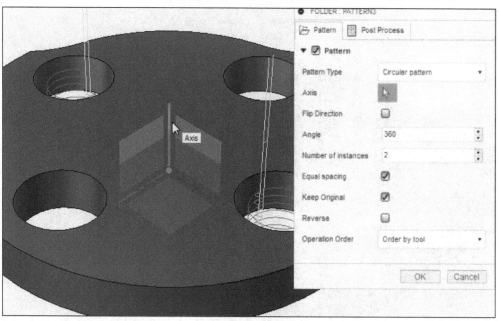

Figure-8. Selecting axis for pattern

- If you want to select the axis from default coordinate system then make it visible from **BROWSER** and select desired axis.
- Select the **Flip Direction** check box of **Pattern** tab to flip the direction of selected edge or axis.
- Click in the **Angle** edit box of **Pattern** tab and specify total angle in which all the instances of pattern will be created.
- Click in the **Number of instances** edit box of **Pattern** tab and specify the value to define number of instances to be created in circular pattern.
- Select the **Equal spacing** check box of **Pattern** tab to equally distribute the instances of selected pattern within total span of specified angle.
- Select the **Keep Original** check box of **Pattern** tab to keep the original toolpaths as well after creating new pattern.
- Select the **Reverse** check box to reverse the pattern order.
- The other tools of the drop-down are same as discussed earlier in this book. After specifying the parameters, click on the **OK** button from **FOLDER : PATTERN** dialog box; refer to Figure-9. The preview of pattern will be created and displayed on the model.

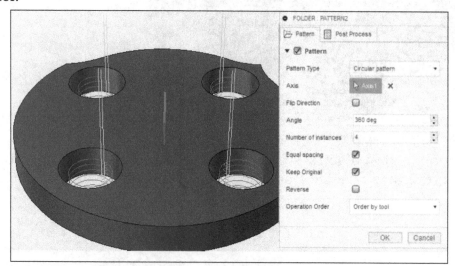

Figure-9. Preview of circular pattern

Mirror Pattern

In this section, we will discuss the procedure of creating the mirror pattern.

- Select the **Mirror Pattern** option from **Pattern Type** drop-down in **FOLDER :
  PATTERN** dialog box for creating mirror copy of selected operation; refer to Figure-10.

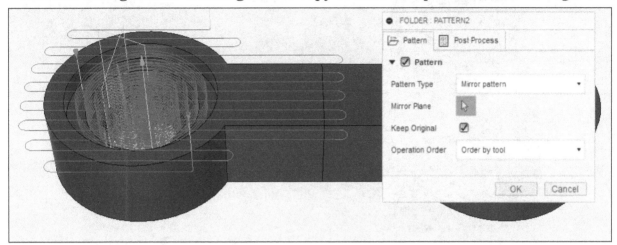

Figure-10. Mirror pattern option

- The selection button of the **Mirror Plane** section in **Pattern** tab is active by
  default. You need to click on the plane or face from model to select.
- The other options of the dialog box are same as discussed earlier.
- After specifying the parameters, click on the **OK** button from the **FOLDER : PATTERN**
  dialog box; refer to Figure-11. The preview of pattern will be created and displayed
  on the model.

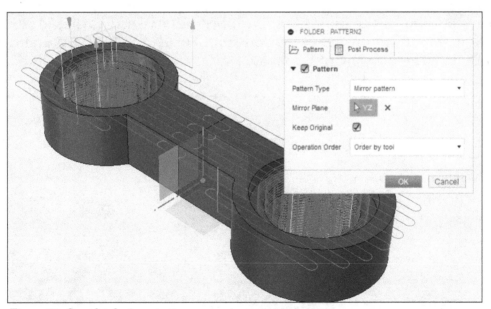

Figure-11. Completed mirror pattern

Duplicate Pattern

In this section, we will discuss the procedure of creating the duplicate copy of selected
operations.

- Click on the **Duplicate Pattern** option from **Pattern Type** drop-down in the
  FOLDER : PATTERN dialog box for creating duplicate pattern; refer to Figure-12.

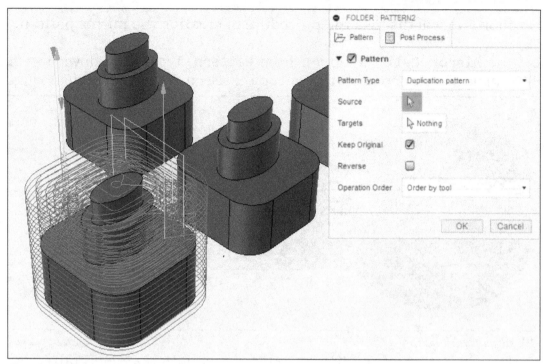

Figure-12. Duplication pattern option

- The selection button for **Source** section in **Pattern** tab is active by default. You need to click on the point from model to be defined as source point for duplicating.
- The selection button for **Targets** section in **Pattern** tab is active by default. You need to click on the target points on the model where duplicate copy of toolpath will be created.
- The other options of the dialog box have been discussed earlier. After specifying the parameters, click on the **OK** button from **FOLDER : PATTERN** dialog box; refer to Figure-13. The preview of duplicate pattern will be created and displayed on the model.

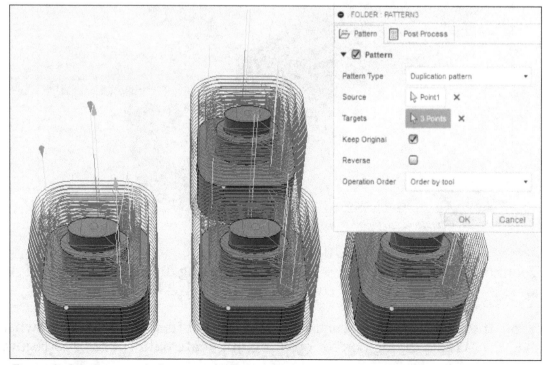

Figure-13. Selecting target body

Component Pattern

The component pattern is used to pattern the toolpaths created on one component onto the other component. Note that to use this pattern, you must have more than one components individually setup for machining in assembly. The procedure to use this option is given next.

- Select the **Component pattern** option from the **Pattern Type** drop-down in the dialog box. The **FOLDER : PATTERN** dialog box will be displayed as shown in Figure-14.

Figure-14. Component Pattern option

- Clear the **Automatic** check box if you do not want to automatically select targets based on selected source components.
- Click on selection button of **Source** option and select the component from where you want to copy the toolpaths.
- Click on the selection button of **Targets** option and select the components on which you want to paste the duplicate copies of toolpaths; refer to Figure-15.

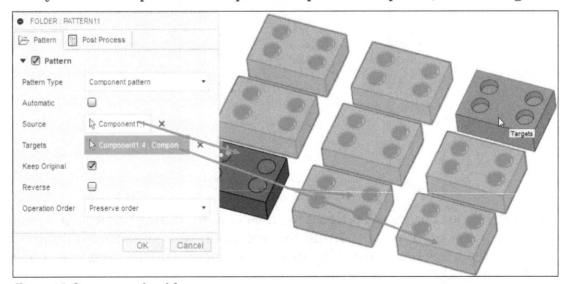

Figure-15. Components selected for pattern

- After specifying desired parameters, click on the **OK** button from the dialog box.

Manual NC

The **Manual NC** tool is used to insert special manual NC entries in the **CAM BROWSER**. The procedure to use this tool is discussed next.

• Click on the **Manual NC** tool from **SETUP** drop-down in **Toolbar**; refer to Figure-16. The **MANUAL NC** dialog box will be displayed; refer to Figure-17.

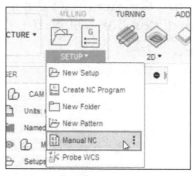

Figure-16. Manual NC tool

Figure-17. MANUAL NC dialog box

• Click on the **Manual Type** drop-down from **MANUAL NC** dialog box and select desired option to specify the type of manual NC operation to be inserted in toolpath.
• Click in the edit box below the drop-down and specify the desired text/code to be output during post processing.
• Click on the **OK** button from the dialog box. The manual nc code will be added in the **SETUP** node of **BROWSER**.

PROBING

Probing is used to measure the geometry of object by sensing various points of object using different types of probes. There are various types of probes like mechanical probes, optical probes, laser probes, and so on. The tools to create NC program for probing are available in the **INSPECTION** tab of **Toolbar**; refer to Figure-18. Various tools in this tab are discussed next.

Figure-18. INSPECTION tab

WCS Probe

The **Probe WCS** tool is used to output NC codes for the probing cycles of your CMM (Coordinate Measuring Machine). The procedure to use this tool is discussed next.

* Click on the **Probe WCS** tool from **SETUP** drop-down in **Toolbar**; refer to Figure-19. The **WCS PROBE** dialog box will be displayed; refer to Figure-20.

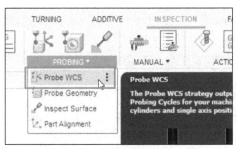

Figure-19. Probe WCS tool

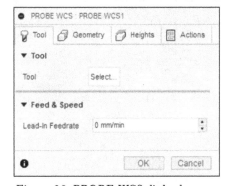

Figure-20. PROBE WCS dialog box

* Click on the **Select** button from **Tool** section in **Tool** tab. The **Select Tool** dialog box will be displayed; refer to Figure-21.

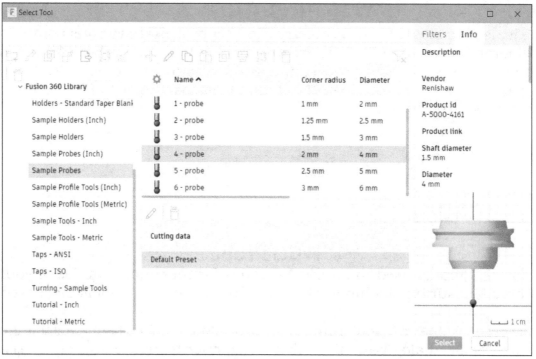

Figure-21. Select Tool dialog box for selecting probe tool

- Select desired probing tool from the table and click on the **Select** button from **Select Tool** dialog box.
- Specify desired lead-in feed rate for the probe in the **Lead-In Feedrate** edit box.

Geometry

- Click on the **Geometry** tab of **WCS PROBE** dialog box. The options will be displayed as shown in Figure-22.

Figure-22. Geometry tab of WCS PROBE dialog box

- Click in the **Probe Mode** drop-down of **Geometry** tab from **WCS PROBE** dialog box and select desired option to define whether you want to probe part/model or stock.
- The **Nothing** button of **Probe Surface(s)** section of **Geometry** tab is active by default. You need to click on the face of model/stock to be checked by probe; refer to Figure-23.

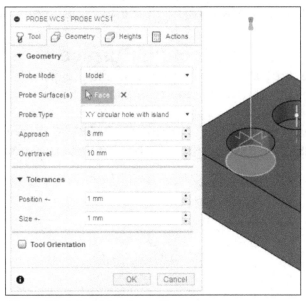

Figure-23. Selected face for probing

- Select desired option from the **Probe Type** drop-down to define what type of probe operation you want to create. If you have selected the **Z surface** option from the drop-down then **Use Selected Point** check box will be displayed. Select this check box if you want to use selected point as reference for probing on the surface. Similarly, you can set parameters for other types as well; refer to Figure-24.

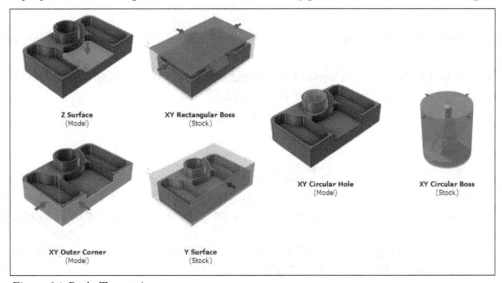

Figure-24. Probe Type options

- Specify desired value of approach distance and over travel of probe in respective edit boxes of the dialog box. The approach distance is the distance of selected point from where probe starts to measure. The over travel distance is the distance from selected point up to which the probe can move past selected point while measuring.
- Specify desired tolerance values for position and size in respective edit boxes of the **Tolerances** section in the dialog box.
- Click on the **Actions** tab of dialog box and select check boxes for actions to be displayed as messages in case of measurement error.
- The other options of the dialog box have been discussed earlier in this book. After specifying the parameters, click on the **OK** button from **PROBE WCS** dialog box. The operation will be created and displaced in the **BROWSER**.

The **Probing Geometry** tool works in the same way as **Probe WCS** tool.

Inspecting Surface

The **Inspect Surface** tool is used to probe multiple points of selected surface by using a machine. The procedure to use this tool is given next.

- Click on the **Inspect Surface** tool from the **PROBING** panel in the **INSPECTION** tab of **Toolbar**. The **INSPECT** dialog box will be displayed; refer to Figure-25.

Figure-25. INSPECT dialog box

- Select desired probing tool and specify feed/speed parameters in the **Tool** tab of dialog box.
- Click on the **Geometry** tab of dialog box and click on the surface at desired locations to specify inspection points to be measured.
- Set the other parameters as discussed earlier and click on the **OK** button from the dialog box.

Part Alignment

The **Part Alignment** tool is used to realign the part in desired orientation for performing measurement of coordinates. The procedure to use this tool is given next.

- Click on the **Part Alignment** tool from the **PROBING** drop-down in **INSPECTION** tab of the **Ribbon**. The **PART ALIGNMENT** dialog box will be displayed; refer to Figure-26.

Figure-26. PART ALIGNMENT dialog box

- Select desired option from the **Method** drop-down to define number of axes about which model can be aligned.
- Click on the **OK** button from the dialog box to activate alignment mode. The **PART ALIGNMENT** tab will be displayed in the **Ribbon**; refer to Figure-27.

Figure-27. PART ALIGNMENT tab

Creating Surface Inspection Feature

- Click on the **Inspect Surface** tool from the **INSPECT SURFACE** panel in the **PART ALIGNMENT** tab of the **Ribbon**. The **INSPECT** dialog box will be displayed.

Figure-28. INSPECT dialog box

- Click on the **Select** button for **Tool** option from **Tool** tab and select desired probe tool.
- Specify desired values of feedrates in the edit boxes of the dialog box.
- Click on the **Geometry** tab in the dialog box. The options to define geometries to be measured will be displayed.
- Click at desired points of the model which are to be measured; refer to Figure-29.

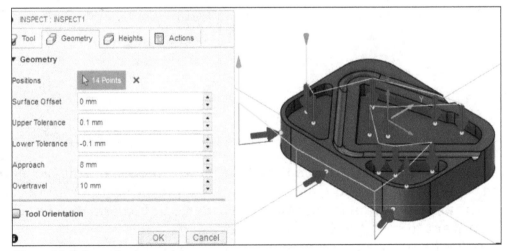

Figure-29. Points to be measured

- Set the other parameters as discussed earlier and click on the **OK** button from the dialog box.

Posting NC Program for Alignment

The **Post For Alignment** tool is used to generate NC program of inspection toolpath for alignment. The procedure to use this tool is given next.

- Click on the **Post For Alignment** tool from the **POST FOR ALIGNMENT** drop-down in the **PART ALIGNMENT** tab of the **Ribbon**. The **NC Program** dialog box will be displayed; refer to Figure 30.

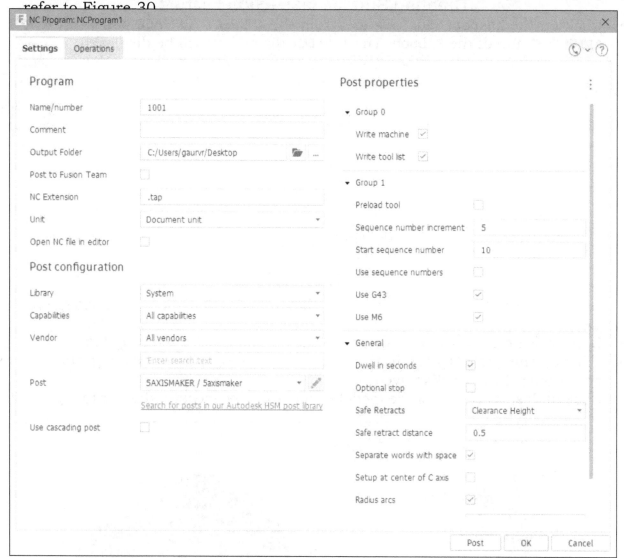

Figure-30. NC Program dialog box

- Specify desired text values in the **Name/number** and **Comment** edit boxes to define program name/number and comments to be displayed in the output file.
- Click on the **Browse** button for the **Output Folder** edit box and specify the location where you want to save the output file generated.
- Select the **Post to Fusion Team** check box and specify location of shared folder if you want to share the output file with your colleagues using Autodesk Fusion Team app.
- The **NC Extension** edit box shows the format in which output file will be generated.
- Select desired option from the **Unit** drop-down to define unit system to be used for generating tool movements in the output file.
- If you have a post processor specific to your machine available in the system library then select the **System** option from the **Library** drop-down and select the post processor from **Vendor** drop-down in the dialog box. If desired post processor is

not available in the drop-down then you can get it from your machine vendor and use the **Browse** option in **Library** drop-down to get it used by Autodesk Fusion. You can also search for desired post processor by using the **Search for posts in our Autodesk HSM post library** link button.

- Specify the parameter as desired in the **Post properties** section of the dialog box based on post processor selected by you.
- Click on the **Operations** tab in the dialog box to include/exclude inspection toolpaths in the output file; refer to Figure-31.

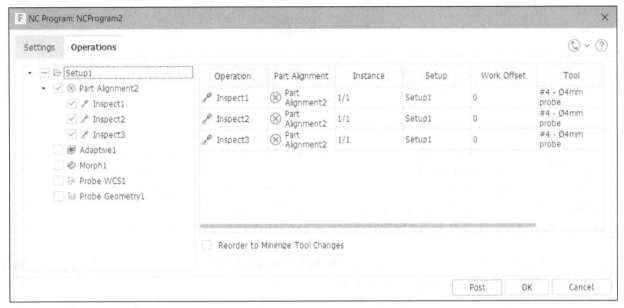

Figure-31. Operations tab

- Select check boxes from the left area in the dialog box to define inspection operations are to be included in the output file.
- After setting desired parameters, click on the **Post** button. The output file will be generated.

Importing Inspection Results

Once you have generated the inspection NC program, upload it on machine and check for the coordinates. If you have not done it yet then a red cross mark is displayed before **Part Alignment** feature in **BROWSER**; refer to Figure-32. The **Import Inspection Results** tool is used to import result files generated by CMM machine and project them on the model in Autodesk Fusion 360. The procedure to use this tool is given next.

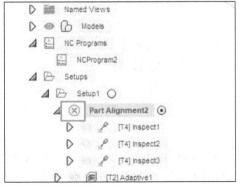

Figure-32. Cross mark for alignment feature

- Click on the **Import Inspection Results** tool from the **GET RESULTS** drop-down in the **PART ALIGNMENT** tab of the **Ribbon**. The **IMPORT INSPECTION RESULTS** dialog box will be displayed; refer to Figure-33.

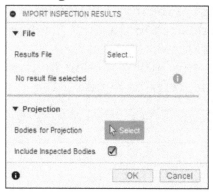

Figure-33. IMPORT INSPECTION RESULTS dialog box

- Select the model on which result points will be projected.
- Click on the **Select** button for **Results File** option and select the file generated by CMM machine after inspection.
- After selecting inspection file, click on the **OK** button from the dialog box. The projected results will be displayed on model along with a green mark before **Part Alignment** feature in the **BROWSER** showing that results have been calculated.

Select the **Show Part Alignment Information** and **Show Results** tools to display respective results in the drawing area.

You can post the output file with alignment using the **Post with Alignment** tool from the **POST WITH ALIGNMENT** drop-down in the **Ribbon**. The procedure is same as discussed earlier.

After performing desired operations, click on the **Exit Part Alignment** tool from the **EXIT ALIGNMENT** panel in the **Ribbon** to exit the alignment mode.

MANUAL INSPECTION REPORT

The tools to create and manage manual inspection reports are available in the **MANUAL** drop-down of the **INSPECTION** tab in **Ribbon**; refer to Figure-34. Using these tools, you can generate inspection report for various features of the model like diameter of holes, length of sides, and so on to be inspected physically using the measuring instruments. The tools in this drop-down are discussed next.

Figure-34. MANUAL drop-down

Creating Manual Inspection Report

The **Create Manual Inspection** tool is used to create a manual inspection report for measuring various features of model using physical instruments. The procedure to use this tool is given next.

- Click on the **Create Manual Inspection** tool from the **MANUAL** drop-down in the **INSPECTION** tab of the **Ribbon**. The **CREATE MANUAL INSPECTION** dialog box will be displayed; refer to Figure-35.

Figure-35. CREATE MANUAL INSPECTION dialog box

- Select two face to measure the distance between them using an instrument; refer to Figure-36.
- Specify desired text in the Name edit box to define name of the dimension.
- Specify desired values in the **Upper Tolerance** and **Lower Tolerance** edit boxes to define range within which the dimension should be, on measuring the part.
- Click in the **Comments** edit box to provide user notes for measuring current dimension.
- Click on the **+** button to add next dimension for inspection. You can repeat these steps to add desired number of dimensions in inspection sheet.

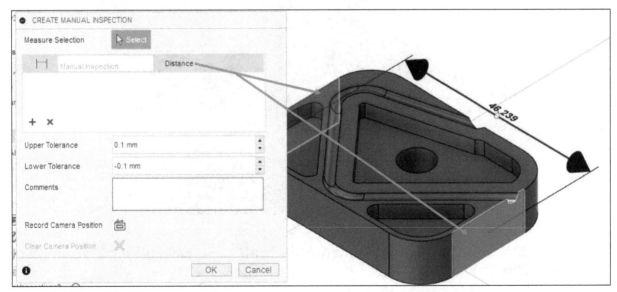

Figure-36. Faces selected for measuring distance

- Click on the **Record Camera Position** button to take a snapshot of the model in current position and orientation. This snapshot will be saved in canvas mode.
- After setting desired parameters, click on the **OK** button from the dialog box. The inspection dimensions will be added in the **Manual Inspection** node; refer to Figure-37.

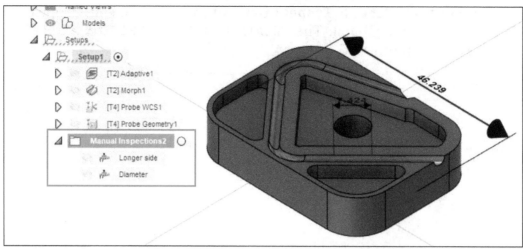

Figure-37. Inspection dimensions added

Recording Manual Inspections

The **Record Manual Inspection** tool is used to record measurements performed on the part in inspection report. The procedure to use this tool is given next.

- Select the inspection dimension to be recorded from the **Manual Inspection** node in the **BROWSER** and click on the **Record Manual Inspection** tool from the **MANUAL** drop-down in the **INSPECTION** tab of the **Ribbon**; refer to Figure-38. The **RECORD MANUAL INSPECTION** dialog box will be displayed; refer to Figure-39.

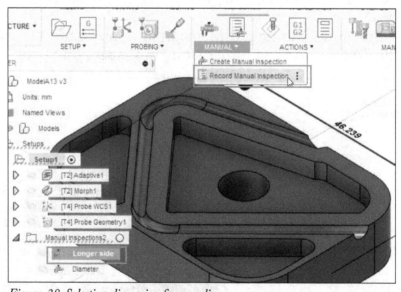

Figure-38. Selecting dimension for recording

Note that you can select multiple dimensions from the **BROWSER** for recording while holding the **CTRL** key.

- Specify the value you get by physically measuring the part in the **Measured** edit box. The deviation from mean value and amount of error will be displayed in the respective fields of the dialog box. After specifying desired parameters, click on the **OK** button from the dialog box to exit.
- If the measured values are within tolerance range then dimensions will be displayed in green color, otherwise they will be displayed in orange color; refer to Figure-40.

Figure-39. RECORD MANUAL INSPECTION
dialog box

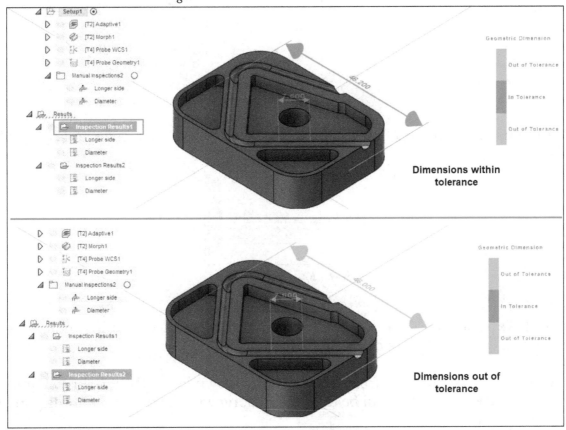

Figure-40. Inspection results

Generating/Regenerating Toolpaths

The **Generate** tool in **ACTIONS** drop-down of **INSPECTION** tab in **Ribbon** is used to generate/regenerate toolpaths after performing modifications. To use this tool, select the toolpaths to be regenerated from the **BROWSER** and click on the **Generate** tool. You can also use **Generate** option from right-click shortcut menu on selected operation to regenerate it. The procedure to generate toolpath is discussed next.

• Click on the operation for which you want to regenerate the toolpath from **Browser Tree**.

- Click on the **Generate** tool of **ACTIONS** drop-down from **Toolbar**; refer to Figure-41.
- You can also select the **Generate** tool from shortcut menu; refer to Figure-42.

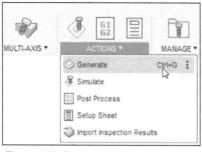

Figure-41. Generate tool

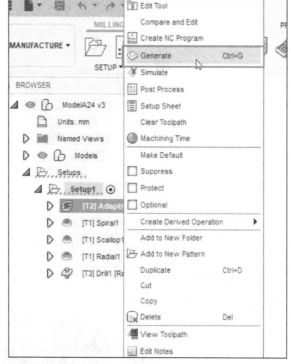

Figure-42. Selecting Generate tool from Shortcut menu

- The toolpath will be regenerated and displayed in **BROWSER**; refer to Figure-43.

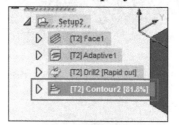

Figure-43. Regenerating toolpath

Simulate

The **Simulate** tool is used to review and simulate the created probing toolpaths. The procedure to use this tool is discussed next.

- Click on the **Simulate** tool from **ACTIONS** drop-down in **Toolbar**; refer to Figure-44. The **SIMULATE** dialog box will be displayed along with model and simulation keys; refer to Figure-45.

Figure-44. Simulate tool

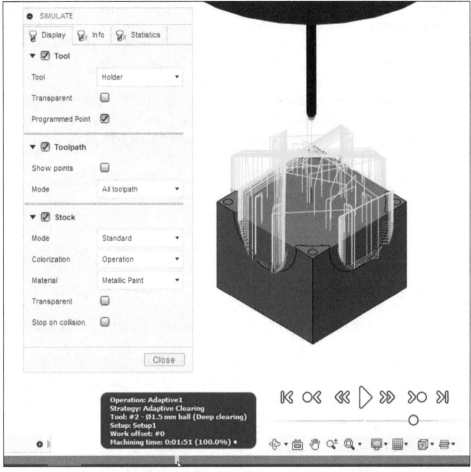

Figure-45. SIMULATE dialog box along with simulation keys

- There is also another way to select **Simulate** tool. Right-click on the specific operation or setup from **BROWSER**. A shortcut menu will be displayed; refer to Figure-46. Click on the **Simulate** tool from the menu. The **SIMULATE** dialog box will be displayed.

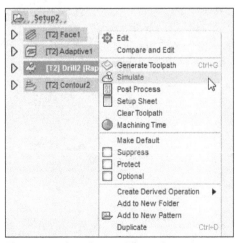

Figure-46. Simulate tool from shortcut menu

Display tab

- Select the **Tool** check box of **Display** tab from **SIMULATE** dialog box to view the tool and holder while machining animation process.
- Click on the **Tool** drop-down of **Display** tab and select the required option to tell software that which segment of the tool to show while machining.

- Select the **Transparent** check box of **Display** tab to make the tool transparent.
- Select the **Programmed Point** check box to display the points where changes occur in toolpath like change in toolpath curve, change in feed & speed, and so on.
- Select the **Toolpath** check box of **Display** tab to display the toolpath on the model. Clear the **Toolpath** check box to hide the toolpath.
- Select the **Show Points** check box of **Toolpath** node to display the points of toolpath on the model. Clear the **Show Points** check box to hide the points.
- Click on the **Mode** drop-down of **Toolpath** node from **Display** tab and select the required option to show the toolpath of the specific part.
- Select the **Stock** check box of **Display** tab to view the stock on the part. Clear the check box to hide the stock from part.
- Select the **Standard** option of **Mode** drop-down from **Stock** check box for simulation of any toolpath including 2D, 3-axis, 3-Axis indexing and multi axis toolpath.
- Select the **Fast (3-Axis only)** option of **Mode** drop-down from **Stock** check box for simulating faster toolpath in large toolpaths.
- Click on the **Colorization** option of **Stock** check box to select the required option to specify how the stock should be colorized.
- Click on the **Material** drop-down of **Stock** check box to specify the material used for visualization.
- Select the **Transparent** check box of **Stock** tab to transparent the stock of model.
- Select the **Stop on collision** option of **Stock** tab to stop the tool on collision with stock of model. When the tool or holder collides with stock then it shows red sign in the animation bar; refer to Figure-47

Info tab

- Click on the **Info** tab of **SIMULATE** dialog box. The **Info** tab will be displayed; refer to Figure-48.

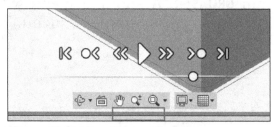

Figure–47. Holder collision with stock

Figure-48. Info tab of SIMULATE dialog box

- In this **Info** tab, the information related to stock, operation, spindle speed, volume, cursor position, and so on will be displayed.

Statistics

- Click on the **Statistics** tab of **SIMULATE** dialog box. The **Statistics** tab will be displayed; refer to Figure-49.

Figure-49. Statistics tab

- In the **Statistic** tab, the information like Machining time, Machining distance, Operations, and Tool change will be displayed.

Simulation Keys

The simulation keys are used to watch the animation of stock removal; refer to Figure-50.

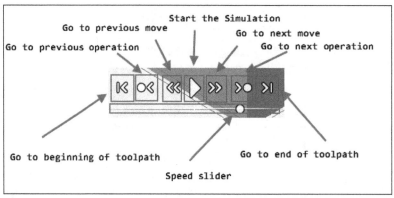

Figure-50. Simulation keys

Post Process

The **Post Process** tool is used to convert the machine-independent cutter location data into machine-specific NC code that can be run directly on CNC machines. The procedure to use this tool is discussed next.

- Click on the **Post Process** tab of **ACTIONS** tab from **Toolbar**; refer to Figure-51. The **Post Process** dialog box will be displayed; refer to Figure-52.

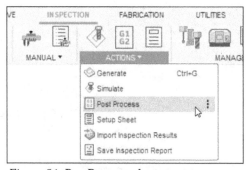

Figure-51. Post Process tool

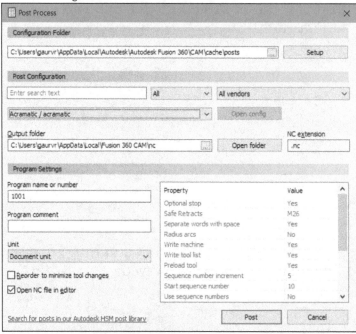

Figure-52. Post Process dialog box

- Click on the **Browse** button of **Configuration Folder** section from **Post Process** dialog box. The **Select post process configuration folder** dialog box will be displayed; refer to Figure-80.

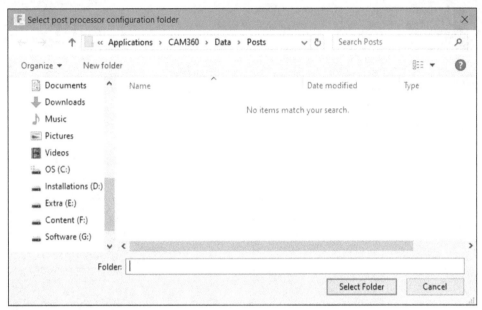

Figure-53. Select post processor configuration folder dialog box

- Create a new folder or select from the existing one to set the post processing configuration folder and click on the **Select Folder** button. Note that all the post processing configurations available in selected folder will be displayed in the dialog box.
- Click on the **Setup** button of **Post Process** dialog box and select desired option to reset the default post processor configuration.
- Select the desired machine type and vendor for the post processor from respective drop-downs in the **Post Configuration** section of the dialog box.
- Click on the **Open config** button of **Post Process** dialog box to open the post configuration in the default word processor application.
- Click in the **Browse** button for **Output folder** option to select the output folder for saving the NC file.
- Click on the **Open Folder** button of **Post Process** dialog box to open the output folder in windows explorer.
- Click in the **Program name or number** edit box of **Post Process** dialog box and enter the desired name or number of the output file.
- Click in the **Program comment** edit box of **Post Process** dialog box and enter the desired text related to the output file.
- Click in the **Unit** drop-down of **Post Process** tab and select the desired unit to specify the output unit. When unit of the output file is set to **Document unit** option then either inch or millimeter will be used.
- Select the **Reorder to minimize tool changes** check box of **Post Process** tab to record the operation between jobs to minimise the number of tool changes.
- Select the **Open NC file in editor** check box of **Post Process** dialog box to open the output NC file in editor.
- If you want to reset a single property or all the properties for output file then right-click on any property from **User defined property** box and select the required option. The property will be reset.

- After specifying the parameters, click on the **Post** button from **Post Process** dialog box. The **Post Process** output location folder will be displayed; refer to Figure-54.

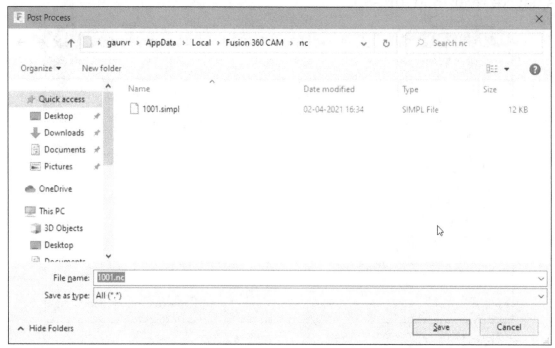

Figure-54. Post Process output location folder

- Click on the **Save** button to save the file. The output file will be save in specified folder.
- If you have selected the option to view the file in editor then the program will be displayed in default word processor application; refer to Figure-55.

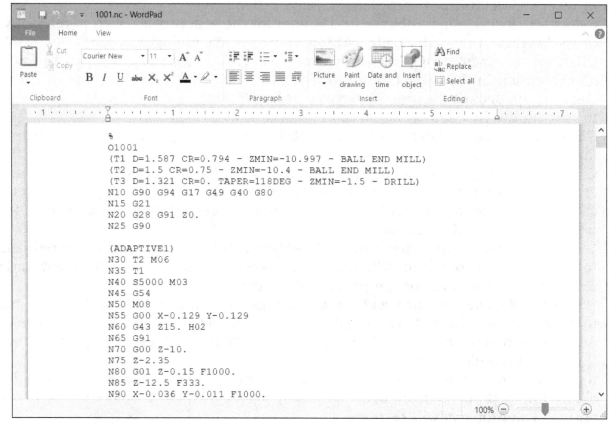

Figure-55. NC file in editor

• You can edit the codes as required by using the options of the editor.

Setup Sheet

The **Setup Sheet** tool is used to generate an overview for the NC program operator. Setup sheet provides the data related to stock, tool data, work piece position, and machine statistics. Before creating the setup sheet, you need to be sure that desired setup is selected from the **BROWSER**.

If you are on a network and the operator has access to a PC on the shop floor, you can save the setup sheet to a folder that the operator can access. This will save paper and ensure the operator always has access to the most current setup information. The default **Setup Sheet** can be viewed in any standard web browser. The procedure to use this tool is discussed next.

• Click on the **Setup Sheet** tool of **ACTIONS** panel from **Toolbar**; refer to Figure-56. The **Select Setup Sheet Output Folder** dialog box will be displayed; refer to Figure-57.

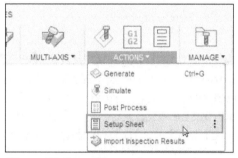

Figure-56. Setup Sheet tool

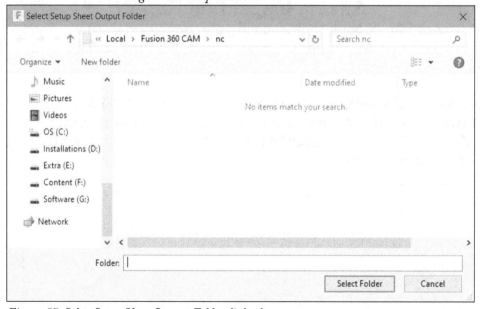

Figure-57. Select Setup Sheet Output Folder dialog box

• Select the location folder as desired and click on the **Select Folder** button from **Select Setup Sheet Output Folder** dialog box. The setup sheet will be created and open in the default internet browser; refer to Figure-58.

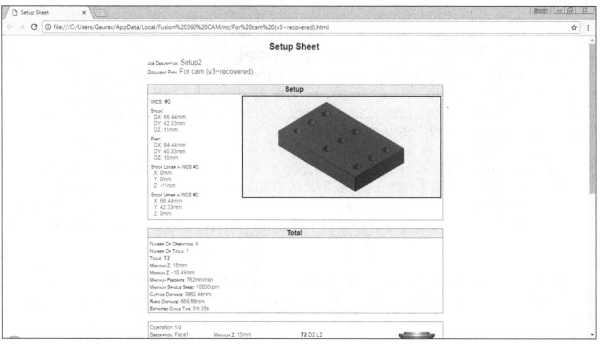

Figure–58. Created setup sheet

Importing Inspection Report

The **Import Inspection Report** tool in **ACTIONS** drop-down of **INSPECTION** tab in **Ribbon** is used to load inspection report generated by CMM machine after measuring the part. You can also project the measured points on the model in drawing area to compare the results. The procedure to use this tool has been discussed earlier.

Saving Inspection Report

The **Save Inspection Report** tool is used to generate PDF file of inspection result report generated by CMM or manually specified. The procedure to use this tool is given next.

• Click on **Save Inspection Report** tool from the **ACTIONS** drop-down in the **INSPECTION** tab of the **Ribbon** after selecting inspection results node; refer to Figure-59. The **Select Inspection Report Save Location** dialog box will be displayed; refer to Figure-60.

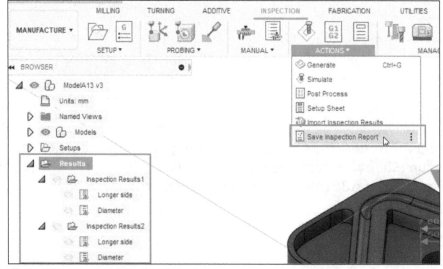

Figure–59. Save Inspection Report tool

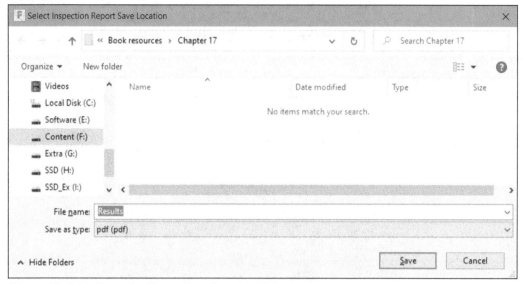

Figure-60. Select Inspection Report Save Location dialog box

- Specify desired name for file in the **File name** edit box and click on the **Save** button. The report file will be saved in specified location. You can open this file using PDF processor; refer to Figure-61.

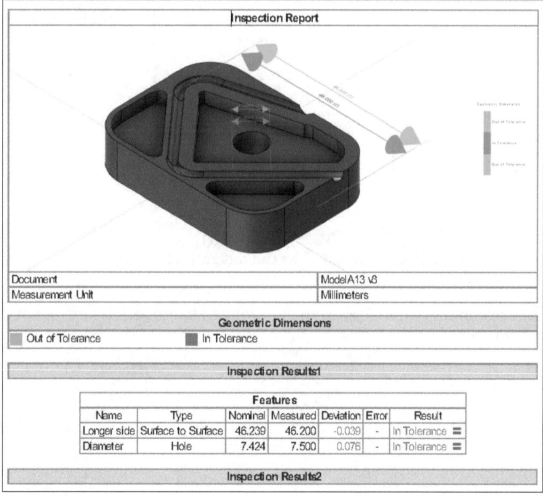

Figure-61. Inspection report generated

TOOLPATH SHORTCUT MENU

On selecting a toolpath from **BROWSER** and right-clicking on it, a shortcut menu will be displayed; refer to Figure-62. Most of the options are same as available in the **Ribbon**. Rest of the options in shortcut menu are discussed next.

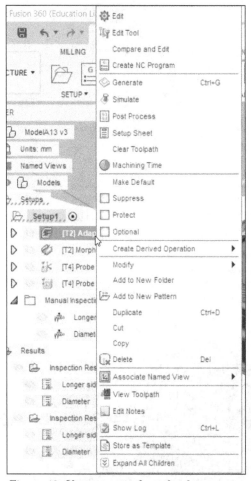

Figure-62. Shortcut menu for toolpaths

Clear Toolpath

The **Clear toolpath** tool is used to clear the toolpath of the selected operation. The procedure to use this tool is discussed next.

• Right-click on the required operation from **BROWSER** and select the **Clear Toolpath** tool from shortcut menu; refer to Figure-63.

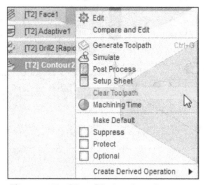

Figure-63. Clear Toolpath tool

- The tool path of selected operation will be deleted. If you want to display this toolpath again then right-click on it and select the **Generate Toolpath** option.

Machining Time

The **Machining Time** tool is used to measure and calculate the machining time for a selected operation with a high accuracy. The procedure to use this tool is discussed next.

- Right-click on the required operation from **BROWSER** and select the **Machining Time** option from shortcut menu; refer to Figure-64.

Figure-64. Machining Time tool

Figure-65. MACHINING TIME dialog box

- Click in the **Feed Scale(%)** edit box of **MACHINING TIME** dialog box and specify desired value of feed rate in percentage to check how machining time will change.
- Click in the **Rapid Feed(mm/min)** edit box of **MACHINING TIME** dialog box and specify desired value to check change in machining time.
- Click in the **Tool change Time(s)** edit box of **MACHINING TIME** dialog box and specify desired value to check change in machining time.
- In the **MACHINING TIME** dialog box, you can also check other information related to operation.

Setting Default for Toolpaths

Using the **Make Default** option in shortcut menu for selected operation, you can make parameters of selected operation default for other similar operations. The procedure to use this option is given next.

- Select the operation which you want to use for setting default parameters from the Setup node in **BROWSER** and right-click on it. A shortcut menu will be displayed.
- Select the **Make Default** option from the shortcut menu; refer to Figure-66. The **Make Default** information box will be displayed; refer to Figure-67.

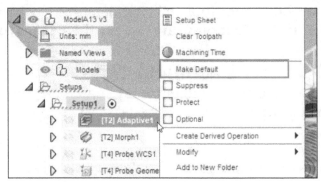

Figure-66. Make Default option

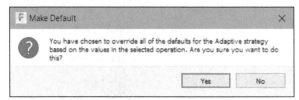

Figure-67. Make Default information box

- Click on the **Yes** button from the information box to apply defaults. For example, if you have selected an adaptive toolpath to make it default then parameters in all the other adaptive toolpaths of current file will be changed according to selected adaptive toolpath.

Suppressing Toolpaths

The **Suppress** check box of shortcut menu is used to make selected toolpath inactive in the setup. Note that on suppressing a toolpath, the stock will be modified as well because effects of suppressed toolpath will be removed from stock. If you want to reactivate the suppressed toolpath then right-click on it and clear the **Suppress** check box.

Protecting Toolpaths

The **Protect** check box of shortcut menu is used to make selected toolpath protected so that accidentally no changes are made in the toolpath. Note that on selecting this check box, a lock sign will be added before toolpath in **BROWSER**; refer to Figure-68.

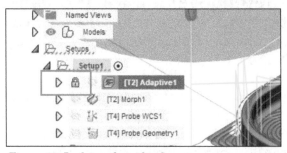

Figure-68. Lock icon for toolpath

Optional Toolpath

The **Optional** check box of shortcut menu is used to mark selected toolpath as optional so that user on machine can include or skip toolpaths from the machining setup.

Creating Derived Operation

The options in the **Create Derived Operation** cascading menu are used to create cutting operation which uses parameters specified in earlier created operation. Note that selected faces, cutting tools, and cutting parameters of earlier created operation will be applied on derived operations. For example, you have performed an Adaptive Clearing operation earlier for roughing pockets and now you want to perform finishing operation then you can select Pocket Clearing operation from **Create Derived Operation** cascading menu of shortcut menu; refer to Figure-69.

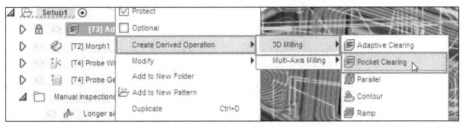

Figure-69. Creating derived operation

Viewing Toolpath Data

The **View Toolpath** option in right-click shortcut menu of an operation is used to check movements of cutting tool in terms of coordinates with feed rate and spindle speed; refer to Figure-70. After checking parameters, click on the **Close** button to exit the information box.

Index	Type	X	Y	Z	Movement	Feedrate	RPM	Compensation
1	Linear	-24.3693	-20.4428	15	Rapid		10000	
2	Linear	-24.3693	-20.4428	5	Rapid		10000	
3	Linear	-24.3693	-20.4428	-4.445	Rapid		10000	
4	Linear	-24.3693	-20.4428	-4.60375	Lead-in	381	10000	
5	Circular	-24.3299	-20.289	-4.7625	Lead-in	381	10000	
6	Circular	-24.1063	-18.7342	-4.7625	Lead-in	381	10000	
7	Linear	-24.1063	-18.6588	-4.7625	Cutting	381	10000	
8	Linear	-24.1964	-18.2175	-4.7625	Cutting	381	10000	
9	Linear	-24.2357	-17.7688	-4.7625	Cutting	381	10000	
10	Linear	-24.2525	-17.3186	-4.7625	Cutting	381	10000	
11	Linear	-24.2596	-16.8682	-4.7625	Cutting	381	10000	
12	Linear	-24.2625	-16.4178	-4.7625	Cutting	381	10000	
13	Linear	-24.2641	-15.5169	-4.7625	Cutting	381	10000	
14	Linear	-24.2575	17.1404	-4.7625	Cutting	381	10000	
15	Linear	-24.2419	17.252	-4.7625	Cutting	381	10000	
16	Linear	-24.204	17.358	-4.7625	Cutting	381	10000	
17	Linear	-24.0724	17.5408	-4.7625	Cutting	381	10000	

Figure-70. Toolpath information box

Editing Notes of an Operation

The **Edit Notes** option for selected operation is used to create user defined notes for the operation. On selecting **Edit Notes** option from the shortcut menu, the **Notes for Adaptive(name of operation)** dialog box will be displayed; refer to Figure-71.

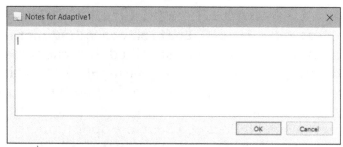

Figure-71. Notes for Adaptive dialog box

Specify desired text comments in the edit box of dialog box and click on the **OK** button.

Similarly, you can use the **Show Log** option from the shortcut menu to check log of events related to selected operation; refer to Figure-72.

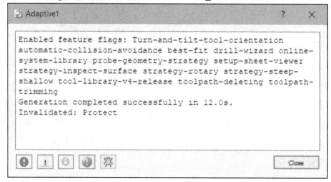

Figure-72. Tool operation log

Creating Form Mill Tool

The **Form Mill** tool in **MANAGE** drop-down is used to create a milling tool of desired shape. The procedure to use this tool is given next.

- Click on the **Form Mill** tool from the **MANAGE** drop-down in **INSPECTION** tab of the **Toolbar**. The **FORM MILL** dialog box will be displayed; refer to Figure-73.

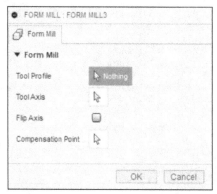

Figure-73. FORM MILL dialog box

- Select the desired sketch profile, axis, and cutting point of the tool; refer to Figure-74.

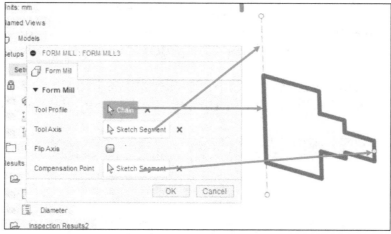

Figure-74. Selection for form tool

- Select the **Flip Axis** check box if you want to reverse the orientation of tool.
- Click on the **OK** button from the dialog box. The new tool will be added in the **Tool Library** dialog box; refer to Figure-75.

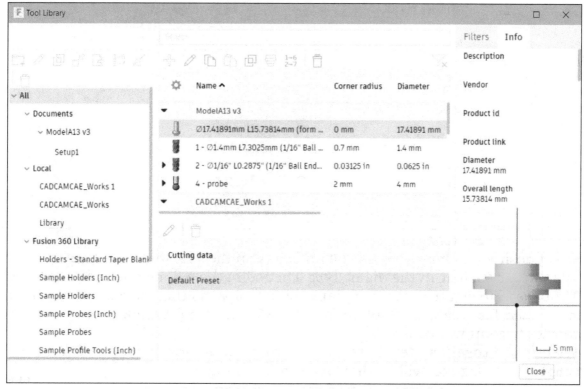

Figure-75. New form tool added in the CAM Tool Library dialog box

Tool Library

The **Tool Library** tool displays the **Tool Library** dialog box where you manage all the tools like milling, lathe, and cutting tools for your individual documents and operations, as well as libraries of predefined tools.

- Click on the **Tool Library** tool of **MANAGE** drop-down from **Toolbar**. The **Tool Library** dialog box will be displayed.
- View or modify the tool as required from the **Tool Library** dialog box. You can also perform common copy-paste functions in the library. The options of this dialog box have been discussed earlier.

Template Library

The **Template Library** tool is used to create and manage templates defined for performing set of specified operations. The procedure to use this tool is given next.

- Click on the **Template Library** tool from the **MANAGE** drop-down in the **INSPECTION** tab of the **Ribbon**. The **Template Library** dialog box will be displayed; refer to Figure-76.

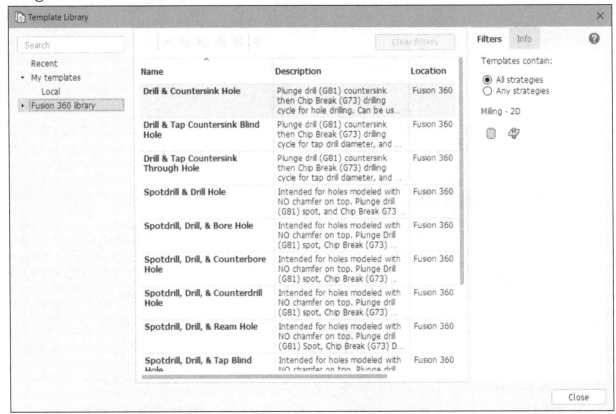

Figure-76. Template Library dialog box

- Select desired strategy from the list and click on the **Use selected template** button from the toolbar in the dialog box. The toolpaths will be added in the **BROWSER**.
- If you want to create a new template using toolpaths then select all the toolpaths to be used for creating template from **BROWSER** and right-click on any of them. A shortcut menu will be displayed.
- Select the **Store as Template** option from the shortcut menu. The **Store as template** dialog box will be displayed; refer to Figure-77.

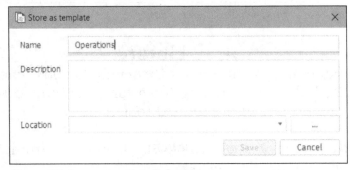

Figure-77. Store as template dialog box

- Specify desired description in the **Description** edit box.
- Click on the **Browse** button [...] for **Location** option to define location for template. The **Template Library** dialog box will be displayed.

- Select the **Local** folder in **My templates** node from left area in the dialog box and click on the **Select folder** button. The **Store as template** dialog box will be displayed again.
- Click on the **Save** button from the dialog box to save the template.

Task Manager

The **Task Manager** tool is used for controlling toolpath generation. CAM allows you to continue working inside Fusion 360 while generating toolpaths in the background. The main interface for controlling toolpath generation is the **CAM Task Manager**.

- Click on the **Task Manager** tool of **MANAGE** drop-down from **Toolbar**; refer to Figure-78. The **CAM Task Manager** dialog box will be displayed; refer to Figure-79.

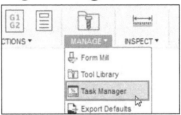

Figure-78. Task Manager tool

Figure-79. CAM Task Manager dialog box

- Now, generate a toolpath of any operation and you can see the progress in the **CAM Task Manager**; refer to Figure-80.

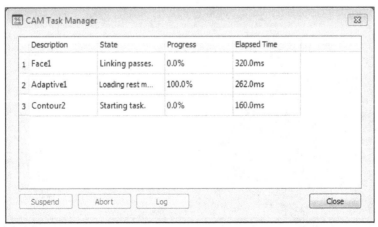

Figure-80. Running multiple operations simultaneously

Select the **Export Defaults** tool from the **MANAGE** drop-down in the **Toolbar** to export all the defaults settings of current manufacturing setup to a file.

Select the **Import Defaults** tool from the **MANAGE** drop-down in the **Toolbar** to import settings of manufacturing setup from a file.

Select the **Reset Defaults** tool from the **MANAGE** drop-down in the **Toolbar** to reset settings of current machining setup.

ADDITIVE MANUFACTURING

Additive manufacturing is a separate extension of Autodesk Fusion Manufacture workspace. You need to purchase it using credits from the **Extensions** dialog box; refer to Figure-81. So, make sure you have access to this extension. The procedure to activate tools for additive manufacturing is discussed next.

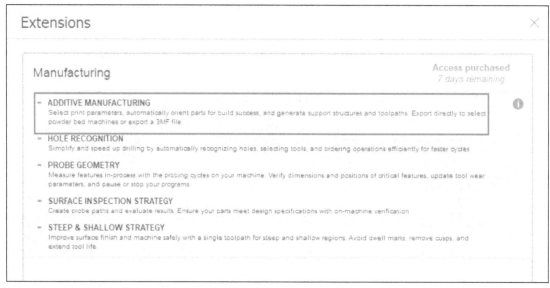

Figure-81. Extensions dialog box

- Click on the **New Setup** tool from the **SETUP** drop-down in the **Toolbar** of **MANUFACTURE** workspace. The **SETUP** dialog box will be displayed as discussed earlier.
- Select the **Additive** option from the **Operation Type** drop-down in the **Setup** tab of dialog box. The options will be displayed as shown in Figure-82.

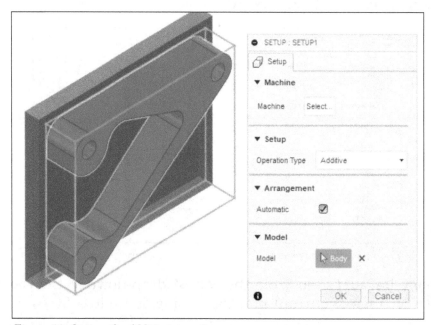

Figure-82. Options for Additive operation setup

- Click on the **Select** button for **Machine** option and select desired machine for 3D printing (Additive Manufacturing machine).
- On selecting the machine, **Print Settings** option will be displayed in the dialog box; refer to Figure-83. Click on the **Select** button for this option. The **Print Setting Library** dialog box will be displayed; refer to Figure-84. Select desired material and nozzle from the list and click on the **Select** button. Print settings will be applied.

Figure-83. Print Settings option

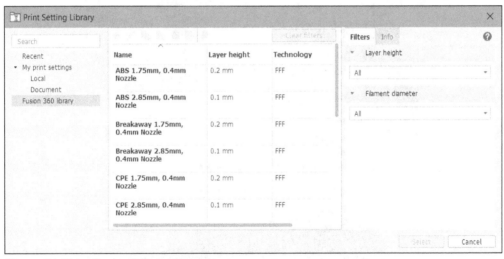

Figure-84. Print Setting Library dialog box

- Click on the **OK** button from the **SETUP** dialog box to apply operation setup.

Now, click on the **ADDITIVE** tab in the **Toolbar**. The options for additive manufacturing will be displayed; refer to Figure-85.

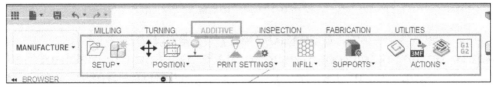

Figure-85. ADDITIVE tab in Toolbar

Most of the tools in this tab have been discussed earlier like Move, Task Manager, Measure, and so on. Rest of the tools are discussed next.

Minimize Build Height

The **Minimize Build Height** tool is used to orient the part in such a way that it occupies minimum height on the machine bed and more parts can be 3D printed on bed stacked one over another if needed. The procedure to use this tool is given next.

- Click on the **Minimize Build Height** tool from the **POSITION** drop-down in the **ADDITIVE** tab of **Toolbar**. The **MINIMIZE BUILD HEIGHT** dialog box will be displayed; refer to Figure-86.

Figure-86. MINIMIZE BUILD
HEIGHT dialog box

- Select the component/components for 3D printing, specify desired value of clearance from platform in the **Platform Clearance** edit box, and click on the **OK** button. The part will be oriented; refer to Figure-87.

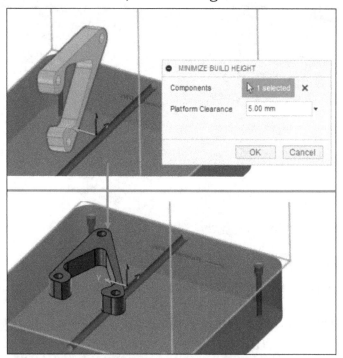

Figure-87. Setting orientation based on minimum build height

Automatic Orientation

The **Automatic Orientation** tool is used to place parts on the machine bed based on manufacturing requirements like minimum support structures and minimum build height parameters. The procedure to use this tool is given next.

- Click on the **Automatic Orientation** tool from the **POSITION** drop-down in the **ADDITIVE** tab of **Toolbar**. The **AUTOMATIC ORIENTATION** dialog box will be displayed; refer to Figure-88.

Figure-88. AUTOMATIC ORIENTATION dialog box

- Select the desired object to be reoriented.
- Click in the **Critical Angle** edit box and specify angle at which support structures will be created.
- Select the **Support Bottom Surface** check box to include bottom surface of model for generating support structures.
- Click in the **Distance to Platform (Z)** edit box and specify distance from the base bed that should be maintained while 3D printing.
- Click in the **Smallest Rotation** edit box and specify the minimum rotation with respect to original orientation within which automatic orientation will work for finding most suitable orientation of model.
- Select desired option from the **Rotation Axis** drop-down to define axis to be used for orienting the model. By default, the **Arbitrary** option is selected in the drop-down so axis will be selected automatically based on geometry of model.
- Click on the **Ranking** tab and select the ranking for various parameters to be considered based on priority level for deciding orientation of model; refer to Figure-89. For example, if you want to set minimum part height as priority then select the **Very high** option from the **Part Height** drop-down in the dialog box.

Figure-89. Ranking tab

- After setting desired parameters, click on the **OK** button from the dialog box after setting desired parameters.

Placing Parts on Platform

The **Place parts on platform** tool is used to place parts directly on platform or at some clearance distance from the platform. The procedure to use this tool is given next.

- Click on the **Place parts on platform** tool from the **POSITION** drop-down in the **ADDITIVE** tab of **Toolbar**. The **PLACE PARTS ON PLATFORM** dialog box will be displayed; refer to Figure-90.

Figure-90. PLACE PARTS ON PLATFORM dialog box

- If you select the **Flat Face** option from the **Type** drop-down then you can select the face to be directly placed on the platform.
- Select desired part and specify parameters as discussed earlier.
- After specifying parameters, click on the **OK** button.

Collision Detection

The **Collision Detection** tool is used to check whether two components on machine bed are interfering. The procedure to use this tool is given next.

- Click on the **Collision Detection** tool from the **POSITION** drop-down in the **ADDITIVE** tab of **Toolbar**. The **COLLISION DETECTION** dialog box will be displayed; refer to Figure-91.

Figure-91. COLLISION DETECTION dialog box

- Select the components whose interference is to be checked and click on the **Compute** button. The status will be displayed. If there is an interference then you can click on the **Auto Arrangement** button to automatically arrange components.
- Click on the **OK** button to exit the dialog box.

Selecting Print Settings

The **Select** tool in the **PRINT SETTINGS** drop-down of the **ADDITIVE** tab is used to select desired 3D printing settings. On selecting this tool, the **Print Setting Library** dialog box will be displayed. Select desired setting as discussed earlier.

Editing Print Settings

The **Print Setting Editor (Edit)** tool is used to modify parameters related to material and machining used by 3D printer. The procedure to use this tool is given next.

- Click on the **Print Setting Editor (Edit)** tool from the **PRINT SETTINGS** drop-down in the **ADDITIVE** tab of **Ribbon**. The **Print Setting Editor** dialog box will be displayed; refer to Figure-92.

Figure-92. Print Setting Editor dialog box

- Specify desired name for the print setting in **Name** edit box at the top in the dialog box.

Basic Tab

The options in the **Basic** tab of the dialog box are used to define basic parameters like height of each layer, fill density, nozzle speed, and so on. Various parameters of this tab are discussed next.

- Click in the **Layer Height** edit box and specify height of each layer while 3D printing.
- Click in the **Extrusion Width** edit box and specify width between two consecutive passes of 3D printer while creating the object.
- Click in the **Sparse Infill Density (%)** edit box and specify density in terms of percentage of general Infill density for regions which are non exposed.
- Select desired option from the **Sparse Infill Pattern** drop-down to define pattern in which nozzle will move to fill non exposed regions.
- Click in the **Travel Speed** edit box to specify at what speed the nozzle will move while 3D printing. If you find the part getting less dense than what is needed then you should check **Layer Height**, **Extrusion Width**, and **Travel Speed** parameters.
- Select the **Nozzle Priming** check box if you want to extrude some material before starting to 3D print the model so that there is a smooth flow of material through nozzle.

- Select the **Enable Raft** check box to create a layer on the build plate for better adhesion of model.
- Select the **Support Enabled** check box to use FFF specific support structures for supporting model while 3D printing.
- Select the **Randomize Perimeter Start Point** check box to start each layer pass from different location on the plane so that weld bead type structure is not formed on the side of model.
- Select the **Enable Nozzle Wipe** check box to perform a clean run through the nozzle to wipe material left in the machine if supported by your machine.
- Click in the **Nozzle Wipe Distance** edit box to specify distance within which nozzle need to be wiped.

Extruder Tab

The options in the **Extruder** tab are used to define speed and temperature of material in the extruder of 3D printer while generating the part; refer to Figure-93. Various options of this tab are discussed next.

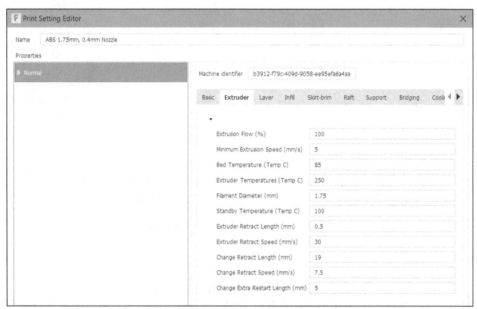

Figure-93. Extruder tab

- Click in the **Extrusion Flow** edit box and specify the percentage of material to be used for 3D printing in terms of total material stores in extruder.
- Specify desired value in the **Minimum Extrusion Speed** edit box to define minimum speed with which extruder can move for successful extrusion.
- Specify desired value in the **Bed Temperature** edit box to define temperature of bed on which model will be created.
- Specify desired value in the **Extruder Temperatures** edit box to define temperature of material in extruder. This should be equal to melting point of material used for creating 3D print.
- Specify desired value in the **Filament Diameter** edit box to define diameter of raw material filament being used for creating model.
- Specify desired value in the **Standby Temperature** edit box to define temperature at which material will be kept in extruder when machine is not in use.
- Specify desired value in **Extruder Retract Length** edit box to define length of filament which will be extruded before nozzle of machine is closed.

- Specify desired value in the **Extruder Retract Speed** edit box to define speed at which filament will move back in extruder from nozzle after completing operation.
- Similarly, specify desired values in **Change Retract Length**, **Change Retract Speed**, and **Change Extra Restart Length** edit boxes to define respective parameters when nozzle of machine is changed using a turret.

Layer Tab

The options in **Layer** tab are used to define properties of layers by which 3D print model is generated; refer to Figure-94. These options are discussed next.

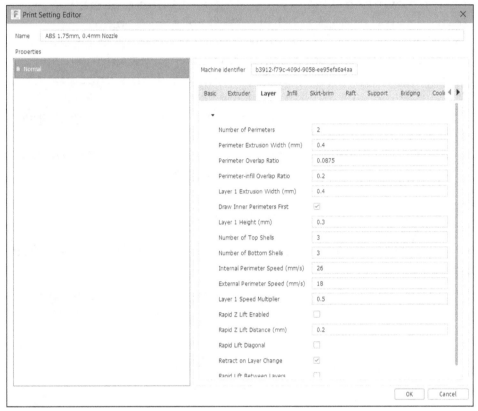

Figure-94. Layer tab

- Specify desired value in **Number of Perimeters** edit box to define boundary beads created on the bed before printing the features of model. This ensures firm base for features to be 3D printed.
- Specify desired value in **Perimeter Extrusion Width** edit box to define width of each strand of boundary bead.
- Specify desired value in **Perimeter Overlap Ratio** edit box to define the amount of overlap between two consecutive perimeter beads.
- Specify desired value in the **Perimeter-infill Overlap Ratio** edit box to define amount of material of printed model that will overlap with innermost perimeter bead.
- Specify desired value in the **Layer 1 Extrusion Width** edit box to define extrusion width of first layer being printed after creating perimeter bead on the bed.
- Select the **Draw Inner Perimeters First** check box to print inner most layer of perimeter beads first.
- Specify desired value in **Layer 1 Height** edit box to define thickness of first layer.
- Specify desired value in **Number of Top Shells** edit box to define number of top layers to be created by solid infill on 3D printed model.

- Specify desired value in **Number of Bottom Shells** edit box to define number of layers to be created at the bottom of 3D print with solid infill.
- Specify desired value in **Internal Perimeter Speed** edit box to define speed at which infill will be applied on internal structures of perimeter bead.
- Specify desired value in **External Perimeter Speed** edit box to define speed at which infill will be applied on external structures of perimeter bead.
- Specify desired value in the **Layer 1 Speed Multiplier** edit box to define factor by which speed for printing first layer will be multiplied.
- Select the **Rapid Z Lift Enabled** check box to allow rapid movement of nozzle in Z direction for moving to next layer or return to home position.
- Specify desired value in **Rapid Z Lift Distance** edit box to define distance up to which nozzle will move in filling mode before switching to rapid movement.
- Select the **Rapid Lift Diagonal** check box to allow rapid movements of nozzle in diagonal direction while moving to fill next layer. If this check box is not selected then nozzle will move directly upward for generating next layer.
- Select the **Retract on Layer Change** check box to force retraction speed movement before going to print next layer.
- Select the **Rapid Lift Between Layers** check box to lift extruder at the end of layer and lower the extruder at the beginning of next layer.

Infill Tab

The options in **Infill** tab of dialog box are used to define parameters related to how material will be filled in various areas of model; refer to Figure-95. The options in this tab are discussed next.

Figure-95. Infill tab

- Specify desired value in the **Sparse Infill Angle** edit box to define the angle at which nozzle will approach non-exposed areas of model while printing.
- Specify desired value in the **Solid Infill Angle** edit box to define the angle at which nozzle will approach non-exposed areas of model while printing.
- You can specify parameters in other edit boxes of this tab as discussed earlier.
- Select the **Connect Single Pass Only** check box to create single printing pass when moving to print next layer of model.

Skirt-brim Tab

The options in the **Skirt-brim** tab are used to define parameters for printing side walls of the model. The options in this tab are similar to those discussed earlier.

Raft Tab

The options in the **Raft** tab are used to define how rafts which are base parts of 3D printed object will be created. The options in this tab are similar to those discussed for Layer and Infill tab.

Support Tab

The options in the **Support** tab of dialog box are used to define how support structures will be filled during 3D printing.

Bridging Tab

The options in the **Bridging** tab are used to define how gap between support structures and other permeable features of model will be bridged (connected).

Cooling Tab

The options in the **Cooling** tab are used to define time required for filling one layer and number of starting layers up to which fan will not be running which indirectly define cooling time for creating 3D printed model.

G-Code Tab

The options in the **G-Code** tab are used to define parameters related to generation of G-Codes for 3D printer machine; refer to Figure-96. The options in this tab are discussed next.

Figure-96. G-Code tab

- Select the **Arc Fit Optimization** check box to enable arc interpolations using G2 and G3 codes.
- Select the **Start at Home Position** check box to always start 3D printing at home position before beginning toolpath.
- Select the **End at Park Position** check box to move nozzle to safe park position at the end of toolpath.
- Select the **Verbose G-code** check box to output comments along with toolpaths in output file.

After setting desired parameters, click on the **OK** button from the dialog box to apply settings.

Infill Settings

The **FFF Infill** tool of **INFILL** drop-down in **Ribbon** is used to define settings for infill of material. Here FFF refers to Fused Filament Fabrication, a type of 3D printing machine. The procedure to use this tool is given next.

* Click on the **FFF Infill** tool from the **INFILL** drop-down in the **ADDITIVE** tab of the **Ribbon**. The **INFILL** dialog box will be displayed; refer to Figure-97.

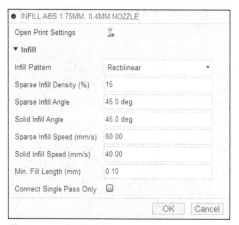

Figure-97. INFILL dialog box

* Select desired option from the **Infill Pattern** drop-down to define how nozzle will move while performing 3D printing. Some common infill patterns are shown in Figure-98.

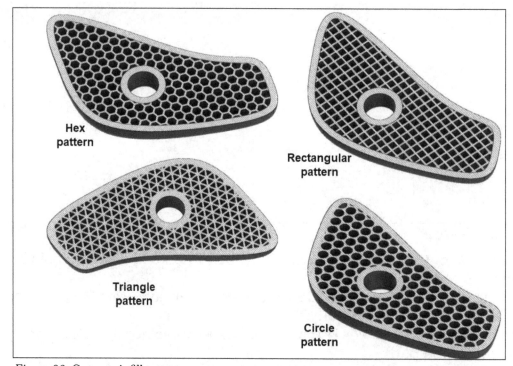

Figure-98. Common infill patterns

* Other options in this dialog box have been discussed earlier. After setting desired parameters, click on the **OK** button from the dialog box.

Support Settings

The **FFF Supports** tool is used to define settings related to creating support structures in the 3D printed model; refer to Figure-99.

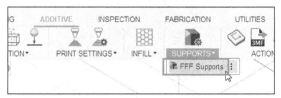

Figure-99. FFF Supports tool

The procedure to modify support settings is given next.

- Click on the **FFF Supports** tool from the **SUPPORTS** drop-down in the **ADDITIVE** tab of the **Ribbon**. The **SUPPORTS** dialog box will be displayed; refer to Figure-100.

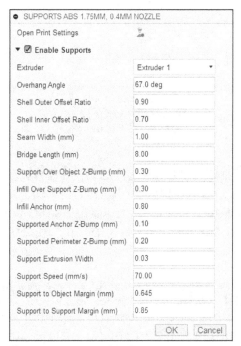

Figure-100. SUPPORTS dialog box

- The options in this dialog box are same as discussed earlier in printing settings. After setting desired parameters, click on the **OK** button from the dialog box. The support settings will be modified accordingly.

Generating Toolpath for Additive Manufacturing

After setting desired parameters, click on the **Generate** button from the **ACTIONS** drop-down in the **ADDITIVE** tab of **Toolbar**. The toolpath will be generated and displayed in the **BROWSER**.

After creating toolpath, click on the **Simulate additive toolpath** tool from the **ACTIONS** panel and check the simulation as discussed earlier.

Using **Print Statistics** tool in **ACTIONS** drop-down, you can check the information related to 3D printing like total time required, filament length used, and so on; refer to Figure-101.

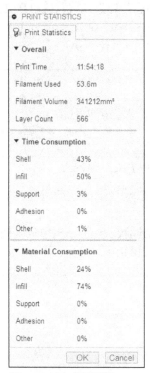

*Figure-101. PRINT
STATISTICS dialog box*

Exporting Toolpath and Printing Data

The **Export Toolpath** and **Export 3MF** tools in the **ACTIONS** panel are used to export the toolpath and printing setup for use by machines or other postprocessing software. Select the desired tool from the panel and specify location where you want to save the exported file.

ADDITIVE BUILD EXTENSION

Extensions are advanced capability apps that can be added to main interface of software to perform additional tasks. One such extension for additive manufacturing is **Additive Build** available in **Extension Manager**; refer to Figure-102. Note that the procedure to add extensions has been discussed in first chapter of the book. The general version of Autodesk Fusion 360 allows to use only FFF type 3D printing usually used for plastic parts. But if you want to 3D print metals and other materials that cannot be printed by FFF method then you can activate the Additive Build extension to use Selective Laser Melting (SLM) method and capable machines. In this method powder of metal is selectively melted to form layers. The tools and options related to SLM additive manufacturing are discussed next.

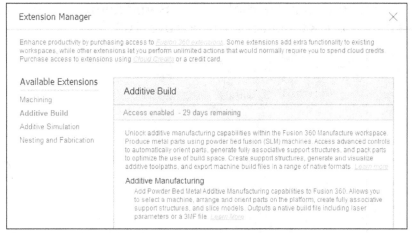

Figure-102. Additive Build extension

Selecting an SLM Machine

The procedure to select an SLM machine for additive manufacturing is similar to earlier discussed procedure to selecting FFF machine. The procedure to setup SLM machine is discussed next.

- After loading/create a model to be 3D printed, click on the **New Setup** tool from the **SETUP** drop-down in the **ADDITIVE** tab of the **Ribbon**. The **SETUP** dialog box will be displayed as discussed earlier.
- Click on the **Select** button for **Machine** option from **Machine** section of the dialog box. The **Machine Library** dialog box will be displayed.
- Select the **Additive** check box from the **Capabilities** section and **SLM** check box from the **Technologies** section of the dialog box. The SLM capable machines will be displayed; refer to Figure-103.

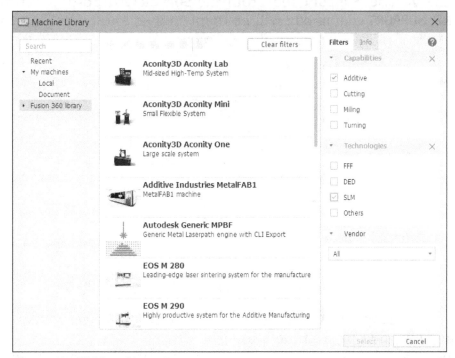

Figure-103. SLM machines

- Select desired SLM machine from the dialog box and click on the Select button. The printing bed for selected machine will be displayed.

- Click on the **OK** button from the dialog box. Some additional tools related to SLM will be displayed in the **Ribbon**; refer to Figure-104.

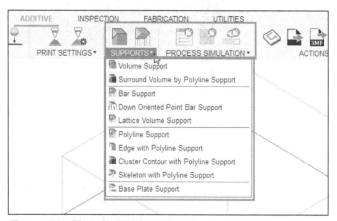

Figure-104. Tools for SLM 3D Printing

These tools are discussed next

Applying Supports

Various tools to create supports for 3D printing a model are available in the **SUPPORTS** drop-down of **ADDITIVE** tab in the **Ribbon**. The tools of this drop-down which are related to SLM are discussed next.

Creating Volume Support

The **Volume Support** tool is used to create support material for the components with specified structure. The procedure to create volume support is given next.

- Click on the **Volume Support** tool from the **SUPPORTS** drop-down in the **Toolbar**. The **VOLUME SUPPORT** dialog box will be displayed as shown in Figure-105.

Figure-105. VOLUME SUPPORT dialog box

- By default, the **Support Target** selection button is active. Select the faces on which you want to apply volume supports.
- Specify the desired values of angles from horizontal line in **Support Angle** and **Critical Support Angle** edit boxes to define width of support structures in normal and critical areas respectively.

- If you do not want to select the faces individually and want to use a window selection then you can use the options in **Advanced Area Filter** section of the dialog box to refine your selected of faces.
- Specify the minimum distance between new support and older support beam in the **Distance to other support** edit box.
- Click on the **General** tab in the dialog box to define general shape of volume support structures; refer to Figure-106.

Figure-106. General tab in VOLUME SUPPORT dialog box

- Select the desired option from the **Filling Type** drop-down to define the general shape of volume support. Select the **Hollow** option from the drop-down to create hollow supports. If you want to create structure of wired wall, punch plate, or solid structure then select the **Structured** option from the drop-down.
- Set the desired parameters in various edit boxes of the tab.
- Click on the **Volume Properties** tab to modify parameters of support structure. The options will be displayed as shown in Figure-107. Note that this tab will be displayed only when **Structured** option is selected in the **Filling Type** drop-down.

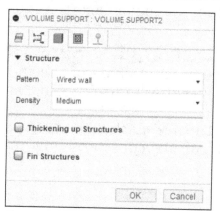

Figure-107. Volume Properties tab

- Select the desired options from the **Pattern** and **Density** drop-downs to define shape and density of structure.

- Specify the desired values in **Top Connections** and **Bottom Connections** edit boxes of the **Thickening up Structures** section to increase/decrease thickness of structure for top or bottom connection points of the structure. If you want to increase or decrease thickness of the main structure then specify desired value in the **Main Structure** edit box.
- Similarly, you can specify parameters for fin structure in the **Fin Structures** section of the dialog box.
- Click on the **Raster and Contour** tab in the dialog box to modify parameters related to net structure of rasters and contours. Contours are the boundary passes and rasters are the infill passes in 3D printing. The options will be displayed as shown in Figure-108.

Figure-108. Raster and Contour tab

- Set desired parameters in this tab and similarly, specify parameters related to different connection points in the **Connections** tab of dialog box.
- Click on the **OK** button from the dialog box to create support structure.

Creating Bar Support

The **Bar Support** tool is used to create bars of specified shape and size to support the printed structure. The procedure to use this tool is given next.

- Click on the **Bar Support** tool from the **SUPPORTS** panel in the **Toolbar**. The **BAR SUPPORT** dialog box will be displayed; refer to Figure-109.

Figure-109. BAR SUPPORT dialog box

- Select the faces that you want to be supported and specify desired parameters in the **Geometry** tab as discussed earlier.
- Click on the **General** tab to define density of base points of support, type of tree structure used for support, and projection angle for support structures in respective edit boxes; refer to Figure-110. You can also add bar supports to contour skeleton while printing layer by layer if the model is unstable at some point of printing.

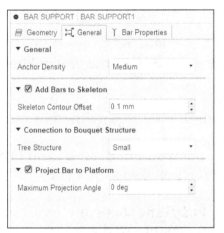

Figure-110. General tab in BAR SUPPORT dialog box

- Click on the **Bar Properties** tab in the dialog box and set the desired shape & size of bar in the dialog box.
- Click on the **OK** button from the dialog box. The support will be created; refer to Figure-111.

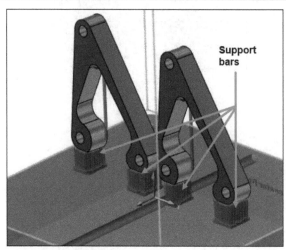

Figure-111. Support bars created

You can use the **Polyline Support**, **Lattice Volume Support**, and other tools of **SUPPORTS** panel in **Toolbar** in same way as other support tools have been discussed.

The **Surround Volume by Polyline Support** tool is used to apply additional polyline support after creating volume support for the model. Note that polyline supports will be added to contours of support only.

The **Down Oriented Point Bar Support** tool is used to apply bar support with its pointed side on the bed.

The **Lattice Volume Support** tool is used to create supporting structure in the form of scaffolding like structures generally found in civil construction sites.

The **Polyline Support** tool is used to create supporting structures in the form of a net line structure using small diameter polyline bar.

The **Edge with Polyline Support** tool is used to add support structures to edges of steep walls of model along with polyline supports.

The **Cluster Contour with Polyline Support** tool is used to add polyline cluster near contours of polyline supports created earlier.

The **Skeleton with Polyline Support** tool is used to create a skeleton of model support and then add polyline support to connect it with bed.

The **Base Plate Support** tool is used to create support between bed and bottom of part's support structures.

AM Process Simulation Settings

The **AM Process Simulation Settings** tool is used to define settings for simulation additive manufacturing process. Note that you need to activate Additive Simulation extension as discussed earlier. The procedure to use this tool is given next.

- Click on the **AM Process Simulation Settings** tool from the **PROCESS SIMULATION** drop-down in the **ADDITIVE** tab of the **Ribbon**. The **SETTINGS** dialog box will be displayed; refer to Figure-112.

Figure-112. SETTINGS dialog box

- Move the **Solve Accuracy** slider to define the accuracy at which you want to check the simulation of 3D printing process.
- Specify desired value in the **Layers per element** edit box to define size of the voxel in mesh to be 3D printed layer by layer. A high value in this edit box generated a coarse mesh.
- Specify desired value in the **Coarsening Generations** edit box to define number of layers to be created between two deposition layers for smooth transition of structure.
- Specify desired value in **Max Adaptivity Level** edit box to define maximum number of times by which coarsening generations can be performed for build plate.
- Select desired option from the **Build Plate Heating** drop-down and define related values if you want to heat the build plate. Heating build plate ensures that during machining cracks are not formed in the model due to large temperature difference and 3D printed model is easily detachable from base plate.
- Select the **Auto Calculate Support Volume Fraction** check box to automatically decide size of volume supports during simulation. If you clear this check box then **Manual Volume Fraction** edit box will be displayed and you need to specify the size fraction for volume supports.
- Click on the **OK** button from the dialog box to apply the settings specified in the dialog box.

After specifying settings, click on the Generate Mesh tool from the from the **PROCESS SIMULATION** drop-down in the **ADDITIVE** tab of the **Ribbon**. The result of meshing will be displayed after the process is complete showing how model will be generated in the drawing area; refer to Figure-113 and **RESULTS** tab will be activated in the **Ribbon**. Click on the **Finish Results** tool from the **Ribbon** to exit the mode and click on the **Close** button from **CAM Task Manager** to exit the dialog box.

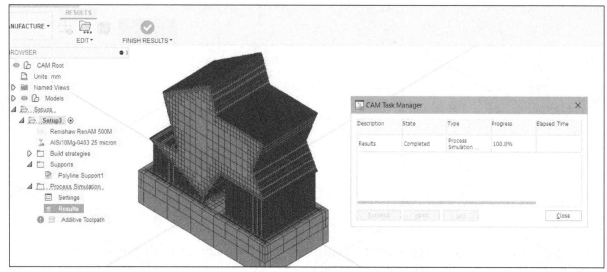

Figure-113. Additive manufacturing meshing results

SELF- ASSESSMENT

Q1. Which of the following tool is used to group together various cutting operations?

a. Setup Sheet   b. Add to New Pattern
c. Add to New Folder   d. Post Process

Q2. Which of the following pattern option is used to copy all the toolpaths created on one body onto another independent body?

a. Circular Pattern   b. Mirror Pattern
c. Duplicate Pattern   d. Component Pattern

Q3. Which of the following NC parameter is not available for Manual NC entry?

a. Start   b. Stop
c. Dwell   d. Optional Stop

Q4. What is the difference between the use of **Probe WCS** tool and **Inspect Surface** tool available in **PROBING** panel of **Toolbar** in **MANUFACTURE** workspace?

Q5. The **Simulate** tool is used to check how tools cuts through material in the form of an animation. (T/F)

Chapter 20

Introduction to Simulation in Fusion 360

Topics Covered

The major topics covered in this chapter are:

- *Introduction.*
- *Types of Analyses performed in Fusion 360.*
- *FEA*
- *User Interface of Fusion 360 Simulation.*
- *Material Properties*

INTRODUCTION

Simulation is the study of effects caused on an object due to real-world loading conditions. Computer Simulation is a type of simulation which uses CAD models to represent real objects and it applies various load conditions on the model to study the real-world effects. Fusion 360 is a CAD-CAM-CAE software package. In Fusion 360 Simulation, we apply loads on a constrained model under predefined environmental conditions and check the result (visually and/or in the form of tabular data). The types of analyses that can be performed in Autodesk Fusion are given next.

TYPES OF ANALYSES PERFORMED IN FUSION 360 SIMULATION

Fusion 360 Simulation performs almost all the analyses that are generally performed in Industries. These analyses and their uses are given next.

Static Analysis

This is the most common type of analysis we perform. In this analysis, loads are applied to a body due to which the body deforms and the effects of the loads are transmitted throughout the body. To absorb the effect of loads, the body generates internal forces and reactions at the supports to balance the applied external loads. These internal forces and reactions cause stress and strain in the body. Static analysis refers to the calculation of displacements, strains, and stresses under the effect of external loads, based on some assumptions. The assumptions are as follows.

- All loads are applied slowly and gradually until they reach their full magnitudes. After reaching their full magnitudes, load will remain constant (i.e. load will not vary against time).
- Linearity assumption: The relationship between loads and resulting responses is linear. For example, if you double the magnitude of loads, the response of the model (displacements, strains and stresses) will also double. You can make linearity assumption if:

1. All materials in the model comply with Hooke's Law that is stress is directly proportional to strain.
2. The induced displacements are small enough to ignore the change is stiffness caused by loading.
3. Boundary conditions do not vary during the application of loads. Loads must be constant in magnitude, direction, and distribution. They should not change while the model is deforming.

If the above assumptions are valid for your analysis, then you can perform **Linear Static Analysis**. For example, a cantilever beam fixed at one end and force applied on other end; refer to Figure-1.

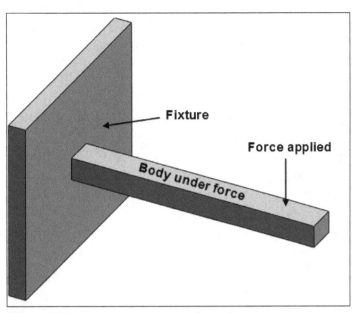

Figure-1. Linear static analysis example

If the above assumptions are not valid, then you need to perform the **Non-Linear Static analysis**. For example, force applied on an object attached with a spring; refer to Figure-2.

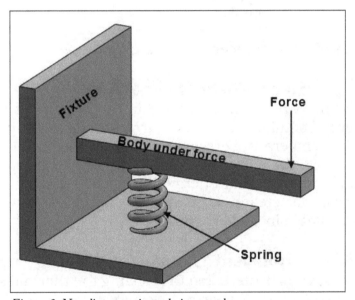

Figure-2. Non-linear static analysis example

Modal Analysis (Vibration Analysis)

By its very nature, vibration involves repetitive motion. Each occurrence of a complete motion sequence is called a "cycle." Frequency is defined as so many cycles in a given time period. "Cycles per second" or "Hertz". Individual parts have what engineers call "natural" frequencies. For example, a violin string at a certain tension will vibrate only at a set number of frequencies, that's why you can produce specific musical tones. There is a base frequency in which the entire string is going back and forth in a simple bow shape.

Harmonics and overtones occur because individual sections of the string can vibrate independently within the larger vibration. These various shapes are called "modes".

The base frequency is said to vibrate in the first mode, and so on up the ladder. Each mode shape will have an associated frequency. Higher mode shapes have higher frequencies. The most disastrous kinds of consequences occur when a power-driven device such as a motor, produces a frequency at which an attached structure naturally vibrates. This event is called "resonance." If sufficient power is applied, the attached structure will be destroyed. Note that armies, which normally marched "in step," were taken out of step when crossing bridges. If the beat of the marching feet align with a natural frequency of the bridge, then it could fall down. Engineers must design in such a way that resonance does not occur during regular operation of machines. This is a major purpose of Modal Analysis. Ideally, the first mode has a frequency higher than any potential driving frequency. Frequently, resonance cannot be avoided, especially for short periods of time. For example, when a motor comes up to speed it produces a variety of frequencies. So, it may pass through a resonant frequency.

Thermal analysis

There are three mechanisms of heat transfer. These mechanisms are Conduction, Convection, and Radiation. Thermal analysis calculates the temperature distribution in a body due to some or all of these mechanisms. In all three mechanisms, heat flows from a higher-temperature medium to a lower temperature one. Heat transfer by conduction and convection requires the presence of an intervening medium while heat transfer by radiation does not.

There are two modes of heat transfer analysis.

Steady State Thermal Analysis

In this type of analysis, we are only interested in the thermal conditions of the body when it reaches thermal equilibrium, but we are not interested in the time it takes to reach this status. The temperature of each point in the model will remain unchanged until a change occurs in the system. At equilibrium, the thermal energy entering the system is equal to the thermal energy leaving it. Generally, the only material property that is needed for steady state analysis is the thermal conductivity. This type of analysis is available in Fusion 360.

Transient Thermal Analysis

In this type of analysis, we are interested in knowing the thermal status of the model at different instances of time. A thermos designer, for example, knows that the temperature of the fluid inside will eventually be equal to the room temperature(steady state), but designer is interested in finding out the temperature of the fluid as a function of time. In addition to the thermal conductivity, we also need to specify density, specific heat, initial temperature profile, and the period of time for which solutions are desired. Till the time of writing this book, the transient thermal analysis was not available in Fusion 360.

Thermal Stress Analysis

The Thermal Stress Analysis is performed to check the stresses induced in part when thermal and structural loads act on the part simultaneously. Thermal Stress Analysis is important in cases where material expands or contracts due to heating or cooling of the part to certain temperature in irregular way. One example where thermal stress analysis finds its importance is two material bonded strip working in a high temperature environment.

Structural Buckling Analysis

Slender models tends to buckle under axial loading. Buckling is defined as the sudden deformation which occurs when the stored membrane (axial) energy is converted into bending energy with no change in the externally applied loads. Mathematically, when buckling occurs, the stiffness becomes singular. The Linearized buckling approach, used here, solves an eigenvalue problem to estimate the critical buckling factors and the associated buckling mode shapes.

In a laymen's language, if you press down on an empty soft drink can with your hand, not much will seem to happen. If you put the can on the floor and gradually increase the force by stepping down on it with your foot, at some point it will suddenly squash. This sudden scrunching is known as "buckling."

Event Simulation

The Event Simulation analysis is used to study the effect of object velocity, initial velocity, acceleration, time dependent loads, and constraints in the design. The results of this analysis include displacements, stresses, strains, and other measurements throughout a specified time period. You can perform this analysis when you need to check the effect of throwing a phone from some height or similar cases where motion is involved.

Shape Optimization

The Shape Optimization in Fusion 360 is not an analysis but a study to find the shape of part which utilizes minimum material but sustains the applied load up to required factor of safety.

Till this point, you have become familiar with the analyses that can be performed by using Fusion 360. But, do you know how the software analyze the problems. The answer is FEA.

FEA

FEA, Finite Element Analysis, is a mathematical system used to solve real-world engineering problems by simplifying them. In FEA by Fusion 360, the model is broken into small elements and nodes. Then, distributed forces are applied on each element and node. The cumulative result of forces is calculated and displayed in results. Note that Fusion 360 uses **Linear Tetrahedron** element with 4 nodes, Parabolic Tetrahedron element with 10 nodes and Parabolic Tetrahedron elements with Curved Edges having 10 nodes to mesh 3D solids. The **Line** element for bolt connectors (only available in Fusion 360 Ultimate). 2D and Planar (shell) elements are not supported in Autodesk Fusion 360 till the time we are writing this book.

As we are ready with some basic information about simulation in Fusion 360. Let's get started with initiating the simulation environment of Fusion 360.

STARTING SIMULATION IN FUSION 360

In Fusion 360, every workspace is available in a seamless manner. To start simulation in Fusion 360, click on the **Change Workspace** drop-down and select the **SIMULATION** option; refer to Figure-3. The **Simulation Workspace** will become active; refer to Figure-4. Also, the **New Study** dialog box will be displayed asking you to select the analysis type you want to perform.

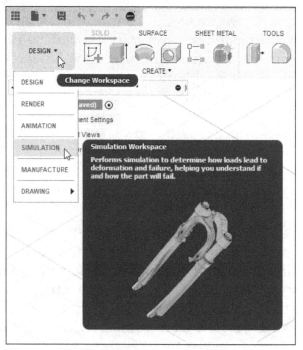

Figure-3. SIMULATION option

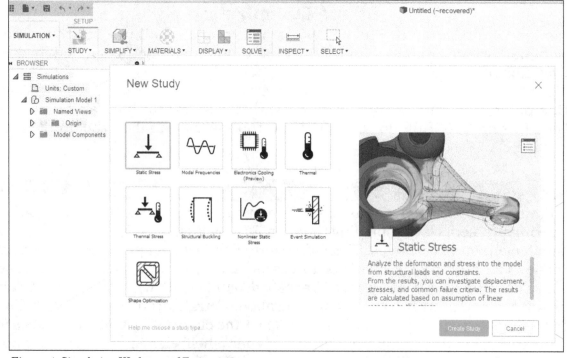

Figure-4. Simulation Workspace of Fusion 360

PERFORMING AN ANALYSIS

The static stress analysis is performed when the load is stable and the object deforms according to Hooke's Law. The procedure to start static stress analysis is given next. You can apply the same procedure for starting other analyses too.

- Double-click on the **Static Stress** button from the **New Study** dialog box. The tools required to perform static stress analysis will be displayed in the **Toolbar**; refer to Figure-5.

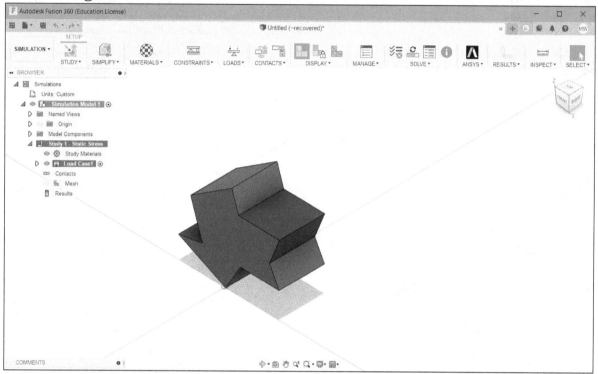

Figure-5. Static stress analysis tools in toolbar

Various tools and options of **Toolbar** in Simulation environment are discussed next.

STARTING NEW SIMULATION STUDY

While you are working on an analysis if you need to start another analysis then you can do so by using the **New Simulation Study** button. The procedure is given next.

- Click on the **New Simulation Study** tool from **STUDY** panel in the **Toolbar**; refer to Figure-6. The **New Study** dialog box will be displayed as discussed earlier. You can also press **N** key from keyboard while in Simulation environment to do the same.

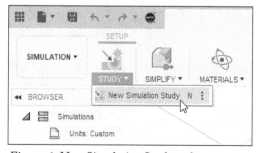

Figure-6. New Simulation Study tool

- Double-click on the desired button to perform respective analysis.

SIMPLIFYING MODEL FOR ANALYSIS

Most of the parts and assemblies have features that are irrelevant to analysis. These features hardly affect the result of analysis but take too much of processing resources. Such features should be removed before performing analysis. Examples of these features can be chamfers, fillets, unnecessary components which have no role in analysis. Note that sometimes these features are important part of calculations so make sure not to remove such features. One such example can be an assembly where load is acting on chamfer so you should not remove such chamfer from model. In Autodesk Fusion, there is a simple way to simplify your model using **Simplify** tool.

- Click on the **Simplify** tool from **SIMPLIFY** panel in the **Toolbar**; refer to Figure-7. The **SIMPLIFY Toolbar** will be displayed; refer to Figure-8.

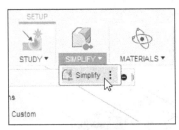

Figure-7. Simplify tool

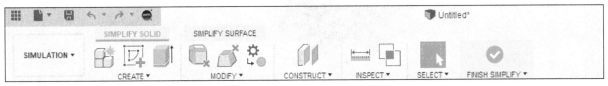

Figure-8. SIMPLIFY toolbar

If you go through various drop-downs and panels in this **Toolbar** then you will find that most of the tools are same as discussed for Modeling and Surface designing in earlier chapters. So, we will skip those tools and discuss the other tools.

Removing Features

The **Remove Features** tool is used to remove selected feature from the model in Simulation environment only. The procedure to use this tool is given next.

- Click on the **Remove Features** tool from **MODIFY** drop-down in the **SIMPLIFY** toolbar; refer to Figure-9. The **REMOVE FEATURES** dialog box will be displayed; refer to Figure-10.

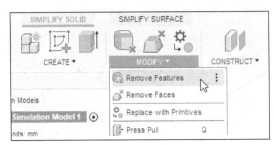

Figure-9. Remove Features tool

Figure-10. REMOVE FEATURES dialog box

- Select the bodies whose extra features are to be removed for performing analysis. The features being removed based on your selection in the **REMOVE FEATURES** dialog box, will be highlighted; refer to Figure-11.

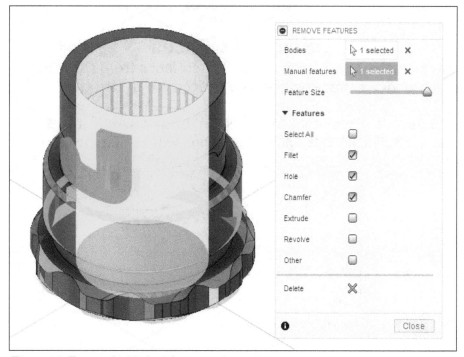

Figure-11. Features highlighted for removal

- Select the desired check boxes from the dialog box to remove respective features.
- Move the **Feature Size** slider left or right to consider smaller or larger features, respectively for removal.
- Although most of the features get selected automatically based on specified conditions in the dialog box but if you want to manually add/remove the features then click on the **Select** button of **Manual features** option in the dialog box and select the desired features.
- After selecting the features, click on the **Delete** button (⊠) at the bottom of the dialog box and close the dialog box; refer to Figure-12.

Figure-12. Selected features removed

Removing Faces

The **Remove Faces** tool is used to remove selected faces from the model and close the open sections of model automatically. The procedure to use this tool is given next.

- Click on the **Remove Faces** tool from **MODIFY** drop-down in the **SIMPLIFY** toolbar; refer to Figure-13. The **REMOVE FACES** dialog box will be displayed; refer to Figure-14.

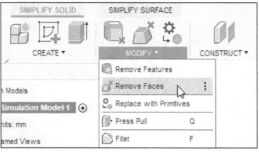

Figure-13. Remove Faces tool

Figure-14. REMOVE FACES dialog box

- The **no selection** button of **Select Faces** section is active by default. Click on the face to be remove; refer to Figure-15.

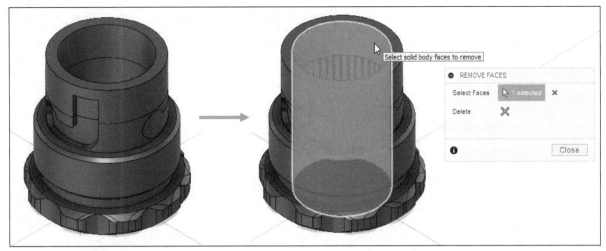

Figure-15. Selection of face to be remove

- Click on the **Delete** button from the dialog box. The selected faces will be deleted; refer to Figure-16.

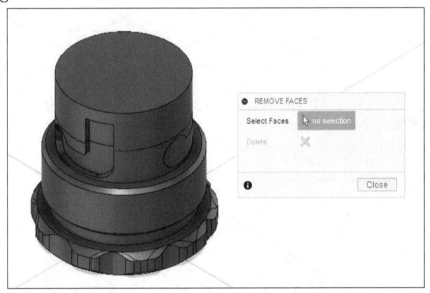

Figure-16. Selected face removed

- Click on **Close** button to exit the tool.

Replace With Primitives

The **Replace with Primitives** tool is used to replace selected body/component by primitive shapes like box, cylinder, and sphere. The procedure to use this tool is given next.

- Click on the **Replace with Primitives** tool from the **MODIFY** drop-down in the **SYMMETRY** toolbar; refer to Figure-17. The **REPLACE WITH PRIMITIVES** dialog box will be displayed; refer to Figure-18.

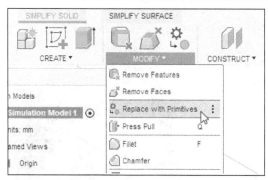

Figure-17. Replace with Primitives tool

Figure-18. REPLACE WITH PRIMITIVES dialog box

- Select the desired object type from the **Object type** drop-down.
- Select the **Bodies** option if you want to replace bodies with primitives and select the **Components** option if you want to replace assembly components with primitives for analysis.
- The **Select** button of **Selection** section is active by default. Click on the body/ component to be selected; refer to Figure-19.

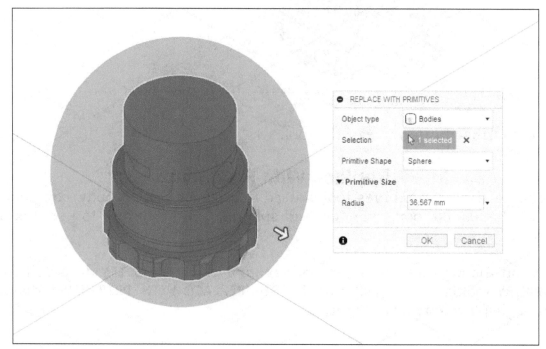

Figure-19. Selection of body to replace

- Select the desired shape from **Primitive Shape** drop-down and specify the related size parameters in the edit boxes of dialog box.
- Click on the **OK** button from the dialog box to replace the body/component; refer to Figure-20.

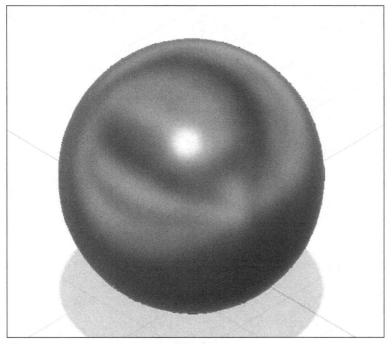

Figure-20. Selected body replaced with sphere

After performing desired simplification operations, click on the **FINISH SIMPLIFY** button from the **Toolbar** to exit Simplify mode.

STUDY MATERIAL

Study material is the material applied to model with all the physical properties so that you can check the effect of load on actual material conditions. The tools related to study material are available in the **MATERIALS** drop-down of the **Toolbar**; refer to Figure-21. These tools are discussed next.

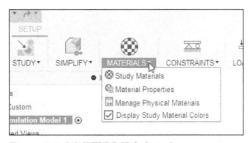

Figure-21. MATERIALS drop down

Applying Study Material

Study material is applied to assign physical properties of material to the model. The procedure to apply study material is given next.

- Click on the **Study Materials** tool from the **MATERIALS** drop-down in the **Toolbar**; refer to Figure-22. The **Study Materials** dialog box will be displayed; refer to Figure-23.

Figure-22. Study Materials tool

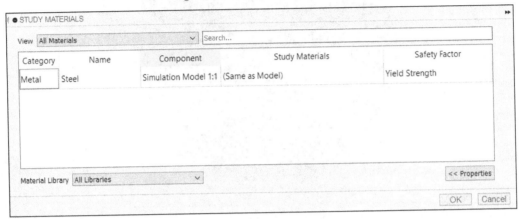

Figure-23. Study Materials dialog box

- Click in the drop-down of **Study Materials** column in the dialog box for the current component and select desired material; refer to Figure-24.

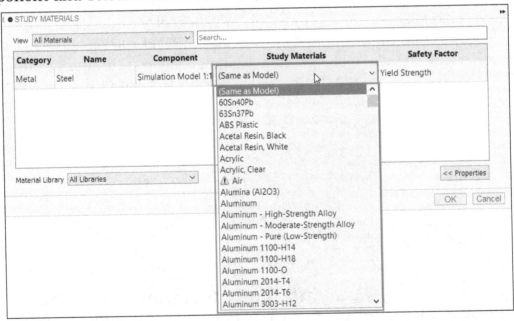

Figure-24. Study Materials drop down

- Select desired safety factor criteria from the drop-down in **Safety Factor** column of the dialog box. There are two options in this drop-down; **Yield Strength** and **Ultimate Tensile Strength**. Yield Strength is the point where metal starts to permanently deform. Ultimate Tensile Strength is the point after which the metal becomes so weak that it can break. Generally Yield Strength is used as standard for determining safety factor.
- After specifying all the desired parameters, click on the **OK** button. The material will be applied.

Displaying Material Properties

All the properties of different materials in material library can be checked by using the **Material Properties** button. The procedure is discussed next.

- Click on the **Material Properties** tool from the **MATERIALS** drop-down in the **Toolbar**; refer to Figure-25. The **Material Properties** dialog box will be displayed; refer to Figure-26.

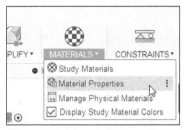

Figure-25. Material Properties tool

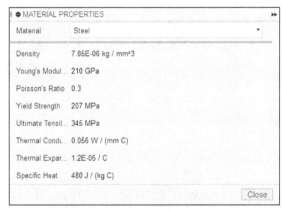

Figure-26. MATERIAL PROPERTIES dialog box

- Select the material from the **Material** drop-down at the top in the dialog box to check the material properties.
- Click on the **Close** button to exit the dialog box.

Managing Physical Material

The **Manage Physical Materials** tool is used to manage physical properties of material. If you want to edit any parameter of material before using it in analysis then you can do so by using this tool. The procedure is given next.

- Click on the **Manage Physical Materials** tool from the **MATERIALS** drop-down in the **Toolbar**; refer to Figure-27. The **Material Browser** dialog box will be displayed; refer to Figure-28.

Figure-27. Manage Physical Materials tool

Figure-28. Material Browser dialog box

- Select the desired material library and category from the left area of the **Material Browser** dialog box.
- Hover the cursor on the material you want to edit from the right area and click on the **Adds material to favorites and displays in editor** button; refer to Figure-29. The editing options will be displayed at the right in dialog box; refer to Figure-30.

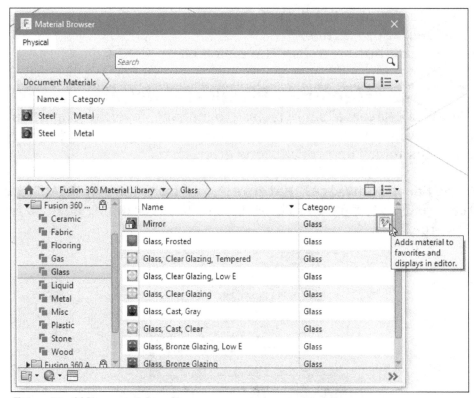

Figure-29. Adding material to edit

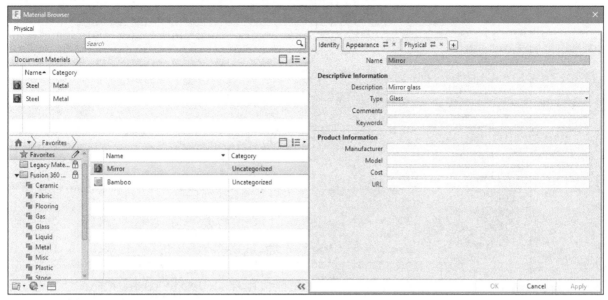

Figure-30. Material editing area

- Click at the desired tab in the right area of the dialog box and specify the parameters related to material.
- Click on the **Apply** button from the right area to apply the changes and click Cancel button to exit editing or click on the **OK** button to apply changes and exit editing mode.
- Click on the **Close** button from top right corner to exit the **Material Browser** dialog box.

Creating New Material

If you want to create a new material then follow the procedure given next.

- Click on the **Create New Library** tool from the **Creates, opens, and edits user-defined libraries** drop-down at the bottom left corner of the **Material Browser** dialog box; refer to Figure-31. The **Create Library** dialog box will be displayed; refer to Figure-32.

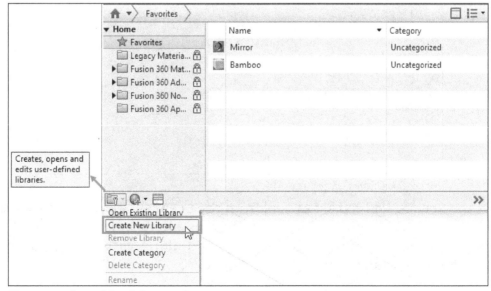

Figure-31. Create New Library tool

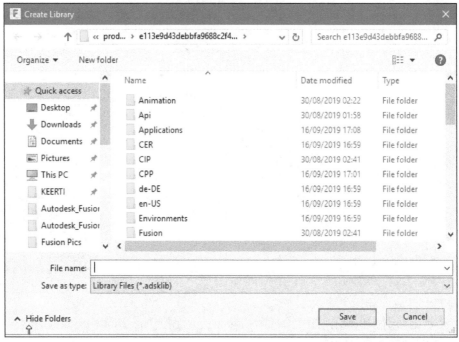

Figure-32. Create Library dialog box

- Specify the desired name of the library in **File name** edit box and click on the **Save** button to save the library file at desired location. A new library will be added.
- Select the newly added library and click on the **Create Category** tool from the **Creates, opens, and edits user-defined libraries** drop-down at the bottom left corner of the **Material Browser** dialog box; refer to Figure-33. A new category will be added to the library. Right-click on the category and select **Rename** if you want to rename it as desired.

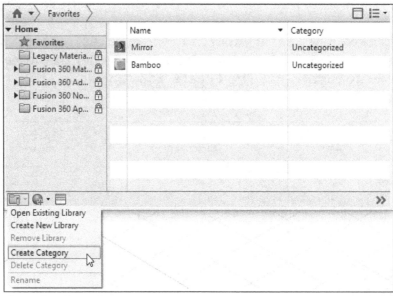

Figure-33. Create Category tool

- Click on the **Create New Material** tool from the **Creates and duplicates materials** drop-down at the bottom in the dialog box as shown in Figure-34. The **Select Material Browser** dialog box will be displayed along with editing options in the **Material Browser** dialog box; refer to Figure-35.

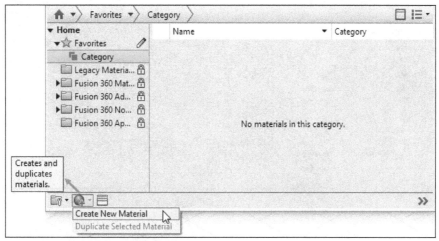

Figure-34. Create New Material tool

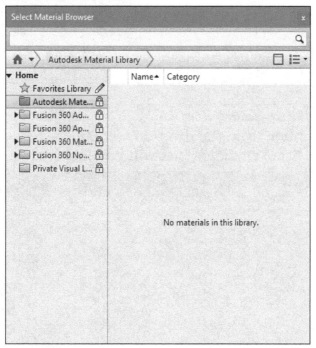

Figure-35. Select Material Browser dialog box

- Close the **Select Material Browser** dialog box and specify desired parameters of material in the right area.
- To apply physical or appearance properties to material, click on the **+** sign next to **Identity** tab in the editing area of the dialog box. A drop-down will be displayed; refer to Figure-36.

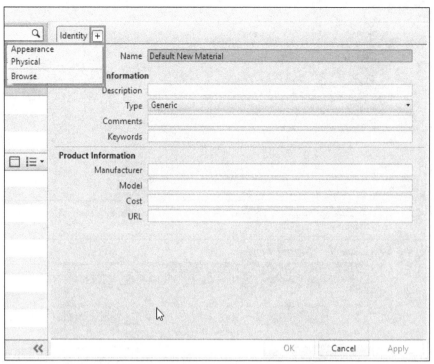

Figure-36. Adding properties of material

- Select the desired option from the drop-down (like, we have selected the **Physical** option). The **Asset Browser** dialog box will be displayed; refer to Figure-37.

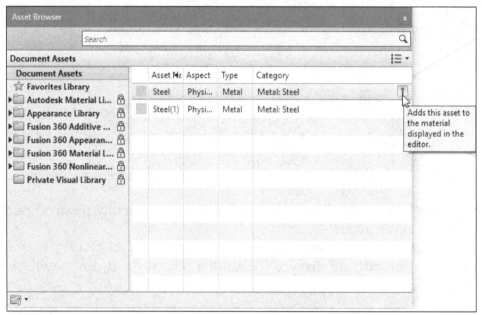

Figure-37. Asset Browser dialog box

- Click on the **Adds this asset to the material displayed in the editor** button as shown in Figure-37 to copy the physical properties of material.
- Close the **Asset Browser** dialog box. The physical properties have been assigned to the new material. Similarly, you can apply **Appearance** properties to the material.
- Click on the **Apply** button. The material will be added in the library.
- Add more materials as required and then close the dialog box. Various parameters related to material are discussed in next topic.

General Parameters of Materials in Autodesk Fusion 360

There are three categories in which properties of materials are defined in Autodesk Fusion 360: Identity, Appearance, and Physical. These categories are available as tabs in the **Material Browser** dialog box when you edit a material or create a new material; refer to Figure-38. The parameters specified in these tabs are discussed next.

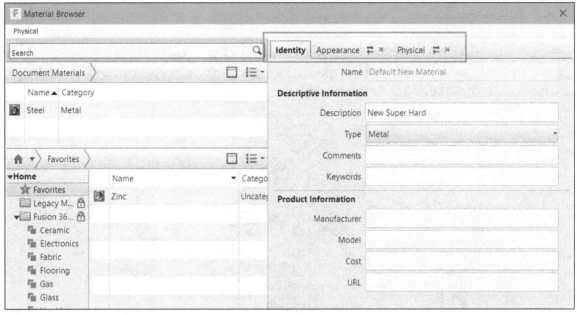

Figure-38. Categories of material properties

Identity Parameters (common for all type of materials)

The parameters of **Identity** tab are used to define description and type of material, manufacturer of material, cost of material and other related data. These parameters are discussed next.

- Specify desired text in the **Description** edit box to add user-define description of material which can be used to identify general use of the material.
- Select desired option from the **Type** drop-down to define type of the material. Note that the type selected here will define the category in which material will be placed in Material Library.
- Specify desired text in the **Comments** edit box to define comments about the material meant to warn or notify the user of material. Like, material is fragile, material is explosive, and so on.
- Specify desired identification keywords in the **Keywords** edit box separated by comma (,) to enable fast filtering of material in the browser based on specified keywords.
- The options in **Product Information** section of this tab are used when generating reports for production or performing cost calculations. Specify desired text in the **Manufacturer** edit box to define the name of vendor from whom your organization purchases the material.
- Specify desired value in the **Model** edit box to define model number of material if provided by your manufacturer.
- Specify desired value in the **Cost** edit box to define cost of material. Note that value specified in this edit box is in the form of text which is not used for any mathematical formula.

- Specify desired value in the **URL** edit box to define website link for the material.

In Autodesk Fusion 360, all the materials fall into 5 categories: Metal, Transparent, Opaque, Glazing, and Layered in terms of their appearance. Any modification in appearance of material does not affect the results of analysis but they can make huge difference when generating rendered images of model discussed in Chapter 10 of this book. Some important parameters of all five categories of material appearances are discussed next.

Appearance Tab (For metal type materials)

If you are defining parameters of appearance for a metal type material then options in editing mode of **Material Browser** will be displayed as shown in Figure-39.

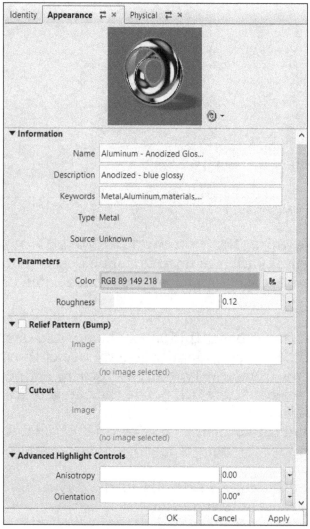

Figure-39. Appearance tab for metal type

- The options in the **Information** section of dialog box are used to define general information about the appearance of material so that user can easily identify the type of material. In this section, you need to specify name, description, and keywords for the material.
- The options in the **Parameters** section are used to define color and roughness of surface of material. Click in the field for **Color** option from **Parameters** section. The **Color Picker** dialog box will be displayed. Select desired color by clicking

in the color board or by specifying RGB values; refer to Figure-40. After setting color, click on the **OK** button from the dialog box. Using the slider for **Roughness** option, you can define the roughness of surface of material for reflectivity. A high roughness value means less reflective surface.

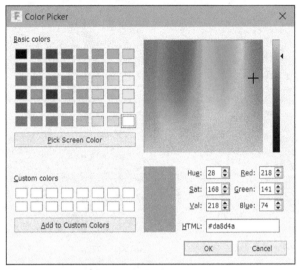

Figure-40. Color Picker dialog box

- Select the **Relief Pattern (Bump)** check box to add bumps in the texture of material. Bumps create an illusion of irregularities generally found on surfaces of real objects; refer to Figure-41. On selecting check box, the **Material Editor Open File** dialog box will be displayed and you will be asked to select an image file of bumps. Select desired image to be projected on material surface for bumps and click on the **Open** button. The appearance of material will be modified accordingly.

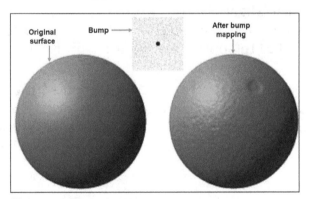

Figure-41. Bump mapping

- Select the **Cutout** check box to make net like texture of material where some portion of material will be transparent and some portion will be opaque; refer to Figure-42. After selecting check box, you need to select an image file with black area to define transparent section of texture as discussed earlier for bump mapping.

Figure-42. Cutout texture

- The options in **Advanced Highlight Controls** section are used to anisotropy, orientation of anisotropy effects, highlight color and so on. Specify desired value using slider of **Anisotropy** option to define enlargement of highlight texture in material. For example, if highlight texture is in the form of circles then anisotropy will transform them into ellipses. Specify desired value using slider of **Orientation** option to define angle at which anisotropy enlargement will occur. Specify desired value in the **Color** selection box to define color by which material will be highlighted. Generally, white color is used to give realistic highlight effect. Select desired option from the **Shape** drop-down to define how reflectivity and highlight effect of material. Select the **Long Falloff** option to produce smoother highlights and select the **Short Falloff** option to produce sharp highlights in material.

Appearance Tab (For Opaque type materials)

If you are defining parameters of appearance for opaque type material then options in editing mode of **Material Browser** will be displayed as shown in Figure-43. Most of the options in this tab are same as discussed for metals. The other options of this tab are discussed next.

- Specify desired value in **Reflectance** field using slider to define amount of light that will be reflected from surface of material. Note that for opaque materials, reflection will be highest at glazing angle.
- Select the **Translucency** check box to define parameters for free transmission of light through the material. After selecting check box, specify desired value of depth up to which there will be free transmission of light though the material using **Depth** slider. Click in the **Weight** selection box and specify desired color for translucency. This color acts as tinting color for material. You can also select pantone color using **Pantone Color Libraries** button next to **Weight** selection box.
- Select the **Emissivity** check box to apply self illumination of material under light. After selecting check box, set desired value using the **Luminance** slider to define brightness of light emitting from material in candelas per meter square unit. Set desired value in **Filter Color** selection box to define color of self illuminance.

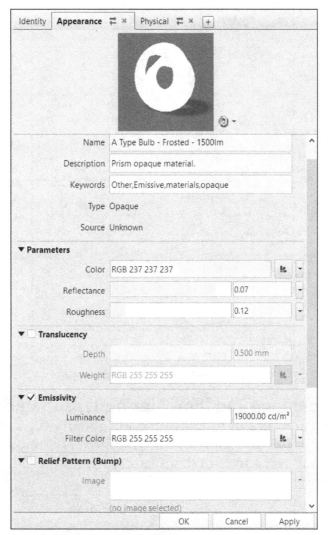

Figure-43. Appearance tab for opaque type

Appearance Tab (For Transparent type materials)

If you are defining parameters of appearance for transparent type material then options in editing mode of **Material Browser** will be displayed as shown in Figure-44. Most of the options in this tab are same as discussed for metals. The other options of this tab are discussed next.

- Specify desired value in the **Absorption Distance** edit box to define distance up to which transmission color will reach through the model after applying transparent material.
- Select desired option from the **Index of Refraction** drop-down to define reference material to be used for defining refraction level of material. After selecting desired option, specify the value of refraction index in the edit box next to drop-down.

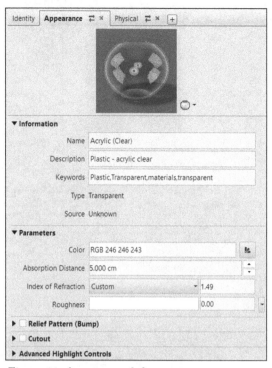

Figure-44. Appearance tab for transparent type

Appearance Tab (For Glazing type materials)

If you are defining parameters of appearance for glazing type material then options in editing mode of **Material Browser** will be displayed as shown in Figure-45. Most of the options in this tab are same as discussed for metals. The other options of this tab are discussed next.

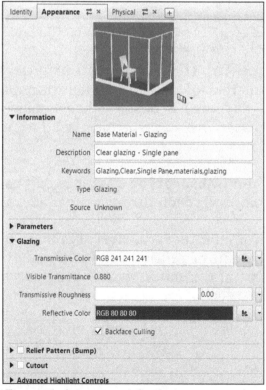

Figure-45. Appearance tab for glazing type

- Specify desired color in the **Transmissive Color** selection box of **Glazing** section to define color which will transmit through the material. Note that other colors will display material as opaque.
- Specify desired value using the **Transmissive Roughness** slider to define how much light transmission will be restricted through the material. A higher roughness value will make the material lesser transmissive.
- Set desired color in **Reflective Color** selection box to define the color which will be reflecting from the surface of objects to which material has been applied.
- Select the **Backface Culling** check box to make only front side of material reflect or transmit light. Rest of the sides of material will be completely transparent.

Appearance Tab (For Layered type materials)

If you are defining parameters of appearance for layered type material then options in editing mode of **Material Browser** will be displayed as shown in Figure-46. Most of the options in this tab are same as discussed for metals. The other options of this tab are discussed next.

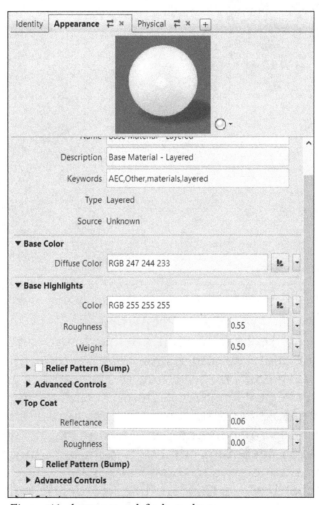

Figure-46. Appearance tab for layered type

- Specify desired colors for base layer, base highlight, and top coat of material. Set the other parameters as discussed earlier.

Physical Tab

The options in Physical tab are used to define properties that affect the physical data of material used by various simulations (analyses). Parameters like thermal conductivity, yield strength, young's modulus are some example of physical properties. On selecting this tab in editing mode of **Material Browser**, the options will be displayed as shown in Figure-47. The options in Information section are same as discussed earlier. The other options are discussed next.

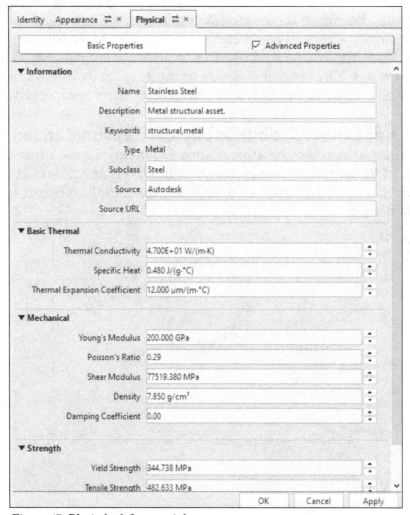

Figure-47. Physical tab for materials

Basic Properties

- The options in **Basic Thermal** section are used to define how much heat can transfer through the material and related parameters. Specify desired value in **Thermal Conductivity** edit box to define the amount of heat in W that can be transferred through the unit length of material at unit temperature. Specify desired value in **Specific Heat** edit box to define the amount of heat energy required by unit mass of material to raise its temperature by 1 unit. Specify desired value in the **Thermal Expansion Coefficient** edit box to define the amount of length increase in material due to unit increase in temperature.

- The options in **Mechanical** section are used to define parameters like Young's modulus, Poisson's ratio and so on. Specify desired value in **Young's Modulus** edit box to define the amount of stress required for per unit strain in the material. This parameter is also called **Modulus of elasticity** and defines relationship between stress and strain of the material. Specify desired value in **Poisson's Ratio** edit box to define relationship between compression applied on one direction of material causing expansion in other perpendicular direction. You can check the role of this parameter by compression a rectangular piece of sponge. Specify desired value in **Shear Modulus** edit box to define amount of shear stress required for unit shear strain to occur in material. Shear forces cause objects to tilt while their base is fixed. Specify desired value in **Density** edit box to define mass of per unit volume of material. This parameter is used to determine mass of the model. Specify desired value in the **Damping Coefficient** edit box to define a ratio by which oscillations in the material are dissipated. Note that if there is no damping in a spring placed in vacuum then it will keep on oscillating forever once stretched and released. The damping coefficient of material directly affects results of Modal analysis and other frequency related analyses.

- The options in **Strength** section are used to define strength parameters of material up to which the material will be useful for mechanical applications. Specify desired value in the **Yield Strength** edit box to define the amount of stress at which permanent deformation will occur in material. Specify desired value in the **Tensile Strength** edit box to define amount of stress required to break off the material. Tensile strength is also called ultimate tensile strength and ultimate strength of material.

Advanced Properties

Select the **Advanced Properties** check box in **Physical** tab to activate advanced properties of the material and then click on the **Advanced Properties** tab. The options will be displayed as shown in Figure-48. The options in this tab define whether material is linear, non-linear, or hyper elastic. These options are discussed next.

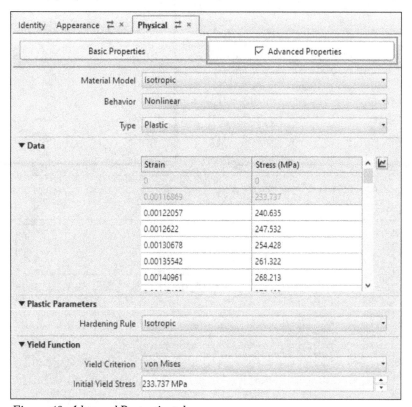

Figure-48. Advanced Properties tab

Isotropic Materials

- Select the **Isotropic** option from the **Material Model** drop-down to define that properties of the material are same in all the direction. It means you can apply 100 N load in any direction of material and it will cause same stress in the material.

- For isotropic materials, three options are available in the Behavior drop-down to define whether material behaves linearly to stress or it behaves non-linearly. Select the Linear option from the drop-down if your material follows Hooke's Law for given range of analysis loads. On selecting **Linear** option, no other parameters need to be specified. Select the **Nonlinear** option from the **Behavior** drop-down if various in stress causes non-linear variation in strain of material; refer to Figure-49. After selecting this option, you need to specify stress and strain data of material in the table of **Data** section. Note that non-linearity can be of three types available in **Type** drop-down which are Elastic non-linearity, Plastic non-linearity, and Elastic-Plastic (Bi-linear) non-linearity; refer to Figure-50. Select the **Elastic** option from **Type** drop-down if your material generates non linear strain for given range of stresses but returns to its original shape when load is removed. Select the **Plastic** option from the **Type** drop-down if your material has non-linear stress-strain curve and do not return to original shape after removing load. Select the **Elastic-Plastic (Bi-linear)** option if material behaves elastically up to specified yield stress value and then behaves plastically up to specified tangent modulus value. Note that Tangent Modulus is equal to Young's Modulus in elastic range of material. Specify the other parameters as needed for non-linearity of material. Note that one of the cause for using Non-linear static analysis in Autodesk Fusion 360 is non-linearity of material selected for model.

A

σ

Linear Material

$E = \dfrac{\sigma}{\epsilon}$

ϵ

B

σ

Nonlinear Material

$E(\epsilon) = \dfrac{d\sigma}{d\epsilon}$

ϵ

Figure-49. Linear vs non linear materials

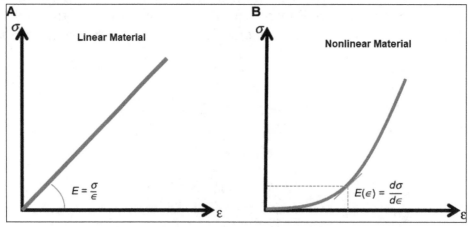

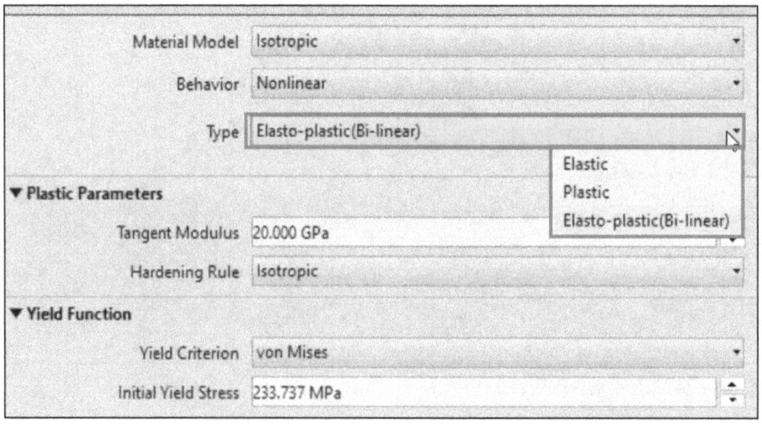

Figure-50. Nonlinearity types

- A material can be non-linear due to change in temperature as well. For such materials, select the **Temperature Dependent** option from the **Behavior** drop-down after selecting Isotropic material model option. The options in dialog box will be displayed as shown in Figure-51. Double-click in the empty fields of tables to specify the behavior of material for different temperature values.

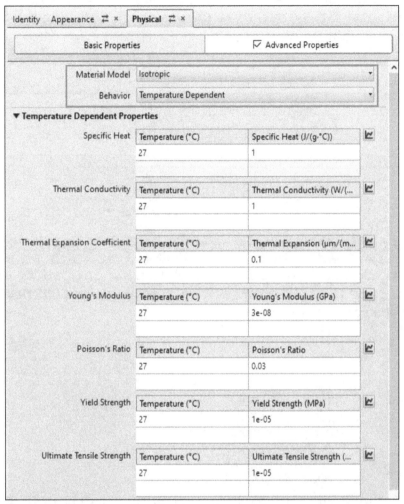

Figure-51. Temperature dependent isotropic material

- Select the **Hyperelastic** option from the **Material Model** drop-down to define that material does not follow standard stress-strain relationship. A hyperelastic material gets its stress-strain relationship from strain energy density function. More detail about this material is beyond the scope of this book but you can assume rubber like materials to be hyperelastic and follow Mooney Rivlin equation for strain less than 100% ; refer to Figure-52.

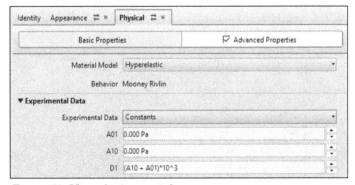

Figure-52. Hyperelastic material type

- After specify desired parameters, click on the **OK** button to apply modifications to material.

Displaying Study Material Colors

By default, appearance assigned to the part in the **Design workspace** is displayed in the **Simulation workspace**. If you want to display appearance of study material then select the **Display Study Material Colors** check box from the **MATERIALS** drop-down; refer to Figure-53.

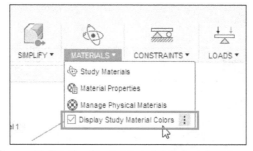

Figure-53. Display Study Material Colors check box

APPLYING CONSTRAINTS

Constraints are used to restrict motion of part when load is applied to form equilibrium. The tools to apply constraints are available in **CONSTRAINTS** drop-down in the **Toolbar**; refer to Figure-54.

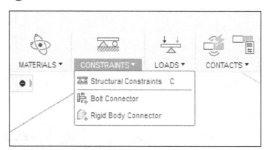

Figure-54. CONSTRAINTS drop down1

The procedure to apply different type of constraints are discussed next.

Applying Structural Constraints

The structural constraints are used to apply different type of structural constraints like fixed, pin, frictionless and so on. The procedure to apply structural constraint is given next.

- Click on the **Structural Constraints** tool from the **CONSTRAINTS** drop-down in the **Toolbar**; refer to Figure-55. The **STRUCTURAL CONSTRAINTS** dialog box will be displayed; refer to Figure-56.

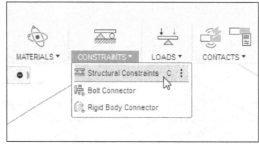

Figure-55. Structural Constraints tool

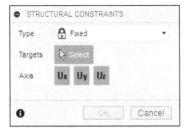

Figure-56. STRUCTURAL CONSTRAINTS dialog box

Fixed Constraint

- Select **Fixed** option from **Type** drop-down in the **STRUCTURAL CONSTRAINTS** dialog box if you want to fix selected faces/edges/vertices of the part.
- Select the face/edge/vertex that you want to be fixed.
- Select desired axis button from the **Axis** section. Like, select the **Ux** button if you want to restrict movement along X axis. By default, all the three buttons are selected and hence the movement along all the three axes is restricted. Refer to Figure-57.

Figure-57. Faces selected for fixed constraint

- Click on the **OK** button from the dialog box to fix selected geometries.

Pin Constraint

The Pin constraint is used to restrict radial, axial, and tangential movement of a cylindrical part. The procedure to use this constraint is given next.

- Select the **Pin** option from the **Type** drop-down in the **STRUCTURAL CONSTRAINTS** dialog box. The options in the dialog box will be displayed as shown in Figure-58.

Figure-58. STRUCTURAL CONSTRAINTS
dialog box with Pin option selected

- Select the cylindrical face on which you want to apply pin constraint; refer to Figure-59.

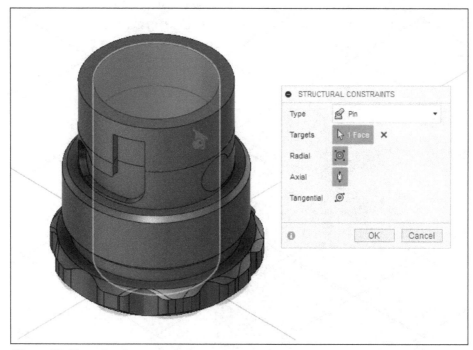

Figure-59. Face selected to apply pin constraint

- Select desired buttons to restrict the respective motion. For example, select the **Radial** button to restrict radial motion.
- Similarly, specify other parameters as required and click on the **OK** button to complete the process.

Frictionless Constraint

The Frictionless constraint is used to restrict the movement of object perpendicular to selected face. However, the object is free to move in the plane as if it is sliding on the face. The procedure to apply this constraint is given next.

- Select the **Frictionless** option from the **Type** drop-down in the **STRUCTURAL CONSTRAINTS** dialog box. The options in the dialog box will be displayed as shown in Figure-60.

Figure-60. STRUCTURAL CONSTRAINTS
dialog box with Frictionless option selected

- Select desired face on which you want to apply the frictionless constraint.

Figure-61. Face selected to apply frictionless constraint

- Click on the **OK** button to exit.

Prescribed Displacement Constraint

The **Prescribed Displacement** constraint is used to apply fixed constraint at specified displacement. The procedure to use this constraint is given next.

- Select the **Prescribed Displacement** option from **Type** drop-down in the **STRUCTURAL CONSTRAINTS** dialog box. The options in the dialog box will be displayed as shown in Figure-62 and you will be asked to select the face/edge/vertex.

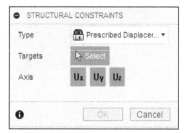

Figure-62. STRUCTURAL CONSTRAINTS dialog
box with Prescribed Displacement option selected

- Select desired geometry on which you want to apply constraint. The **STRUCTURAL CONSTRAINTS** dialog box will be updated; refer to Figure-63.

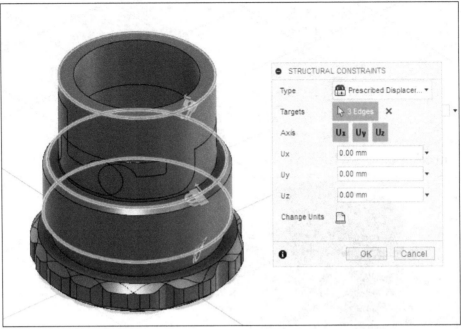

Figure-63. Edges selected to apply prescribed displacement constraint

- Specify the desired parameters and click on the **OK** button to exit the tool.

Remote Constraint

The **Remote** constraint is used to apply fixed constraint linked to a remote location. The procedure to use this constraint is given next.

- Select the **Remote** option from the **Type** drop-down of the **STRUCTURAL CONSTRAINTS** dialog box and you will be asked to select faces to be constrained.
- Select desired face from the model. The options will be displayed as shown in Figure-64.

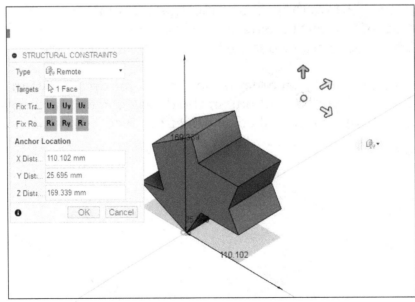

Figure-64. Remote constraint options

- Select desired buttons for **Fix Translation** and **Fix Rotation** options to constraint movements for respective translation and rotation axes.
- Click in the **X Distance**, **Y Distance**, and **Z Distance** edit boxes to define the anchor point location from where the constraint is being applied on selected face/body.
- After specifying desired parameters, click on the **OK** button from the dialog box.

Applying Bolt Connector Constraint

Bolt connector is used to apply connection similar to bolt fastener connection in assemblies. Note that bolt connector represents the nut-bolt connection or threaded nut connection mathematically. The procedure to apply bolt connector is given next.

- Click on the **Bolt Connector** tool from the **CONSTRAINTS** drop-down in the **Toolbar**; refer to Figure-65. The **BOLT CONNECTOR** dialog box will be displayed; refer to Figure-66.

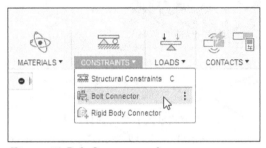

Figure-65. Bolt Connector tool

Figure-66. BOLT CONNECTOR dialog box

- The **Select** button of **Location for Bolt Head** option is active by default.
- Select the round edge of the part where you want the bolt head to be placed. The **BOLT CONNECTOR** dialog box will be updated; refer to Figure-67.

Figure-67. Updated BOLT CONNECTOR dialog box after selecting location for bolt head

Bolt Fastener with Nut

- Select the **With Nut** option from the **Bolt Subtype** drop-down if you want to create a bolt-nut fastener constraint. You will be asked to select the round edge where nut will be placed.
- Select the desired edge. The **BOLT CONNECTOR** dialog box will be updated with preview of bolt-nut fastener; refer to Figure-68.

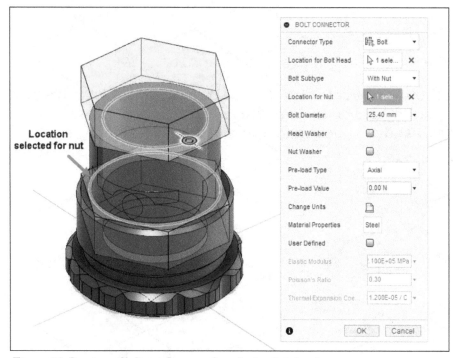

Figure-68. Preview of bolt nut fastener after selecting location for nut

- Specify the parameters like bolt diameter, bolt washer, nut washer, pre-load etc. and click on the **OK** button to create the constraint.

Bolt with Threaded Hole

- Select the **Threaded Hole** option from the **Bolt Subtype** drop-down if you want to create threaded bolted connection. You will be asked to select face to be threaded.
- Select the desired face. Preview of the bolt will be displayed; refer to Figure-69.

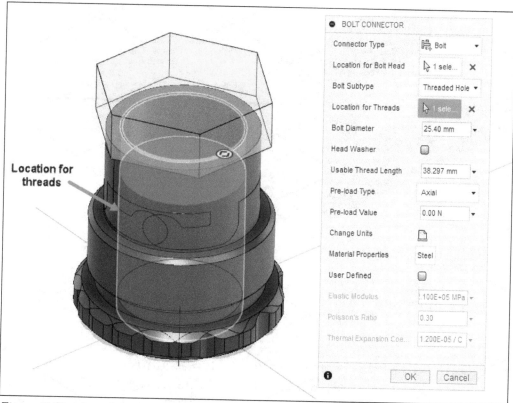

Figure-69. Preview of threaded bolt

- Specify the parameters as required and click on the **OK** button to exit the tool.

Rigid Body Connector Constraint

The Rigid Body Connector constraint is used where a vertex of one component is to be rigidly connected with face, edge, or vertex of other body. The procedure to use this constraint is given next.

- Click on the **Rigid Body Connector** tool from the **CONSTRAINTS** drop-down in the **Toolbar**; refer to Figure-70. The **RIGID BODY CONNECTOR** dialog box will be displayed; refer to Figure-71.

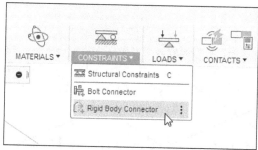

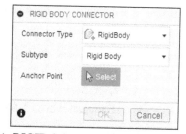

Figure-71. RIGID BODY CONNECTOR dialog box

Figure-70. Rigid Body Connector tool

- Select desired vertex that you want to be anchor point for connection on the first body/component. You will be asked to select dependent entities.

- Select the points/faces/edges that are dependent on the anchor point for movement; refer to Figure-72.

Figure-72. Rigid Body Connector constraint

- Similarly, you can use the **Interpolation** option from the **Subtype** drop-down to create rigid connection with translational and rotational constraining.

APPLYING LOADS

Loads in Fusion 360 are the representation of forces and loads applied on the part in real world. The tools to apply loads are available in the **LOADS** drop-down in the **Toolbar**; refer to Figure-73. Various tools in this drop-down are discussed next.

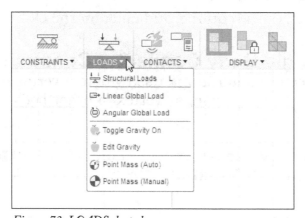

Figure-73. LOADS drop down

Applying Structural Loads

There are various structural loads that can be applied on the object like force, pressure, moment, remote force, bearing load, and hydrostatic pressure. The procedure to apply different loads are discussed next.

- Click on the **Structural Loads** tool from the **LOADS** drop-down in the **Toolbar**. The **STRUCTURAL LOADS** dialog box will be displayed; refer to Figure-74.

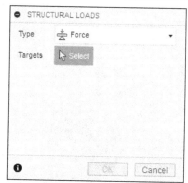

Figure-74. STRUCTURAL LOADS dialog box

Applying Force

- By default, **Force** option is selected in the **Type** drop-down of **STRUCTURAL LOADS** dialog box. If not selected by default in your case then select the **Force** option from the **Type** drop-down. You will be asked to select face/edge/point on which force is to be applied.
- Select desired face/edge/point. The options in the **STRUCTURAL LOADS** dialog box will be modified according to geometry selected; refer to Figure-75.
- Select desired **Direction Type** button from the dialog box. If you have selected the **Normal** button ⬚ then force will be applied perpendicular to the selected face. You can use the **Flip** ⬚ button below it to reverse direction of force; refer to Figure-76.

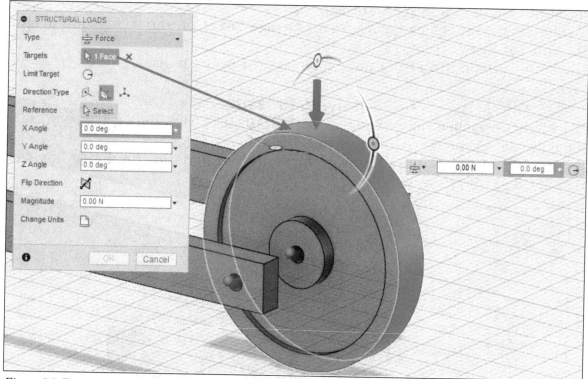

Figure-75. Face selected to apply force

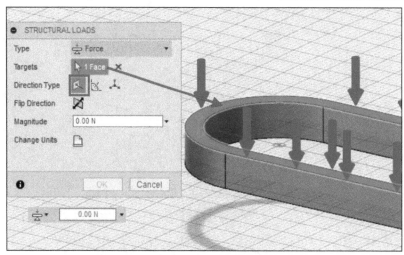

Figure-76. Force applied with Normal button selected

- Select the **Angle (delta)** button 🔲 from the **Direction Type** section of the dialog box if you want to apply force at some angle; refer to Figure-75. Specify the desired angle values in the **X Angle**, **Y Angle**, and **Z Angle** edit boxes. Select the **Flip Direction** button to reverse the direction if required. Note that the **Limit Target** button is also available in the dialog box. Select this button and specify the radius range in which the force will be applied.
- Select the **Vectors** button from the **Direction Type** section if you want to specify force value along each vector direction.
- To change the unit of load, click on the **Change Units** button if you want to change the unit for load.
- Click on the **OK** button from the dialog box to apply the load.

Applying Pressure

- Select the **Pressure** option from the **STRUCTURAL LOADS** dialog box. You will be asked to select the faces to apply pressure.
- Select the face(s) to apply pressure force. The options in the dialog box will be displayed as shown in Figure-77.

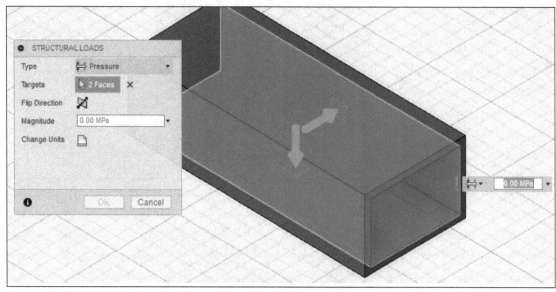

Figure-77. STRUCTURAL LOADS dialog box with pressure option

- Specify desired value of pressure in the **Magnitude** edit box. If you want to change the unit then select the **Change Units** button and specify the desired value in the edit box displayed.
- Click on the **OK** button from the dialog box to apply the pressure load.

Applying Moment

- Select the **Moment** option from the **STRUCTURAL LOADS** dialog box. You will be asked to select the faces to apply moment.
- Select desired face. The options in the dialog box will be updated; refer to Figure-78.

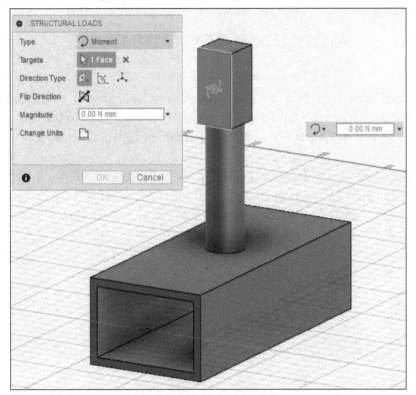

Figure-78. STRUCTURAL LOADS dialog box with moment option

- Set desired value of moment in the **Magnitude** edit box.
- Set other parameters as discussed earlier and click on the **OK** button to apply moment load.

Applying Bearing Load

Bearing load is the force exerted by bearing on round face of the part. The procedure to apply bearing load is given next.

- Select the **Bearing Load** option from the **Type** drop-down in the **STRUCTURAL LOADS** dialog box. You will be asked to select the round face on which bearing load is to be applied.
- Select the desired face. The options in the dialog box will be displayed as shown in Figure-79.

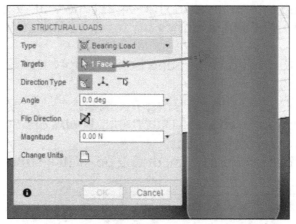

Figure-79. STRUCTURAL LOADS dialog box with Bearing Load option

- Specify desired parameters as discussed earlier. Note that bearing load is a directional force and applicable on only half of the full 360 cylindrical face.
- After specifying the parameters, click on the **OK** button to apply bearing load.

Applying Remote Force

The remote force is used to represent effect of load on selected faces which was applied at different location; refer to Figure-80. The procedure to apply load is given next.

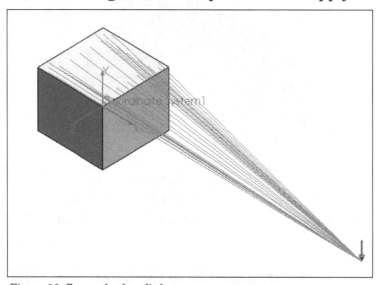

Figure-80. Remote load applied

- Select the **Remote Force** option from the **Type** drop-down in the **STRUCTURAL LOADS** dialog box. You will be asked to select a location to apply force.
- Select the desired face/edge/point. The options in the dialog box will be displayed as shown in Figure-81.

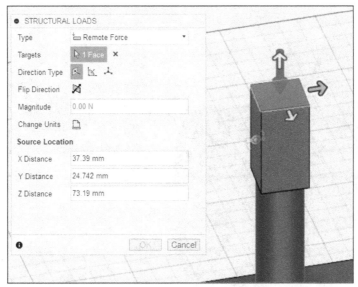

Figure-81. STRUCTURAL LOADS dialog box with Remote Force option

- Set the X, Y, and Z distances of the load location in the **X Distance**, **Y Distance**, and **Z Distance** edit boxes of the dialog box, respectively.
- Specify other parameters as discussed earlier.
- Click on the **OK** button to apply remote force.

Similarly, you can apply remote moment by using the **Remote Moment** option from the **Type** drop-down in the **STRUCTURAL LOADS** dialog box.

Applying Hydrostatic Pressure

Hydrostatic pressure is a linearly varying pressure exerted by fluid on the surface of part. This force is applicable when the part is in contact with high volume of fluid. The procedure to apply this load is given next.

- Select the **Hydrostatic Pressure** option from the **Type** drop-down in the **STRUCTURAL LOADS** dialog box.

Note that if you are applying hydrostatic pressure for the first time then a message box will be displayed prompting you to activate gravity. Activate the gravity by clicking on the **OK** button.

- The **Hydrostatic Pressure** option will be selected in the dialog box. You will be asked to select the face(s)to apply hydrostatic pressure.
- Select the desired face(s). The options in the dialog box will be displayed as shown in Figure-82.

Figure-82. STRUCTURAL LOADS dialog box with Hydrostatic Pressure option

- Click on the **Select** button for **Select Surface Point** section in the dialog box and select the desired point up to which the fluid is filled in the system.
- Specify the desired offset value for fluid surface point in the **Offset Distance** edit box.
- Select the desired fluid type from the **Fluid Type** drop-down. If you have a different fluid that the options available then select the **Custom** option and specify the density of fluid in the **Density** edit box.
- Click on the **OK** button after specifying the desired values to apply load.

Applying Linear Global Load (Acceleration)

Linear Global Load is the applied when the whole system is under acceleration like an object placed in an accelerating car. The procedure to apply linear global load is given next.

- Click on the **Linear Global Load** tool from the **LOADS** drop-down in the **Toolbar**. The **LINEAR GLOBAL LOAD** dialog box will be displayed as shown in Figure-83 and you will be asked to select a reference for acceleration direction.

Figure-83. LINEAR GLOBAL LOAD dialog box

- Select the desired face/edge to specify the direction of acceleration; refer to Figure-84.

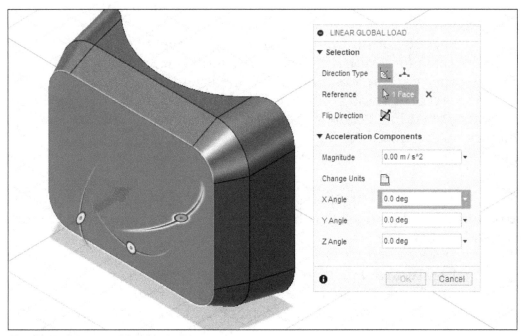

Figure-84. Selection of face for acceleration

- Specify desired value of acceleration in the **Magnitude** edit box.
- Similarly, specify desired angle values for X, Y, and Z in the **X Angle**, **Y Angle**, and **Z Angle** edit boxes.
- Specify other parameters as required and click on the **OK** button to apply the linear global load.

Applying Angular Global Load

The Angular Global Load is applied to give angular velocity or angular acceleration to the system. The procedure to apply angular global load is given next.

- Click on the **Angular Global Load** tool from the **LOADS** drop-down in the **Toolbar**. The **ANGULAR GLOBAL LOAD** dialog box will be displayed; refer to Figure-85.

Figure-85. ANGULAR GLOBAL LOAD dialog box

- The **Select** button of **Location Reference** section is active by default. Select the desired location for applying velocity or acceleration. The input boxes will be displayed to apply angular velocity and specify the location of the exerting point along X direction.
- Click on the **Select** button from the **Direction Reference** section and select the face to define axis for angular velocity/acceleration.
- Specify the other parameters as discussed earlier.
- Click on the **Acceleration** button if you want to specify the acceleration also from the **Acceleration Components** rollout in the dialog box. Specify the related parameters as discussed earlier.
- Click on the **OK** button to apply angular velocity/acceleration.

Toggling Gravity On/Off

Anyone who has passed high school should be knowing what is gravity! The procedure to activate and de-activate gravity is given next.

- Click on the **Toggle Gravity On** button from the **LOADS** drop-down in the **Toolbar** if the gravity is off and you want to activate it for simulation.
- Click on the **Toggle Gravity Off** button from the **LOADS** drop-down in the **Toolbar** if the gravity is on and you want to de-activate it for simulation; refer to Figure-86.

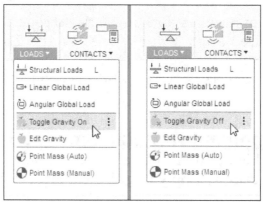

Figure-86. Toggle Gravity On or Off tool

Editing Gravity

The **Edit Gravity** tool is used to edit the value and direction of gravity acting in the simulation. The procedure is given next.

- Click on the **Edit Gravity** tool from the **LOADS** drop-down in the **Toolbar**. The **EDIT GRAVITY** dialog box will be displayed; refer to Figure-87.

Figure-87. EDIT GRAVITY dialog box

- The **Select** button of **Reference** section is active by default. Select the desired direction reference (face/edge) to specify the direction of gravity.
- Specify the desired value of gravity in the **Magnitude** edit box.
- Click on the **OK** button to apply the edited Gravity.

Apply Point Mass (Auto)

The **Point Mass (Auto)** tool is used to replace the real component with a point mass. This phenomena is used to simplify simulation calculations. The procedure to use this tool is given next.

- Click on the **Point Mass (Auto)** tool from the **LOADS** drop-down in the **Toolbar**. The **POINT MASS (AUTO)** dialog box will be displayed; refer to Figure-88.

Figure-88. POINT MASS (AUTO) dialog box

- The **Select** button of **Bodies** section is active by default. Select the object that you want to be replaced by point mass.
- Click on the **Select** button for **Geometries** section and select the face on which you want to place the mass.
- Specify the desired mass value in the **Mass** edit box of the dialog box; refer to Figure-89.

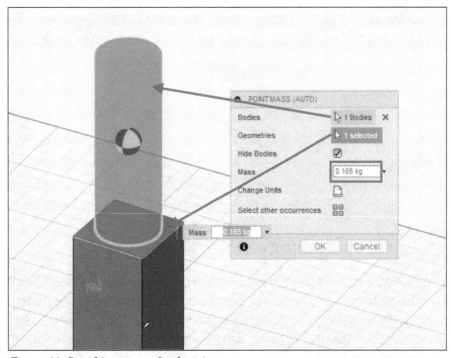

Figure-89. Specifying mass value for point mass

- Specify other parameters as required and click on the **OK** button to apply point mass.

Apply Point Mass (Manual)

The **Point Mass (Manual)** tool works in the same way as **Point Mass (Auto)**. The only difference between the two is that in case of **Point Mass (Manual)** tool, you do not need to select a body to be replaced by mass but specify the location where point mass will be placed.

APPLYING CONTACTS

Contacts are applied when there are two or more bodies/components in contact with each other and load is transferred between them during simulation. The tools to apply contact are available in the **CONTACTS** drop-down; refer to Figure-90. These tools are discussed next.

Figure-90. CONTACTS drop down

Applying Automatic Contacts

If you are performing analysis on an assembly with multiple components then applying automatic contact is a very important steps. Without applying automatic contact, you can not perform analysis of assembly in Fusion 360. Based on the joints applied to components and gap between the components, contacts are applied between the components automatically. The procedure to apply automatic contact is given next.

- Click on the **Automatic Contacts** tool from the **CONTACTS** drop-down in the **Toolbar**. The **AUTOMATIC CONTACTS** dialog box will be displayed; refer to Figure-91.

Figure-91. AUTOMATIC CONTACTS dialog box

- Specify the desired value of tolerance in **Solids** edit box of the **AUTOMATIC CONTACTS** dialog box. The tolerance specified here is the maximum gap up to which the software will apply contacts. If the gap between two components is more than the specified value then Fusion will not apply any contact automatically.
- Click on the **Generate** button. The contacts will be generated automatically.

Modifying Contacts

The **Automatic Contacts** tool applies the same contact to the components in the assembly. The procedure to check and modify automatically applied contacts is given next.

- To check the automatically applied contacts, click on the **Edit** button displayed on hovering the cursor over **Contacts** node of the **BROWSER**; refer to Figure-92. The **CONTACTS MANAGER** dialog box will be displayed; refer to Figure-93. Here, you can check the contacts automatically applied.

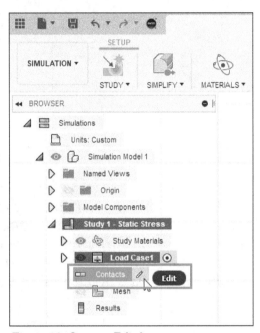

Figure-92. Contacts Edit button

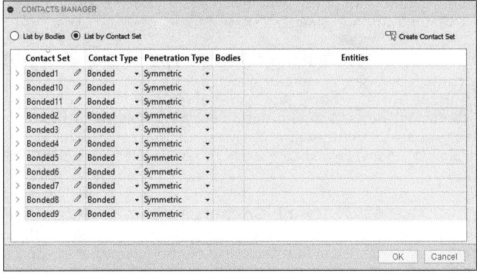

Figure-93. CONTACTS MANAGER dialog box

- Select the contact that you want to edit from the **Contact Set** column in the **CONTACTS MANAGER** dialog box. The contact will be highlighted in the model.
- To modify the contact, click in the **Contact Type** column for the selected contact. List of different available contacts will be displayed; refer to Figure-94.

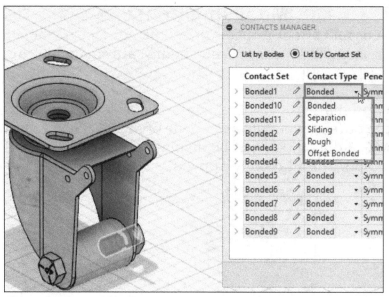

Figure-94. Contact Type list

- Select desired contact type to change. Various contact types are discussed next.

Bonded Contact Type

The bonded contact is used when there is no relative displacement between two connected solid bodies. This type of contact is used to glue together different solids of an assembly. The two surfaces that are in contact are classified as master and slave. Every node in the slave surface(slave nodes) is tied to a node in the master surface(master node) by a constraint. You will learn about master surface and slave surface in the next topic.

Separation Contact Type

The separation contact is applied when separation between parts is allowed but prohibits part penetration.

Sliding Contact Type

The sliding contact is a type of contact which allows displacement tangential to the contacting surface but no relative movement along the normal direction. This type of contact constraint is used to simulate sliding movement in the assembly. The two surfaces that are in contact are classified as master and slave. Every node in slave surface(slave nodes) is tied to a node in the master surface(master node) by this constraint.

Rough Contact Type

The rough contact is used when two parts cannot slide over each other as friction between them is very high. Note that the parts cannot penetrate in each other if this contact type is selected.

Offset Bonded Contact Type

The offset bonded contact is used when two parts are at a distance in assembly but you want them to be bonded as bonded contact type.

Applying Manual Contacts

Applying automatic contacts is the first step for performing analysis on the assembly but **Automatic Contacts** apply the same contact to all the assembly joints which can be changed by using **CONTACTS MANAGER**. But what to do if automatic contacts are not generated for required faces. The **Manual Contacts** tool is used to apply these contacts. The procedure to use this tool is given next.

*   Click on the **Manual Contacts** tool from the **CONTACTS** drop-down in the **Toolbar**. The **MANUAL CONTACTS** dialog box will be displayed as shown in Figure-95. Also, you will be asked to select the master body.

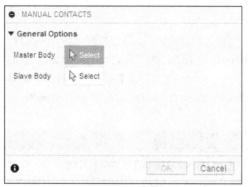

Figure-95. MANUAL CONTACTS dialog box

*   The **Select** button of **Master Body** section is active by default. Select the first body. You will be asked to select the slave body.
*   Select the second body. You will be asked to select the face/edge on the first body.
*   Select the desired face/edge at which the body is in contact with other body.
*   Click on the **Select** button of **Selection Set 2** section in the dialog box and select the contacting face/edge on the second body. The options in the dialog box will be displayed as shown in Figure-96.

*Figure-96. Options in MANUAL
CONTACTS dialog box*

- Select the desired contact type from the **Contact Type** drop-down in the dialog box.
- Select the desired option from the **Penetration Type** drop-down. If you have selected the **Symmetric** option then both master component and slave component cannot penetrate into each other. If the **Unsymmetric** option is selected then the master component can penetrate the slave component.
- Specify the desired maximum activation distance in the **Max. Activation Distance** edit box. This parameter is useful when parts are not coincident and a small gap is present between them. Choose a small value to prevent conflicting contact interactions.
- Similarly, specify the other parameters as required.
- Click on the **OK** button to create the contact.

Manage Contacts Tool

The **Manage Contacts** tool in the **CONTACTS** drop-down is used to edit contacts earlier applied. On clicking this tool, the **CONTACTS MANAGER** will be displayed. The options in the **CONTACTS MANAGER** have already been discussed.

SOLVING ANALYSIS

Once you have applied all the information required to perform analysis, you need to perform a check whether you have specified the required information or not. Once the system says, it has the required information then you are good to go for analysis. The tools to perform pre-check and analysis are available in the **SOLVE** drop-down of the **Toolbar**; refer to Figure-97. These tools are discussed next.

Figure-97. SOLVE drop down

Performing Pre-check

Pre-checking is an important step before performing analysis. Although, the tool will not tell you that you have specified load at wrong place or other design faults but the tool will warn you that you have not specified load, constraint, contact like parameters which need to be specified before performing analysis. The procedure to perform pre-check is given next.

- Click on the **Pre-check** tool from the **SOLVE** drop-down in the **Toolbar**. If there is any parameter left to be specified then **Cannot Solve** dialog box will be displayed; refer to Figure-98.

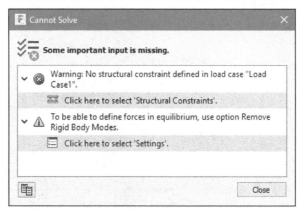

Figure-98. Cannot Solve dialog box

- Apply the parameters which are not specified. If all the parameters are specified then **Ready to Solve** dialog box will be displayed; refer to Figure-99.

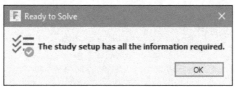

Figure-99. Ready to Solve dialog box

- Click on the **OK** button and perform meshing.

Meshing

Meshing is the base of FEM. Meshing divides the solid/shell models into elements of finite size and shape. These elements are joined at some common points called nodes. These nodes define the load transfer from one element to other element. Meshing is a very crucial step in design analysis. The automatic mesher in the software generates a mesh based on a global element size, tolerance, and local mesh control specifications. Mesh control lets you specify different sizes of elements for components, faces, edges, and vertices.

The software estimates a global element size for the model taking into consideration its volume, surface area, and other geometric details. The size of the generated mesh (number of nodes and elements) depends on the geometry and dimensions of the model, element size, mesh tolerance, mesh control, and contact specifications. In the early stages of design analysis where approximate results may suffice, you can specify a larger element size for a faster solution. For a more accurate solution, a smaller element size may be required.

Meshing generates 3D tetrahedral solid elements and 1D beam elements. A mesh consists of one type of elements unless the mixed mesh type is specified. Solid elements are naturally suitable for bulky models. Shell elements are naturally suitable for modeling thin parts (sheet metals), and beams and trusses are suitable for modeling structural members.

The procedure to create the mesh of the solid is given next.

- Click on the **Generate Mesh** tool from the **SOLVE** drop-down in the **Toolbar**. The system will start creating mesh and progress bar will be displayed. Once the operation is complete. The mesh will be displayed; refer to Figure-100.

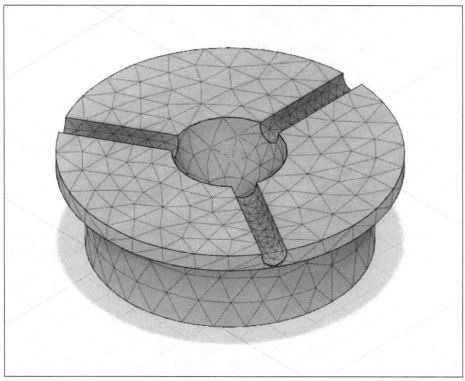

Figure-100. Mesh automatically created

- If you want to change the automatic mesh settings then click on the **Edit** button displayed on hovering the cursor over **Mesh** in the **BROWSER**; refer to Figure-101. The **Mesh Settings** dialog box will be displayed; refer to Figure-102.

Figure-101. Edit button of mesh

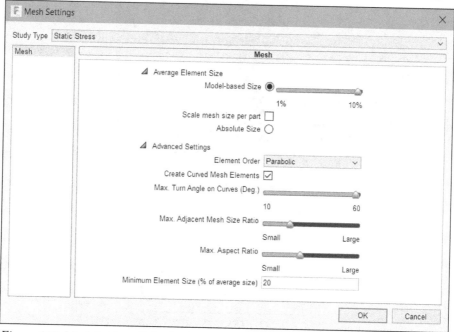

Figure-102. Mesh Settings dialog box

- Move the **Model-based Size** slider towards left to decrease the average size of mesh. Note that decreasing the size will increase the analysis solution time.

- Select the **Scale mesh size per part** check box if you want system to scale mesh size of each individual part in assembly based on its size. In other words, if there are 10 parts in assembly with different sizes then mesh elements of each part will have different size. The bigger the part size, the bigger will be mesh element size.

- If you want to specify a value to define size of all mesh elements then select the **Absolute Size** radio button and specify desired value for element size in the edit box next to it.

- Expand the **Advanced Settings** node to define advanced parameter of mesh. Select desired element order from the **Element Order** drop-down in the **Advanced Settings** node of the dialog box. Select the **Parabolic** option element order for complex parts which require higher degree of elements. For simple parts, select the **Linear** option from the drop-down.

- Select the **Create Curved Mesh Elements** check box if you want the mesh elements to follow curvature of round/curved faces of the part. Note that boundaries of mesh elements will be distorted to get curves of part.

- Similarly, specify other parameters in the **Advanced Settings** node and click on the **OK** button to apply mesh settings.

Applying Local Mesh Control

Local mesh control is used when you need to increase or decrease the size of elements in a finite area of the part. The procedure to apply local mesh control is given next.

- Click on the **Local Mesh Control** tool from the **MANAGE** drop-down in the **Toolbar**; refer to Figure-103. The **LOCAL MESH CONTROL** dialog box will be displayed; refer to Figure-104. You will be asked to select a face/edge.

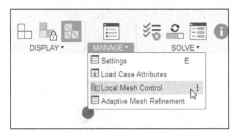

Figure-103. Local Mesh Control tool

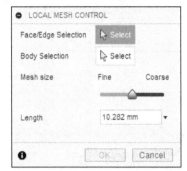

Figure-104. LOCAL MESH CONTROL dialog box

- The **Select** button of **Face/Edge Selection** section is active by default. Select the face/edge(s) for which you want to increase/decrease the element size. If you are working on an assembly then you can select the body after clicking on the **Select** button from **Body Selection** section of the dialog box.
- After selecting desired geometries, move the slider towards coarse or fine to change the mesh size.
- Click on the **OK** button to apply the change.

Note that changes in mesh will not be reflected automatically. To update mesh, right-click on the **Mesh** in the **Browser** and select the **Generate Mesh** option from the shortcut menu; refer to Figure-105.

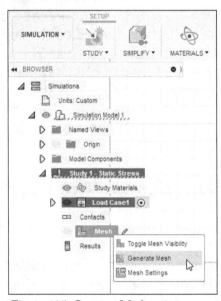

Figure-105. Generate Mesh option

Adaptive Mesh Refinement

Adaptive mesh refinement is used when dynamic refinement of mesh is required at the stress-strain locations to increase accuracy. The procedure to apply adaptive mesh refinement is given next.

- Click on the **Settings** tool from the **MANAGE** drop-down in the **Toolbar**. The **Settings** dialog box will be displayed. Click on the **Adaptive Mesh Refinement** option in the left side of the **Settings** dialog box. The **Adaptive Mesh Refinement** page will be displayed; refer to Figure-106.

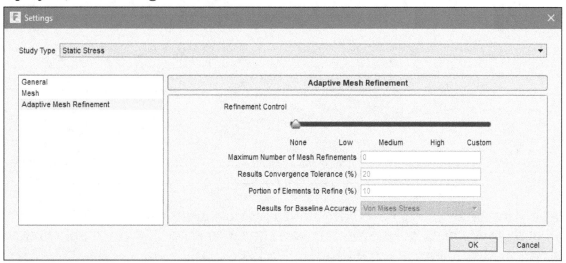

Figure-106. Adaptive Mesh Refinement page in Settings dialog box

- Move the **Refinement Control** slider towards right to increase refinement level of meshing at high stress areas.
- Click on the **OK** button to apply refinement.

Solving Analysis

Once you have specified all the parameters then it is time to solve the analysis. The procedure to do so is given next.

- Click on the **Solve** tool from the **SOLVE** drop-down in the **Toolbar**. The **Solve** dialog box will be displayed; refer to Figure-107.

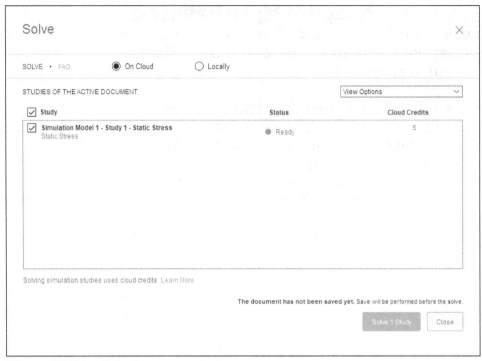

Figure-107. Solve dialog box

- Make sure **Ready** is displayed in the **Status** column for study to be performed.
- Select desired check boxes from the **View Options** drop-down to display respective objects in the dialog box; refer to Figure-108.

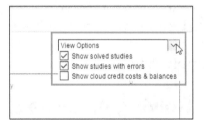

Figure-108. View Options drop-down

- Click on the **Solve 1 Study** button. The results will be displayed; refer to Figure-109.

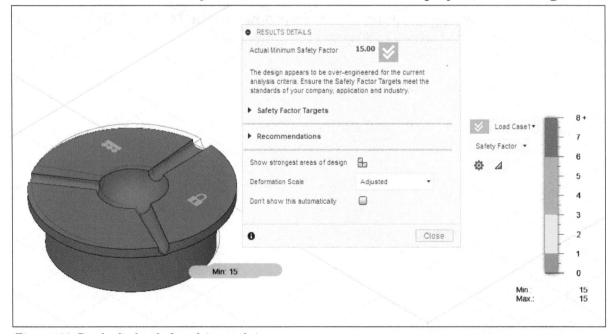

Figure-109. Results displayed after solving analysis

To check the strongest areas of design, click on the **Show strongest areas of design** button from the **RESULTS DETAILS** dialog box. The strongest areas of design will be displayed; refer to Figure-110.

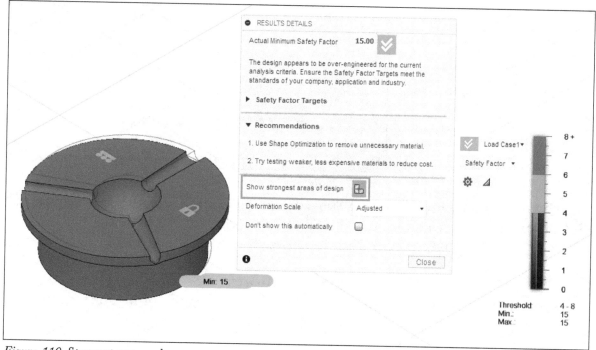

Figure-110. Strongest area results

Move the sliders on scale to change the threshold to be counted as weak area.

Preparing and Managing the Results

Once the analysis is complete, the next step is to prepare and manage the results as required. The tools to prepare results are available in the **RESULT TOOLS** drop-down from **RESULTS** tab in the **Toolbar**; refer to Figure-111. The tools in this drop-down are discussed next.

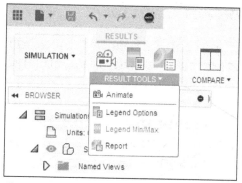

Figure-111. RESULT TOOLS drop down

Animating the Results

Animation is used to represent analysis results in dynamic motion. The procedure to animate the analysis result is given next.

• Click on the **Animate** tool from the **RESULT TOOLS** drop-down in the **Toolbar**. The **ANIMATE** dialog box will be displayed; refer to Figure-112.

Figure-112. ANIMATE dialog box

- Select the **One-way** check box or **Two-way** check box to repeat the animation.
- Set other parameters as required and click on the **Play** button.
- Click on the **OK** button to exit the dialog box.

Legend Options

The **Legend Options** tool is used to modify the appearance of legends displayed in the results. The procedure to use this tool is given next.

- Click on the **Legend Options** tool from the **RESULT TOOLS** drop-down in the **Toolbar**. The **LEGEND OPTIONS** dialog box will be displayed; refer to Figure-113.

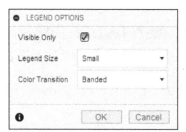

Figure-113. LEGEND OPTIONS dialog box

- Select desired legend size and color transition method from the **Legend Size** drop-down and **Color Transition** drop-down, respectively.
- Click on the **OK** button to apply changes.

Generating Reports

Once you find the analysis results as expected, it is the time to generate reports. The procedure to generate report is given next.

- Click on the **Report** tool from the **RESULT TOOLS** drop-down in the **Toolbar**. The **Report** dialog box will be displayed; refer to Figure-114.

Figure-114. Report dialog box

- Click on the **Save** button to save the report. The **Save Report As** dialog box will be displayed; refer to Figure-115.

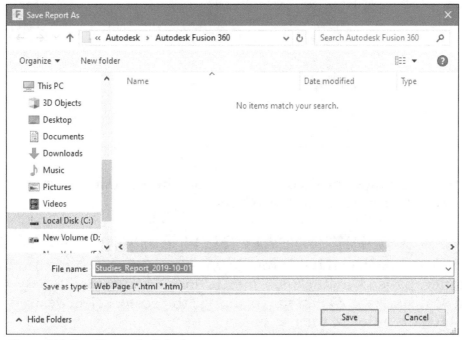

Figure-115. Save Report As dialog box

- Specify desired name of the report and click on the **Save** button. The file will be saved and displayed in the web browser.

Comparing Results of Analyses

The **Compare** tool in **COMPARE** drop-down is used to compare results of two analyses along with parameters specified for those analyses. The procedure to use this tool is given next.

- After perform at least two analyses to be compared, click on the **Compare** tool from the **COMPARE** drop-down in the **RESULTS** tab of the **Toolbar**. The results of analyses will be displayed in two windows of application; refer to Figure-116.

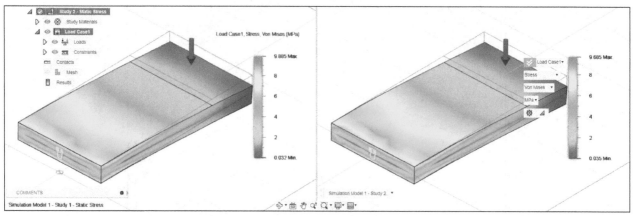

Figure-116. Comparing results

- Click on desired side in result window and set desired result parameters like select **Stress** option from result type drop-down of the legends bar; refer to Figure-117.

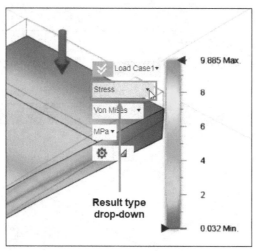

Figure-117. Result type drop-down

- Check other results as desired and click on the **Finish Compare** tool from the **FINISH COMPARE** drop-down in the **Toolbar**.

DEFORMATION Drop-down

The options in the **DEFORMATION** drop-down are used to scale up or scale down the deformation caused in the part due to load in the results. By default, the deformation scale is set to **Adjusted**. To change the scale, select desired option from the **DEFORMATION** drop-down; refer to Figure-118.

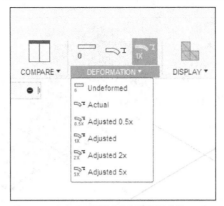

Figure-118. DEFORMATION drop down

DISPLAY Options

The options of **DISPLAY** drop-down are used to switch between various display styles of the model in results. The display commands are accessible through the **DISPLAY** panel in the **Toolbar**; refer to Figure-119.

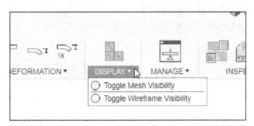

Figure-119. DISPLAY drop down

- The **Toggle Mesh Visibility** option is used to enable or disable the display of the mesh in results.
- The **Toggle Wireframe Visibility** option is used to toggle the wireframe display of the model in results.

Note that if you are not in **Results** mode then two more options are displayed in the **DISPLAY** drop-down; refer to Figure-120.

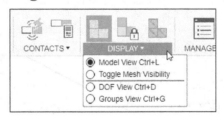

Figure-120. DISPLAY drop down

- Select the **DOF View** radio button to check whether the model is fully fixed, partially fixed or free. The model will be displayed in respective color code; refer to Figure-121.

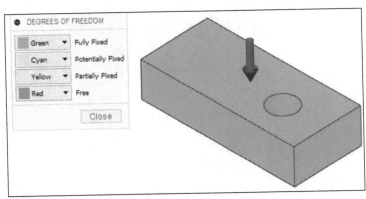

Figure-121. Degrees of freedom

- Select the **Groups View** radio button from the drop-down if you want to display components of different groups in different colors.

Load Case Attributes

The **Load Case Attributes** tool is used to check and modify loads applied in the analysis. The procedure to use this tool is given next.

- Click on the **Load Case Attributes** tool from the **MANAGE** drop-down in the **RESULTS** tab of **Toolbar**. The **LOAD CASE ATTRIBUTES** dialog box will be displayed; refer to Figure-122.

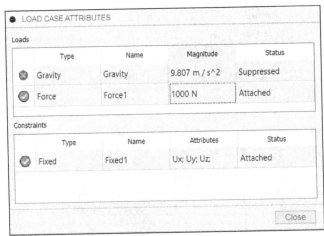

Figure-122. LOAD CASE ATTRIBUTES dialog box

- Double-click on desired parameter in the dialog box to modify. Related dialog box will be displayed. For example, double-click on **Force** parameter in the dialog box then **EDIT STRUCTURAL LOAD** dialog box will be displayed.
- After setting desired parameters, click on the **Close** button.

Inspecting Result Parameters

The options in the **INSPECT** drop-down are used to check various result parameters of the analysis performed on the model; refer to Figure-123. These tools are discussed next.

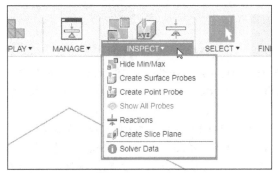

Figure-123. INSPECT drop-down

Showing/Hiding Minimum and Maximum Result Value

Select the **Show Min/Max** option from the **INSPECT** drop-down if you want to show minimum and maximum value of result parameter in the results. If the values are shown in the results and you want to hide them then select **Hide Min/Max** option.

Creating Surface Probe

The **Create Surface Probes** tool is used to check analysis results on various surface points of the model. The procedure to use this tool is given next.

- Click on the **Create Surface Probes** tool from the **INSPECT** drop-down in the **RESULTS** tab of the **Ribbon**. The **CREATE SURFACE PROBES** dialog box will be displayed and analysis result will be displayed when you hover cursor on the surface of model; refer to Figure-124.

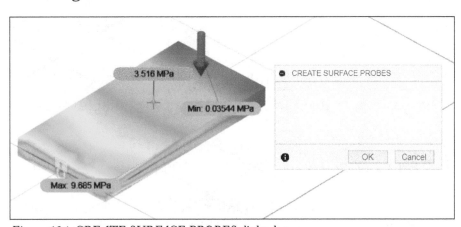

Figure-124. CREATE SURFACE PROBES dialog box

- Click at desired locations on the surface of model to check the probe results; refer to Figure-125.

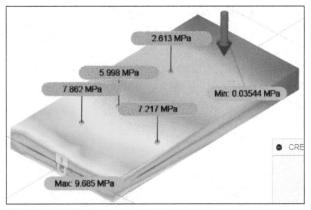

Figure-125. Surface probe results

*   Click on the **OK** button from the dialog box to exit.

Creating Point Probe

The **Create Point Probe** tool is used to check analysis results at desired point (location) on the model. The procedure to use this tool is given next.

*   Click on the **Create Point Probe** tool from the **INSPECT** drop-down in the **RESULTS** tab of the **Toolbar**. The **POINT PROBE** dialog box will be displayed and as you hover cursor on the model, the result parameter will be displayed along with coordinates of the point; refer to Figure-126.

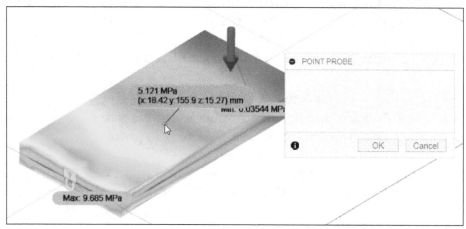

Figure-126. Result on a point

*   Click at desired location to display results at selected locations and click on the **OK** button.

Hiding Probes

The **Hide All Probes** tool is used to hide all the probes displayed on the model.

Showing Probes

The **Show All Probes** tool is used to display all the probes again if hidden.

Deleting All Probes

The **Delete All Probes** tool is used to delete all the probes earlier applied on the model.

Checking Reaction Forces and Moments

The **Reactions** tool in **INSPECT** drop-down is used to check reaction forces and moments caused in the model due to application of forces. On clicking this tool, the **REACTIONS** dialog box will be displayed and you will be asked to select faces to be checked; refer to Figure-127. Select the constrained face on which you want to check reaction forces & moments; refer to Figure-128.

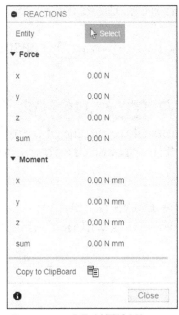

Figure-127. REACTIONS dialog box

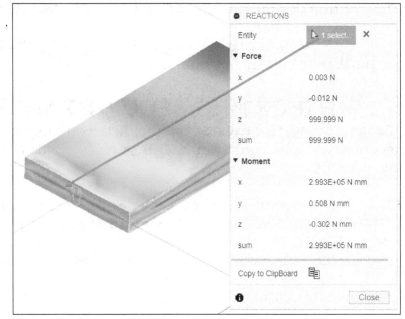

Figure-128. Face selected for reaction

- Click on the **Close** button from the dialog box to exit.

Creating Slice Plane

The **Create Slice Plane** tool is used to check result of analysis at desired level in model using clipping plane. The procedure to use this tool is given next.

- Click on the **Create Slice Plane** tool from the **INSPECT** drop-down in the **RESULTS** tab of the **Toolbar**. The **SLICE PLANE** dialog box will be displayed.
- Select desired face of the model to be used as reference for creating clipping plane. The **EDIT SLICE PLANE** dialog box will be displayed; refer to Figure-129.

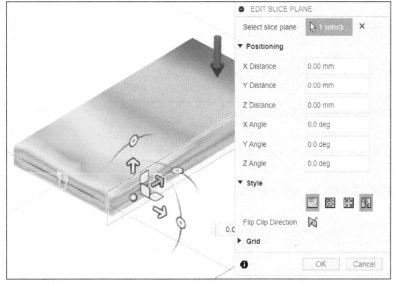

Figure-129. EDIT SLICE PLANE dialog box

- Move the slice plane to desired location using handles displayed on the model or specify desired parameters in the dialog box to set location of clipping plane.
- Using the buttons in **Style** section, you specify the style of displaying results at clipping plane.

- After setting desired parameters, click on the **OK** button from the dialog box.

Click on the **Finish Result** tool from the **FINISH RESULTS** drop-down in the **RESULTS** tab of **Toolbar** to exit the result mode.

EXPORTING STUDY TO ANSYS SETUP

The **Export Study to Ansys Setup** tool is used to export complete setup of analysis like geometry, loads, constraints, and so on to a .sdz file. The procedure to use this tool is given next.

- Click on the **Export Study to Ansys Setup** tool from the **ANSYS** drop-down in the **SETUP** tab of the **Toolbar**. The **Export to Ansys Setup** dialog box will be displayed; refer to Figure-130.

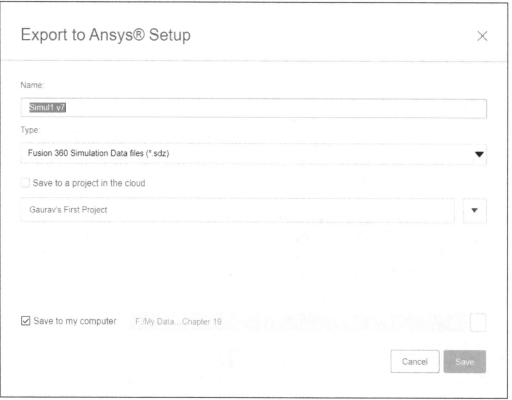

Figure-130. Export to Ansys Setup dialog box

- Specify desired value in the **Name** edit box to define name for simulation file.
- Select desired option from the **Type** drop-down to define format in which the data related to analysis will be saved.
- Select the **Save to my computer** check to save the file at desired location in computer.
- Click on the **Save** button from the dialog box to save the file. You can later load this file in Ansys software to perform advanced analyses.

In this chapter, we have worked on a Static Stress analysis. We have also gone through the basic process of analysis in Fusion 360. In the next chapter, we will work through the other types of analysis available in Autodesk Fusion 360.

SELF-ASSESSMENT

Q1. An analysis fulfills following conditions:

- All loads are applied slowly and gradually until they reach their full magnitudes. After reaching their full magnitudes, load will remain constant (i.e. load will not vary against time).
- Linearity assumption: The relationship between loads and resulting responses is linear. For example, if you double the magnitude of loads, the response of the model (displacements, strains and stresses) will also double.

Which of the following analyses should be used to check the design?

a. Linear Static Analysis   b. Non-linear Static Analysis
c. Linear Dynamic Analysis   d. Buckling Analysis

Q2. What is the different between steady state thermal analysis and transient thermal analysis?

Q3. Which of the following analysis is used to study the effect of object velocity, initial velocity, acceleration, time dependent loads, and constraints in the design in Autodesk Fusion 360?

a. Shape Optimization   b. Structural Buckling Analysis
c. Event Simulation   d. Modal Analysis

Q4. Which of the following is not an element type for 3D objects in Autodesk Fusion 360 simulation?

a. Linear Tetrahedron   b. Parabolic Tetrahedron
c. Parabolic Tetrahedron with Curved edges   d. Wedge

Q5. Which of the following tool is used to apply pin constraint?

a. Structural Constraint   b. Bolt Connector
c. Rigid Body Connector   d. Joint Origin

Q6. A linearly varying pressure exerted by fluid on surface of a part is called ....

Q7. The **Pre-check** tool is used to warn if constraint is applied on wrong face of model. (T/F)

FOR STUDENT NOTES

Chapter 21

Simulation Studies
in Fusion 360

Topics Covered

The major topics covered in this chapter are:

- *Introduction*
- *Nonlinear Static Stress Analysis*
- *Modal Frequencies Analysis*
- *Buckling Analysis*
- *Thermal Analysis*
- *Event Simulation*
- *Shape Optimization*
- *Electronics Cooling*

INTRODUCTION

In the previous chapter, you have learned about the basics of the analysis. In this chapter, you will learn the procedure of applying different analyses on the part.

NONLINEAR STATIC STRESS ANALYSIS

Non-linear static stress analysis is used to check the effect of load on part when three common forms of nonlinearity like material, geometric, and boundary conditions nonlinearity are applicable in analysis. The procedure to apply non-linear static stress analysis is given next.

- Click on the **New Simulation Study** button from **STUDY** drop-down in the **Toolbar**. The **New Study** dialog box will be displayed. You can also press **N** which in **Simulation** workspace to do the same.
- Double-click on the **Nonlinear Static Stress** button from the dialog box. The analysis environment will be displayed.
- Click on the **Settings** button from the **MANAGE** drop-down in the **Toolbar**; refer to Figure-1. The **Settings** dialog box will be displayed; refer to Figure-2.

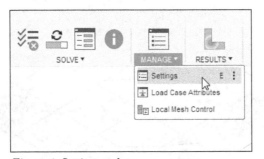

Figure-1. Settings tool

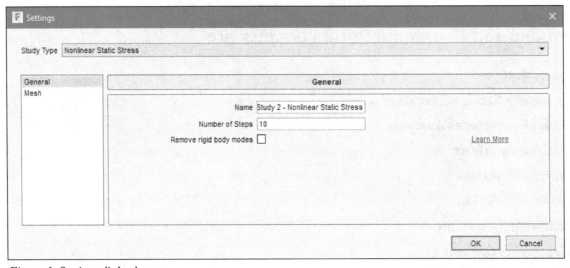

Figure-2. Settings dialog box

- In the **Number of Steps** edit box, specify the desired number of steps in which the total load will be applied.
- Select the **Remove rigid body modes** check box to exclude linear component of force effects. Click on the **OK** button to apply the parameters.

- Apply the material, load, and constraint as required on the model; refer to Figure-3. Note that for nonlinear materials, you need to select **Fusion 360 Nonlinear Material Library** option from the **Material Library** drop-down in the **STUDY MATERIALS** dialog box; refer to Figure-4.

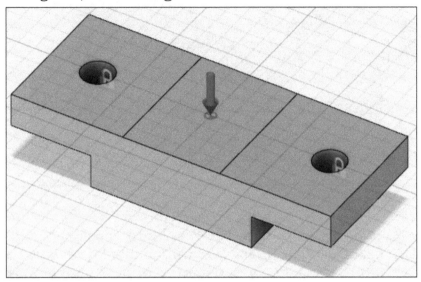

Figure-3. Load and constraint applied

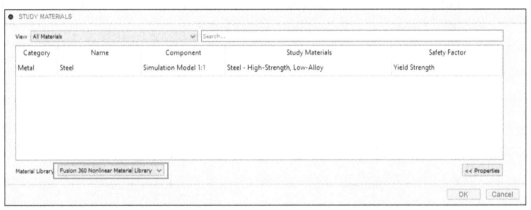

Figure-4. Non linear material library option

- Click on the **Solve** button from the **SOLVE** drop-down in the **Toolbar**. The **Solve** dialog box will be displayed.
- Click on the **Solve** button from the dialog box. Make sure **Cloud** radio button is selected while solving the analysis as you cannot solve non-linear analyses locally in Autodesk Fusion 360. The result will be displayed in the screen; refer to Figure-5.

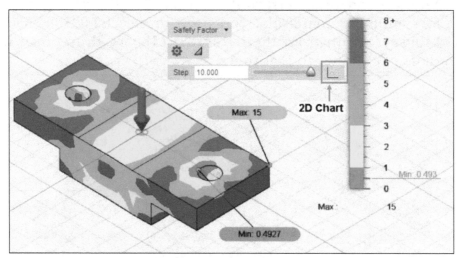

Figure-5. Result of non linear static analysis

- Click on the **2D Chart** button □ in the results to check the transient behavior. The **TRANSIENT RESULTS PLOT** dialog box will be displayed with the results; refer to Figure-6. Note that this plot gives insight of what is happening in the model at each time step. There can be some cases where maximum stress will be occurring at the middle of time steps and value at last time step will be minimal. If you miss such value then your part may fail in the assembly.

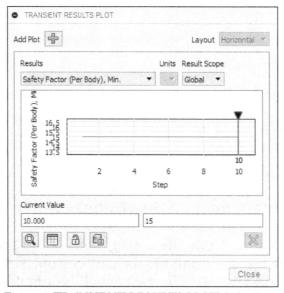

Figure-6. TRANSIENT RESULTS PLOT dialog box

- Move the slider in the graph to check the results of load at different steps. Select the desired result type from the **Results** drop-down in the dialog box.

You can generate the report as discussed earlier.

STRUCTURAL BUCKLING ANALYSIS

Slender models tend to buckle under axial loading. Buckling is defined as the sudden deformation which occurs when the stored membrane (axial) energy is converted into bending energy with no change in the externally applied loads. Mathematically, when buckling occurs, the stiffness becomes singular. The Linearized buckling approach, used here, solves an eigenvalue problem to estimate the critical buckling factors and the associated buckling mode shapes.

A model can buckle in different shapes under different levels of loading. The shape the model takes while buckling is called the buckling mode shape and the loading is called the critical or buckling load. Buckling analysis calculates a number of modes as requested in the Buckling dialog. Designers are usually interested in the lowest mode (mode 1) because it is associated with the lowest critical load. When buckling is the critical design factor, calculating multiple buckling modes helps in locating the weak areas of the model. The mode shapes can help you modify the model or the support system to prevent buckling in a certain mode.

A more vigorous approach to study the behavior of models at and beyond buckling requires the use of nonlinear design analysis codes. In a laymen's language, if you press down on an empty soft drink can with your hand, not much will seem to happen. If you put the can on the floor and gradually increase the force by stepping down on it with your foot, at some point it will suddenly squash. This sudden scrunching is known as "buckling."

Models with thin parts tend to buckle under axial loading. Buckling can be defined as the sudden deformation, which occurs when the stored membrane (axial) energy is converted into bending energy with no change in the externally applied loads. Mathematically, when buckling occurs, the total stiffness matrix becomes singular.

In the normal use of most products, buckling can be catastrophic if it occurs. The failure is not one because of stress but geometric stability. Once the geometry of the part starts to deform, it can no longer support even a fraction of the force initially applied. The worst part about buckling for engineers is that buckling usually occurs at relatively low stress values for what the material can withstand. So they have to make a separate check to see if a product or part thereof is okay with respect to buckling.

Slender structures and structures with slender parts loaded in the axial direction buckle under relatively small axial loads. Such structures may fail in buckling while their stresses are far below critical levels. For such structures, the buckling load becomes a critical design factor. Stocky structures, on the other hand, require large loads to buckle, therefore buckling analysis is usually not required.

Buckling almost always involves compression. In civil engineering, buckling is to be avoided when designing support columns, load bearing walls and sections of bridges which may flex under load. For example an I-beam may be perfectly "safe" when considering only the maximum stress, but fail disastrously if just one local spot of a flange should buckle! In mechanical engineering, designs involving thin parts in flexible structures like airplanes and automobiles are susceptible to buckling. Even though stress can be very low, buckling of local areas can cause the whole structure to collapse by a rapid series of 'propagating buckling'.

Buckling analysis calculates the smallest (critical) loading required for buckling a model. Buckling loads are associated with buckling modes. Designers are usually interested in the lowest mode because it is associated with the lowest critical load. When buckling is the critical design factor, calculating multiple buckling modes helps in locating the weak areas of the model. This may prevent the occurrence of lower buckling modes by simple modifications.

USE OF BUCKLING ANALYSIS

Slender parts and assemblies with slender components that are loaded in the axial direction buckle under relatively small axial loads. Such structures can fail due to buckling while the stresses are far below critical levels. For such structures, the buckling load becomes a critical design factor. Buckling analysis is usually not required for bulky structures as failure occurs earlier due to high stresses. The procedure to use buckling analysis in Fusion 360 is given next.

- Open/create the part on which you want to perform buckling analysis. Click on the **New Simulation Study** button from **STUDY** drop-down in the **Toolbar**. The **New Study** dialog box will be displayed.
- Double-click on the **Structural Buckling** tool from the **New Study** dialog box. The analysis environment will be displayed.
- Apply material, constraint, load, and other parameters as required and solve the study as discussed earlier. The results of buckling will be displayed; refer to Figure-7.

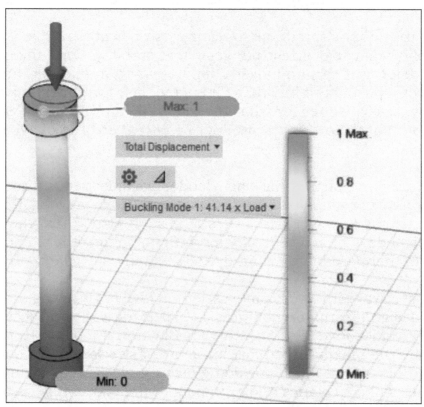

Figure-7. Result of buckling analysis

- Click in the **Buckling Mode** drop-down in the results and select the desired buckling mode to check the effect; refer to Figure-8.

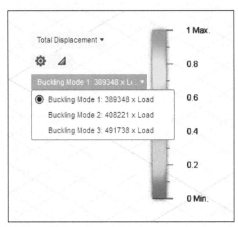

Figure-8. Buckling Mode drop down

MODAL FREQUENCIES ANALYSIS

Every structure has the tendency to vibrate at certain frequencies, called **natural or resonant frequencies**. Each natural frequency is associated with a certain shape, called **mode shape**, that the model tends to assume when vibrating at that frequency. When a structure is properly excited by a dynamic load with a frequency that coincides with one of its natural frequencies, the structure undergoes large displacements and stresses. This phenomenon is known as **Resonance**. For undamped systems, resonance theoretically causes infinite motion. **Damping**, however, puts a limit on the response of the structures due to resonant loads.

A real model has an infinite number of natural frequencies. However, a finite element model has a finite number of natural frequencies that are equal to the number of degrees of freedom considered in the model. Only the first few modes are needed for most purposes.

If your design is subjected to dynamic environments, static studies cannot be used to evaluate the response. Frequency studies can help you design vibration isolation systems by avoiding resonance in specific frequency band. They also form the basis for evaluating the response of linear dynamic systems where the response of a system to a dynamic environment is assumed to be equal to the summation of the contributions of the modes considered in the analysis.

Note that resonance is desirable in the design of some devices. For example, resonance is required in guitars and violins.

The natural frequencies and corresponding mode shapes depend on the geometry, material properties, and support conditions. The computation of natural frequencies and mode shapes is known as modal, frequency, and normal mode analysis.

When building the geometry of a model, you usually create it based on the original (undeformed) shape of the model. Some loads, like the structure's own weight, are always present and can cause considerable effects on the shape of the structure and its modal properties. In many cases, this effect can be ignored because the induced deflections are small.

Loads affect the modal characteristics of a body. In general, compressive loads decrease resonant frequencies and tensile loads increase them. This fact is easily demonstrated by changing the tension on a violin string. The higher the tension, the higher the frequency (tone).

You do not need to define any loads for a frequency study but if you do their effect will be considered. By having evaluated natural frequencies of a structure's vibrations at the design stage, you can optimize the structure with the goal of meeting the frequency vibro-stability condition. To increase natural frequencies, you would need to add rigidity to the structure and (or) reduce its weight. For example, in the case of a slender object, the rigidity can be increased by reducing the length and increasing the thickness of the object. To reduce a part's natural frequency, you should, on the contrary, increase the weight or reduce the object's rigidity.

Note that the software also considers thermal and fluid pressure effects for frequency studies.

The procedure to perform the frequency analysis is given next.

- Open/create the part on which you want to perform modal analysis. Click on the **New Simulation Study** button from **STUDY** drop-down in the **Toolbar**. The **New Study** dialog box will be displayed.
- Double-click on the **Structural Buckling** tool from the **New Study** dialog box. The analysis environment will be displayed.
- Click **Settings** tool from the **MANAGE** drop-down in the **Toolbar**. Select **Modal Frequencies** option from **Study Type** drop-down in the dialog box. The **Settings** dialog box for **Modal Frequencies** option will be displayed; refer to Figure-9.

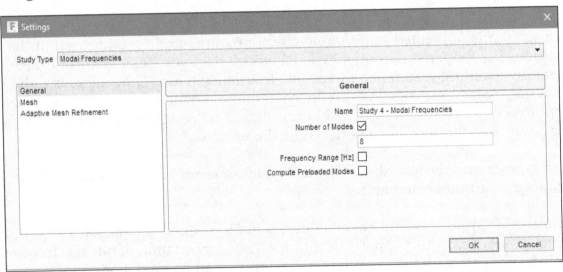

Figure-9. Settings dialog box for modal frequencies option

- Select the **Number of Modes** check box and specify the total number of frequencies that you want to test for resonance.
- Select the **Frequency Range (Hz)** check box to specify the minimum & maximum frequencies within which the natural frequencies are to be found.
- Select the **Compute Preloaded Modes** check box if you include the effect of structural loads in modal frequency analysis.

- After specifying all the parameters, click on the **OK** button.
- Apply material, constraint, load, and other parameters as required and solve the study as discussed earlier. Note that you can perform modal analysis without applying any load but if you apply a load then its effects will also be counted in the results if **Compute Preloaded Modes** check box is selected in the **Settings** dialog box. The results of modal analysis will be displayed; refer to Figure-10.

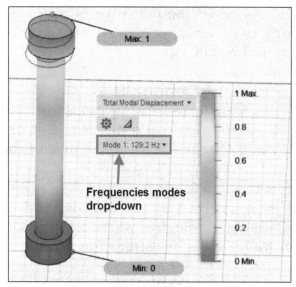

Figure-10. Frequencies drop-down

- Click on the **Mode** drop-down in **Results** area to check different natural frequencies of the model.

THERMAL ANALYSIS

Thermal analysis is a method to check the distribution of heat over a body due to applied thermal loads. Note that thermal energy is dynamic in nature and is always flowing through various mediums. There are three mechanisms by which the thermal energy flows:

- Conduction
- Convection
- Radiation

In all three mechanisms, heat energy flows from the medium with higher temperature to the medium with lower temperature. Heat transfer by conduction and convection requires the presence of an intervening medium while heat transfer by radiation does not.

The output from a thermal analysis can be given by:
1. Temperature distribution.
2. Amount of heat loss or gain.
3. Thermal gradients.
4. Thermal fluxes.

This analysis is used in many engineering industries such as automobile, piping, electronic, power generation, and so on.

Important terms related to Thermal Analysis

Before conducting thermal analysis, you should be familiar with the basic concepts and terminologies of thermal analysis. Following are some of the important terms used in thermal analysis:

Heat Transfer Modes

Whenever there is a difference in temperature between two bodies, the heat is transferred from one body to another. Basically, heat is transferred in three ways: Conduction, Convection, and Radiation.

Conduction

In conduction, the heat is transferred by interactions of atoms or molecules of the material. For example, if you heat up a metal rod at one end, the heat will be transferred to the other end by the atoms or molecules of the metal rod.

Convection

In convection, the heat is transferred by the flowing fluid. The fluid can be gas or liquid. Heating up water using an electric water heater is a good example of heat convection. In this case, water takes heat from the heater.

Radiation

In radiation, the heat is transferred in space without any matter. Radiation is the only heat transfer method that takes place in space. Heat coming from the Sun is a good example of radiation. The heat from the Sun is transferred to the earth through radiation.

Thermal Gradient

The thermal gradient is the rate of increase in temperature per unit depth in a material.

Thermal Flux

The Thermal flux is defined as the rate of heat transfer per unit cross-sectional area. It is denoted by q.

Bulk Temperature

It is the temperature of a fluid flowing outside the material. It is denoted by Tb. The Bulk temperature is used in convective heat transfer.

Film Coefficient

It is a measure of the heat transfer through an air film.

Emissivity

The Emissivity of a material is the ratio of energy radiated by the material to the energy radiated by a black body at the same temperature. Emissivity is the measure of a material's ability to absorb and radiate heat. It is denoted by e. Emissivity is a numerical value without any unit. For a perfect black body, e = 1. For any other material, e < 1.

Stefan–Boltzmann Constant

The energy radiated by a black body per unit area per unit time divided by the fourth power of the body's temperature is known as the Stefan-Boltzmann constant. It is denoted by s.

Thermal Conductivity

The thermal conductivity is the property of a material that indicates its ability to conduct heat. It is denoted by K.

Specific Heat

The specific heat is the amount of heat required per unit mass to raise the temperature of the body by one degree Celsius. It is denoted by C.

PERFORMING THERMAL ANALYSIS

- Click on the **New Simulation Study** button from **STUDY** drop-down in the **Toolbar**. The **New Study** dialog box will be displayed.
- Double-click on the **Thermal** button from the dialog box. The thermal analysis environment will be displayed.
- Apply the desired material to the mode as discussed earlier.
- Click on the **Thermal Loads** tool from the **LOADS** panel in the **Toolbar**; refer to Figure-11. The **THERMAL LOADS** dialog box will be displayed; refer to Figure-12.

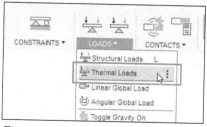

Figure-11. Thermal Loads tool

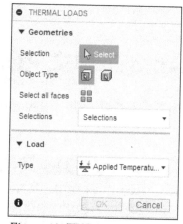

Figure-12. THERMAL LOADS dialog box

- Select the desired thermal load type from the **Type** drop-down in the **Load** section of the dialog box. Select the **Applied Temperature** option from the drop-down and enter the desired temperature if you want to apply a fix temperature to the selected geometry. Select the **Heat Source** option if you want to specify the amount of heat energy to be applied on selected geometry. Select the **Radiation** option if heat is transferred through radiation to the selected face. Select the **Convection** option if heat is transferred through convection. Select the **Internal Heat** option from the drop-down if heat is generated inside the model.

- Select the face/edge/vertex on which you want to apply thermal load. If you want to select a body then click on the **Bodies** button from the **Object Type** section of the dialog box and then select the body.

- Specify the desired values of thermal load in edit boxes as per the option selected in the **Type** drop-down.

- Click on the **OK** button from dialog box to apply the settings.

- Specify the contacts as discussed earlier for heat flow between different components of the assembly.

- Click on the **Settings** button from the **MANAGE** drop-down in the **Toolbar**. The **Settings** dialog box will be displayed. Specify the desired value of atmospheric temperature in the **Global Initial Temperature** edit box. Set the other parameters as discussed earlier. Click on the **OK** button from the dialog box to apply settings.

- After specifying all the parameters, click on the **Solve** button. The **Solve** dialog box will be displayed. Click on the **Solve Study** button in the dialog box. The results of analysis will be displayed; refer to Figure-13. You can use probes to check results at different locations as discussed earlier.

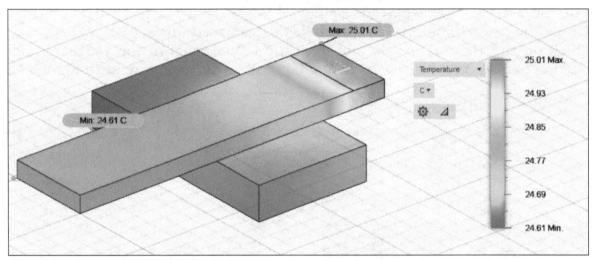

Figure-13. Result of thermal analysis

THERMAL STRESS ANALYSIS

The thermal stress analysis is used to check the effect of thermal and structural loads on the model. The procedure to perform the analysis same as discussed earlier for thermal and stress analyses.

EVENT SIMULATION

Event Simulation is used to study the effect of motion/load on different parts/bodies in model. This analysis is similar to dynamic non-linear analysis, you may have studied in engineering. In this example, we will simulate the collision of a ball on the plate. The procedure to perform event simulation is given next.

- Click on the **New Simulation Study** button from **STUDY** drop-down in the **Toolbar**. The **New Study** dialog box will be displayed. Double-click on the **Event Simulation** button from the dialog box. The environment to solve event simulation will be displayed.

- Specify the material, constraints, and load as applied earlier.
- Click on the **Prescribed Translation** tool from **CONSTRAINTS** drop-down in the **Toolbar**; refer to Figure-14 and specify the displacement, velocity, or acceleration of the selected body as required; refer to Figure-15.

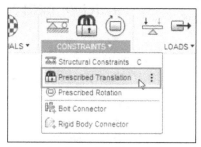

Figure-14. Prescribed Translation tool

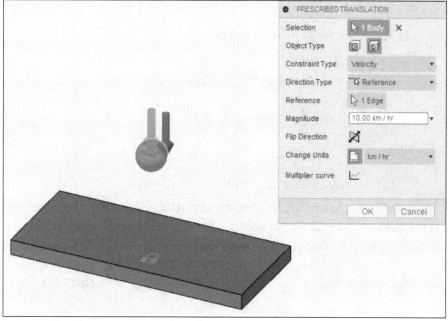

Figure-15. PRESCRIBED TRANSLATION dialog box

- Click on the **Solve** tool from **SOLVE** drop-down in the **Toolbar** and then click on the **Solve** button from the dialog box displayed. The results of analysis will be displayed. Set the slider to desired step to check the analysis result at intermediate steps.

Note that you can also use **Prescribed Rotation** tool from the **CONSTRAINTS** drop-down in **Toolbar** to apply rotation constraint to the model.

SHAPE OPTIMIZATION

The Shape optimization study is used to reduce the mass of part while satisfying all the design requirements of the part. The procedure to use shape optimization is given next.

- Create/open the model on which you want to perform shape optimization study. Click on the **New Simulation Study** tool from **STUDY** drop-down in the **Toolbar**. The **New Study** dialog box will be displayed.
- Double-click on the **Shape Optimization** button from the dialog box. The environment to perform shape optimization will be displayed.

• Set the material, constraint, load, and contacts as required.

Preserve Region

• Click on the **Preserve Region** tool from the **SHAPE OPTIMIZATION** drop-down in the **Toolbar**; refer to Figure-16. The **PRESERVE REGION** dialog box will be displayed; refer to Figure-17.

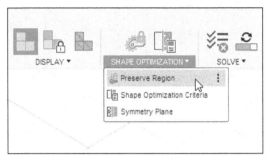

Figure-16. Preserve Region tool

Figure-17. PRESERVE REGION dialog box

• Select the face that you want to be preserved after shape optimization. The options for preserving region based on selected faces will be displayed in the dialog box.
• Set the desired options like if you have select a cylindrical face then specify the radius up to which you want to preserve the region; refer to Figure-18.
• Click on the **OK** button from the dialog box to apply the parameters.

Figure-18. PRESERVE REGION dialog box with cylindrical face selected

Setting Shape Optimization criteria

- Click on the **Shape Optimization Criteria** tool from the **SHAPE OPTIMIZATION** drop-down in the **Toolbar**; refer to Figure-19. The **SHAPE OPTIMIZATION CRITERIA** dialog box will be displayed; refer to Figure-20.

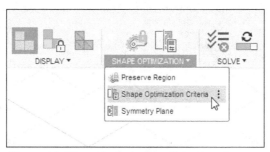

Figure-19. Shape Optimization Criteria tool

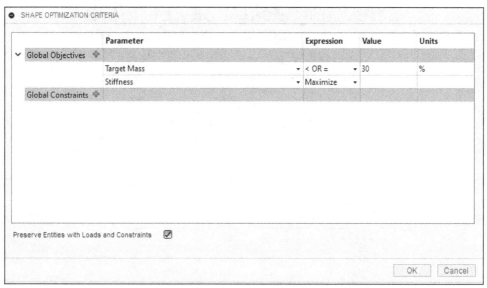

Figure-20. SHAPE OPTIMIZATION CRITERIA dialog box

- Set the desired conditions in the dialog box like, Target Mass less than or equal to 40%.
- Click on the **OK** button to apply the settings.
- Create desired load and constraint conditions for the model.
- Click on the **Solve** tool from **SOLVE** drop-down in the **Toolbar**. The **Solve** dialog box will be displayed.
- Click on the **Solve** button. The result will be displayed; refer to Figure-21.

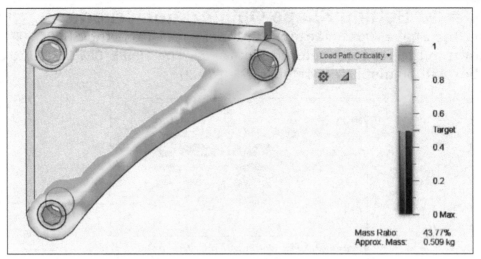

Figure-21. Result of shape optimization

Generating Mesh Body of Result

*   Click on the **Promote** tool from **RESULT TOOLS** drop-down in **Toolbar**; refer to Figure-22. The **PROMOTE** dialog box will be displayed; refer to Figure-23.

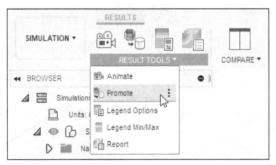

Figure-22. Promote tool

Figure-23. PROMOTE dialog box

*   Select the **Model Workspace** option from the **Add mesh object to** drop-down to generate mesh model as mesh object in model. Select the **Existing Simulation Model** option from the drop-down to add the mesh object in selected simulation model. Select the **Clone Current Simulation Model** option to create a copy of simulation model with mesh body. Select the **Clone Studies** check box while making copy to also copy simulation study parameters.
*   Click on the **OK** button from the dialog box. The mesh body will be displayed which can be used to modify the part; refer to Figure-24.

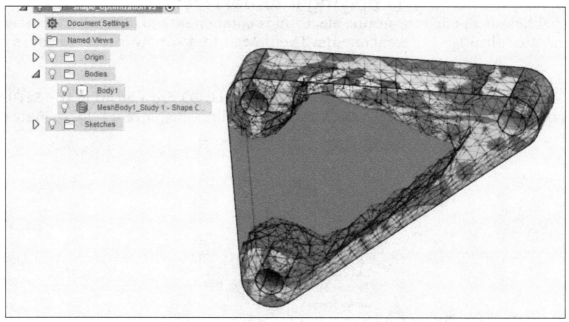

Figure-24. Mesh model generated in Model workspace

ELECTRONICS COOLING

The Electronics Cooling study is performed to check whether temperature of selected bodies exceed the specified values or not due to internal heat loads on the PCB. While writing this edition of book, the Electronics Cooling study was in Preview mode. So, you expect more changes in this study. The procedure to perform this study is given next.

• Create/open the model on which you want to perform electronic cooling study. Click on the **New Simulation Study** tool from **STUDY** drop-down in the **Toolbar**. The **New Study** dialog box will be displayed.
• Double-click on the **Electronics Cooling** button from the dialog box. The environment to perform electronics cooling study will be displayed; refer to Figure-25.

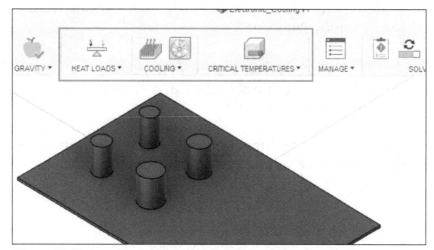

Figure-25. Tools specific to electronics cooling study

• Apply the materials as discussed earlier. Generally, PCB is made of epoxy resin with fiberglass layers.

Applying Internal Heat

Internal heat is applied to various electronics components to define how much heat is generated through the components. The procedure to apply internal heat is given next.

- Click on the **Internal Heat** tool from the **HEAT LOADS** drop-down in the **SETUP** tab of the **Toolbar**. The **THERMAL LOADS** dialog box will be displayed; refer to Figure-26.

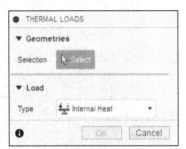

Figure-26. THERMAL LOADS dialog box

- Select all the bodies to which you want to apply thermal loads and specify desired value of heat in the **Internal Heat Value** edit box; refer to Figure-27.

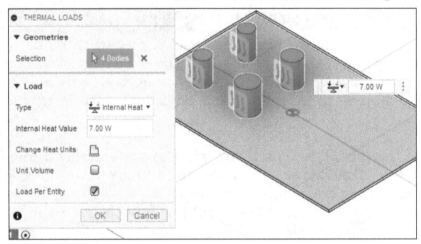

Figure-27. Applying internal heat load

- After setting desired parameters, click on the **OK** button from the dialog box. Internal heat loads will be generated.

Applying Heat Transfer

The **Heat Transfer** tool is used to apply effect of a heat sink on selected body. The procedure to apply heat sink is given next.

- Click on the **Heat Transfer** tool from **COOLING** drop-down in the **SETUP** tab of the **Toolbar**. The **HEAT TRANSFER** dialog box will be displayed; refer to Figure-28.

Figure-28. HEAT TRANSFER dialog box

- Select desired body to be used as heat sink and click on the **OK** button from the dialog box.

Applying Forced Flow

The **Forced Flow** tool is used to apply volumetric air or fluid flow to selected bodies. The procedure to apply force flow is given next.

- Click on the **Forced Flow** tool from the **COOLING** drop-down in the **SETUP** tab of the **Toolbar**. The **FORCED FLOW** dialog box will be displayed; refer to Figure-29.

Figure-29. FORCED FLOW dialog box

- Select desired body to be used as target for applying cooling fan volume; refer to Figure-30. Note that we have create a box with air material applied to it as forced flow body.

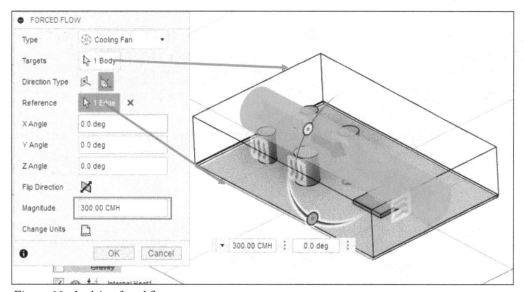

Figure-30. Applying forced flow parameters

- After setting desired parameters, click on the **OK** button from the dialog box to apply the forced flow.

Specifying Critical Temperature

Critical Temperature is defined as maximum safe temperature up to which selected electronic components can function properly. The procedure to specify critical temperature is given next.

- Click on the **Critical Temperatures** tool from the **CRITICAL TEMPERATURES** drop-down in the **SETUP** tab of the **Ribbon**. The **CRITICAL TEMPERATURES** dialog box will be displayed; refer to Figure-31.

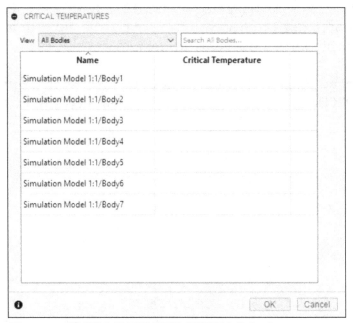

Figure-31. CRITICAL TEMPERATURES dialog box

- Specify desired values in the edit boxes under **Critical Temperature** column. After setting desired parameter, click on the **OK** button from the dialog box.

After setting desired parameters, click on the **Solve** tool from the **SOLVE** drop-down in the **SETUP** tab of **Toolbar** and solve the analysis as discussed earlier.

SELF ASSESSMENT

Q1. Which of the following is not a parameter of non linearity for non-linear static analysis?

a. Material   b. Geometry
c. Load   d. Time

Q2. Using transient results plot, you can check the effect of load with respect to time. (T/F)

Q3. When the stored membrane (axial) energy is converted into bending energy with no change in the externally applied loads then you can perform structural buckling analysis. (T/F)

Q4. The buckling mode displays shape of model under buckling load. (T/F)

Q5. When a structure is properly excited by a dynamic load with a frequency that coincides with one of its natural frequencies, the structure undergoes large displacements and stresses. (T/F)

Q6. The phenomena by which a structure goes large deformation when excited at natural frequency is called ....

Q7. Which of the following is a parameter does not apply for structural analysis?

a. Film Coefficient   b. Pressure
c. Density   d. Force

PRACTICE 1

Consider a rectangular plate with cutout. The dimensions and the boundary conditions of the plate are shown in Figure-32. It is fixed on one end and loaded on the other end. Under the given loading and constraints, plot the deformed shape. Also, determine the principal stresses and the von Mises stresses in the bracket. Thickness of the plate is 0.125 inch and material is **AISI 1020**.

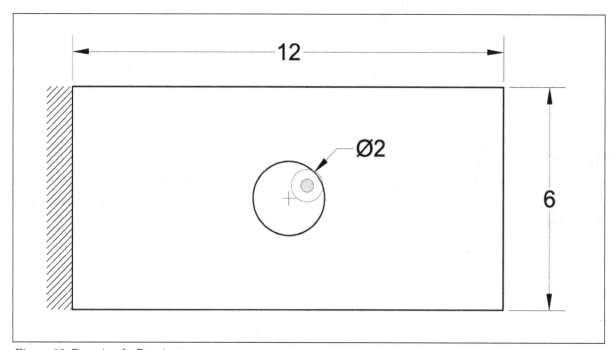

Figure-32. Drawing for Practice1

PRACTICE 2

Open the model for this exercise from the resource kit and perform the static analysis using the conditions given in Figure-33. Find out the **Factor of Safety** for the model.

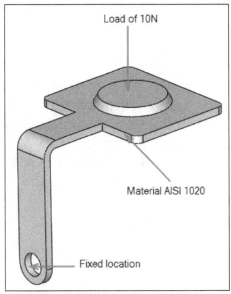

Figure-33. Model for Practice2

PRACTICE 3

Check out what happens to the fork under the conditions specified in Figure-34. Note that the material of fork is **Alloy Steel** and it is in the hands of a nasty kid. (You know kids!! they don't exert linear forces.)

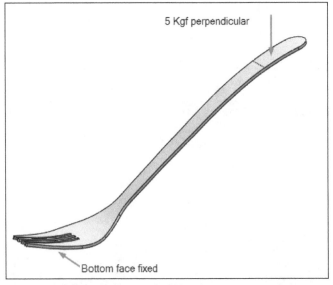

Figure-34. Model for Practice3 of Non-linear static analysis

PROBLEM 1

A metal sphere of diameter d = 35mm is initially at temperature Ti = 700 K. At t=0, the sphere is placed in a fluid environment that has properties of T∞ = 300 K and h = 50 W/m2-K. The properties of the steel are k = 35 W/m-K, ρ = 7500 kg/m3, and c = 550 J/kg-K. Find the surface temperature of the sphere after 500 seconds.

PROBLEM 2

A flanged pipe assembly; refer to Figure-35, made of plain carbon steel is subjected to both convective and conductive boundary conditions. Fluid inside the pipe is at a temperature of 130°C and has a convection coefficient of hi = 160 W/m²-K. Air on the outside of the pipe is at 20°C and has a convection coefficient of ho = 70 W/m²-K. The right and left ends of the pipe are at temperatures of 450°C and 80°C, respectively. There is a thermal resistance between the two flanges of 0.002 K-m²/W. Use thermal analysis to analyze the pipe under both steady state and transient conditions.

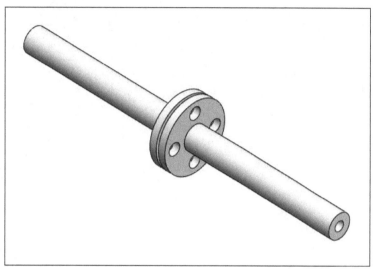

Figure-35. Flanged pipe assembly

FOR STUDENT NOTES

Chapter 22

Sheetmetal Design

Topics Covered

The major topics covered in this chapter are:

- *Introduction to Sheet Metal*
- *Sheetmetal Rules*
- *Creating Flanges*
- *Unfolding Sheetmetal Part*
- *Convert to Sheet Metal*
- *Flat Pattern*
- *Exporting DXF*
- *Practical and Practice*

INTRODUCTION

Sheet metal is used when you need a component of thickness in the range of 0.16 mm to 12.70 mm and do not require conventional cutting machines. The components that can be created by Punch-press and bending machines are designed in Sheet Metal environment. In Autodesk Fusion, the tools to design sheet metal components are available in **SHEET METAL** tab of **DESIGN** workspace; refer to Figure-1. Various tools and parameters of **SHEET METAL** tab are discussed next.

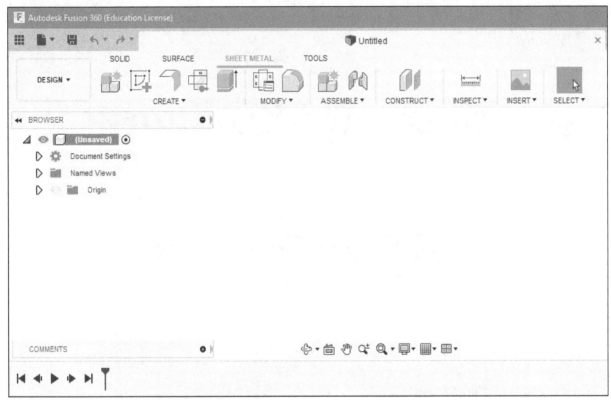

Figure-1. SHEET METAL toolbar in DESIGN Workspace

SHEET METAL RULES

The Sheet Metal rules are used to specify various parameters related to sheet metal design like bend radius, corner conditions, K Factor, thickness of sheet, and so on. The procedure to set sheet metal rules in Autodesk Fusion are discussed next.

*   Click on the **Sheet Metal Rules** tool from the **MODIFY** drop-down in **Toolbar**; refer to Figure-2. The **SHEET METAL RULES** dialog box will be displayed; refer to Figure-3.

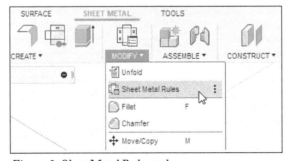

Figure-2. Sheet Metal Rules tool

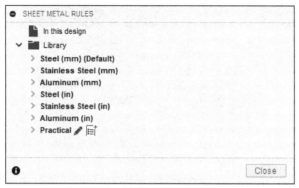

Figure-3. SHEET METAL RULES dialog box

- Expand desired category from the dialog box to check the parameters; refer to Figure-4.

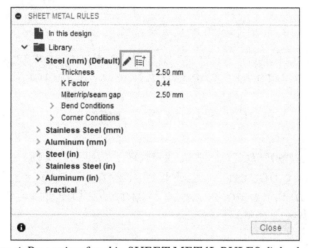

Figure-4. Properties of steel in SHEET METAL RULES dialog box

- Hover the cursor on the name of property. Two buttons will be displayed next to it as shown in Figure-4.

Modifying the rules

- Click on the **Edit Rule** button for desired sheet metal rule from **SHEET METAL RULES** dialog box; refer to Figure-5, if you want to modify the property. The **Edit Rule** dialog box will be displayed; refer to Figure-6.

Figure-5. Edit Rule button

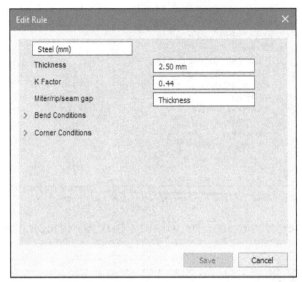

Figure-6. Edit Rule dialog box

- Specify desired name for the rule in the **Rule Name** edit box.
- Set the other parameters like sheet thickness, K factor, seam gap, etc. in their respective edit boxes.
- Click on the **Save** button after setting parameters to save the rule.

Creating New Sheet Metal Rule

- Click on the **New Rule** button from **SHEET METAL RULES** dialog box if you want to create a new sheet metal rule; refer to Figure-7. The **New Rule** dialog box will be displayed; refer to Figure-8.

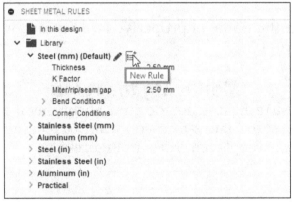

Figure-7. New Rule button

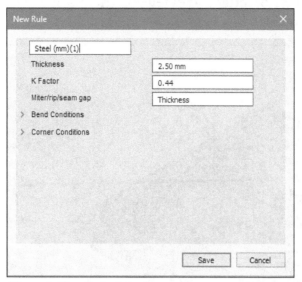

Figure-8. New Rule dialog box

- Set desired parameters as discussed earlier and click on the **Save** button to save new sheet metal rule.

CREATING FLANGES

Flanges are the building blocks of sheet metal parts. Flanges act as base and walls of sheet metal parts. The procedure to create flange is given next.

- Click on the **Flange** tool from the **CREATE** drop-down in the **Toolbar**; refer to Figure-9. The **FLANGE** dialog box will be displayed; refer to Figure-10. You will be asked to select a sketch or edge to create flange.

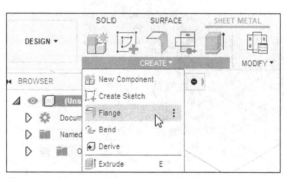

Figure-9. Flange tool

Figure-10. FLANGE dialog box

- Select desired sketch or edge. Preview of the flange will be displayed; refer to Figure-11 and Figure-12.

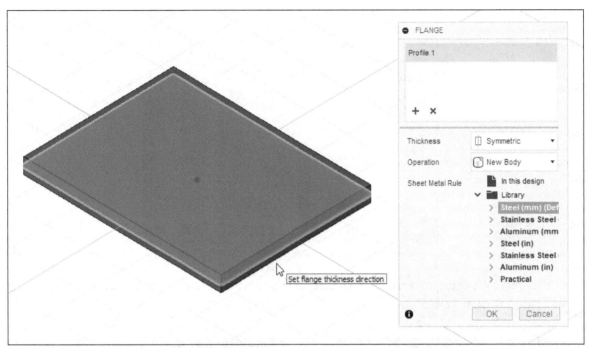

Figure-11. Preview of flange by using sketch

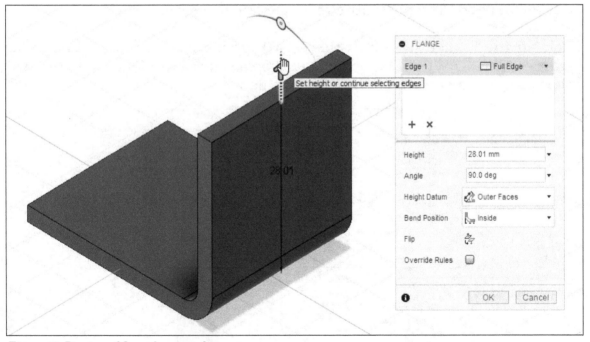

Figure-12. Preview of flange by using edge

- If you have used a closed sketch to create flange then set the thickness direction in the **Thickness** drop-down.
- If you have selected an edge to create flange then the options of the dialog box will be displayed as shown in Figure-12. Set desired **Height** and **Angle** values in the respective edit boxes in the dialog box.

- Select desired option from the **Height Datum** drop-down. Select the **Inner Faces** option if you want to set the height of flange from the inner face. Select the **Outer Faces** option if you want to set the height of flange from the outer face. Select the **Tangent To Bend** option if you want to set the height from the edge of bend created by flange.
- Select the **Inside** option from the **Bend Position** drop-down if you want the bend to be created starting from the inner edge of the flange. Select the **Outside** option if you want the bend to be created from outer edge of the flange. Select the **Adjacent** option if you want to create bend ahead of the outer edge of bend equal to the thickness of sheet. Select the **Tangent** option if you want the bend to start tangent to the face of the selected flange.
- To flip the direction of flange, click on the **Flip** button.
- By default, the bend rules are applied on the flange to be created but if you want to change any of the rule specific to the flange then select the **Override Rules** check box. The options below it will be displayed; refer to Figure-13.

Figure-13. Override Rules options in FLANGE dialog box

- Select desired check box and change the parameter as required.
- Click on the **OK** button to create the flange.

Creating Contour Flange

Using the **Flange** tool, you can also create contour flanges. The procedure is given next.

- Create a sketch of desired shape and size for contour flange; refer to Figure-14.

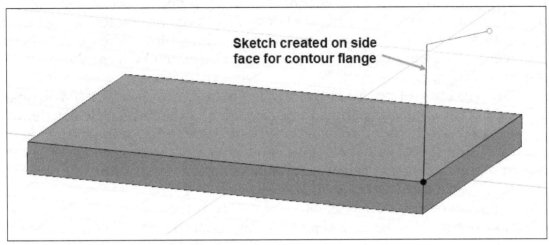

Figure-14. Sketch created for contour flange

- Click on the **Flange** tool from the **CREATE** drop-down in the **Toolbar** and select the newly created sketch for creating contour flange; refer to Figure-15.

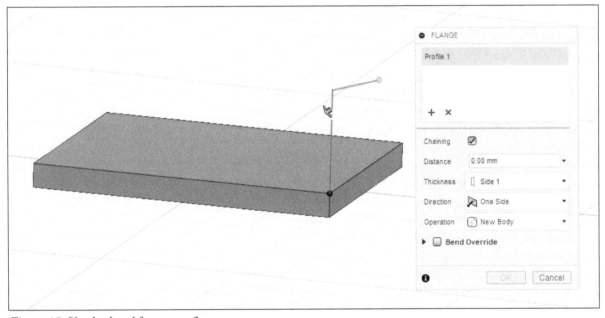

Figure-15. Sketch selected for contour flange

- Now, select all the edges one by one by clicking **Add New Selection** button from **FLANGE** dialog box to define shape of flange. Preview of the contour flange will be displayed; refer to Figure-16.

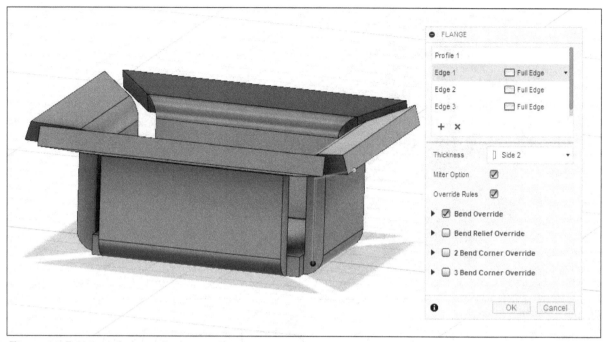

Figure-16. Preview of contour flange

- Select the **Miter Option** check box if you want to create miter cut in flanges.
- Click on the **OK** button from the dialog box to create the flanges.

UNFOLDING SHEETMETAL PART

The **Unfold** tool is used to unbend all the selected bends of the part. The procedure to unfold is given next.

- Click on the **Unfold** tool from **MODIFY** drop-down in the **Toolbar**; refer to Figure-17. The **UNFOLD** dialog box will be displayed; refer to Figure-18.

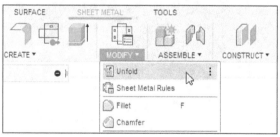

Figure-17. Unfold tool

Figure-18. UNFOLD dialog box

- Select desired face to be made stationary. All the bends will be highlighted; refer to Figure-19.

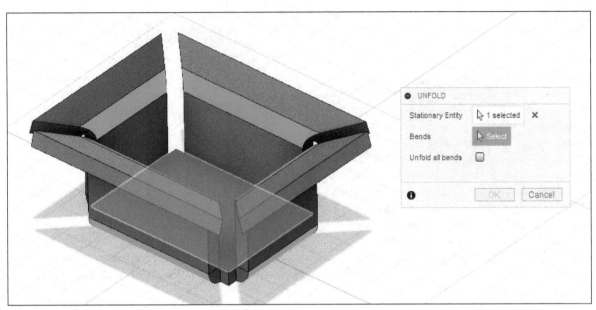

Figure-19. Bends highlighted

- Select the bends that you want to be unfolded. Preview of unfolding will be displayed; refer to Figure-20.

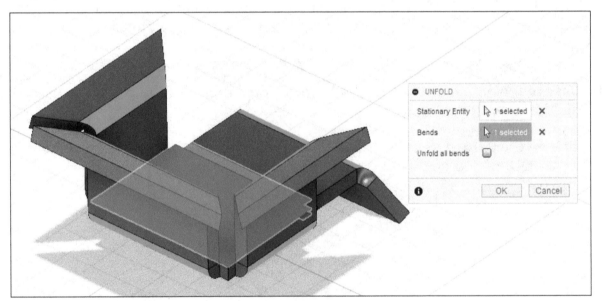

Figure-20. Preview of unfolding

- If you want to unfold all the bends then select the **Unfold all bends** check box.
- Click on the **OK** button to create the unfold feature.

Refolding Faces

The **Refold** tool is used to refold the unfolded faces. To do so, click on the **Refold** tool from **REFOLD FACES** drop-down in the **Toolbar**; refer to Figure-21. The **Refold** tool will become active when the faces are being unfolded.

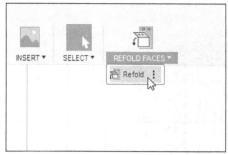

Figure-21. Refold tool

Some people may ask why to unfold and refold the faces of sheet metal parts. The answer is editing between unfolding and refolding steps. After unfolding, you can create holes or other modifications at the bend locations and then you can refold them.

CONVERT TO SHEET METAL

The **Convert to Sheet Metal** tool is used to convert a thin part into sheet metal part. The procedure to use this tool is given next.

- Click on the **Convert to Sheet Metal** tool from the **CREATE** panel in the **SHEET METAL** tab of **Toolbar**; refer to Figure-22. The **CONVERT TO SHEET METAL** dialog box will be displayed; refer to Figure-23 and you will be asked to select a thin part created as solid.

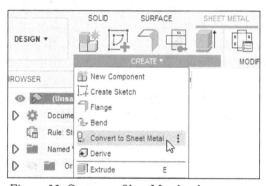

Figure-22. Convert to Sheet Metal tool

Figure-23. CONVERT TO SHEET METAL dialog box

- Click on the face of part to be converted into sheet metal. The updated **CONVERT TO SHEETMETAL** dialog box will be displayed; refer to Figure-24.

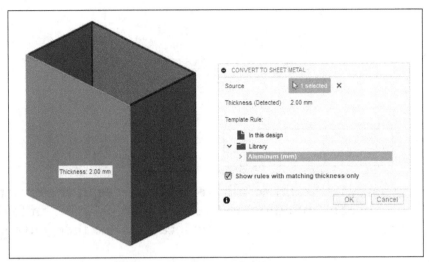

Figure-24. Updated CONVERT TO SHEET METAL dialog box

- By default the **Show rules with matching thickness only** check box is selected and hence only the rules that match with the model thickness are displayed. Clear this check box to select other rules.
- Click on the **OK** button from the dialog box. The part will be converted to a sheet metal part.

FLAT PATTERN

The **Flat Pattern** tool is used to create flat pattern using the sheet metal part so that the part can be cut from the sheet and bent to form desired product. The procedure to use this tool is given next.

- Click on the **Create Flat Pattern** tool from **CREATE** drop-down in the **Toolbar**; refer to Figure-25. The **CREATE FLAT PATTERN** dialog box will be displayed; refer to Figure-26.

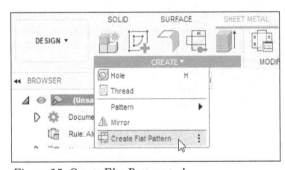

Figure-25. Create Flat Pattern tool

Figure-26. CREATE FLAT PATTERN dialog box

- Select the face that you want to be stationary and click on the **OK** button. The flat pattern will be created; refer to Figure-27.

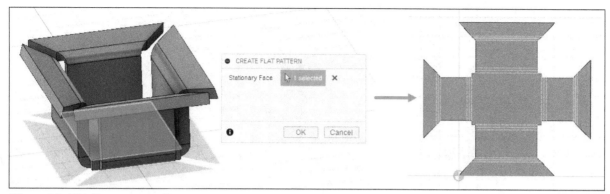

Figure-27. Flat Pattern created

EXPORTING DXF

The **Export Flat Pattern as DXF** tool is used to export the created flat pattern. The procedure is given next.

- Click on the **Export Flat Pattern as DXF** tool from **EXPORT** panel in the **Toolbar** after creating the flat pattern; refer to Figure-28. The **EXPORT FLAT PATTERN AS DXF** dialog box will be displayed; refer to Figure-29.

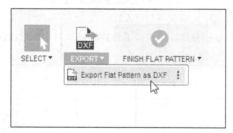

Figure-28. Export Flat Pattern as DXF tool

Figure-29. EXPORT FLAT PATTERN AS DXF dialog box

- Select the **Convert Splines to Polylines** check box if you want to convert all splines in the flat pattern to polylines. Specify the **Tolerance** value in the edit box as desired. This option is useful when your laser cutting machine do not support splines.
- Click on the **OK** button from the dialog box. The **Save As DXF** dialog box will be displayed; refer to Figure-30.

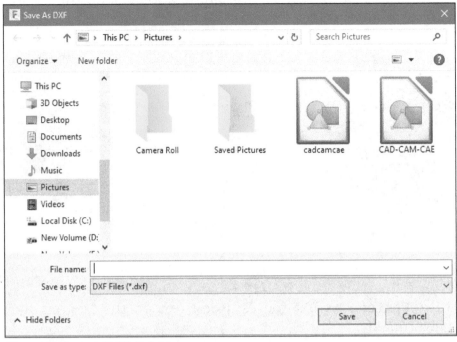

Figure-30. Save As DXF dialog box

- Specify desired name and save the file at desired location.

PRACTICAL

Create the sheet metal model as shown in Figure-31. Dimensions are given in Figure-32. Also, create flat pattern and annotate the drawing.

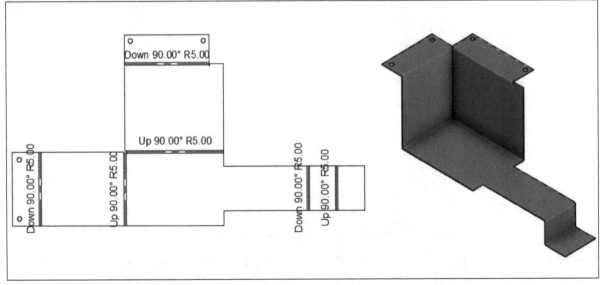

Figure-31. Sheet Metal model

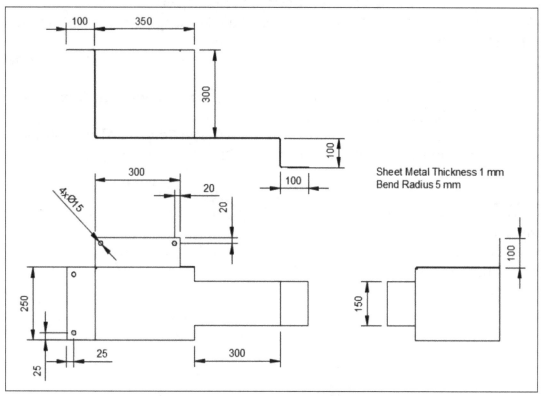

Figure-32. Dimensions for sheetmetal model

Starting Sheet Metal and Creating Base Sheet

- Start Autodesk Fusion and select the **SHEET METAL** tab in **DESIGN** workspace. The tools related to Sheet Metal design will become active.
- Click on the **Create Sketch** tool from **CREATE** drop-down in the **Toolbar** and create a sketch on the **XY** plane as shown in Figure-33.

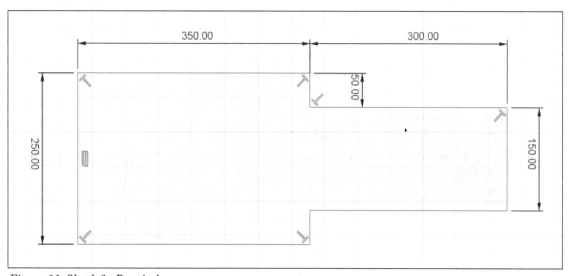

Figure-33. Sketch for Practical

- Click on the **FINISH SKETCH** tool from the **Toolbar** to exit the sketching environment.

Setting Sheet Metal Rules

- Click on the **Sheet Metal Rules** tool from **MODIFY** drop-down in the **Toolbar**. The **SHEET METAL RULES** dialog box will be displayed.
- Select any of the sheet metal rule and click on the **New Rule** button. The **New Rule** dialog box will be displayed.
- Specify the name of rule as **Practical** in the **Rule Name** edit box. Click in the **Thickness** edit box and specify the value as **1 mm**.
- Expand the **Bend Conditions** node and specify the **Bend radius** as **5 mm**.
- Click on the **Save** button and **Close** the dialog box.

Creating Flanges

- Click on the **Flange** tool from **CREATE** drop-down in the **Toolbar** and select the newly created sketch.
- Select the **Practical** option from the **Sheet Metal Rule** area of the dialog box; refer to Figure-34 and click on the **OK** button.

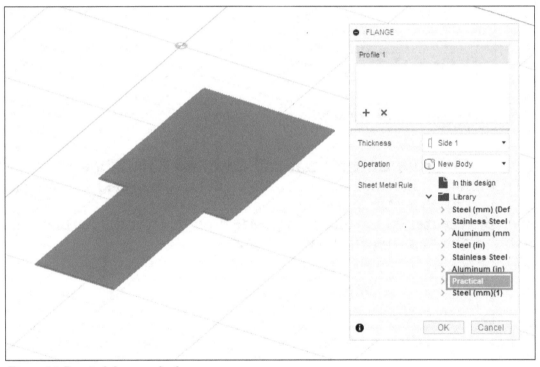

Figure-34. Practical sheet metal rule

- Click on the **Flange** tool from **CREATE** drop-down in the **Toolbar**. Select the edges as shown in Figure-35 and specify the height as **300**.

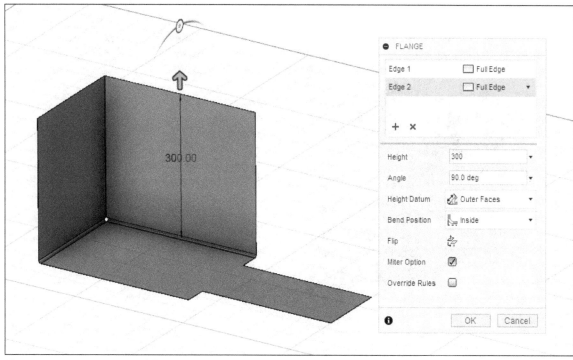

Figure-35. Edges selected for Flange1

- Click on the **OK** button to create vertical flanges.
- Click again on the **Flange** tool and select the outer edges of the newly created vertical flanges.
- Set the height of flanges as **100** mm and clear the **Miter Option** check box. Preview of the flanges will be displayed; refer to Figure-36.

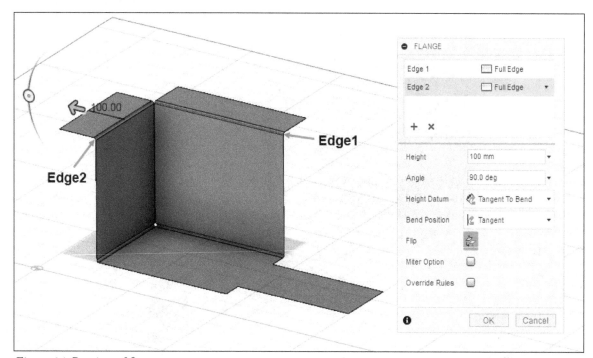

Figure-36. Preview of flanges

- Click on the **Full Edge** option of **Edge 1** in the selection box and select the **Two Offset** option. Now, you will be able to edit the length of flange.
- Move the drag handle to **50** mm backward; refer to Figure-37.

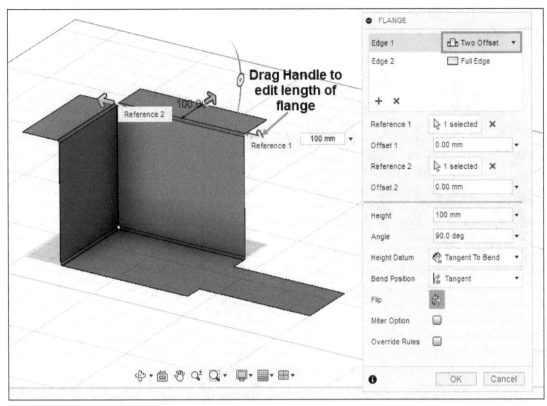

Figure-37. Editing length of flange

- Click on the **OK** button from the dialog box. Similarly, you create other flanges to form a part as shown in Figure-38.

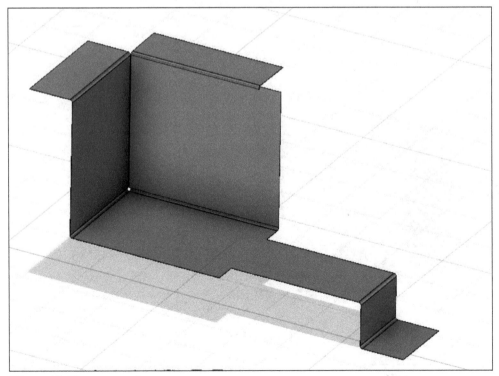

Figure-38. Flanges created

Creating Holes

- Click on the **Create Sketch** tool from **CREATE** drop-down in the **Toolbar**. You will be asked to select a face/plane.
- Select the face as shown in Figure-39. Create four points as shown in Figure-40 and exit the sketching environment.

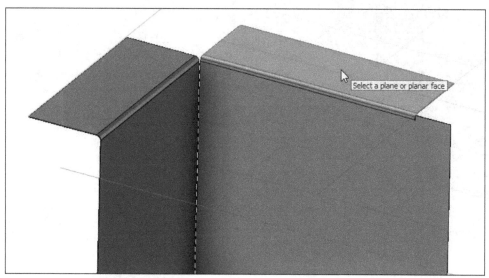

Figure-39. Face selected for sketching

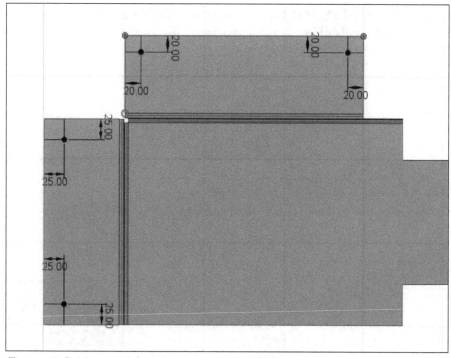

Figure-40. Points created for hole

- Click on the **Hole** tool from **CREATE** drop-down in the **Toolbar**. The **HOLE** dialog box will be displayed.
- Select the **From Sketch (Multiple Holes)** button from **Placement** section in the dialog box. You will be asked to select the sketch points.
- Select all the sketch points created earlier and specify depth as 5 mm and diameter as 15 mm in the respective edit boxes of the dialog box. Preview of the holes will be displayed; refer to Figure-41.

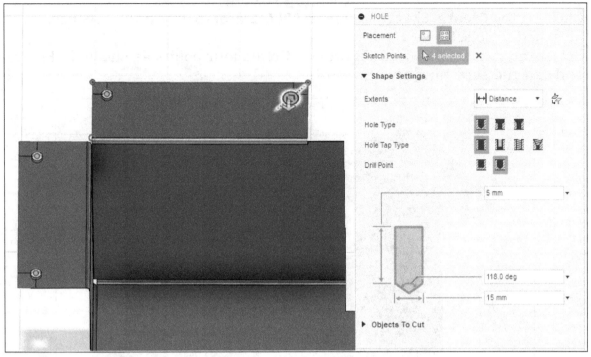

Figure–41. Preview of holes

- Click on the **OK** button from the dialog box to create the holes. The final part will be created.
- Click on the **Create Flat Pattern** tool from **CREATE** drop-down in the **Toolbar**. You will be asked to select a stationary face. Select the base flange and click on the **OK** button. The flat pattern will be displayed; refer to Figure-42.

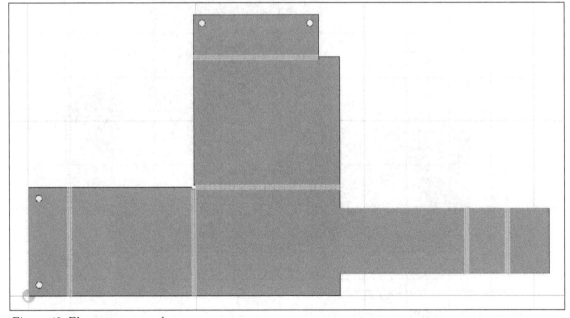

Figure–42. Flat pattern created

Creating Drawing for Flat Pattern

- Click on the **From Design** tool from **DRAWING** cascading menu in the **Change Workspace** drop-down. The **Save** dialog box will be displayed.
- Specify desired name and location and then click on the **Save** button. The **CREATE DRAWING** dialog box will be displayed.

- Set desired parameters and click on the **OK** button. You will be asked to place the drawing view.
- Set desired parameters in the **DRAWING VIEW** dialog box and place the **TOP** view of flat pattern; refer to Figure-43.

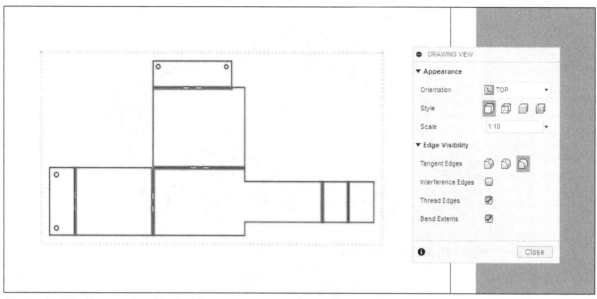

Figure-43. Flat Pattern placed in drawing

Annotating Flat Pattern in Drawing

- Click on the **Leader** tool from **TEXT** drop-down in the **Toolbar**. You will be asked to select the edges to be annotated.
- One by one, select the center lines of all the bends in the flat pattern. The annotations will be displayed as shown in Figure-44. You can also use bend identifiers with table to annotate bends in a sheet metal part.

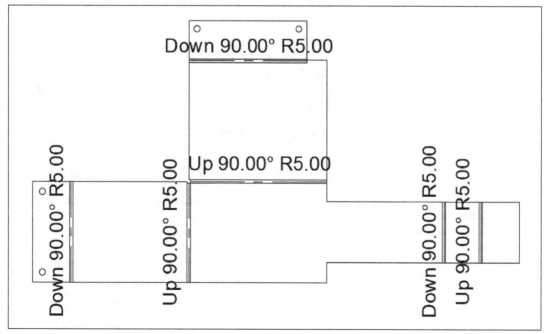

Figure-44. Annotations applied for bends

- Apply the other dimensions and create the other views as discussed earlier in the book.

PRACTICE

Create the sheet metal model as shown in Figure-45. Dimensions are given in Figure-46.

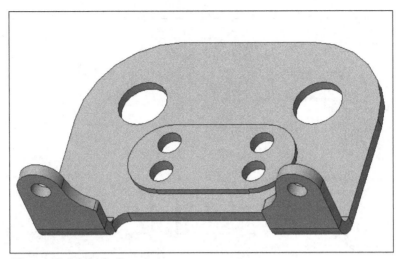

Figure-45. Model for Practice1

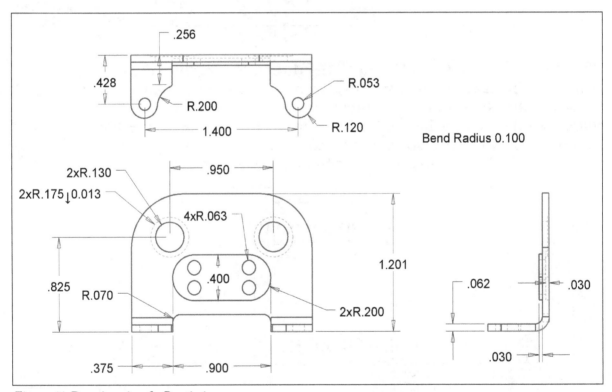

Figure-46. Drawing view for Practice1

SELF ASSESSMENT

Q1. In which areas, **Sheet Metal Rules** are used?

Q2. If you want to modify the property of sheet metal rule then click on the **Edit Rule** button from **SHEET METAL RULES** dialog box. T/F

Q3. Flanges act as .... and .... of sheet metal parts.

Q4. Which of the following options should be selected from **Bend Position** drop-down in the **FLANGE** dialog box to create bend ahead of the outer edge of bend equal to the thickness of sheet?

a) Adjacent
b) Outside
c) Inside
d) Tangent

Q5. The **Refold** tool will become active when the faces are being unfolded. T/F

Q6. Select the .... check box from **EXPORT FLAT PATTERN AS DXF** dialog box if you want to convert all splines in the flat pattern to polylines.

FOR STUDENTS NOTES

Chapter 23

Generative Design

Topics Covered

The major topics covered in this chapter are:

- *Introduction to Generative Design*
- *Generative Study Settings*
- *Design Space Parameters*
- *Setting Design Criteria*
- *Running Generative Study*
- *Exploring Generative Design Results*

INTRODUCTION

Generative Design in Autodesk Fusion 360 is used to modify the shape of components under specified load conditions so that minimum material is used for getting same optimum results of analysis. Following are the general features of Generative Designing in Autodesk Fusion 360:

- Consolidating parts by reducing the overall number of components;
- Reducing the weight of parts and components by using the least amount of material to make the element as effective as possible;
- Increasing performance by designing stronger parts and components.

Generative Design is sometimes confused with Optimization Study. Note that in optimization study, the material is removed from material to get the optimum model for given analysis but in generative design study, shape of the model is also modified as required to get optimum model. The tools to perform Generative Design study are available in the **Generative Design** workspace; refer to Figure-1. The procedure to switch workspace has been discussed earlier. Various tools of Generative Design workspace are discussed next.

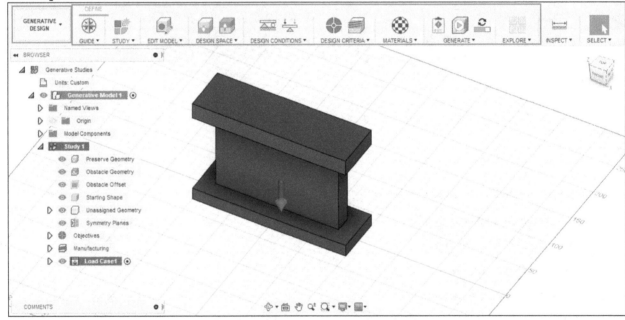

Figure-1. Generative Design workspace

LEARNING PANEL

The **Learning Panel** is used to access interactive help on using Generative Design and other features of the software. The tool is available in **GUIDE** drop-down of the **Ribbon**. Click on the **Toggle Learning Panel** tool from the **GUIDE** drop-down in the **DEFINE** tab of the **Ribbon**. The **Learning Panel** will be displayed; refer to Figure-2. Select the Interactive toggle button to display help interactively. Click on the Close button at top right corner of the **Learning Panel** to close it.

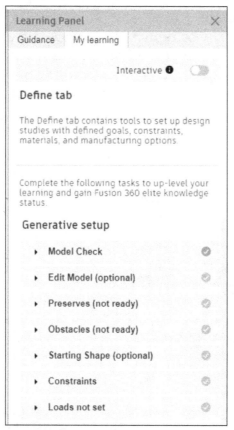

Figure-2. Learning Panel

WORKFLOW OF GENERATIVE DESIGN

Every analysis study follows a predefined strategy to get the result output. For Generative Design, the workflow can shown summarized by Figure-3.

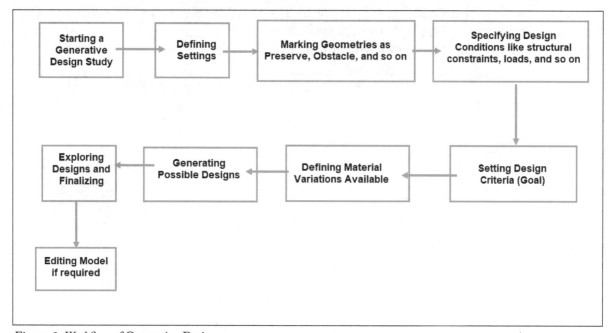

Figure-3. Workflow of Generative Design

Various tools available in Generative Design workspace are discussed next sequentially.

CREATING NEW GENERATIVE DESIGN STUDY

By default when you switch workspace to Generative Design, a generative design study is created automatically. If you want to create a new generative design study, then click on the **New Generative Study** tool from the **STUDY** drop-down in the **DEFINE** tab of **Ribbon**. A new generative design study will be added in the **BROWSER**; refer to Figure-4.

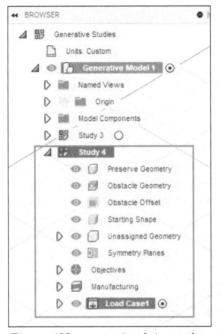

Figure-4. New generative design study

Note: For some unknown reasons, Generative Design workspace did not pick the unit system applied to the model automatically till writing this book. One workaround for this is, click on the **Edit** button next to **Units** in **BROWSER**; refer to Figure-5. The **Units Settings** dialog box will be displayed; refer to Figure-6. Set desired units or unit system as discussed earlier and click on the **OK** button.

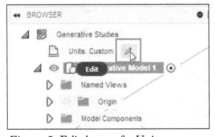

Figure-5. Edit button for Unit

Figure-6. Units Settings dialog box

DEFINING DESIGN SPACE REGIONS

By Design space, we mean categorizing the model into different regions based on whether the region can be modified, preserved, avoided, or marked as starting point for generating designs. The tools to perform these tasks are available in **DESIGN SPACE** drop-down of the **Ribbon**; refer to Figure-7. These tools are discussed next.

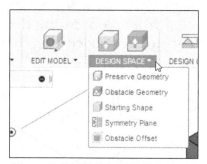

Figure-7. DESIGN SPACE drop-down

Preserving Geometry

The **Preserve Geometry** tool is used to keep selected portion of the model as it is in all variations of the model generated by study. Generally, bodies of the model which are referenced in assembly are preserved like cap of a bottle, speaker housing of a headphone, and so on. It happens because in assembly line, some components are manufactured by vendors and they fit in many models of the same product. The procedure to use this tool is given next.

• Click on the **Preserve Geometry** tool from the **DESIGN SPACE** drop-down in the **DEFINE** tab of the **Ribbon**. The **PRESERVE GEOMETRY** dialog box will be displayed; refer to Figure-8.

Figure-8. PRESERVE GEOMETRY dialog box

• Select the body(s) that you want to preserve in variations of model; refer to and click on the **OK** button. The selected bodies will be marked in green color denoting that they are preserved geometries.

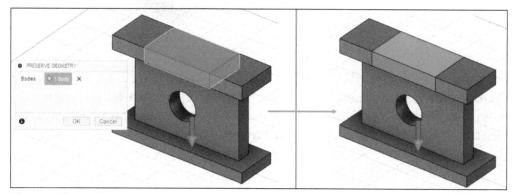

Figure-9. Preserving geometry

Note: If you have single body model and want to preserved some regions of the model then you can use **Split Body** tool of **Design** workspace discussed earlier in the book.

Obstacle Geometries

The obstacle geometries are those which cannot be modified by generative design and do not allow generative design variations to expand past them. For example, roof and floor of a car are obstacles when generating designs of car seat. A car seat cannot go below the floor and cannot go above the roof. The procedure to define obstacle geometries is given next.

- Click on the **Obstacle Geometry** tool from the **DESIGN SPACE** drop-down in the **DEFINE** tab of the **Ribbon**. The **OBSTACLE GEOMETRY** dialog box will be displayed; refer to Figure-10.

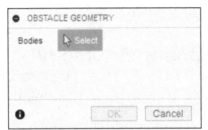

Figure-10. OBSTACLE GEOMETRY dialog box

- Select desired bodies from the model which you want to use as limiting boundaries and click on the **OK** button. Select objects will be marked in red color denoting obstacle geometries; refer to Figure-11.

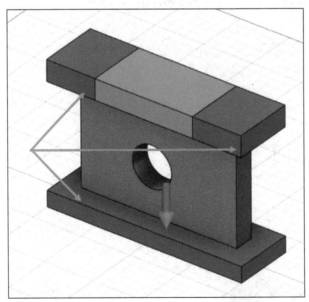

Figure-11. Obstacle geometries marked

Defining Starting Shape

The **Starting Shape** tool is used to define initial shape from which generative design outcomes will be generated. Note that it is not compulsory to select starting shape. The procedure to use this tool is given next.

- Click on the **Starting Shape** tool from the **DESIGN SHAPE** drop-down in the **DEFINE** tab of the **Ribbon**. The **STARTING SHAPE** dialog box will be displayed and you will be asked to select the body.
- Select desired body and click on the **OK** button. Select body will be marked as starting body; refer to Figure-12.

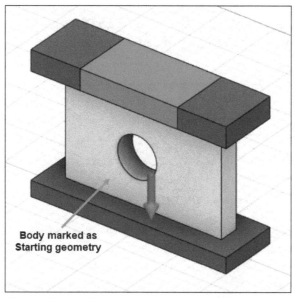

Figure-12. Marking starting body

Applying Symmetry about Plane

The **Symmetry Plane** tool is used to reduce the computing load of generative design study as system will need to perform calculations for only half of the model and mirror copy it for other half. Note that you can apply symmetry only if the model is mirror image about selected plane. The procedure to use tool is given next.

- Click on the **Symmetry Plane** tool from the **DESIGN SPACE** drop-down in the **DEFINE** tab of the **Ribbon**. The **SYMMETRY PLANE** dialog box will be displayed and you will be asked to select planes about which the model is symmetric.
- Select desired plane(s); refer to Figure-13 and click on the **OK** button. The symmetry about plane(s) will be applied.

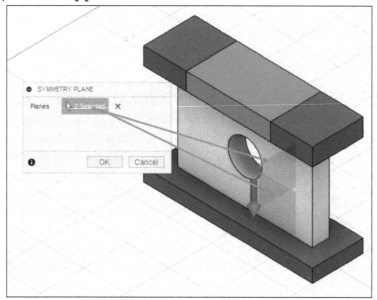

Figure-13. Selecting symmetric planes

Obstacle Offset

The offset word in CAD is used to define the distance by which a body is enlarged or diminished. Obstacle offset is used to expand the size of obstacle body by specified distance value so that this new imaginary expanded body is considered as limiting boundary for generative design. The procedure to apply obstacle offset is given next.

- Click on the **Obstacle Offset** tool from the **DESIGN SPACE** drop-down in the **DEFINE** tab of the **Ribbon**. The **OBSTACLE OFFSET** dialog box will be displayed; refer to Figure-14.

Figure-14. OBSTACLE OFFSET dialog box

- Select desired body(s) to be used as obstacle and specify desired offset distance value in the **Distance** edit box. Preview of the offsetted obstacle will be displayed; refer to Figure-15.

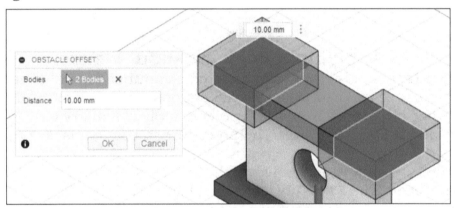

Figure-15. Obstacle offset applied

- Click on the **OK** button to apply the feature.

After defining design space requirements, define the design conditions using tools available in the **DESIGN CONDITIONS** drop-down of **Ribbon**. The tools in this drop-down have been discussed earlier in previous chapters.

Defining Objectives for Generative Design

The **Objectives** tool in **DESIGN CRITERIA** drop-down of **Ribbon** is used to define objectives for performing generative design study. Generally, the objective of generative design study is to either reduce the mass or increase the stiffness of model but in Autodesk Fusion Generative Design, you can also specify limiting factors for design simulations like safety factor, starting modal frequency, maximum displacement allowed, and so on. The procedure to use this tool is given next.

- Click on the **Objectives** tool from the **DESIGN CRITERIA** drop-down in the **DEFINE** tab of the **Ribbon**. The **OBJECTIVES AND LIMITS** dialog box will be displayed; refer to Figure-16.

- Select the **Minimize Mass** radio button to reduce maximum possible mass while keeping model strength enough to get safety factor specified in **Safety Factor** edit box of this dialog box.
- Select the **Maximize Stiffness** radio button to increase the strength of model while keeping mass of model near the value specified in **Mass Target** edit box of this dialog box; refer to Figure-17.

Figure-16. OBJECTIVES AND LIMITS dialog box

Figure-17. Mass Target edit box

- Select the **Modal Frequency** check box from the **Limits** section of the dialog box to define lowest value of natural frequency to be achieved in the generated models. After selecting this check box, specify the frequency value in the **Min. First Mode Frequency** edit box.
- Select the **Displacement** check box from the dialog box to define the limit up to which the model can deform due to loads. On selecting check box, various options to define displacement in different directions will be displayed. Set the parameters as discussed earlier in the book.
- Select the **Buckling** check box to define limiting safety factor up to which model strength can reduce.
- After setting desired parameters, click on the **OK** button from the dialog box.

Defining Manufacturing Criteria

The **Manufacturing** tool in **DESIGN CRITERIA** drop-down of **Ribbon** is used to define the manufacturing processes by which the model will be manufactured after generating designs. The design alternatives generated by Generative Design will be based on the manufacturing processes selected. For example if Milling process is selected then software will make sure to include only those features in design which can be manufactured by milling. You can select multiple processes to explore design alternatives for respective processes. The procedure to use this tool is given next.

- Click on the **Manufacturing** tool from the **DESIGN CRITERIA** drop-down in the **DEFINE** tab of the **Ribbon**. The **MANUFACTURING** dialog box will be displayed; refer to Figure-18.
- Select the **Costing** check box and specify desired value in the **Production Volume** edit box to define the number of pieces to be produced per year. Note that this value will be used to get production cost of model. This value does not affect the manufacturing method used for production.

Figure-18. MANUFACTURING dialog box

- Select the **Unrestricted** check box to tell software that manufacturing constraints do not apply when generating different shapes. Selecting this check box is not advised when you want to generate a manufacturable model for production because giving software total freedom in generating shapes can produce design alternatives which may not be possible to manufacture using conventional machines causing the cost of production to go high.

- Select the **Additive** check box to use 3D printing for manufacturing of model. This will make software to generate designs which are manufacturable by Additive manufacturing processes. After selecting check box, define the direction in which nozzle of 3D printer will create layers. Specify desired values in the **Overhang Angle** and **Minimum Thickness** edit boxes as discussed earlier.

- Select the **Milling** check box to use Milling as manufacturing process for producing model. By default, a 3-axis machine configuration is added in the machine list. You can add as many machining configurations as available in your machine shop. For example, if you have three machines available in your machine shop viz. 2.5 axis, 3-axis, and 5-axis mill machines then you can add the new configurations by using **Add new configuration** button from the **Milling** section; refer to Figure-19. Specify desired values for minimum tool diameter, tool shoulder length, and head (shaft and holder) diameter in respective edit boxes.

- Similarly, you can select **2-axis Cutting** and **Die Casting** check boxes if you want to include laser/waterjet/plasma cutting and die casting manufacturing processes for generating designs. Die casting is the process in which molten material is filled in closed pocket impression of final part and extracted after cooling to get final product. Note that if you have selected multiple manufacturing processes in this dialog box then designs for all selected processes will be generated separately.

- After setting desired parameters in the dialog box, click on the **OK** button from the dialog box.

Figure-19. Adding new configurations

SELECTING STUDY MATERIALS

In Autodesk Fusion 360 Generative Design, you can select multiple materials to check different variations of the design. Generally, material cost takes a fair share in total manufacturing cost of product. If you can find a cheaper material that fulfills all the design requirements then it will reduce the final cost of production. The procedure to select materials for design study is given next.

• Click on the **Study Materials** tool from the **MATERIALS** drop-down in the **DEFINE** tab of the **Ribbon**. The **STUDY MATERIALS** dialog box will be displayed; refer to Figure-20.

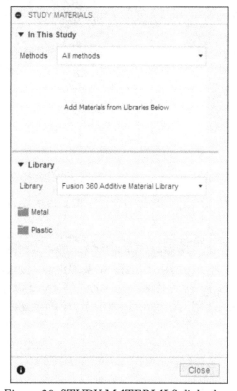

Figure-20. STUDY MATERIALS dialog box

- If you have selected additive manufacturing process along with other processes then by default, `Fusion 360 Additive Material Library` option will be selected in the **Library** drop-down of the **Library** section in the dialog box. This happens because here the limiting process for manufacturing is 3D printing, the other manufacturing processes can handle almost all the available materials in Fusion 360 Material Library. So, it is advised to choose materials from additive material library only. Expand desired material folder from Library section, select the material you want to add in study, and then drag it to **In This Study** section of the dialog box; refer to Figure-21. You can add up to 7 different materials (including the material applied in design environment) in the study.
- If you want to remove a material from the list then right-click on it and select the **Delete** option from shortcut menu displayed; refer to Figure-22.

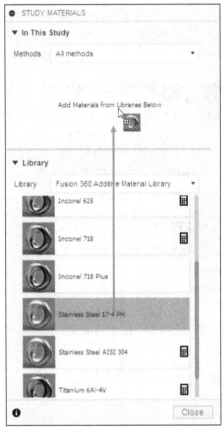

Figure-21. Adding study material

Figure-22. Removing material from list

- After adding desired materials, click on the **Close** button to exit dialog box.

PREFORMING PRE-CHECK

Click on the **Pre-check** tool from the **GENERATE** drop-down to check whether all the requirements to perform study are met or not. If everything is in place then **Ready to Generate** dialog box will be displayed; refer to Figure-23. Click on the **OK** button to exit the dialog box.

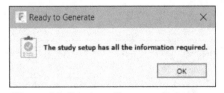

Figure-23. Ready to Generate dialog box

CHECKING PREVIEW OF GENERATIVE DESIGN

The **Previewer** tool in **GENERATE** drop-down of **Ribbon** is used to check preview of study results by solving the study setup without including manufacturing and material settings; refer to Figure-24. Previewer performs calculates locally to display result preview. This tool helps in finding which sections of model are getting modified by study. If you find unwanted areas of model being modified then you can change the setup before finally solving the study. After checking preview, click on the **Stop Previewer** tool from **GENERATE** drop-down of **Ribbon** to exit preview mode.

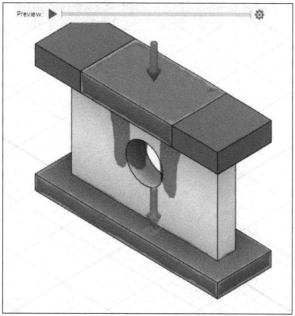

Figure-24. Preview of generative design study

STUDY SETTINGS

The study settings are used to define the scope within which study will be performed. You have learned about study settings earlier in Chapters related to Simulation. To access the settings of Generative Design study, click on the **Study Settings** tool from the **STUDY** drop-down in the **Ribbon** or press **E** from keyboard. The **Study Settings** dialog box will be displayed; refer to Figure-25. Using the **Resolution** slider, specify the smoothness level in model structure up to which you want to generate the designs. Select the **Alternative Outcomes** check box to get additional designs for selected manufacturing processes using experimental methods being created by Autodesk Fusion development team. Note that this option is in Preview mode while writing this book. Select the **Remove rigid body modes** check box to perform study on setup where model is not fully constrained. After setting desired parameter, click on the **OK** button from the dialog box.

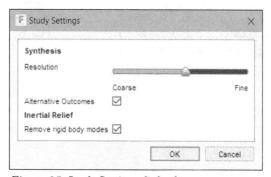

Figure-25. Study Settings dialog box

GENERATING DESIGNS

The **Generate** tool in **GENERATE** drop-down of **Ribbon** is used to solve study for getting design variations. The procedure to use this tool is given next.

- Click on the **Generate** tool in **GENERATE** drop-down of **Ribbon**. The **Generate** dialog box will be displayed similar to **Solve** dialog box discussed earlier in Simulation; refer to Figure-26.
- Click on the **Generate Study** button from the bottom of dialog box to perform study. The status of study will be displayed in **Job Status** dialog box; refer to Figure-27. Once the study is complete, you can check the results in Explore environment displayed automatically.

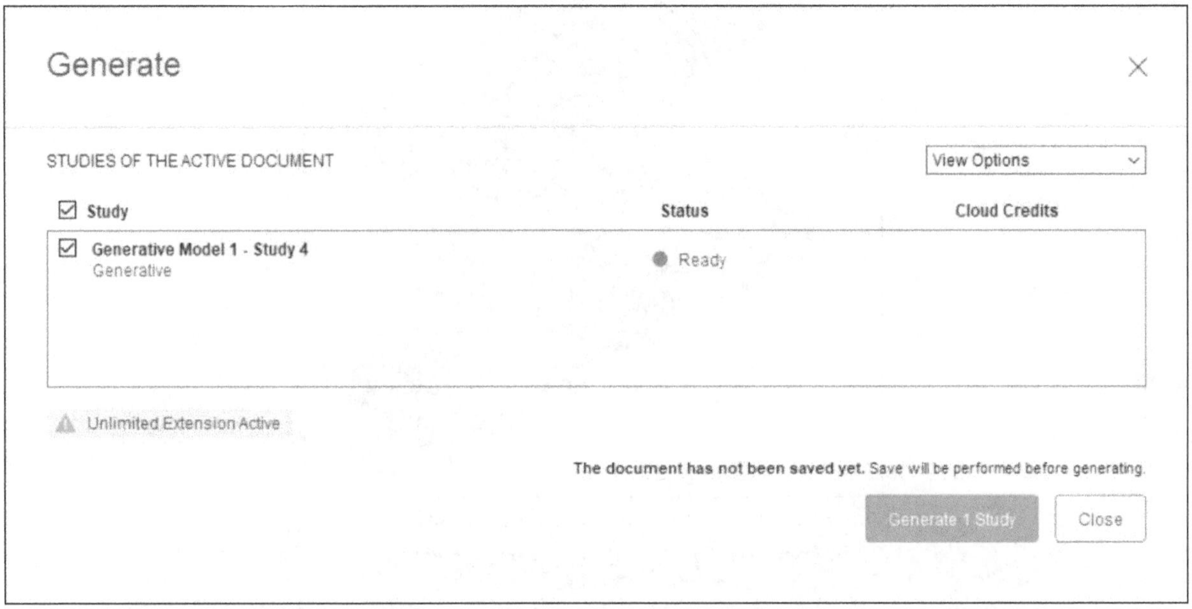

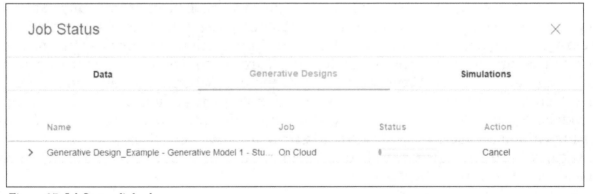

Figure-26. Generate dialog box

Figure-27. Job Status dialog box

- You can filter the designs based on specified criteria in **Outcome Filters** section of the application window; refer to Figure-28.

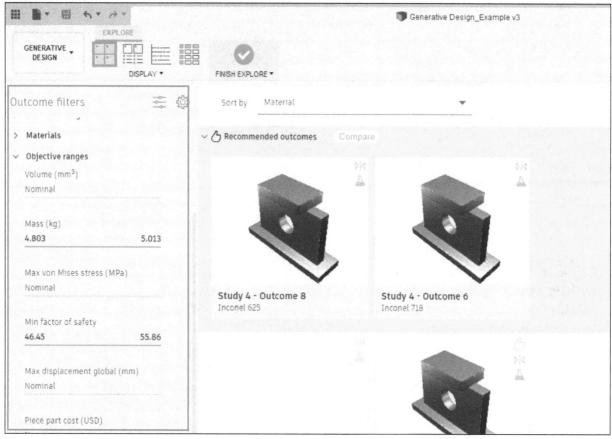

Figure-28. Outcome filters

- Double-click on the desired design alternative which you find most suitable for your design. The **OUTCOME VIEW** tab will be displayed in **Ribbon** and selected design alternative will be displayed in the drawing area; refer to Figure-29.

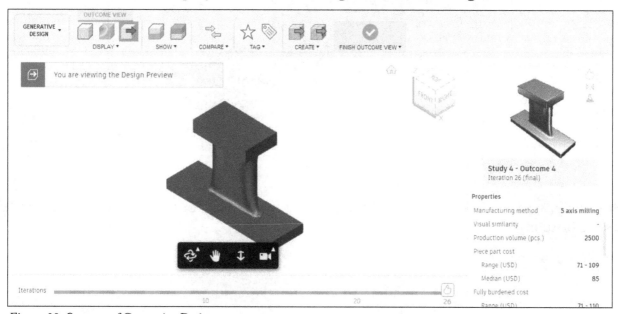

Figure-29. Outcome of Generative Design

- Check various results of outcome using the tools available in **DISPLAY** and **SHOW** drop-downs of the **Ribbon**. You can use the **Compare** tool of **COMPARE** drop-down to compare two outcomes.

- To create design from the outcome, click on the **Design from Outcome** tool from the **CREATE** drop-down in the **OUTCOME VIEW** tab of the **Ribbon**. The **Create a design from this outcome** dialog box will be displayed; refer to Figure-30.

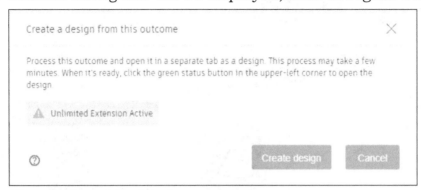

Figure-30. Create design dialog box

- Click on the **Create design** button from the dialog box. The design model file will be generated using cloud server services and after the process is complete, model will be displayed in **DESIGN** workspace; refer to Figure-31. You can modify the design as discussed in previous chapters.

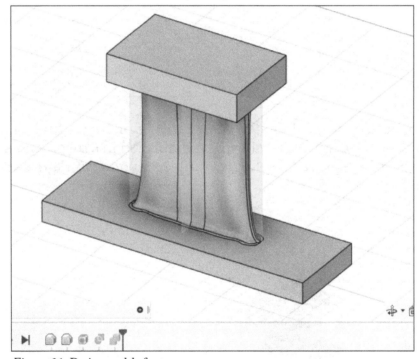

Figure-31. Design model of outcome

Similarly, you can use the **Mesh Design from Outcome** tool to generate mesh model using outcome.

Index

Ethics of an Engineer

- Engineers shall hold paramount the safety, health and welfare of the public and shall strive to comply with the principles of sustainable development in the performance of their professional duties.

- Engineers shall perform services only in areas of their competence.

- Engineers shall issue public statements only in an objective and truthful manner.

- Engineers shall act in professional manners for each employer or client as faithful agents or trustees, and shall avoid conflicts of interest.

- Engineers shall build their professional reputation on the merit of their services and shall not compete unfairly with others.

- Engineers shall act in such a manner as to uphold and enhance the honor, integrity, and dignity of the engineering profession and shall act with zero-tolerance for bribery, fraud, and corruption.

- Engineers shall continue their professional development throughout their careers, and shall provide opportunities for the professional development of those engineers under their supervision.

OTHER BOOKS BY CADCAMCAE WORKS

Autodesk Inventor 2021 Black Book

Autodesk Revit 2021 Black Book

Autodesk Fusion 360 Black Book (V 2.0.10027)

AutoCAD Electrical 2021 Black Book
AutoCAD Electrical 2020 Black Book

SolidWorks 2021 Black Book
SolidWorks 2020 Black Book
SolidWorks 2019 Black Book

SolidWorks Simulation 2021 Black Book
SolidWorks Simulation 2020 Black Book

SolidWorks Flow Simulation 2021 Black Book
SolidWorks Flow Simulation 2020 Black Book

SolidWorks Electrical 2021 Black Book
SolidWorks Electrical 2020 Black Book

Mastercam X7 for SolidWorks 2014 Black Book
Mastercam 2017 for SolidWorks Black Book

Creo Parametric 7.0 Black Book
Creo Parametric 6.0 Black Book
Creo Parametric 5.0 Black Book

Creo Manufacturing 4.0 Black Book

Autodesk CFD 2018 Black Book

Basics of Autodesk Inventor Nastran 2020
Basics of Autodesk Inventor Nastran 2021

ETABS 2016 Black Book
ETABS 2018 Black Book

Solid Edge 2021 Black Book